Natural Resources in Afghanistan

Natural Resources in Afghanistan

Geographic and Geologic Perspectives on Centuries of Conflict

John F. Shroder

Senior Research Scholar
Center for Afghanistan Studies

Emeritus Professor of Geography and Geology
University of Nebraska
Omaha, NE 68182

ELSEVIER

Amsterdam • Boston • Heidelberg • London • New York • Oxford • Paris
San Diego • San Francisco • Singapore • Sydney • Tokyo

Elsevier
525 B Street, Suite 1900, San Diego, CA 92101-4495, USA
225 Wyman Street, Waltham, MA 02451, USA

Library of Congress Cataloging-in-Publication Data
Shroder, John F., 1939-
Natural resources in Afghanistan : geographic and geologic perspectives on centuries of conflict / John F. Shroder. -- First edition, 2014.
pages cm.
Includes bibliographical references and index.
ISBN 978-0-12-800135-6 (alkaline paper) 1. Natural resources--Afghanistan. 2. Economic geography--Afghanistan. 3. Afghanistan--Economic conditions. 4. Agricultural geography–Afghanistan. 5. Water-supply--Afghanistan. I. Title.
HC417.5.S57 2014
333.709581--dc23
2014005299

British Library Cataloguing in Publication Data
A catalogue record for this book is available from the British Library

ISBN: 978-0-12-800135-6

For information on all Elsevier publications visit our web site at store.elsevier.com

Afghan carpet picture courtesy of Richard Pratt, Emmett Eiland's Oriental Rug Company, Berkeley, CA

Contents

Dedication

To the young children of Afghanistan—may you grow up into a decent and healthy environment of expanding horizons brought about by the wise and beneficial exploitation of your tremendous natural resource base. Allah seems to have given your region (Afghanistan) a great and good quantity and quality of natural resources to make your lives better, providing that your elders have the wisdom to exploit the resources properly.

Professor John F. Shroder—October 2013.

Bismullah!

Dedication from Abdul Yaseer and Sher Jan Ahmadzai in Dari

The people of Afghanistan have been blessed by Allah with a huge natural resource base of minerals and oil and gas, which if exploited properly with modern methods while protecting the natural surface environment, would make them rich beyond belief. Wise selling of their own resources to raise their level of living must have been the plan that Allah had for the people of Afghanistan. Bismullah, that they be allowed this blessing.

خداوند بالای مردم افغانستان مرحمت فرموده است که منابع بزرگ طبعی مینرالها، نفت و گازو غیره را دراختیارشان قرار داده است. اگرآنها در استخراج این منابع از روشهای عصری با درنظرداشت حفاظهٔ محیط استفاده نمایند، این امرایشان را بیشتر از آنچه تصور میشود، ثروتمند تر خواهد ساخت.

فروش با درایت این منابع جهت بلند بردن سطح زندگی شان، طرحیست که ازجانب خداوند به انها تفویض شده است.

خداوند ایشان را توفیق نصیب کند تا بتوانند از این نعمت بصورت درست استفاده کنند.

Acknowledgments

This book is a product of fascination with the enigmatic but enormously hospitable people of Afghanistan, coupled with the unusual historical circumstances that have led to the special conditions of the modern Afghanistan. Many people and institutions over the years have helped to accrue the necessary knowledge on Afghanistan and motivate the author to keep at it even when the situation was not particularly conducive to continuation. Afghanistan is a difficult place to work, in large part because of threats from polluted water, lurking pathogenic organisms, baksheesh-ridden and crooked bureaucracies, as well as life-threatening driving on tortured roads, not to mention the great danger from the murderous terrorists who imagine themselves to be some sort of religious freedom fighters. But such difficulties aside, the effort to do this book was well worth it because of the opportunities to show possibly realistic opportunities for things to actually get better for the people who value life, rather than the more death-loving Islamicist dimwits who operate so much in the region. Mankind seems to have many possible paths to failure and Afghanistan has more than most, while it also slyly teases at the possibilities of some success. Great thanks are due to the many people and institutions over the past four decades for the opportunity to work on this fascinating place.

A number of institutions gave considerable encouragement and funding to study Afghanistan. This especially includes the US Agency for International Development (USAID) who first provided the author direct (Atlas of Afghanistan) funding though the US National Foundation, and most recently indirectly (Hindu Kush and Himalayan glaciers) through the US National Academy of Sciences. Unusual open-ended funding from the National Geographic Society also helped recently to contribute to the author's plentiful travel and stays in South Asia to gather some of the information and make contact with a few of the people needed. Early on, the American Geological Institute which publishes the Geotimes/Earth had asked for Afghanistan resource information within the context of the Cold War. The author had responded then, and a number of times since, as political conditions in Afghanistan have changed (Shroder, 1987, 2003, 2004, 2009, 2013). The most recent two of these articles were under the guidance of Managing Editor, Megan Server, at Earth, which are the publication opportunities for which great thanks are due. Many thanks are also extended to personnel at various times in the US Geological Survey (Jack Medlin, Ritchie Willams, Jane Ferrigno, Bruce Molnia, Jeff Kargel, Cal Ruleman, Hugh Kieffer, Craig Wandry, and John SanFilipo), who all provided help at various times. Financing from NASA to Michael Bishop and the author for the Global Land Ice Measurements from Space (GLIMS) Regional Center for Southwest Asia (Afghanistan and Pakistan) was most welcome.

Other encouragement and funding has long come from the Center for Afghanistan Studies (CAS) at the University of Nebraska at Omaha (UNO), which has been the author's main place of employment since first coming there in 1969 from employment at the University of Malawi in East Africa. In 1972, Christian Jung and the author had started the CAS at UNO (Figure 1), in part because of recognition of the great potential that Afghanistan had to provide interesting research for young professors. Little did the author realize! But Chris died of a rare blood disease a year later and the author realized that help was needed if the CAS was to become a lasting entity. With the accompanying support of the UNO administration, great help came in the person of Thomas Gouttierre, a former Peace Corps teacher in Afghanistan in the 1960s, who was the director of AFAMEC (Afghan American Educational Commission) in Kabul in the 1970s, where he ran the Fulbright program. Although he took considerable convincing to move to the middle of flatland USA, he was finally prevailed upon to run the nascent program. The result was transformative as Tom was able to transfuse high energy and plentiful Kabul and Washington associations into the CAS. This he did to bring UNO into active participation just as the country of Afghanistan was achieving a peak of Cold War competition between the USA and the USSR. Chancellor Del Weber and the late Vice Chancellor Otto Bauer at UNO also gave the author great support for years, even while their successors in later years did not quite seem to realize or care that Afghanistan really did matter as much as it many thought that it did. With my retirement to Emeritus status in 2011 and reappointment as the Senior Research Scholar at the CAS enabled continuation of the work through renewed grant monies.

Drafting of the various graphics in this book came from many sources but special thanks goes to former students, Mr Brandon Weihs, now a doctoral candidate at Kansas State University. Professor Breckle gave permission to republish a number of his excellent graphics in Chapter 4 as well, for which he is greatly thanked.

The Afghans who helped along the way are too numerous to mention but a few who are especially noteworthy are former student Lutfullah Safi, as well as Professor (later Rector), Hamidullah Amin, Drs Ismael Burhan and Abdul Yaseer, and Sher Jan Ahmadzai. Ms Sahista Wahab, the head of our Arthur Paul Afghanistan Library collection, has looked after the author's Afghanistan maps and files very well for years.

Finally, dear wife, Susie Nye, has supported me through the thick and thin on production of this book, and I truly must say that had she not given me all the uncompromising business acumen and culinary expertise and effort to keep me contented, this work could not have been done. I am madly grateful! Grandson John Ficenec seems to have bitten into the fruit of knowledge too, as he seeks to become a photographer/writer in the mode of his grandfather. And finally, to my little namesake grandson, Henry Shroder Stubbe,

I wish you such a bright life because you kept me laughing all through the writing of this work.

References

Shroder Jr., J.F., March 1987. Afghanistan resources and the Soviet Union. Geotimes 32, 4–5.
Shroder Jr., J.F., October 5, 2003. Reconstructing Afghanistan: nation building or nation failure? Geotimes.
Shroder, J., October 2004. Afghanistan redux: better late than never? Geotimes, 34–8.
Shroder Jr., J.F., 2009. Saving Afghanistan: redevelopment one resource at a time. Earth 54 (7), 38–47.
Shroder, J.F., September 2013. Building resource corridors in Afghanistan: a solution to an interminable war? Earth, 25–31.

Foreword

Timeliness, content, context are essential to any study of a nation as storied and complex as Afghanistan. John (Jack) Shroder effectively employs each of these qualities in his book, *Natural Resources of Afghanistan*.

Following decades of war and destruction, it is the right time for a penetrating examination of Afghanistan's natural resources as that nation continues its difficult reconstruction with anticipated reduced assistance from international partners. *Natural Resources of Afghanistan* provides the historical and social context that enables both the scholar and the casual reader the means to understand the problems and opportunities facing Afghans at this crucial juncture.

The depth and length of Professor Shroder's commitment to the study of a singular country's natural resources are most uncommon. The result is a book rich in content regarding what only a few scholars and intelligence personnel understood—Afghanistan possesses natural resources substantial enough, if properly exploited and used, to underwrite significant portions of its future rebuilding and development.

I first met Jack Shroder in the winter of 1973–1974. A member of a delegation from the University of Nebraska Omaha (UNO), he and three colleagues arrived in snow-bound Kabul several days behind schedule, after a rather harrowing journey. (Jack describes this trip in the Preface.)

I was then the Executive Director of the Afghan-American Education Commission, which oversaw Fulbright and other education exchanges between the USA and Afghanistan. I was in charge of coordinating the schedule of the UNO delegation, which had been reduced to 3 days due to the delayed arrival. One of the delegation's objectives, in which it was successful, was to persuade me to move to UNO to head up that institution's newly created Center for Afghanistan Studies. Thus was born a professional and personal association between Jack and myself that has now spanned an uninterrupted four decades.

Jack's challenging personal introduction to Afghanistan must have appealed to his geoscience nature. He set aside previous focuses on the geology of Malawi and tree rings in Utah and invested himself fully into research regarding Afghanistan.

He pursued his commitment to Afghanistan Studies with what at times seemed a curmudgeonly dedication. His efforts contributed to the growth, reputation, and reach of the Center for Afghanistan Studies; and, Jack became an internationally recognized authority on Afghanistan's earth sciences.

Jack's second trip to Afghanistan was no less inauspicious than his first. The recipient of a Fulbright Fellowship, he was obliged to leave Kabul, like many other Americans, due to a pro-Soviet coup and the changed political climate. He returned to the area on another Fulbright Fellowship in 1983–1984 to work with Kabul University colleagues in exile in Pakistan. Jack took full advantage of these new surroundings to begin research on the Karakoram Range and regional glacial and water studies. Since then, he has made numerous trips to South Asia to collaborate with Pakistani, Indian, and Nepalese scholars.

Since 9/11, Jack has returned several times to Afghanistan to reestablish connections and to begin collaboration on research with a new generation of Afghan earth scientists. Due to his experiences in South Asia, Jack has also been instrumental in facilitating research dialogue and activities between Afghans and other scholars in South Asia.

The expanse and diversity of these experiences have enhanced the breadth of knowledge Jack has been able to incorporate into his research efforts and is now able to share with others in the *Natural Resources of Afghanistan*.

The scholar will appreciate the sections that focus on Afghanistan's environment and resource base, and their relationship to development; others will appreciate how Jack has tied the resources together with history and global politics. All will be interested in the scenarios, both optimistic and pessimistic, that Jack paints regarding Afghanistan's utilization of its natural resources in future development.

Natural Resources of Afghanistan is an engaging scholarly read, containing interesting facts and anecdotes, as well as critical scientific information. Jack's writing style is sometimes brusque, at other times, delightfully colorful; it is always entertaining, fully cited, and scientifically sound.

Natural Resources of Afghanistan is a major contribution to the study of Afghanistan. It affords all readers an informed look into a country whose natural and human resources are generally underestimated, into a nation often misrepresented and misunderstood, into a proud and independent people who take pride in their reputation for the hospitality they eagerly provide their friends.

Thomas E. Gouttierre
Senior Officer for Global Engagement
Dean, International Studies and Programs
Director, Center for Afghanistan Studies
University of Nebraska Omaha

Preface

Capt Henry Drummond, a member of the British 3rd Bengal Light Cavalry who was part of the invasion force into Afghanistan in 1839, described his geological rambles through the wildest parts of the country to conduct a first Western mineral survey (Drummond, 1841). He found abundant green stains and different ores of copper at Aynak and elsewhere, some of which rivaled the great deposits of Chile at the time, as well as rich veins of iron ore, both deposits of which in 2013 have new mines finally contracted and planned after more than a century and three quarters later. Although many of his countrymen viewed Afghanistan as an unruly place, where a man could not stir many yards from his home or tent without the risk of being assassinated, Drummond wandered without arms and lived among the people on many occasions. He was of the opinion that in any attempt to develop the resources of the country, an acquaintance with the character of the inhabitants was a matter of serious consideration. He thought that given the keen commercial spirit of the nation, if the Afghans could be convinced of British friendship, then no end of good things could happen. Mining, he felt—not the gun—offered the best hope to pacify the territory and win over the Afghans.

> *"Give them, however, but constant employment, with good wages and regular payment; encourage a spirit of industry, both by precept and example; let strict justice be dealt out to them without respect of persons; and we shall shortly see their swords changed into plowshares, industry take place of licentiousness, and these people be converted into peaceable and useful subjects,"*

Drummond opined.

But the Afghans were not keen on the idea of handing over their minerals to occupiers, or for that matter, on tolerating the British occupation itself for much longer. A year later the Afghans massacred most of the British army who had remained in country after Drummond had gone back to India.

If more people had paid attention to Drummond long ago, things might be very different today. As Karr (1849) first noted at about the same time, "Plus ça change, plus c'est la même chose." The more things change the more they remain the same. Indeed, in Afghanistan the irony is that the issue of its natural resources appears to have been an item of some attention for a rather long time. This book is an assessment of many of the aspects of Afghanistan that are responsible for the situation in which the country finds itself today. It is possible that these resources, which as can be seen from Drummond's early words, may be at least a partial solution to Afghanistan's future betterment.

References

Drummond, H., 1841. On the mines and mineral resources of Northern Afghanistan. Journal Asiatic Society of Bengal 10, 74–93.
Karr, J.-B.A., 1849. Les Guêpes. Paris.

List of Acronyms[1]

AAF	Afghan Air Force
ACI	Afghan Cartographic Institute
ADKN	Afghanistan Disaster Knowledge Network
ADP	Asian Development Bank
AFG-FHM	Afghanistan Flood Hazard Map
AGE	Advisory Group on the Environment (Afghanistan)
AGS	Afghanistan Geological Survey
AID	Agency for International Development (United States)
AJP	Afghanistan Justice Project
APRP	Afghanistan Peace and Reintegration Program
ALP	Afghanistan Local Police
ANAAC	Afghan National Army Air Corps
AOI	Area of interest
AST	ASTER satellite spectral signatures
AVIRIS	Airborne Visible Infra-Red Imaging Spectrometer
CE	Common Era (substitute for AD—Anno Domini)
CIA	Central Intelligence Agency (United States)
BCE	Before Common Era (substitute for BC—Before Christ)
Bsh	Warm steppe (climate)
Bsk	Cold steppe (climate)
Bwh	Warm desert (climate)
Bwk	Cold desert (climate)
Csb	Mediterranean (climate)
Csa	Humid subtropical (climate)
CAPTF	Combined Air Power Transition Force
CB	Badakhshan (fault)
CH	Chaman (fault)
CHKRCS	Cross Hindu Kush Resource Corridor Segment
D	Humid continental (climate)
DC	Development Corridor
DEM	Digital elevation model
DOD	Department of Defense (U.S.)
DTED	Digital terrain elevation data
DTSS	Digital topographic support system
E	East
EaPaHiKu	Eastern Pamir - Hindu Kush Geomorphic Province
EI	Extractive industries
EIA	Environmental Impact Assessment
EITI	Extractive Industries Transparency Initiative
EIVC	Extractive Industries Value Chain
EP	Equator Principles
ET	Tundra and ice (climate)
EW	East–west
FAO	Food and Agriculture Organization (United Nations)

[1] *Chemical formulas and element symbols not included.*

GEO	Geophysics (gravity and magnetics)
GDP	Gross domestic product
GFDL	Geophysical Fluid Dynamics Laboratory (Princeton University)
GIRoA	Government of the Islamic Republic of Afghanistan
GLASOD	Global assessment of soil degradation
GLIMS	Global Land Ice Measurements from Space
GoA	Government of Afghanistan
HIV/AIDS	Human immunodeficiency virus infection/acquired immunodeficiency syndrome
HKH	Hindu Kush—Himalaya (mountains)
HR	Hari Rud (Rod) (fault)
HSD	HyMap spectral data
HVA	Helmand Valley
HVP	Helmand Valley Project
IAP	International Cooperation Panel (Government of Afghanistan)
ICA	International Cooperation Administration
ICRC	International Committee of the Red Cross
IED	Improvised explosive device
IPB	Intelligence preparation of the battlefield
IPCC	Intergovernmental Panel on Climate Change (U.N.)
ISAF	International Security Assistance Force
ISI	Directorate of Inter-services Intelligence (Pakistan military)
ITCZ	Intertropical Convergence Zone
IWA	Integrity Watch Afghanistan
JOG	Joint Operations Graphic (U.S. DOD map)
KGB	Komitet gosudarstvennoy bezopasnosti; Committee for State Security (U.S.S.R.)
KGPS	Kilometric global positioning system
KU	Kabul University
LiDAR	Light Detecting and Ranging system
LLOF	Landslide lake outburst flood
LREE	Light rare earth elements
LTERA	Land Titling and Economic Restructuring Activity (USAID)
METI	Ministry of Economy Trade, and Industry (Japan)
MDG	Millennium development goals
MK	Morrison Knudsen Co.
MoAIL	Ministry of Agriculture, Irrigation and Livestock
MoCI	Ministry of Commerce and Industries (Afghanistan)
MoEP	Ministry of Energy and Power (Afghanistan)
MoF	Ministry of Finance (Afghanistan)
MoM	Ministry of Mines (Afghanistan)
MoPW	Ministry of Public Works (Afghanistan)
MoTCA	Ministry of Transport and Civil Aviation (Afghanistan)
mya	Million years ago
NAPA	National Adaptation Programme of Action to Climate Change (Afghanistan)
NASA	National Aeronautics and Space Administration (U.S.)
NATO	North Atlantic Treaty Organization
NDM	Northern Distribution Network
NDMC	National Disaster Management Commission (Afghanistan)
NDMP	National Disaster Management Plan (Afghanistan)
NDRRP	National Disaster Risk Reduction Plan (Afghanistan)
NEPA	National Environmental Protection Agency (Afghanistan)
NHRCS	Northern Hydrocarbon Resource Corridor segment
NIMA	National Intelligence Mapping Agency (United States)

NOAA	National Oceanic and Atmospheric Administration (United States)
NPP	National Priority Program
NRRCP	National and Regional Resource Corridors Program
NS	North–south
NW	Northwest
NWFP	North-West Frontier Province (Pakistan)
P	Pressure (atmospheric)
PAH	Polyaromatic hydrocarbons
PIA	Pakistan International Airlines
PIP	Public Investment Program for Environmental Management (Afghanistan)
PDPA	People's Democratic Party Afghanistan
PPP	Private-public-partnerships
RAAF	Royal Afghan Air Force
RC	Resource Corridor
RGA	Royal Government of Afghanistan
RLAP	Rural Land Administration Project
SECRCS	Southeast Copper Resource Corridor Segment
SERCLS	Steel and Energy Resource Corridor Link Segment
SSE	South–southeast
SME	Small and Medium enterprise
SMR	Soil-moisture regime
STR	Soil-temperature regime
T	Temperature (atmospheric)
TAP	Trans-Afghanistan Pipeline
TAPI	Turkmenistan-Afghanistan-Pakistan-India (pipeline)
TDA	Tactical decision aids
TFBSO	Task Force for Business and Stability Operations (U.S. military)
UK	United Kingdom
UNAMA	U.N. Assistance Mission to Afghanistan
UNEP	U.N. Environmental Program
UNHRC	U.N. High Commissioner for Refugees
UNICEF	U.N. Children's Fund
UNO	University of Nebraska at Omaha
US	United States
USA	United States of America
USACE	U.S. Army Corps of Engineers
USAID	United States Agency for International Development
USDA	United States Department of Agriculture
USDOD	U.S. Department of Defense
USGS	U.S. Geological Survey
USICA	U.S. International Cooperation Administration
USSR	Union of Soviet Socialist Republics
VTU	Soviet MilitaryTopographic Directorate
W	West
WSW	West–southwest
WWII	World War II

In this Chapter

1

Introduction: Historical Overview of Afghanistan at War

Abstract

This introductory chapter to the book about Afghanistan's resources and environment explains the author's initial introduction and travel to Afghanistan more than 40 years ago. Subsequent multiple returns to the country and long-term extensive research enabled production of this book, as divided into four parts: I—geology and geography; II—resources; III—environment; and IV—pessimistic and optimistic future scenarios of possibility for the country. The historical overview of Afghanistan details chief points of the last two centuries of violent warfare and chief political occurrences in the country, as well as a chronology of the main historical events from the 1830s to the present day. This sets the stage for evaluating Afghanistan's resources and environment from different perspectives concerning certain geographic and geologic aspects that have contributed to the centuries of conflict in the region.

Keywords: Abdur Rahman Khan; Afghanistan history; Amanullah; Anglo–Afghan Wars; Daoud; Mujahideen; NATO (ISAF) withdrawal plans; Soviet–Afghan War; Taliban; US-led invasion.

This book on the resources and environment of Afghanistan, a country upon which I have expended considerable mental effort for over four decades, is an enigma wrapped in a South Asian cocoon of romance and my scientific fascination with its landscapes and resources, coupled with no little disgust at the so-dreadful, but also heartwarming human conditions exhibited there. I became fascinated with the place as an ardent young associate professor when first arriving in Afghanistan in the deep and snowy winter of 1973 on a flight from New Delhi to Kabul that instead was diverted to Kandahar by bad weather. After hours of wrangles with immigration and customs officials in the middle of the night, we then proceeded to drive in a bus into the teeth of a Hindu Kush winter gale toward Kabul with only half of a front windshield partially covered with a blanket, and a corpse in a pine box on the roof with a wailing brother on the seat below. This was indeed, a fitting first entry into a country that was itself practically a moribund relic of ancient civilizations seeking entry into a more

Natural Resources in Afghanistan. http://dx.doi.org/10.1016/B978-0-12-800135-6.00001-5

idyllic future; even if it was to be only a dodgy great powers game being played out by a small, hopelessly incompetent buffer state, trapped forever between a heartless Cold War bear and a powerful Uncle Sam. This was not an auspicious beginning, and it grew worse thereafter, but still the "oriental romance" of the place had a certain staying power that enticed the dogged scientist in me to stay on and try to comprehend, even perhaps to the point of foolishness in the face of implacable problems and people.

But Afghanistan has a way of sticking like a Nebraska cocklebur beneath the saddle of an earnest geomorphologist and resource specialist. The ride was not smooth, with numerous bucks and twists, but staying on board for the long count was ultimately highly rewarding to me, as also emerged the undoubted considerable resource rewards possible to a sore-beleaguered people who have little choice but to endure their mortal ride in that intractable place. When I was later deported from Kabul just before the Soviets invaded Afghanistan, I surely felt done with the place, yet within a few years it seemed reasonable for another dose, this time to help the Mujahideen in Peshawar as they fought the Soviet juggernaut. I did not return as a soldier or a spy, but rather as an educator with the belief that maintaining the Afghan educational system in exile was better than the alternatives of nothing for their youth. I and my Afghan colleagues failed in the early 1980s to produce an *Azad Afghan Pohuntoon* (Free Afghanistan University) but collateral efforts by the University of Nebraska at Omaha (UNO) to keep some form of primary and secondary schooling continuous in what passed into decades of war, did become successful (Shroder, 2007). At one time in the mid-1980s, UNO's efforts in Peshawar employed more people inside Pakistan and Afghanistan than in the entire statewide University of Nebraska system at the time.

Because I became *persona non grata* in Afghanistan throughout the 1980s during the Soviet–Afghan War, I had to be careful in Peshawar. This became acute when Chancellor Weber, Dean Gouttierre, and I were singled out for newspaper notice by Tudeh, the Iranian Communist Party in league with the Soviet Union's 'Komitet gosudarstvennoy bezopasnosti' (KGB), because we three purportedly were really bad people who were trying to mislead young Muslim minds. Consequently, recognition dawned on me that my life might be a trifle vulnerable on a bicycle surrounded by people and standing by the public tandoor to obtain my nightly *naan*. When both the Pakistani Interservices Intelligence (ISI) and the USA Central Intelligence Agency (CIA) asked to come to have *chai* and request that I seek greater defendable housing and a less visible presence in public in Peshawar, worry became a habit. I retreated into work on glaciers and landforms in the Karakoram and on Nanga Parbat, which made the crevasse danger and earthquake landslides seem quite safe by comparison. But then later from afar, as it were, I and my friends kept eyes on Afghanistan anyway, and when possible, new satellite imagery and more maps of Afghanistan allowed progressive understandings. The original project in the country in the 1970s as Director of the National

Atlas of Afghanistan never quite went away; it just retreated into the background with the map collections waiting for an opportunity when they might be needed again. Now is such a time.

This book is an attempt to bring together a somewhat disparate mélange of historical fact, together with information on the geological substrate and its landforms, the despoiled environment, and an unclear future with its rich resource base, into a coherent whole that can serve as a partial guide to perhaps help bring Afghanistan a little way back from its long nightmare. With any kind of good luck, strong politics, and military protection to keep the lid on the pot, the country is poised on the brink to take off economically with its resources, which after more than 30 years of war, could not happen a moment too soon. This is especially poignant because the Afghans and Russians knew quite clearly about most of these resource facts more than 30 years ago, but it has taken the outside world a bit more time to catch on. Now is the question of whether the people of Afghanistan will be allowed to prosper like so many other Asians are doing at the moment, or will instead so many more Afghans be forced again to live a life of abject poverty and fear of reprisal or death from people who would like to control Afghanistan for their own financial benefit. It is a zero-sum game all around, with many different kinds and political flavors of spoilers trying to get in on some of the resource action. Only strong pressure against such machinations with linkages amongst the people outside, such as artful agreements between different combinations of China, India, and Pakistan, and possibly Iran, can the violence be stemmed enough to allow sufficient peaceful progress for the resources to produce sufficient revenues for the economy to jumpstart.

After this introduction with an historic overview, the book is divided into five parts: Part I is a look at the natural physical environment of bedrock, landforms, climate, water, soils, and vegetation as the foundation for the ethnic diversity, long-term travel along the multistranded Silk Road through Afghanistan, and hazards and disasters that periodically assault the Afghan people with horrendous earthquakes, landslides, dust storms, floods, and droughts. Part II is on the now internationally recognized, rich resources of Afghanistan and their possible uses to rebuild a war-shattered nation. Part III is a look at the despoiled environment of Afghanistan that has been under natural and human stress for centuries and now with the incipient resource boom, is threatening to get even worse, if such can be imagined. Development and redevelopment issues are examined, as well as environmental degradation and the new efforts and laws about environmental protection. Excellent new environmental legislation has been implemented but we will have to wait to see whether or not this will work in a milieu where corruption is so pervasive. The last chapter in this section is an examination of new climate models that make predictions of increased moisture in the north and decreased precipitation in the south, which is likely to make the political situation in the country even more contentious, because the south is where the bellicose Pashtun tribes hold sway. Already the increasing drought stress has driven more and more farmers deeper into the clutches of the drug mafias as they have to plant drought-tolerant opium poppies to survive. Additional aridity stress piled onto an already volatile situation is obviously not a preferable situation. Finally, Part IV is an either-or look at positive and negative potential futures for examination; a negative set of scenarios that occur in a fashion recognizable from the history of Afghanistan wherein outsiders continue to foment troubles for the country to further their own agendas designed to gain power over territory and resources; and a positive approach in which the excellent new ideas are put forth to lift the country by its resource bootstraps and set it on a firm course of development with the new revenues so generated. The positive scenarios are thought to occur potentially because all of the countries neighboring Afghanistan recognize that there is a good deal to gain and much more benefit for them to suppress the violence and get the resources out of the ground, than there is for the alternatives. What would be an intelligent thing to do in that part of the world is not however, what too often occurs; the partition of the British Raj into India and Pakistan in 1947 is an example, as is the abortive excursion into communism for Afghanistan in the 1980s. Many other negative scenarios are always possible in that volatile stew-pot of people and ideas.

This book is meant to provide a compendium of information about the Afghanistan resources and physical environment, together with enough historical and general geographical information to understand some of the difficult situations and effects on complex human behavior that play out in the diverse country. The likelihood is high that Afghanistan will continue to need foreign advisors on into the foreseeable future; this book is meant to serve as a partial guide to some of what many of them will need to know to do their jobs more effectively. The reference lists also bring in a great many less well-known sources, particularly the considerable "gray" literature of so much of the Internet, which because of the difficulties and time-consuming nature of such web searches, is gathered here as a new web tool, although I make no pretenses of an exhaustive web search, if such even be possible.

Historical Overview

This historical overview of Afghanistan at war is provided to give the interested reader a simplified timeline account of the historical events, places, and personages who figured most prominently in establishing the situation in which the country now finds itself. This account is not intended as a serious historical analysis to find motivations and reasons for everything that occurred, but just to provide basic understandings to the reader who may wish to know more about why or how certain actions were taken or behaviors occurred in the past, as well as to better understand how resources and the environment fit into the historical picture.

As an overview then, Afghanistan history, the "graveyard of empires", appears superficially at least, to have played out as centuries of warfare leavened by some brief periods of quiet. Only the last two centuries of Afghan history are presented in summary form here, mainly to give some limited sense of the difficulties under which Afghans have had to live for so long, to those readers who may require that background for better understanding. Most newcomers to Afghanistan are immediately struck by the seemingly incessant war in the place, which does demand some overview and explanation, if however cursory.

In the nineteenth and early twentieth centuries, Afghanistan fought three major wars with the British, the first one of which in the early 1840s, ended with the massacre of almost all of the British troops and their families. The reprisals by the British thereafter and a second foray by the British again in the last quarter of the nineteenth century only cemented the dislike of the Afghans against the foreign invader as the "Great Game" unfolded. The expanding British Raj faced off against the growing Russian Empire, with Afghanistan caught between as a constricted buffer state whose very ability to have a foreign policy with other nations was not allowed. In the early twentieth century, the Afghans having finally had quite enough, invaded British India just for a few kilometers and with a weak enough response to be driven back. This did, however, initiate a new dialogue with the World War I-weary British so that the Afghans finally won their freedom to engage with the world outside their borders. The recently Sovietized Russians arrived in Kabul soon thereafter and began a new dynamic that decades down the line would not work out so well for either side. In the meantime, modernization attempts by the Pashtun Afghan monarchy led to a first Afghan Civil War during which a Tajik took over for a short while before the Pashtun monarchy was reestablished by Nadir Shah. The so-called "Golden Age" of the Pashtun Musahiban Dynasty lasted throughout the 1930s to the end of the 1970s when Afghan communists took over, mismanaged everything into a complete shambles, and the Soviets finally fully invaded in 1979, mainly to prop up the failing government. A decade of horrific war followed throughout the 1980s, filled with atrocities from both sides, before the Soviets were forced to withdraw in the face of a stalemate as the USA poured in weapons and financial support for the Mujahideen resistance. During this period the Islamist supporters from elsewhere who came to fight the Soviet juggernaut imported a particular brand of a fundamentalist or Salafist view of a vastly more restrictive and warped fundamentalist Islam, where such had never existed in Afghanistan to any real extent before. This brought an entirely new and most difficult framework of oppression into the country.

The second Afghanistan Civil War erupted in the decade of the 1990s when the Soviets retired from the country, essentially in defeat, and the largely Pashtun Taliban first rose to dominance and quickly gained support from Pakistani intelligence. Founded against the many forms of depravity and criminality that existed in the country, the largely uneducated, illiterate, and ignorant Taliban soon took over most of the country by extreme force, unrelenting violence, and genocide against the other ethnic groups with whom they had long been in competition. Their oppression of women and anyone who did not espouse their fundamentalist and ill-informed opinions, as well as providing sanctuary for terrorists who committed violence in the name of their warped view of Islam elsewhere in the world, was opposed by the Northern Alliance of non-Pashtun Tajiks, Uzbeks, Hazara, and other ethnic groups of the country. Things finally came to a head and boiled over when Osama bin Laden and his al Qaeda organization hiding in Afghanistan began to unleash attacks against the US embassies in East Africa, the USS Cole warship, and the 9/11 attacks with jet aircraft into buildings in New York and Washington.

War with the USA and its North Atlantic Treaty Organization (NATO) and other allies went on for more than a decade after their invasion of Afghanistan in late 2001. At first regarded as liberators of a sorely oppressed nation, within a few years in the face of being largely ignored by the invading powers, Taliban resistance returned and grew. As the Western Coalition troops instead went blithely off to a new and largely unwarranted war in Iraq, things began to steadily fall apart in Afghanistan, until finally it was quite clear that the withdrawal of yet another invading empire was required. That withdrawal is underway as this book goes to press.

An abbreviated timeline of wars in Afghanistan (Anon, 2013) is included here to further facilitate the reader's efforts to follow the complex history of the country. This whole historical section hereafter can be safely ignored, however, by the reader who is only interested in the resources and environmental issues that will so affect the future of Afghanistan, but because the history of Afghanistan is so replete with recurrent violence, some accounting of these past occurrences is necessary, perhaps for better understanding of future events.

Timeline of Wars in Afghanistan

Although many wars have been fought in Afghanistan far back into antiquity, including those of Alexander the Great, Ghengis Khan, Tamerlane, and Emperor Babur in founding the Mogul Dynasty rule over India, only the wars of the last two centuries are included here. This timeline of the incessant violence and wars in Afghanistan is exemplary of why that fractious country has been such a focal point of contention in Southwest Asia for so long, as well as why the great powers of the day have been so caught up in attempted domination of the nation of Afghanistan, generally to their eventual considerable dismay.

Chronology of Events

1838 to 1842—British forces invaded and installed King Shah Shujah. He was assassinated in 1842. During their forced retreat from Kabul, British and Indian troops were massacred with few survivors. (*See*, Preface about Captain Henry Drummond and his interest in resources in Afghanistan.)

1878 to 1880—British invaded Afghanistan over fears of Russian advances as the Great Game unfolded. A treaty and installation of Iron Amir Abdur Rahman Khan give Britain control of Afghan foreign affairs. (*See*, Chapter 10 concerning Greisbach's plentiful geological and resource investigations.) The Durand Line was subsequently established to divide Pashtuns and enable British rule over the half left in British India.

1919—Emir Amanullah Khan declared independence from British influence. Afghans invaded British India. The third Anglo–Afghan War was brief. (*See*, Chapter 10 and arrival of Russian geologist Obruchev to survey the resources in Afghanistan.)

1926 to 1929—Amanullah tried to introduce social reforms, which however, stirred civil unrest. He fled Afghanistan.

1933—Zahir Shah became the King and Afghanistan remained a monarchy for the next four decades.

1953—General Mohammed Daoud became the prime minister and turned to Soviet Union for economic and military assistance. (*See*, Chapter 11 on Soviet discoveries of rich resources.) Daoud introduced social reforms, such as abolition of purdah (practice of secluding women from public view).

1963—Mohammed Daoud forced to resign as prime minister.

1964—Constitutional monarchy introduced—but led to political polarization and power struggles.

1973—Mohammed Daoud seized power in a coup and declared Afghanistan a republic. He tried to play off the USSR against Western powers.

1978—General Daoud overthrown and killed in a communist coup. Rule began by Tarakhi and Amin. The beginnings of armed revolt were started by the Mujahideen.

Soviet intervention

1979 December—Soviet Red Army invaded and propped up communist government.

1980—Babrak Karmal installed as ruler, backed by Soviet troops. But antiregime resistance intensified with various Mujahideen groups fighting Soviet forces. The US, Pakistan, China, Iran, and Saudi Arabia supplied money and arms.

1985—Mujahideen came together in Pakistan to form official alliance against Soviet forces. Half of Afghan population estimated to be displaced by war, with many fleeing to neighboring Iran or Pakistan.

1986—The US began supplying Mujahideen with Stinger missiles, enabling them to shoot down Soviet helicopter gunships. Babrak Karmal replaced by Najibullah as head of Soviet-backed regime.

1988—Afghanistan, the USSR, the US, and Pakistan signed peace accords and Soviet Union began pulling out troops.

Red Army quits

1989—Last Soviet troops left, but Civil War continued as Mujahideen pushed to overthrow Najibullah.
1992—Najibullah's government toppled but a devastating Civil War followed.
1996—Taliban seized control of Kabul, executed Najibullah, and introduced hard-line version of Islam, banning women from work, and introducing Islamic punishments, which included stoning to death and amputations.
1997—Taliban recognized as legitimate rulers by Pakistan and Saudi Arabia. At that time they controlled about two-thirds of country.
1998—The US launched missile strikes at suspected bases in Afghanistan of militant Osama bin Laden, accused of bombing the US embassies in Africa.
1999—The UN imposed an air embargo and financial sanctions to force Afghanistan to hand over Osama bin Laden for trial.
2001 September—Ahmad Shah Masoud, leader of the main opposition to the Taliban—the Northern Alliance—was assassinated by al Qaeda.

US-led invasion

2001 October—US-led bombing of Afghanistan began following the 9/11 attacks on the United States. The antiTaliban Northern Alliance forces entered Kabul shortly thereafter.
2001 December—Afghan groups agreed to a deal in Bonn, Germany for an interim government. Hamid Karzai was sworn in as head of an interim power-sharing government.
2002 January—Deployment of first contingent of foreign peacekeepers—the NATO-led International Security Assistance Force (ISAF)—marking the start of a protracted fight against the Taliban.
2002 April—Former King Zahir Shah returned, but made no claim to the throne and died in 2007.
2002 June—Loya Jirga, or grand council, elected Hamid Karzai as interim head of state. Karzai picked members of his administration, which was to serve until 2004.
2003 August—NATO took control of security in Kabul, its first-ever operational commitment outside Europe.

New constitution

2004 January—Loya Jirga adopted new constitution which provided for strong presidency.
2004 October–November—Presidential elections. Hamid Karzai was declared winner.
2005 September—Afghans voted in first parliamentary elections in more than 30 years.
2005 December—Parliament opened with warlords and strongmen in most of the seats.

2006 October—NATO (ISAF) assumed responsibility for security across the whole of Afghanistan, taking command in the east from a US-led coalition force.
2007 August—Opium production soared to a record high, the UN reported.
2008 June—President Karzai warned that Afghanistan would send troops into Pakistan to fight militants if Islamabad failed to take action against them.
2008 July—Suicide bomb attack on Indian embassy in Kabul killed more than 50.
2008 September—US President George Bush sent an extra 4500 US troops to Afghanistan, in a move he described as a "quiet surge".
2009 January—US Defense Secretary Robert Gates told Congress that Afghanistan is the new US administration's "greatest test".
2009 February—NATO countries pledged to increase military and other commitments in Afghanistan after the US announced dispatch of 17,000 extra troops.

New US approach

2009 March—US President Barack Obama unveiled new strategy for Afghanistan and Pakistan to combat an increasingly "perilous situation". An extra 4000 US personnel would train and bolster Afghan army and police, and new support for civilian development would also occur.
2009 August—Presidential and provincial elections were marred by widespread Taliban attacks, with a patchy turnout and claims of serious fraud.
2009 October—Karzai declared winner of August presidential election, after second-placed opponent Abdullah Abdullah pulled out before the second round.
2009 December—US President Barack Obama decided to boost US troop numbers in Afghanistan by 30,000, bringing total to 100,000. He said the US would begin withdrawing its forces by 2011. An al Qaeda double agent killed seven CIA agents in a suicide attack on a US base in Khost.
2010 February—NATO-led forces launched major offensive, Operation Moshtarak, in bid to secure government control of southern Helmand province.
2010 July—Whistleblowing website Wikileaks published thousands of classified US military documents relating to Afghanistan. General David Petraeus took command of US and ISAF forces.
2010 August—Dutch troops adverse to combat, withdrew. Karzai said private security firms—accused of operating with impunity—must cease operations. He subsequently watered down the decree.
2010 September—Parliamentary polls marred by Taliban violence, widespread fraud and a long delay in announcing results.
2010 November—NATO—at a summit in Lisbon—agreed to plan to hand control of security to Afghan forces by the end of 2014.

2011 January—President Karzai made first official state visit to Russia by an Afghan leader since the end of the Soviet invasion in 1989.

2011 February—Number of civilians killed since 2001 invasion hit record levels in 2010, Afghanistan Rights Monitor reported.

2011 April—Burning of Koran by a US pastor prompted countrywide protests in which foreign UN workers and several Afghans were killed. Some 500 prisoners, mostly Taliban broke out of jail in Kandahar.

2011 May—Osama bin Laden killed by US Navy Seals in Abbottabad, Pakistan. President Obama indicated the US would withdraw from Afghanistan by 2014.

2011 July—President's half-brother and Kandahar governor Ahmad Wali Karzai killed in Taliban campaign against prominent figures.

2011 September—Ex-President Burhanuddin Rabbani—a go-between in talks with the Taliban—assassinated by a suicide attacker with a bomb in his turban.

2011 October—As relations with Pakistan worsened after a series of attacks, Afghanistan and India signed strategic partnership to expand cooperation in security and development.

2011 November—President Karzai indicated the US needed to stay in Afghanistan for at least a decade beyond the planned withdrawal of 2014.

2011 December—About 58 people were killed in twin attacks at a Shia shrine in Kabul and a Shia mosque in Mazari Sharif. Pakistan and Taliban boycotted the scheduled Bonn Conference on Afghanistan. Pakistan refused to attend after a NATO air strike killed Pakistani soldiers on Afghan border.

2012 January—Taliban agreed to open an office in Dubai as a move towards peace talks with the US and Afghan government.

2012 February—At least 30 people killed in protests about burning Korans at US Bagram airbase. US officials believed Taliban prisoners were using the books to pass messages, and that they were extremist texts but not Korans. Two soldiers are also killed in reprisal attacks.

2012 March—US Army Sgt. Robert Bales was accused of killing 16 civilians in an armed rampage in the Panjwai district of Kandahar. His guilty plea and defense in May–June 2013 claimed posttraumatic stress disorder from multiple deployments in Iraq and Afghanistan, and resulted in a penalty of life in prison, which allowed him to escape the death penalty.

2012 April—Taliban announced "spring offensive" with audacious attack on diplomatic quarter of Kabul. The government blamed the Haqqani Network. Security forces killed 38 militants.

NATO withdrawal plans

2012 May—NATO summit endorsed plan to withdraw foreign combat troops by end of 2014.

New French President Francois Hollande indicated France would withdraw its combat mission by the end of 2012—a year earlier than planned.

Arsala Rahmani of High Peace Council was shot dead in Kabul. A former Taliban minister and he was crucial in reaching out to rebel commanders. The Taliban denied responsibility.

2012 July—Tokyo donor conference pledged $16 billion in civilian aid to Afghanistan up to 2016, with the US, Japan, Germany, and the UK supplying the bulk of funds. Afghanistan agreed to new conditions to counter corruption.

2012 August—US military disciplined six soldiers for accidentally burning copies of Koran and other religious texts in Afghanistan. They would not face criminal prosecution. Three US Marines were also disciplined for a video in which they were seen urinating upon the bodies of dead Taliban fighters.

2012 September—The US handed over Bagram high-security jail to Afghan government, although it retained control over some foreign prisoners until March 2013. The US also suspended training new police recruits in order to carry out checks on possible ties to Taliban following series of attacks on foreign troops by apparent police and Afghan soldiers.

2013 February—President Karzai and Pakistan's Asif Ali Zardari agreed to work for an Afghan peace deal within six months after talks hosted by Britain's Prime Minister David Cameron. They backed opening of Afghan office in Doha and urged the Taliban to do the same for talks to take place.

2013 March—Two former Kabul Bank chiefs, Sherkhan Farnood and Khalilullah Ferozi, jailed for multimillion dollar fraud that almost led to its collapse and that of the entire Afghan banking system in 2010.

President Karzai met emir of Qatar for talks believed to pave the way for the opening of a Taliban office in the Gulf emirate.

2013 May—New Pakistani Prime Minister Nawaz Sharif elected and promised to curtail American drone strikes against militants in tribal areas adjacent to Afghanistan.

2013 June—Afghan Army took command of all military and security operations from ISAF forces.

President Karzai suspended security talks with the US after Washington announced it was planning to hold direct talks with the Taliban. Afghanistan insisted on conducting the talks with the Taliban in Qatar itself.

2013 June—The US drone strikes in Pakistan killed at least seven militants in North Waziristan.

Newly appointed Pakistani Minister for Water and Power, Khwaja Muhammad Asif, was charged with addressing the urgent energy crisis that has caused widespread economic distress and general misery.

July 2013—A 4-h firefight at the Kabul Airport ended with seven militants killed by Afghan forces.

2013 October—Registration deadline for presidential elections to replace Karzai brought out 28 candidates: Zalmay Rassoul, a favorite of Karzai; Abdullah Abdullah, number two against Karzai in 2009; Qayum Karzai, president's older brother but not favored by Karzai; Abdul Rab Rassoul, accused of war crimes; Ashraf Ghani, former World Bank expert; and Gul Agha Shirzai, former provincial governor and accused pedophile (*bacha bazi*).

President Karzai blasted the US role in the Afghan War, saying that NATO coalition had failed to deliver what it promised and had inflicted needless suffering on Afghans.

In Afghanistan the civilian death toll by terrorist attacks was 25 percent higher than last year's toll at the same time. The war in Afghanistan showed no significant diminution in more than a decade of effort against the militancy and promised to go on into the future until conditions changed sufficiently in some fashion to ensure that certain grievances were met. The diffuse grievances, abject poverty, general disenchantment with the government, and pressures from agencies external to the government and territory of Afghanistan seemed to offer little solution to the recurrence of incessant war.

Reference

Anon, 31 March 2013. Afghanistan Profile. BBC News South Asia. http://www.bbc.co.uk/news/world-south-asia-12024253.

Shroder, J., 2007. Afghanistan's development and functionality: Renewing a collapsed state. Geojournal, 70: 91–107.

Part I

Overview of the Geology and Geography of Afghanistan

The land of Afghanistan strikes nearly all outsiders who go there as a place where the more primal forces of much of the natural environment are on all too natural display for the taste of the villagers who have to cope with "Allah's wrath". To a geoscientist, however, the wrath of God is more like the breath of Gaia, the seemingly (to the short-lived human) eternal flame of the planet Earth in ceaseless turmoil, imperceptible to cataclysmic, which endlessly changes the surface upon which we all live. Indeed, in Afghanistan each time there is an earthquake everyone just 'knows' that Allah is angry about something and is sending a message of his unhappiness with human behavior. The problem, of course, is in discerning exactly which set of behaviors or occurrences have produced that particular upset of an apparently quiet surface. Other hazards such as floods or dust storms may be the work of *djinns* or malevolent spirits that love to cause human havoc just for the cruelty of it. But we geoscientists understand things rather differently and seek the patterns of explanation that satisfy educated curiosity of why things happen this particular way and not in that other fashion, and why things look the way they do. Afghanistan is a perfect place to see the results of long-continued forces of crustal plate motion, rafted islands of exotic rocks from other places, sutured onto the soft underbelly of Asia, where tectonic mountain uplifts resulted, with still active rock folds and fault breaks. Massive transpressive flower structures of rock masses grew slowly up out of the Earth, unfolding as province-sized rock masses, which rose slowly to open their ponderous rock petals. Across the surfaces of such massifs, the labile atmosphere exerted its daily forces of unceasing rain and snow, wind and water, to sculpt the rocks and generate the fertile plains or undulating sand dunes. Slowly, the exposure of the rocks and sediments to the rain and the sparse plant life began to add organic humus to build the soils that would enable human agriculture. In the high mountains the glaciers came and went, controlled by world climate changes, themselves the product of other Gaian forces still only dimly understood as scientists seek answers to such global fluctuations. In Afghanistan, huge lakes of fresh water diminished as climates changed, effectively dragging cities over the millennia along with their fluctuating shorelines to stay located near the vital water. The lakes became more salty, the fish died off, and so did the people, until only the *hamun* dry lakes were left as mud flats and salt pans in which weathered lithium from the surrounding bedrock had accumulated to one day perhaps derive a cell phone battery for a new civilization.

Many other geologic changes have occurred in Afghanistan too, with humans entering the picture to exert their changes only within the last few thousands of years. Even so, with the diligent daily effort to wrest a living from the ground, the ceaseless toil of myriads of grazing animals to get enough to eat as they are tended by the shepherds, the irrigation *juie* (jōy) ditches that bring vital water, but salts too to the land, the unending cutting down of trees for houses and fuel; the end result is a human-caused desertification that can be understood as a geologic force in its own right. The result is a land stripped bare, devegetated, salinated, wind-blasted, and merciless to its people, who in turn are merciless to each other. It is a hard place in which to try to live.

These chapters on the bedrock and the sediments, the landforms, climate, soils, and vegetation are the geoscience setup to understanding some of the human uses of this land. The ethnic patchwork of Afghanistan reflects the millennia of human migrations across and through the mountain and desert fastnesses, where some have decided to settle down in isolated patches, or others realized that constant nomadism was a good and free life as well. But all still were oppressed by the constant hazards and disasters that the land also produced. The ever-present earthquakes produced landslides from the steep mountain slopes, elsewhere the windblown dust, the loess that can be so fertile, failed and flowed down as mud to overwhelm village and field. Floods of water and rocks rushed down *nala* or *wadi* gulleys, sand and dust storms striped away farm fields. These chapters set the physical stage for the understanding of the environment with which the Afghan people have long had to contend as they developed their civilization. Their strength of history played out on a dynamic landscape.

In this Chapter

2

Rock and Landform Jigsaw Puzzles

Abstract

The geology of Afghanistan is a jigsaw puzzle of jumbled continental fragments moved across the ancient Tethys Seaway over several hundred million years to dock into faults against the south Asian coasts of geologic time. Four main tectonic regions exist: the transpressional plate boundary along the border with Pakistan; the accreted terranes in the central and western regions of the country; the middle Afghanistan shear zone; and the north Afghanistan platform. All tectonic regions are overlain by various sedimentary basins. The fundamental bedrock geology and overlying sediment cover constitute the materials that control most of the surficial landforms of Afghanistan, the mapping of which was first done by prior invading armies with military terrain analysis. Present-day geomorphological mapping by the author, however, takes most into account the tectonics of the country to achieve a more process-oriented explanation of why Afghanistan has the landforms it does.

Keywords: Accreted terranes; Faults; Geomorphological mapping; Middle Afghanistan shear zone; North Afghanistan platform; Plate tectonics; Sedimentary basins; Tectonic regions; Terrain analysis; Transpressional plate boundary region.

Bedrock Geology, Structure, and Surficial Sediments

To understand the geologic underpinnings of Afghanistan that control the landforms one sees on the surface requires reading a basic geologic history of the rocks there, many of which were originally emplaced elsewhere, billions to millions of years before being brought finally to their present locations. These bedrock amalgamations have been extensively faulted and broken up, before being overlain by various surficial sediments and volcanic rocks, with the whole mass then being eroded into its modern geomorphology or landforms at the Earth's surface. The best and most modern understandings of the controlling processes that generated this basic geologic superstructure involve the theory of plate tectonics, wherein mobile crustal plates on the Earth's surface over geologic time have slowly formed the geologic substrate of Afghanistan.

Natural Resources in Afghanistan. http://dx.doi.org/10.1016/B978-0-12-800135-6.00002-7

Plate Tectonic History of Afghanistan

Afghanistan is a patchwork region of crustal fragments or exotic terranes that have been rafted into place by plate motions over a long time and ultimately sutured together into a geologically jumbled amalgam of highly sheared slivers of rock (Powell, 1979; Şengor, 1984). Basically as Wheeler and Rukstales (2007) have noted, Afghanistan forms a southward-projecting promontory of the Eurasian plate, welded to the Indian plate, which regions are only roughly distinguishable from each other somewhat arbitrarily by locations of modern earthquake seismicity. The North Afghanistan Platform, or Turan Block, has been a part of the Eurasian Plate for some 350–250 million years, mostly as an offshore continental shelf, with typical carbonate and clastic shelf sediments being deposited in the shallow marine environments there. Over about the past 250 million years, several volcanic arcs of islands, as well as various fragments of continental and oceanic crust have collided with, and been added to the southerly edges of the North Afghanistan Platform, and are now part of the Eurasian plate (Wheeler et al., 2005).

The regional tectonic map (Figure 2.1) shows details of the modern configuration of southwest Asia in part from which the past geologic history may be

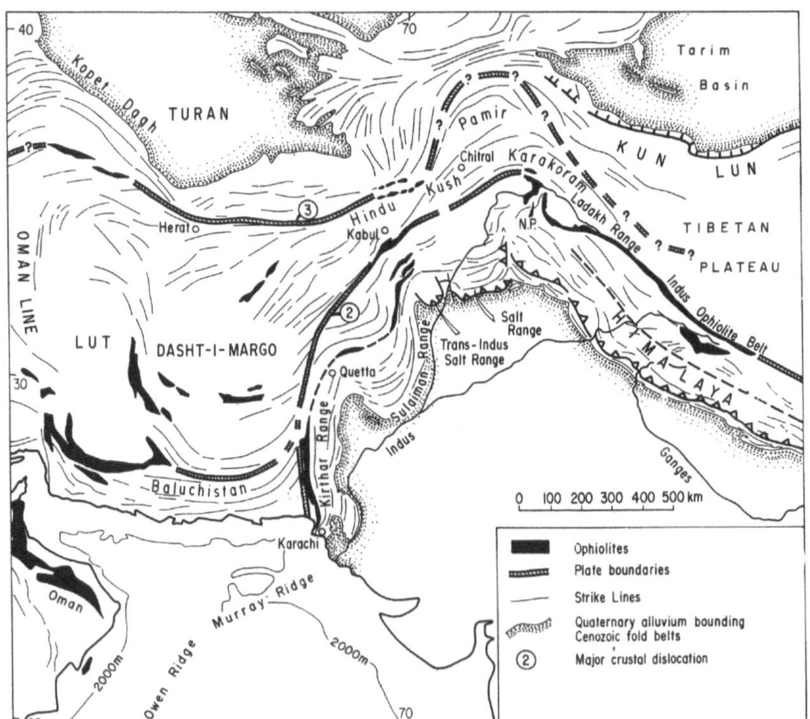

Figure 2.1 Regional tectonic sketch map of Southwest Asia from Iran to India, which focuses in the center on Afghanistan and Pakistan. Circled number 2—Chaman fault zone; circled number 3—Hari Rud fault zone. Ophiolite zones are remnants of ancient igneous rocks from the sea floor now caught up in suture zones as crustal fragments were brought progressively into this region and welded onto the Asian continent. *After Powell (1979).*

deduced. The presence of ophiolites implies that oceanic-basin or sea-floor rocks have been caught up in the collision and welded into the surface crust. Modern understandings of this plate-tectonic process concerning Afghanistan that are mapped, show that at least as far back as Late Triassic time about 210 million years ago (mya) in the early Mesozoic, several recognizable crustal fragments of western Afghanistan (Farad block, Helmand Block) were moving through the Tethys Seaway ocean basin, perhaps as some sort of islands at the time (Figure 2.2). By some 40 million years after that in Middle Jurassic time (170 mya), the Farad block had been rafted up to the Eurasian plate and sutured onto it, while the Helmand Block was still moving across the Tethys on its course to eventual collision and final welding to the plate as well (Figure 2.3).

At a later time the much larger continental fragment of Greater India, or the Indian plate, is seen to have been detached from the great southern

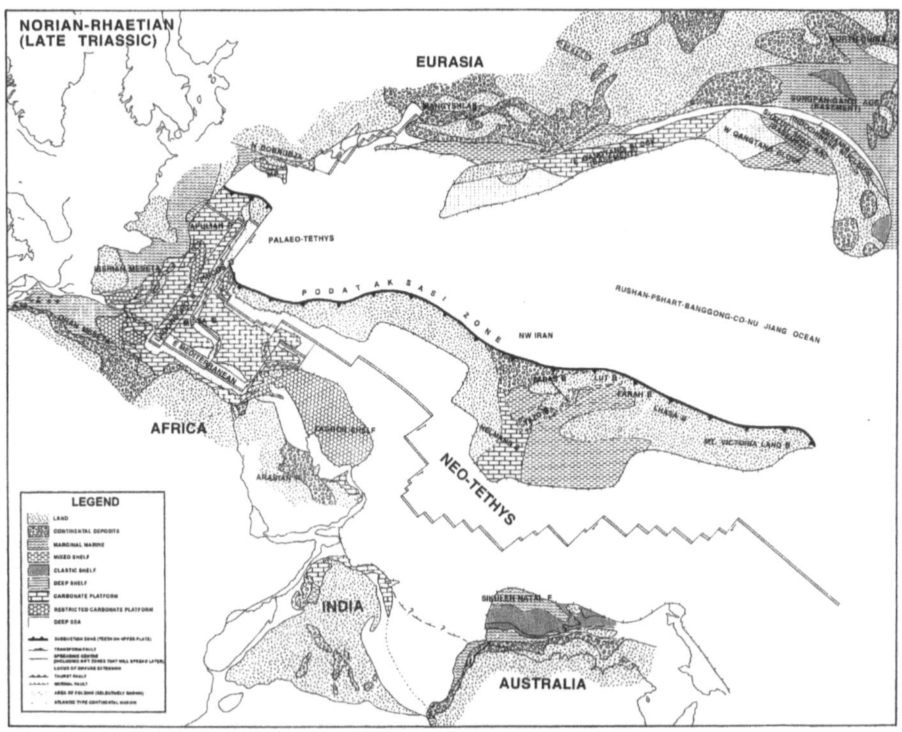

Figure 2.2 Map of Late Triassic time about 210 mya showing possible configuration of continents and continental fragments as islands that were moving through ancient oceans to eventually become parts of the asian continent in more recent times. Note the Helmand B(lock) and Farad B(lock) just above the Neo-Tethys, which is the name for part of the ancient ocean basin there. Both the Helmand and Farad blocks went on to become parts of southwest Afghanistan, once they had migrated across the Neo- and Paleo-Tethys to "dock" with the Eurasian continent much later. *After Şengör, and Natal'in (1996).*

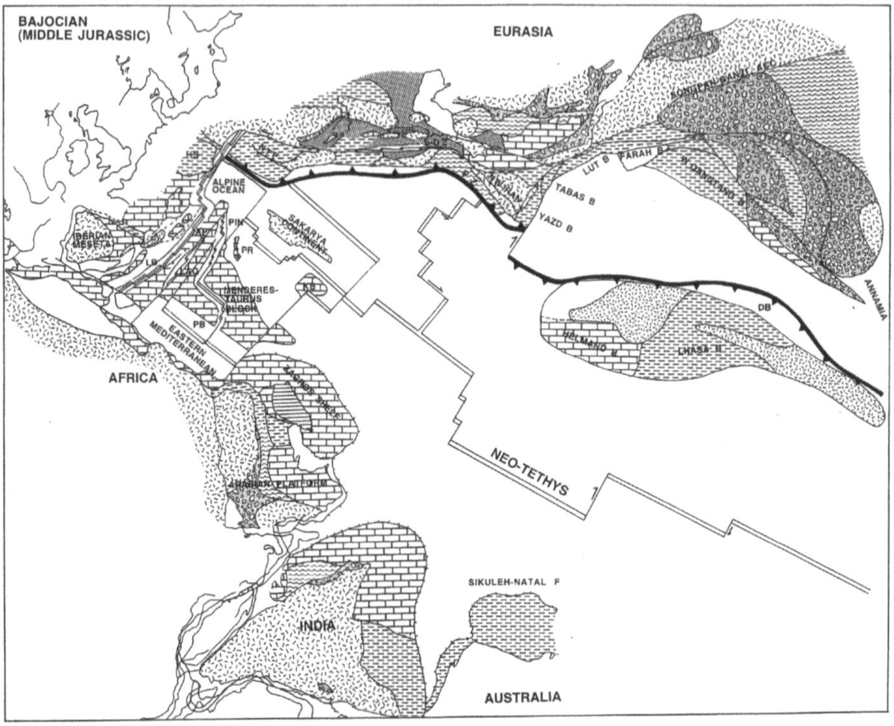

Figure 2.3 Middle Jurassic time about 170 mya showing possible suturing of the Farad block to the Eurasian Plate in the upper right of the figure and the Helmand Block midway in its migration across the Tethys Seaway. *After Şengör and Natal'in (1996).*

supercontinent of Gondwanaland and moved progressively northward over late Mesozoic and Cenozoic time (Figure 2.4). Moreover in greater detail with the crustal fragments of Afghanistan in mind, we can see that in Late Cretaceous time about 70 mya the movement north across the Tethys of the Indian plate had led to the formation of the great volcanic Kohistan—Ladak island arc (now mostly in Pakistan) in the ocean above the subsiding subduction zone (Figure 2.5). This island arc would ultimately become caught between the colliding Indian and Eurasian plates to become partly melted and metamorphosed as it was forced up to become part of the mountains of the Himalaya and Hindu Kush (Figure 2.6). This continued process of convergence and suturing to the present day (Figure 2.7) forced the oceanic subduction zone ultimately to be covered over completely by continental rocks that are now some of the highest mountains in the world. The rocks, mountains, and landforms of Afghanistan are a heritage of this plate-tectonic motion that drove the welding together of such a diverse mélange of materials, as well as shattering the rocks into myriad small fragments bounded by various sorts of faults.

Faults in Afghanistan

In the course of this extensive suturing over the past several hundred million years since Paleozoic time, hundreds or even perhaps thousands of faults

Figure 2.4 **The positions of the Greater India Plate as it moved across the Tethys Seaway from the Late Cretaceous time of the Mesozoic ~80 mya to dock with the Eurasian Plate and force up the Himalaya (cross-hatched pattern) and Hindu Kush in late Cenozoic time.** This is the overall pattern of plate movement in South Asia for the past 65 million years of Cenozoic time, in which the more precise details of events in Afghanistan are shown in the following figures. *After Powell (1979).*

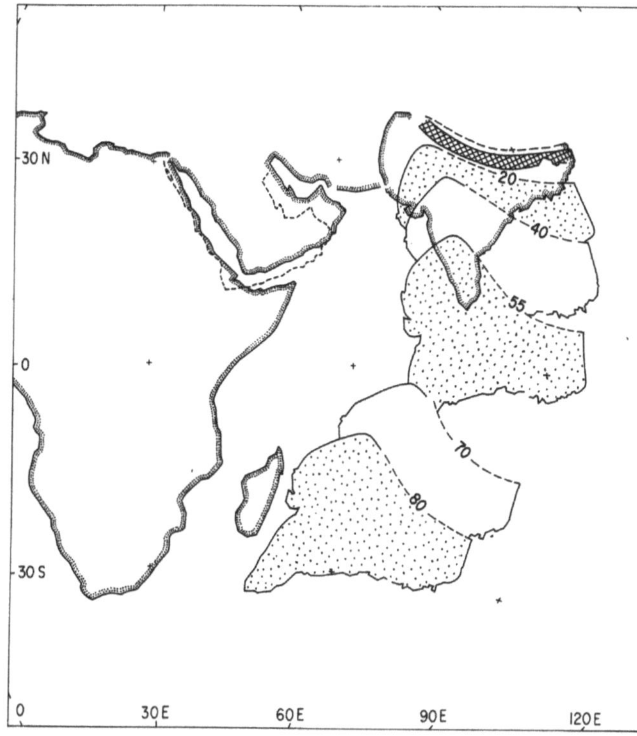

Figure 2.5 **Tectonic sketch map of the Greater India Plate in Late Cretaceous time 70 mya showing the central Iran (C.I.) and Afghanistan (A) microcontinents, which represent small crustal fragments that had migrated across the Tethys Seaway previously to dock with the Eurasian Plate and become sutured to it.** The barbed line with the name "Chitral" along it represents a subduction zone with an oceanic plate diving down to the north that generated volcanoes above it to form an island arc. *After Powell (1979).*

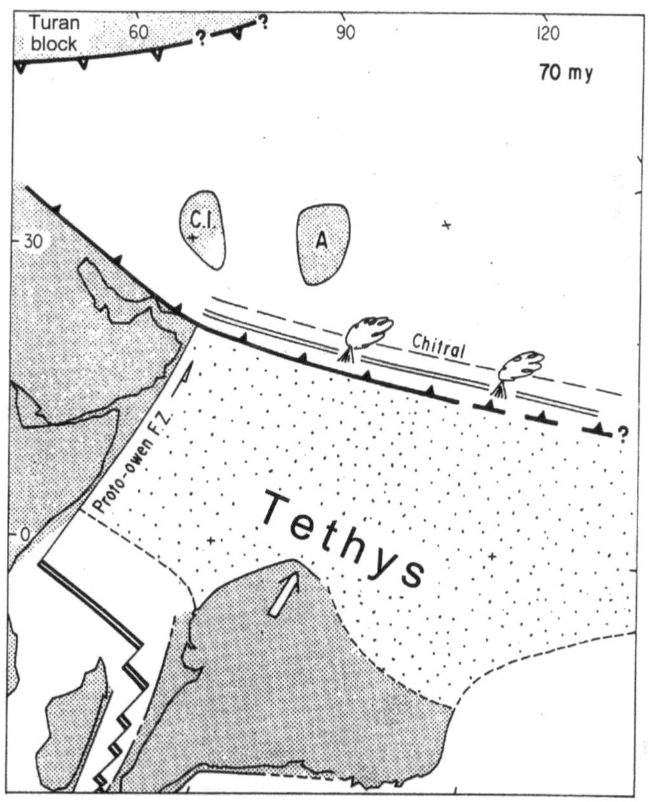

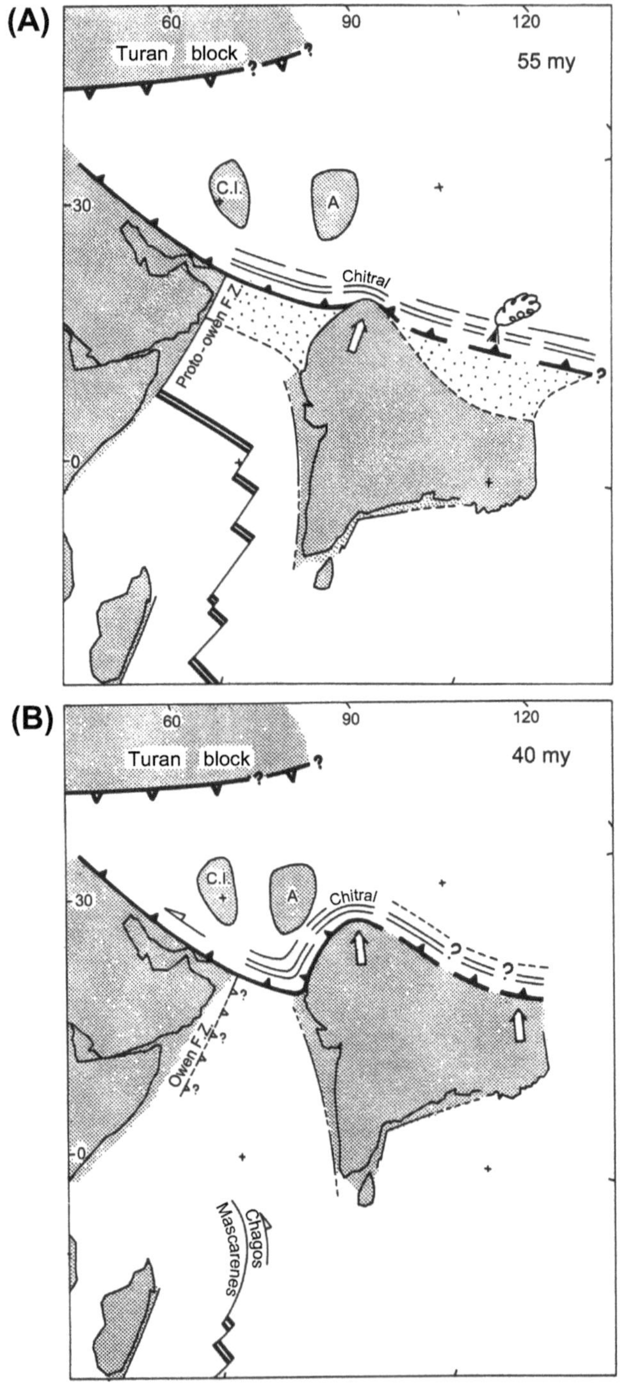

Figure 2.6 Tectonic sketch maps of the Greater India plate at the (A) Paleocene/Eocene boundary (55 mya) and in (B) Late Eocene (40 mya) as it continues to move northward and interact with the Eurasian plate by compressing the volcanic island arc and forcing the Iran (C.I.) and Afghanistan (A) microcontinents closer together. *After Powell (1979).*

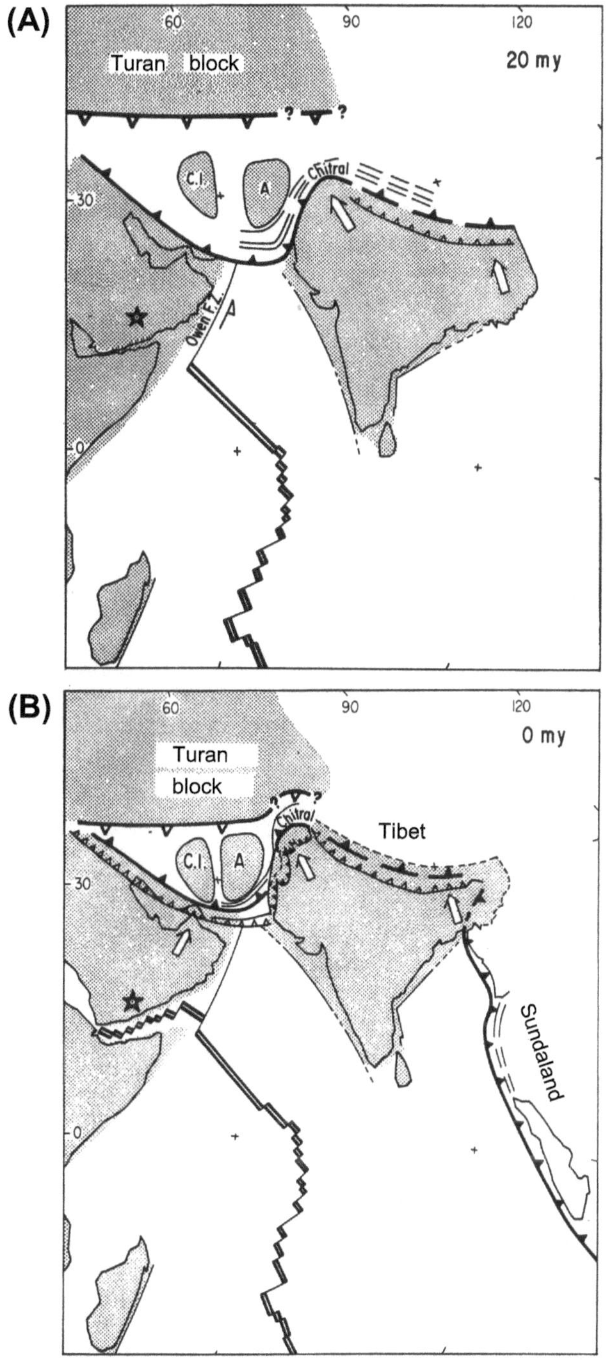

Figure 2.7 Tectonic sketch maps of the Greater India plate in (A) Miocene (20 mya) and the (B) present (0 mya) as it continues to move northward and interact with the Eurasian plate by compressing the volcanic island arc and forcing the Iran (C.I.) and Afghanistan (A) microcontinents ever closer. The great chains of the Hindu Kush and the Himalaya mountains are the result. *After Powell (1979).*

have developed in Afghanistan. Some of the older ones are now more-or-less inactive but others are newly developing and even highly active as seismic sources, as well as constituting useful demarcation lines between one tectonic or geomorphic province or another (Figure 2.8 of main fault zones; Table 2.1 of fault names).

At the present time, the east–west striking Hari Rud fault (Table 2.1; Figure 2.8) constitutes the main boundary between the North Afghanistan platform on the north and the accreted island arcs and continental fragments to the south. The Indian Plate is the most recent of these colliding fragments from Gondwanaland, and seems to be subducting north under the Eurasian plate somewhere between ~2.9–4 cm per year, while the Eurasian plate moves southward at ~1.3 cm per year (Jade, 2004; Bird, 2003; Venant et al., 2004;

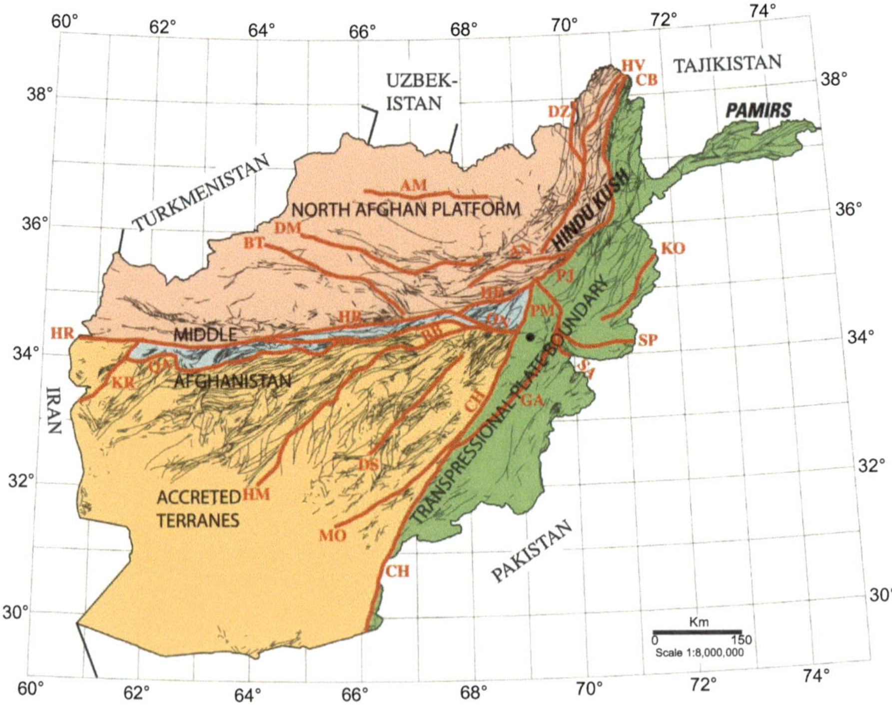

Figure 2.8 The regional tectonic structure of Afghanistan. The red letters designate the faults that are named and described further in Table 2.1. The map shows the four great geological terranes of Afghanistan that are: (1) the transpressional plate boundary (TPB) geomorphic region of eastern Afghanistan; (2) the accreted terranes (AT) geomorphic region of southwestern Afghanistan (yellow); (3) the middle Afghanistan shear zone (MASZ) geomorphic region of central Afghanistan (blue); and (4) the north Afghanistan platform geomorphic region (NAP) (pink). This configuration gives us the basic structural control over the finer geomorphologic subdivision to help better understandings of a complex region. *After Wheeler et al. (2005).*

Ruleman et al., 2007; Sella et al., 2002; Wheeler and Rukstales, 2007) (Figures 2.9 and 2.10). The Indian plate is moving roughly northward past present-day Afghanistan, and as it does so the two are separated by a broad north–northeast trending transpressional (flower structure) plate boundary zone (Wheeler and Rukstales, 2007), which varies in width from 100–400 km.

Table 2.1 Major fault systems of Afghanistan

Fault Name Abbreviation	Fault System Name
AM	Alburz–Marmul fault
AN	Andarab fault
BB	Bandi Bayan fault
BT	Bandi Turkestan fault
CB	Central Badakhshan fault
CH	Chaman fault
DM	Dosi Mirzavalan fault
DS	Darfashan fault
DZ	Darvaz faul
GA	Gardez fault
HM	Helmand fault
HR	Hari Rud (Herat) fault
HV	Hanjvan fault
KO	Kunar (Konar) fault
KR	Kaj Rud fault
MO	Mokur fault
ON	Onay fault
PJ	Panshir fault
PM	Paghman fault
QA	Qarghanaw fault
SA	Sarobi fault
SP	Spin Ghar fault

Each of these faults is actually multiple-shear zones in the Earth's crust, not a single linear break as generally portrayed in Figure 2.8. (Transcription of names from Arabic script in which Afghanistan languages are converted through Russian and other languages into English produces variable spellings, particularly with vowels).
After Wheeler et al. (2005).

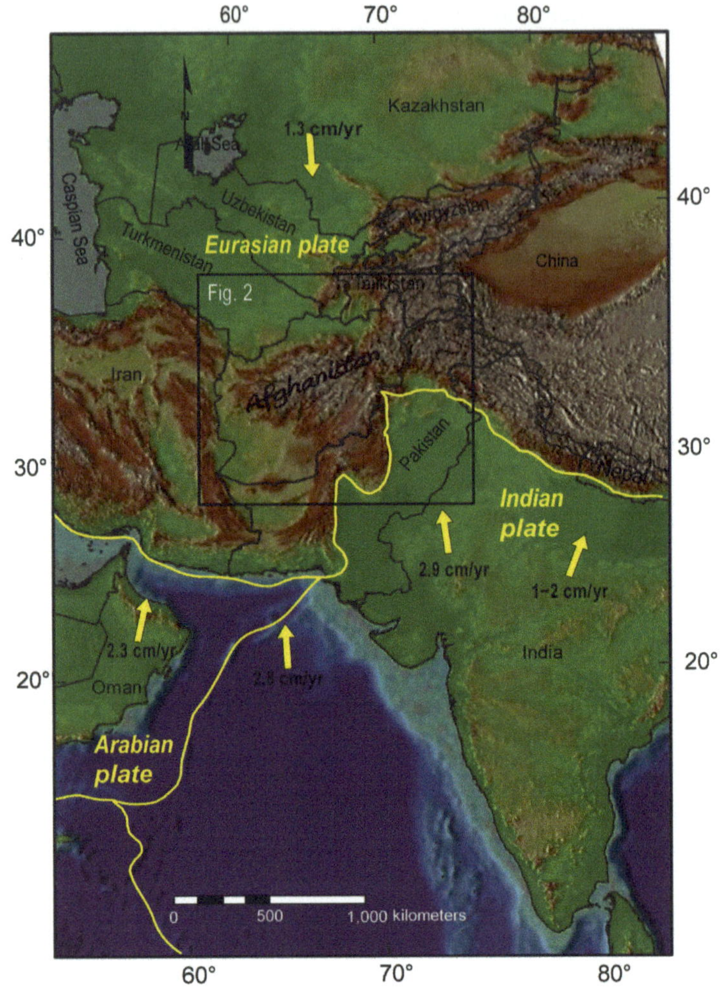

Figure 2.9 Location of Afghanistan in the greater Alpine-Himalayan orogenic belt in Asia. The box shows location of Figure 2.10. Vectors show relative plate motions and velocities between the Indian, Eurasian, and Arabian plates. *After Ruleman et al. (2007).*

Movement in the zone is mainly by left-lateral shear, but individual fault blocks and mountain ranges have been variously broken and folded throughout. In southeast Afghanistan to the northwest end of the Kabul Basin, the Chaman fault constitutes the most active leading edge of this motion (Ruleman et al., 2007). To the southwest of Afghanistan the present-day tectonics are driven by the subduction of the Arabian plate under Eurasia, at ~2.3 cm per year, mainly along the northwest-trending subduction zone that coincides with the coastal edge of the Zagros Mountains in Iran (Figure 2.9). The shallow dip of this subduction zone north beneath Baluchistan, Pakistan, and as far north as the Seistan or Helmand Basin in southwestern Afghanistan has allowed some volcanoes in both countries to be produced from it. In Afghanistan the Dashti

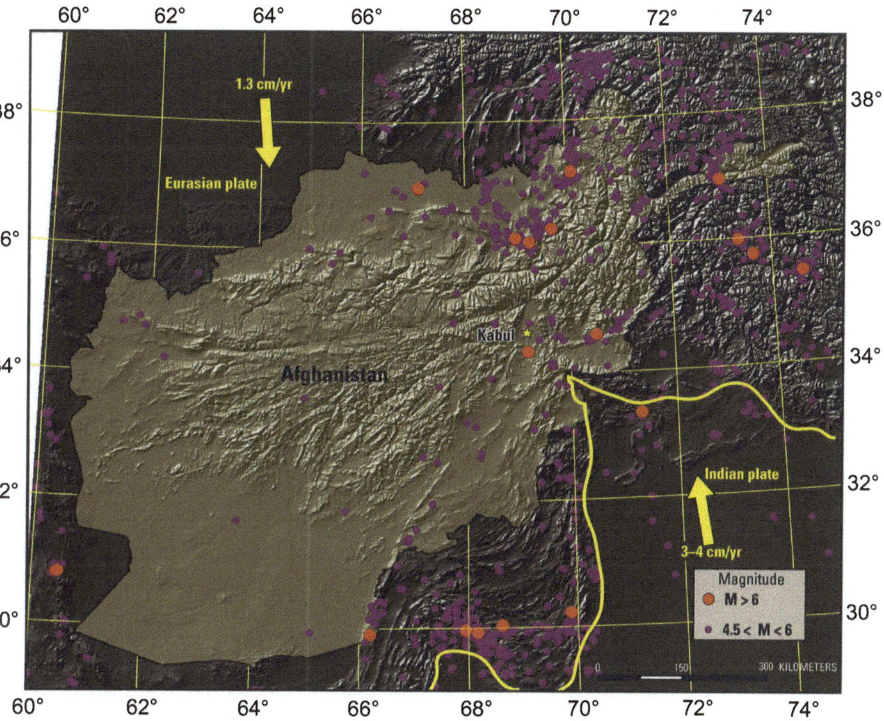

Figure 2.10 Model of relation of Eurasian plate to Indian plate beneath Afghanistan with shallow (<35 km deep) historical earthquakes. Several other possible divisions between the plates could be drawn. The numerous earthquakes and active faulting in Afghanistan are caused by the active plate motion that has uplifted the Hindu Kush, Himalaya, and Pamir mountains in the unlabeled high terrain in the northeastern part of the map. *Dewey (2006).*

Nawar volcanic caldera west of Ghazni is the remnant of just such a volcano, now extinct, and in the southern Helmand, the Kohi Khanneshin mountains are another remnant volcano.

Finally it should be noted that the modern subduction and movement of the Indian plate may be creating a system of new faults that cut across the older faults formed by the plate-tectonic construction of Afghanistan (Wheeler and Rukstales, 2007; Ruleman et al., 2007). The Hari Rud, Chaman, and other faults may be reactivated in the modern stress field, but in many places a northwest-southeast trending set of steeply dipping new faults appears to be being emplaced across previous northeast-southwest-trending fault sheaves.

Tectonic Regions of Afghanistan

The four main tectonic regions of Afghanistan (Figure 2.8), the transpressional plate boundary tectonic region, the accreted terranes tectonic region, the North Afghanistan platform tectonic region, and the Middle Afghanistan shear zone tectonic region help to establish the similarly named, four main geologic and geomorphic regions of the country.

Transpressional Plate Boundary Tectonic Region

The ongoing collision between the northward-moving Indian subcontinent and Eurasia takes place in particular along the Chaman–Paghman–Panjshir–Central Badakhshan fault zones that run approximately southwest to northeast through the eastern quarter of Afghanistan. In the south the left-lateral, strike-slip Chaman fault is likely to have the highest slip rate of any fault in Afghanistan, whereas to the north in the Paghman, Panjshir, and Badakhshan extensions of the fault system, several other faults appear to accommodate the motion in a variety of transpressional and transtensional settings (Ruleman et al., 2007). Transpression, in particular, constitutes a variety of simultaneous occurrences of compression or convergence of the Earth's crust wherein rocks can be faulted upward to form positive flower structures, as well as areas of transtension, in which rocks can drop down to form negative flower structures. The result geomorphologically is a series of upthrust ranges and downwarped basins that are developed into complex terrains. The Kabul block in particular is a separate group of downdropped ancient rocks of Precambrian age set apart by Paghman fault to the northwest and the Gardez fault to the south, and which is overlain by the much younger sediments of the Kabul Basin. Similarly the Jalalabad basin is set off between the Spin Ghar fault on the south and part of the Konar fault to the north.

Accreted Terranes Tectonic Region

In a somewhat similar fashion, the many complex ranges of southwestern Afghanistan are of many diverse rock types and geological ages that are extensively faulted. Many of the largest northeast–southwest trending faults in the region include the Kaj Rod, Helmand-Bande Bayan, Darfashan, and Mokur fault systems between the ranges. A close geological look at the rocks of this region reveals that they have all the earmarks of being formed elsewhere and then being rafted into place on the backs of moving tectonic plates, where they ultimately docked against the Asian plate and were sutured onto it during the following several hundred million years. These isolated small patches of different rocks are what geologists refer to as "suspect terranes" because of their lack of similarity or real association to other nearby rocks with which they are in contact. The extreme southwest of this region to the borders with Iran and Pakistan has subsided sufficiently in the past few million years to serve as an entrapment basin for young sediments of the Seistan that overlie the older bedrock of the accreted terranes (Figure 2.11).

Middle Afghanistan Shear-Zone Tectonic Region

As the rock masses of the accreted terranes region and the transpressional plate boundary region were juxtaposed onto the leading edges of the Asian plate throughout the Phanerozoic over the past half billion years or so, the east-west-trending Hari Rud fault zone came into existence as a suture zone and ultimately became a left-lateral, strike-slip fault zone with a number of other associated faults

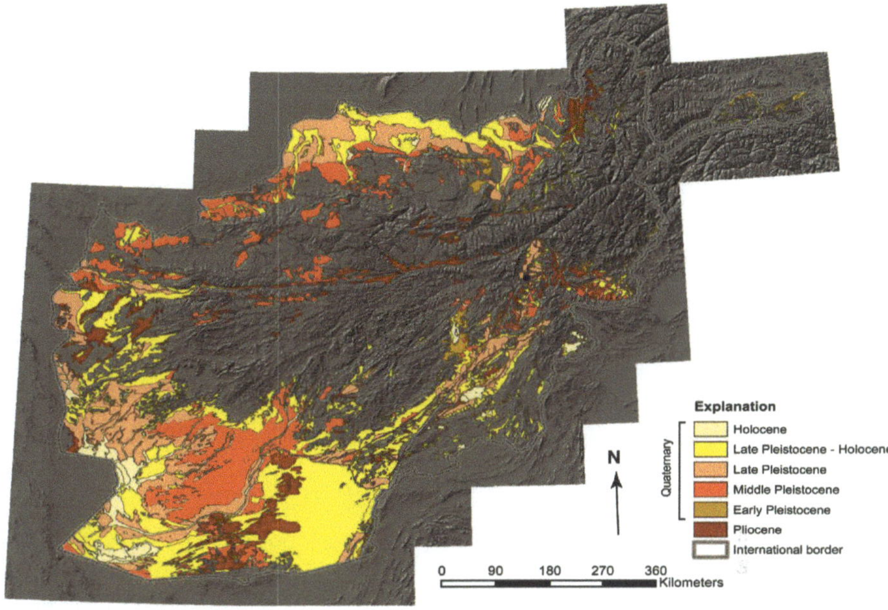

Figure 2.11 Late Cenozoic sedimentary basins in Afghanistan. *After Ruleman et al. (2007).*

of various types splayed out along it. The whole zone is about 780m long east to west, 30–60m-wide in the east, and is now as much as 120km wide at the Iranian border where it opens out. In this region the shear-zone region splays outward with the main Hari Rud fault on the south side of the shear-zone region passing westward into Doruneh fault system at the north end of the Lut structures in Iran, and the northwest-trending part on the north side of the shear-zone region passes northwest into the right-lateral Kopeh Dagh fault system (Berberian, 1981). This designation used and defined here is in contrast to that designated to be the "Middle Afghanistan" tectonic region as first mapped by Wheeler et al. (2005), but is justified herein as a more reasonable viewpoint from the perspectives of underlying tectonics combined with geomorphology and topography.

This Hari Rud fault zone was reactivated in transtension during about Oligocene–Miocene time some 25–20mya, in a series of pull-apart basins with left-lateral strike-slip offsets of as much as 600km that allowed accumulation of clastic sediments, as well as alkaline volcanic activity. Fault-bounded, lens-shaped rock masses trend roughly east-west throughout this region and constitute the most structurally complex part of Afghanistan (Shareq, 1981; Wheeler et al., 2005).

North Afghanistan Platform Tectonic Region

The North Afghanistan Platform region is part of the original coast and continental shelf of formerly flat-lying sediments in southern Asia before they were uplifted out of the former Tethys Ocean to become the high plateaus, mountains, and plains of today. The whole region is crossed by numerous large faults and folds of various types as a result of the collisions of other suspect terranes into it on the south and the northeast. The Bande Turkestan fault system, for example, is thought by some geologists to represent a relatively new shear zone caused by the Hari Rud fault stepping out to the northwest to relieve stress as its western extent toward Herat is at least partially locked and not moving. As a result of the change in direction from east to west of the main right-lateral Hari Rud, to the northwest of the Bande Turkestan fault, the Yakawlang pull-apart basin or graben has resulted as a prominent trench into which the Amir river flows from the lakes of Bande Amir, before breaking out to the northeast through an erosional canyon across the Dosi Mirzavalan fault and the Alburz–Marmul fault and monoclinally uplifted rocks of the North Afghanistan Platform and pouring out onto the vast plains of the Amu Darya near Mazari Sharif. Other major fault systems in the mountains of the North Afghanistan Platform rocks are the Andarab in the southeast of the region and the Darvaz and Henjvan of the northeast.

Sedimentary Basins of Afghanistan

The highlands of the northeast portions of Afghanistan are bounded by the surrounding lowlands, some of which were subsiding basins that accumulated sediments over the past 5 million years or through the Pliocene and Quaternary epochs (Figure 2.11). Such basins also existed as strike-slip, pull-apart features through the Middle Afghanistan shear-zone tectonic region as well. In some places extensive sheets of alluvium and colluvium were spread out by river floods, or rapid, wet debris flows; in others, lakes filled the basins with various kinds of lacustrine sediment. Broad expanses of wind-blown loess dust blanket older bedrock, and sheets of sand dunes were moved across desert floors (Doebrich and Wahl, 2006; Ruleman et al., 2007).

In sum, the tectonically deformed and uplifted, partially sediment-covered substrate of Afghanistan has a full temporal and lithologic suite of rocks that constitute the sedimentary, igneous and metamorphic foundations upon which the rich geomorphology has been developed. The result is an impressively complex terrain over which humans have lived, moved, and fought for many millennia. The exposition of that geomorphic terrain is explained in the following paragraphs.

Geomorphologic Mapping of Afghanistan

The evaluation and mapping of the regional terrain and landforms of the Earth have a long history in which different intentions have been expressed. Pure scientific curiosity about what is there and how it came to be is offset by various desires to use the geology and landforms in different ways. In the case of Afghanistan, it would appear that three main needs would be met by mapping the landforms of the country: (1) for educational purposes to increase understandings; (2) for development in which landform analysis can provide fundamental understandings in civil engineering; and (3) for military purposes where better understandings of terrain increase trafficability and security. In the course of attempting to meet these three main needs, certain factors have become obvious that partly control the inherent ease of access to information.

For example, centuries of rapacious despoliation of the semiarid landscape of Afghanistan, by a people who had little other choice in the face of a generally exceptionally low standard of living, have resulted in vast areas of increasing desertification in that country (See Part III, Chapter 17 on Environmental Degradation). Hand in hand with the natural climate change to the greater dryness in post-Pleistocene (Ice Age) time, has come the modification of the landscape in Afghanistan into one closely conditioned and controlled by processes that reflect their great strength of action. Thus one sees in the strong structurally organized landforms of Afghanistan, overprints of similarly robust wind, ice, water, and gravity to produce obvious palimpsests of aeolian, glacial, fluvial and mass-movement (landslide) actions. Thus the two main procedures of interest to us to help better understand these prominent and complex landscapes of Afghanistan would be terrain analysis and geomorphological exposition.

Terrain Analysis

At one end of the extremes of use are the practicability, evaluation, and mapping of terrain for military purposes, which can be considered as an essential exercise in maintaining control (Mitchell, 1991; Rinker, 1998; Wilson and Gallant, 2000). The best terrain analysis (Table 2.2) is a full discussion of the physical geography and geology of a region in which all the details of weather and climate, vegetation, soils, land use and resources, hydrology, and geology and landforms are presented. That full exposition, however, is easily possible to be done for most developed nations but can be a daunting task where not much is known about an undeveloped place such as Afghanistan, for example. In addition, tactical terrain analysis also consists of interpreting natural and man-made features in a region, together with weather and climate and other details of the physical geography, to determine their effects on military operations. Typically in military applications, data collected for tactical terrain analysis would include obstacles, ground water, surface materials (soil types), surface drainage and its configurations, vegetation types and distributions, transportation systems, and cross-country mobility or trafficability.

Terrain evaluation can also be the complete appraisal of the capabilities and limitations of a region in relation to a particular kind of land use, be it military or engineering applications, for example. Thus the different terrain units recognized from analysis are areas of ground whose recognizable form and observable characteristic relief occurs from the interactions of climate, pedogenesis, hydrology, geology, and geomorphic processes. Land classification in this fashion leads to the recognition of a scale-dependent hierarchy of land categories (Beaven and Lawrence, 1973; Mitchell, 1991) (Table 2.3). At the

Table 2.2 Terrain analysis checksheet

1. Weather and climate
   a. Global circulation region
   b. Climatic region (symbol and descript.)
   c. Climatic controls
   d. Air masses/source region
      - January regime
      - July regime
   e. Temperature
      - Surface temperature region
      - January normal (range, °F or °C)
      - July normal (range, °F or °C)
      - Annual range (range, °F or °C)
      - Peak month
   f. Pressure
      - Dominance pattern (Jan)
      - Dominance pattern (Jul)
   g. Wind
      - Jan direction—Strng/Mod/Wk
      - Jul direction—Strng/Mod/Wk
   h. Precipitation
      - Nov–Apr total
      - May–Oct total
      - Annual total
      - Variability (% from normal annual)
   i. Weather hazards (most frequent)
   j. Climograph
2. Vegetation
   a. Natural vegetation types (symbol and descript.)
   b. Actual vegetation type (environment)
   c. Foliage height/canopy coverage
   d. Seasonal foliage changes
3. Soils
   a. Great soil group (identify)
   b. Pedogenic regime
   c. Dominant factors (short discussion)
      - Climate
      - Vegetation
      - Extent of leaching
      - Specify others
   d. Soil erosion potential or status
   e. Barren soil exposed (est. %)
4. Land use and resources
   a. Land-use classification
      - Predominant economies
      - Agricultural region
      - Other land uses
   b. Natural resources—surface only
   c. Major cultural features reshaping/influencing terrain
5. Hydrology
   a. Rivers and valleys
      - Identify major streams
      - Drainage pattern
      - Drainage density
      - Valley and channel form
      - Major drainage basins (location)
      - Stage of development
      - Flood season(s)
   b. Lakes/ponds/reservoirs
      - Identify major natural and artificial
      - Describe function
   c. Wetlands—swamps/marshes (identify and locate major areas)
   d. Springs/Oases (identify and locate)
   e. Coastal zone
      - Tidal range (hi/mod/low)
      - Surface currents
        (1) Name
        (2) Temperature (warm/cool)
        (3) Flow direction
        (4) Seasonal variations (drift)
   f. Hazards along coast (most frequent)
   g. Glaciers and snowfields
      - Identify
      - Hazards in snow/ice areas
6. Geology and landforms
   a. General rock type (surface geology)
   b. Major geologic structures
   c. Topographic relief (maximum, meters)
   d. Orientation of prominent linear landforms
   e. Landform classification—regional
   f. Physiographic province
   g. Glaciated or nonglaciated
   h. First-order landforms
      - Name of plate
      - Oc eanic or continental
      - Direction of movement
   i. Second-order landforms—describe (including initial and sequential processes and present stage of development)
   j. Third-order landforms—describe (including initial and sequential processes and present stage of development)
   k. Slope characteristics
      - Areas of steep slopes (>30%)
      - Areas of moderate slopes (15–30%)
      - Areas of gentle slopes (0–15%)
   l. Geologic hazards

After EV 104 Terrain Analysis and Evaluation, Department of Earth, Space and Graphic Sciences, U.S. Military Academy, West Point, NY, undated but post-1974.

higher levels one is considering whole climate regions, continents, and other vast regions. At the lower levels a land system would have more-or-less similar weathering under fairly uniform geological conditions of the same geological formation (all one rock type, or fairly uniformly varying rock types in different beds), as well as uniform climatic conditions, and would have a complement of small topographic units, called "land facets" that were all developed by the same geomorphologic process. The land facet is the basic unit of mapping and is generally quite small so it can only be shown at a scale of 1:50,000 or less. It is also defined on the basis of its geology, water regime, and topography, but in a much more restricted way than its parent land system. Land units have simple forms, fitting into one slope class on a slope map, for example, and they have the same parent material for their soils, as well as a single water regime for both surface and ground water. As the name implies, they tend to have planar or uniformly curving topography. In some cases a very small feature on the landscape is too small to be mapped (a small outcrop with glacial straie, a cave mouth, a soil pedon, etc.) but is important enough to be recognized as a land clump, land subfacet, or a land element that is somehow diagnostic of some process or other feature that should be noted (Table 2.3).

Table 2.3 Hierarchical classification of terrain, soil, and ecological units

Terrain Unit	Soil Unit	Vegetation Unit	Mapping Scale
Land zone	Order		<1:50,000,000
Land division	Suborder	Plant panformation *zone écologique*	1:20,000,000– 1:50,000,000
Land province	Great group		1:5,000,000– 1:20,000,000
Land region	Subgroup	Subprovince	1:1,000,000– 1:5,000,000
Land system	Family	*Region écologique*	1:200,000–(simple)[1] 1:1,000,000
Land catena	Association	*Secteur écologique*	1:80,000–1:200,000
Land facet	Series	Subformation *station écologique*	1:10,000–1:80,000
Land clump	Complex	Subformation *station écologique*	1:10,000–1:80,000
Land subfacet	Type		Not mapped
Land element	Pedon	*Element de station Écologique*	Not mapped

[1] A land system is complex if it represents a combination of two or more geomorphologically related simple land systems, or compound if the combination is not geomorphogenetic, or formed through the same process, or at the same time. Complex and compound land systems are appropriate to the mapping scale of the land region.
After Mitchell (1991).

More specifically in direct military applications with what is termed, "intelligence preparation of the battlefield", a digital topographic support system (DTSS) can be created to facilitate terrain analysis and tactical decision aids (TDAs) for battlefield command. The DTSS provides software to produce a variety of mobility, visibility, and special purpose TDAs, such as off-road and on-road speed products, combined obstacle overlays for various wheeled and tracked vehicles, shaded time distance, and maneuver networks and predictions. Various visibility functions can be used to generate masked and visibility areas for point sites, linear routes, and regional or areal analyses. Special purpose products can be built to include helicopter landing zones, bivouac sites, lines of communication, and other such practical military necessities.

In these fashions of hierarchical and special purpose mapping, terrain classification can be useful in the delineation of the land surface for various military purposes. But in the background to all such procedures for terrain analysis would be a deeper look at the underlying geological firmament from which the surficial terrain is derived. Of course the basic geology of any region has also long been known to be quite relevant to military purposes (Leith, 2002; Underwood and Guth, 1998), but in this case only a regional approach to the bedrock geology is required in order to more effectively understand the controls of the basic geomorphology. Geomorphologic subdivision of a region is thus viewed as a fundamental means to understanding the terrain in a more comprehensive and scientific fashion than the pure practicalities of military strategy and tactics.

Historical Terrain Analysis

Terrain analysis by the British in Afghanistan in the nineteenth century in the course of their two wars (1839–1842; 1878–1880) largely involved basic topographic mapping based on simple measurements and triangulations, coupled with hand-drawn hachures to delineate elevations. This was coupled with notations about availability of water, animal fodder, and camping or bivouac sites. This continued well into the twentieth century with periodic map upgrades, conversion to contour lines, and mapping of details to add to the map catalogue of the British Empire (Figure 2.12).

The updating of the topographic maps of the British Raj and its periphery in Afghanistan ceased after partition of India and formation of Pakistan in 1947 and the Cold War began between the communist and capitalist worlds. In the 1950s negotiations between the governments of USA, Afghanistan, and the USSR led to acquisition of aerial photographs by Soviet and American (Fairchild Survey to US Department of Defense (DOD)) aircraft with construction of topographic maps by them and the Afghan Cartographic Institute (ACI) (Glicken, 1960; Shroder, 1983). Maps were produced at scales of 1:50,000 (US DOD, Soviet), 1:100,000 (US DOD, Soviet), 1:200,000 (Soviet), and 1:250,000 (US DOD, ACI) (Figure 2.13). Most of these maps were restricted or classified as low-level state secrets, with the result that they were not easily available.

(A)

Figure 2.12 (A) British topographic map (Zebak J-42X; Quarter Inch Series 1942; 1:253,440) showing the Ishkashim (Afghanistan) border area at the mouth of the Wakhan Corridor, extending from Tajikistan in the northeast across the Noshaq (Naushaq) mountain (24,566 ft, 7492 m) of Afghanistan to the Tirich Mir (25,238 ft, 7690 m) mountain area of Chitral, in what is now Pakistan; (B) US DOD Joint Operations Graphic (Series 1501; Sheet NJ4216; Edition 1 c. 1970 (1:250,000)) from the US Army Topographic Command. Strong similarities in border configuration between the British and American maps show the transition from older ground survey technologies to new aerial photogrammetry, especially where the map sources are noted as being irreconcilable.

(B)

Figure 2.12 —cont'd

Subsequently in the later 1960s and 1970s, progressive takeover of ACI by Soviet cartographers (Shroder, 1983) enabled progressively better map making and the Soviet invasion of Afghanistan in 1979 enabled refinement of these maps. Their VTU (Military—Topographic Directorate) had begun a systematic program of terrain analysis, initially at scales of 1:50,000, with soils, vegetation, climate and other topographic information printed on the back side of each of their 1:200,000-scale map sheets, or some 128 in all, which includes overlapping parts of the neighboring countries (Figure 2.13). These Soviet maps were completed in the 1984–86 period and each one had on its reverse the map divided into its natural soil-type regions, with discussions of terrain characteristics (general topography, hydrology, vegetation, and so forth), but the geomorphology was not well discussed because so little was known about it.

After the defeat and departure of the Soviet troops from Afghanistan in 1988–89, followed soon after by the related collapse of the Soviet Union in the early 1990s, the Russian general staff sold off their map collections to Western agencies as a way to enhance revenues in their economic crisis. These descriptions of map terrains were then translated and published by East View Cartographic (Anonymous, 2003) in Minneapolis, MN, USA, in their *Terrain Analysis*

(A)

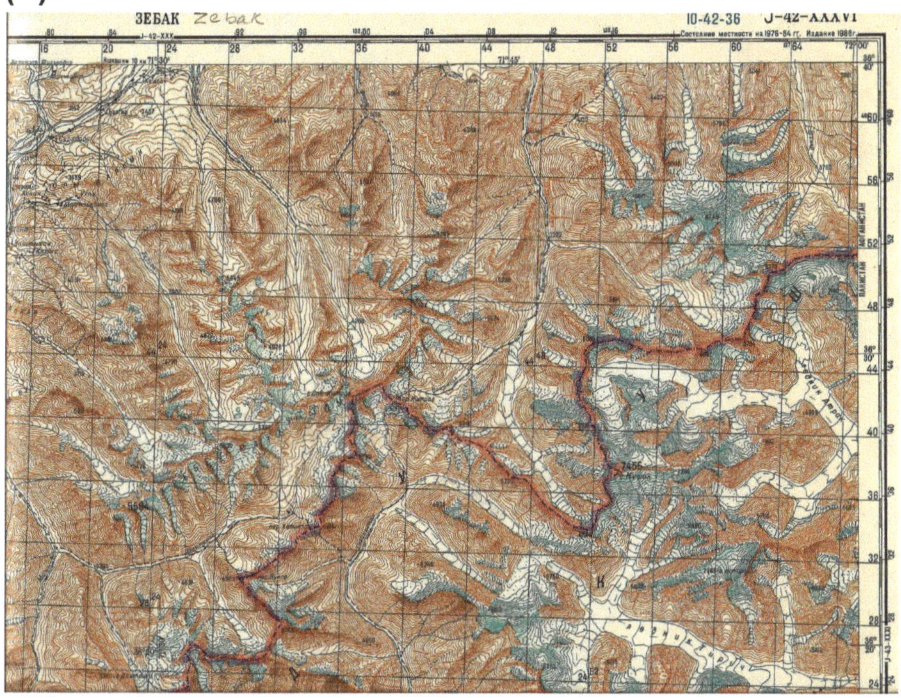

Figure 2.13 (A) Soviet topographic map of part of the Zebak area (Zebak 10-42-36 J-42-XXXVI; 1988; 1:200,000) showing the pronounced border reentrant between Afghanistan and Pakistan that places Afghanistan's tallest mountain, Noshaq (Naushaq) Peak at 7455 m. On the reverse of the original of this map, the Soviet topographers described populated places, roads, topography & soils, hydrology, vegetation, and climates aspects. The accompanying soil map showed generalized distributions of ice, rock, rubble, & loam. (B) Large-scale (1:100,000) Soviet topographic map (10-42-132 J-42–132) with the border reentrant and Noshaq Peak on the lower right at 7455 m.

of Afghanistan in which the 1:200,000-scale maps were featured. All of the terrain text was translated into English, as were all the place names on simplified smaller scale topographic map representations. The texts for each map have sections on populated places, roads, topography and soils, hydrology, vegetation, and climate, which are the most comprehensive such materials

(B)

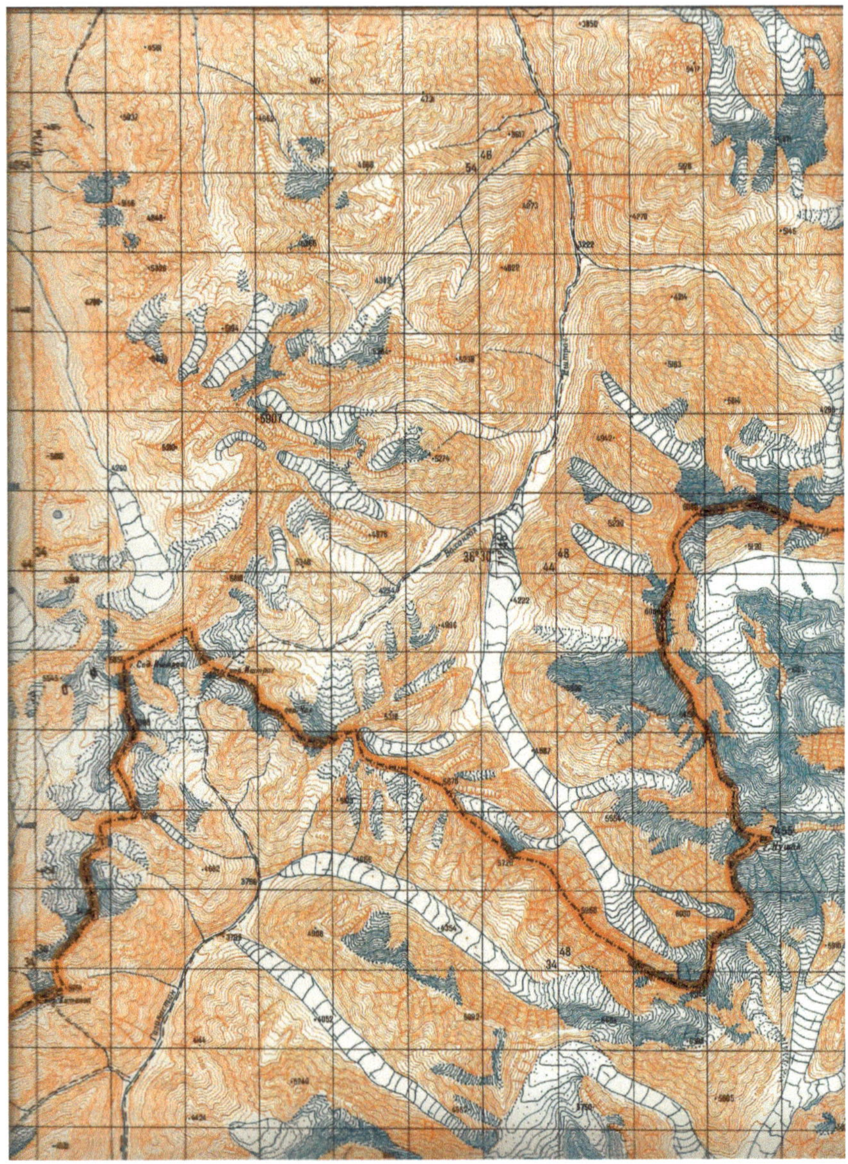

Figure 2.13 —cont'd

available on Afghanistan. These materials are wholly descriptive, however, and lack fundamental genetic or causative elements in them, which retards fundamental understandings. Furthermore, even though they were begun at larger scales of 1:50,000, comparison of some areas of Afghanistan mapped by the Soviets with US DOD maps done at the presumably somewhat less accurate 1:100,000-scale shows that the Soviet versions tended to overly generalize some features, such as glaciers, for example (Figure 2.14). As a springboard for basic geomorphologic analysis, however, the large-scale topographic maps can be combined with the basic geology maps for a more true understanding of the physical situation.

Finally, following the 9/11 horror in the USA when the need to invade Afghanistan by the Americans was seen as the only way to avoid future attacks, the US DOD map catalogue was electronically digitized by the National Intelligence Mapping Agency (NIMA) of the time (now National Geospatial-Intelligence Agency) and reclassified. This then meant that the new computerized versions of these maps were to become unavailable to civilians, Afghanistan nationals, or nongovernment development agencies. As of this writing, this fact is unlikely to change, with the result that the work expressed herein had first to be started with the older paper maps, and only later updated with higher resolution satellite imagery that is available to civilians.

Geomorphological Mapping

Geomorphologists in the nineteenth and early twentieth centuries produced physiographic maps of parts of the United States as exploration and mapping of vast regions enabled understanding of new terrains (Thornbury, 1965). Geomorphologic subdivision and mapping progressed primarily through recognition of underlying and controlling geologic structures, overprinted by surficial processes. After more than a century of effort to map the geomorphology of the United States, roughly from the time of Lesley (1856, 1869) and Powell (1895), through Fenneman (1914, 1917, 1928, 1931, 1938), and ending approximately with Thornbury (1965), the concepts of mapping regional geomorphic situations were fairly well established by the establishment of the major geomorphic divisions, geomorphic provinces, and geomorphic sections. In the course of this work a number of people tried to define what was being done. Hinds (1952) described geomorphic provinces in California as being characterized by a distinguishing geologic record, mainly of the later geological history, and with more-or-less uniform relief features or combinations of such features throughout its region. Thornbury's (1965) teacher had originally taught him that geomorphic units were divisions of the land wherein the topographic elements of altitude, relief, and type of landforms were characteristic and contrasted with, or distinct from those elsewhere. But this definition unfortunately had left out considerations of rock structure or the basic geology. In contrast Dietz (1952) thought that the topography of any area could

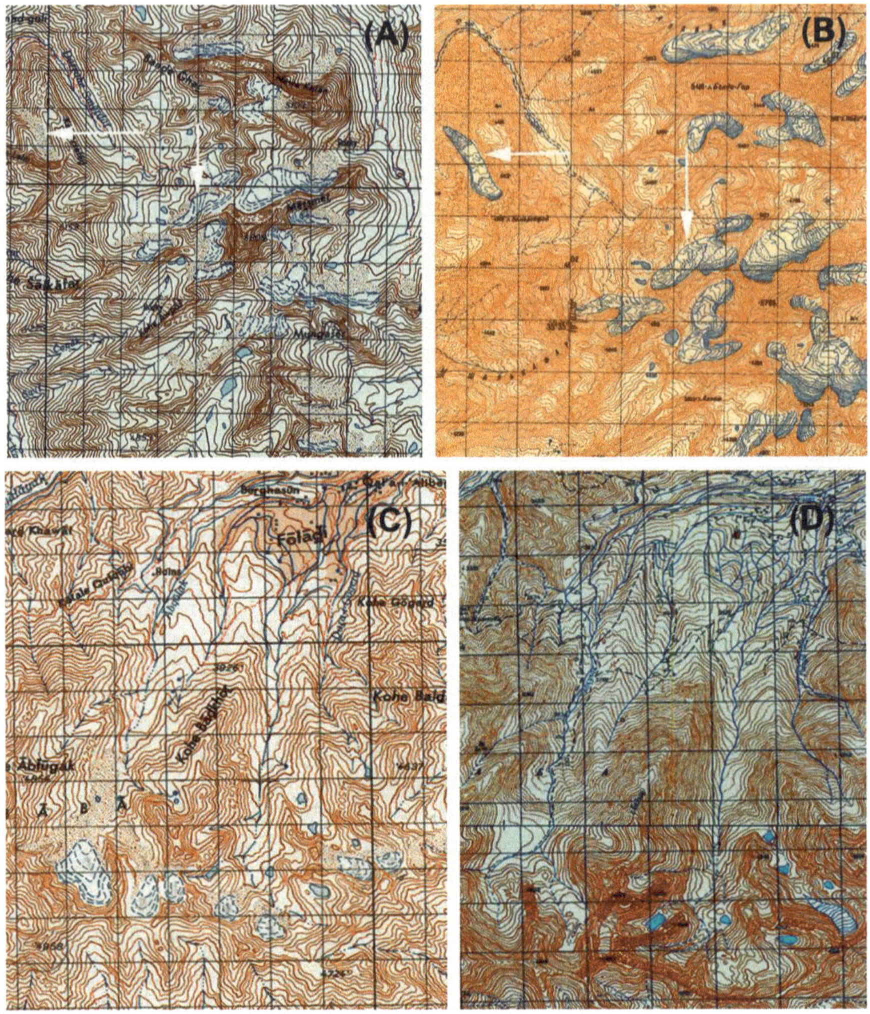

Figure 2.14 Comparison of US DOD and soviet topographic maps of the early 1960s. (A) Glaciers in Mir Samir area of the Panshir Valley published by US DOD at a scale of 1:100,000; (B) Same area published by the Soviet military at a scale of 1:50,000 (Note debris-covered ice pointed out by horizontal arrows that was incorrectly mapped as blue ice by the Soviet cartographer). (C) Kohi Foladi area of the Kohi Baba mountain range south of Bamiyan published originally by US DOD at 1:100,000 showing as many as seven ice glaciers, some with lower debris covers, whereas (D) the Soviet cartography at 1:50,000 scale inaccurately showed only the Foladi Glacier as a pointed feature in blue ink.

be defined in terms of the formative tectonics, the internal bedrock structure, the erosional and sedimentological processes acting on the land, the intensity of those process, and the length of time that they had been operating. Thus today most geoscientists would agree that a geomorphic unit has an individual expression that can be related to controlling factors such as the underlying geology and its structure, as well as the geomorphic processes that have operated in the area throughout certain periods of geologic time (Thornbury, 1965), which collectively influence and produce its topography.

Any scheme of geomorphic division of an area of course is an attempt by humans to constrain and classify reality into the "tyranny of the pigeonhole", wherein some phenomena are not easily so constrained. Clear-cut boundaries in nature are actually rather unusual in many cases and in some places no natural boundaries may exist so that somewhat artificial or arbitrary lines may have to be drawn to separate geomorphic provinces. Outliers of a certain type may have to be included within the geomorphic unit of which they are a part, rather than the one they most resemble. Overall, however, greater clarity and identification of terrain types can be the result of such geomorphic division.

Attention to detailed geomorphologic mapping in the developed world in the later twentieth century enabled better understanding of geomorphic complexities; such mapping, however, was never done for much of Asia, especially remote and little known Afghanistan. Modern attempts at healing the failed state of Afghanistan are critically dependent upon strong security and accurate mapping, including comprehensive delineation of the land surface of Afghanistan for such things as developmental and flood-hazard mapping, better understanding of the physical environment, and seismological zonation. In addition, the analysis of the land has obvious military significance in helping to overcome the terrorist insurgents and maintain peace-making efforts.

Geomorphological subdivision of Afghanistan is based upon modern understandings of the tectonic provinces of the country. Any effort to map the complex geomorphology of Afghanistan has to be based on the details of rock structure, geomorphic process, and the complex topography so that the variable terrain can be more easily understood. The author has further subdivided these regions of Afghanistan into numerous geomorphologic provinces, with an additional group of sections to enhance understanding. In general terms the country is a mosaic of endogenetic fault and fault-line escarpments, active and inactive fold uplifts, a volcanic caldera, and recent volcanoes; and the exogenetic glaciated and glacierized high peaks, with an overall dominance of fluvial and aeolian eroded terrain, coupled with numerous intermontane basins of fluvial and lacustrine fill, overlain with dunes and loess, and dissected by landslides. An understanding of the details of the geomorphology is viewed as one element in the critical rebuilding of the country to a more modern standard to help keep the fundamentalist terrorists at bay.

Earlier recognition that the procedures of geomorphic subdivision might be valuable in the study of Afghanistan has led to several prior attempts at smaller scales. For example, C.H. Halikov et al. (1979) is known to have done a geomorphological map of Afghanistan at a scale of 1:1,000,000 for the Ministry of Water and Power but no copy of it has been seen in some time (US/AID—Nathan Associates & Berger International, 1992). Similarly during communist times, Geokart, an organization in Warsaw, Poland, produced colored maps of geomorphology and types of natural landscape at scales of 1:6,000,000 for the Afghanistan National Atlas (Geokart, 1984) that are discussed further below.

The Polish cartographers of Geokart, who reported (email communication from L. Fronckiel to the author, May 17, 2006) that "elaboration and production [of maps] was not an easy task", no doubt had a good deal of trouble gathering sufficient information to do the job. For the geomorphology map they showed large mountain ridges, a few tectonic scarps along the Chaman–Paghman fault, some volcanic cones and a few water gaps (Figure 2.15). The bulk of the geomorphology was designated as: (1) erosion relief in high mountains, with and without glaciers; (2) denudation–erosion relief as mountains, mountain-cuesta relief, and low mountain and mondanock relief; (3) accumulation relief as piedmonts with undulating plains or alluvial ("deluvial") sediments; (4) alluvial cones that are "Old" Pleistocene, Middle Pleistocene (maxium glaciation), Holocene river valleys, and intermountain basins; and (5) eolian relief. A problem with most of these designations is they are not especially accurate in color differentiation (several shades of green were indistinguishable from each other), genesis, timing, or location, and so are not that useful today.

Similarly the map of "types of natural landscape" by Geokart (1984) included classifications of mountains, uplands, lowlands, and valley bottoms, each with varieties of vegetation that constitute what might be considered to be different physiographic or terrain-analysis classes (Figure 2.16). Thus there are high mountain steppes, subalpine zones that were previously glaciated, mountain woodlands, uplands with juniper, or uplands with shrub-steppe vegetation, or lowlands with sage and salt-tolerant vegetation. Sandy deserts with ephemeral vegetation and flat valley bottoms with willows, poplars, and oleanders are other classes.

J. Al-Absy (2008) produced a thesis and a map of the geomorphology of Afghanistan at a scale of 1:3,000,000 as a student senior thesis for the author. While this work was guided by the author to a certain extent when our schedules overlapped on rare occasions, Mr Alabsy had never been to Afghanistan so he had to rely rather too strongly upon educated guesses about certain phenomena. The result was that many of his interpretations needed revision once the thesis was finished, thus the more detailed work herein.

Geomorphic Subdivision of Afghanistan

Earlier recognitions that the procedures of geomorphic subdivision might be valuable to the study of the physical environment of Afghanistan has led to several prior attempts that were done at small scales that were previously mentioned (Halikov et al., 1979; Geokart, 1987; Alabsy, 2008). Now with the availability of large-scale digital elevation model (DEM) data for the entire country (Figure 2.17), and satellite image mosaics (Figure 2.18) coupled with excellent new satellite imagery from multiple platforms and Google Earth™ views at

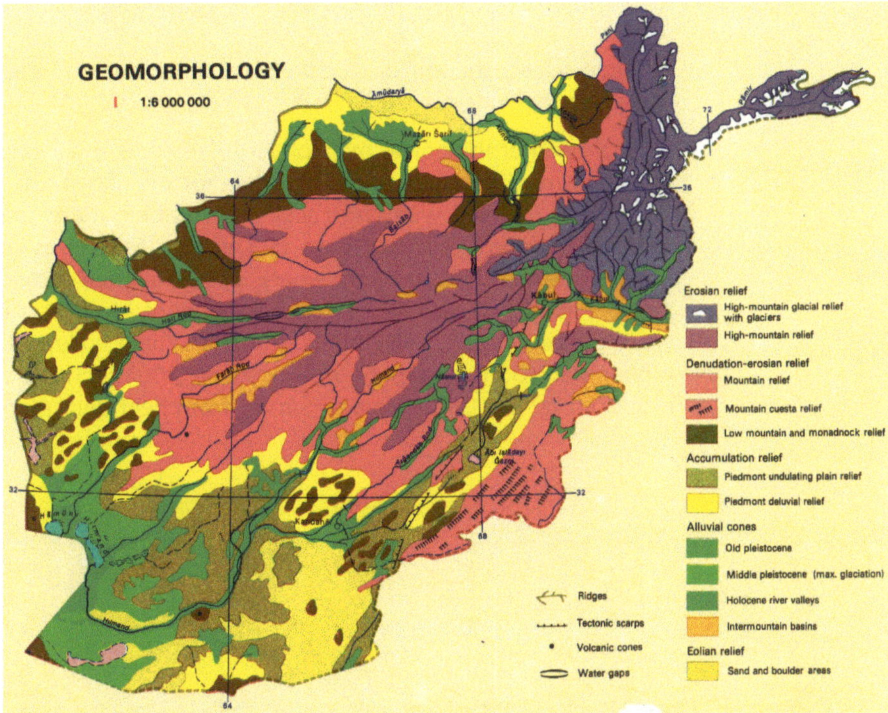

Figure 2.15 Map of the general geomorphology of Afghanistan published originally at a scale of (1:6,000,000) by the Geokart™ atlas of Afghanistan.

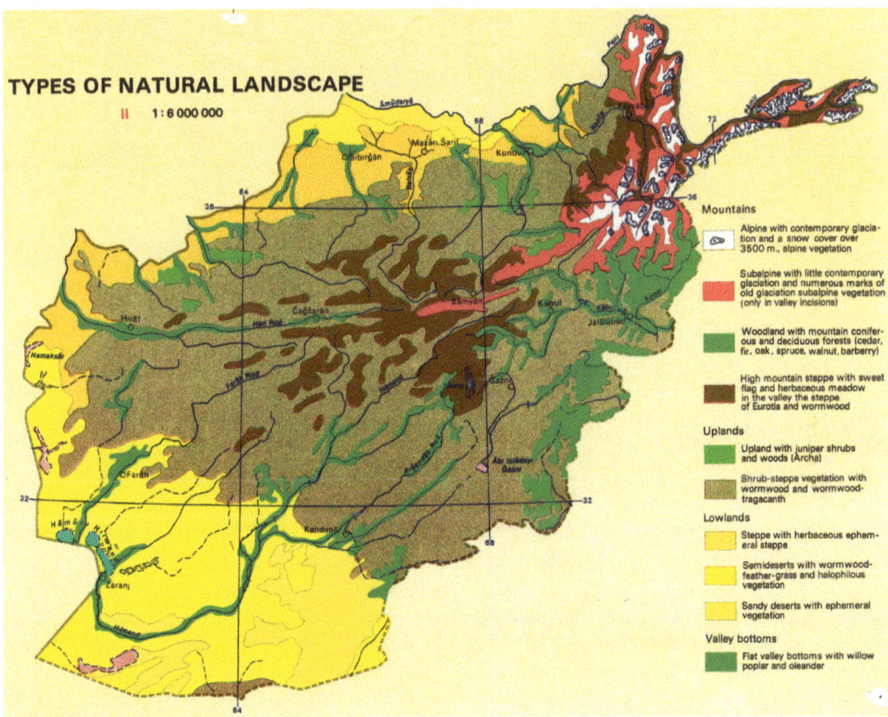

Figure 2.16 Map of types the basic natural landscape regions of Afghanistan (1:6,000,000) from the Geokart™ atlas of Afghanistan.

multiple scales, the ability to develop a more detailed geomorphological map was undertaken here.

Thus a geomorphological map of Afghanistan is now made possible with the availability of: (1) a large-scale DEM data for the entire country (Chirico and Barrios 2005), coupled with large scale topographic maps made from aerial photographs by the US Department of Defense, as well as the Soviet Union; (2) excellent new ASTER satellite imagery, as well as previously classified Corona and Keyhole imagery, and other satellite imagery of Afghanistan; (3) Google Earth™ views of the landscape of Afghanistan at multiple scales; (4) the soils regions of Afghanistan (US Department of Agriculture, 2001); (5) large scale geology maps produced by the US Geological Survey from field data mapped by many Soviet

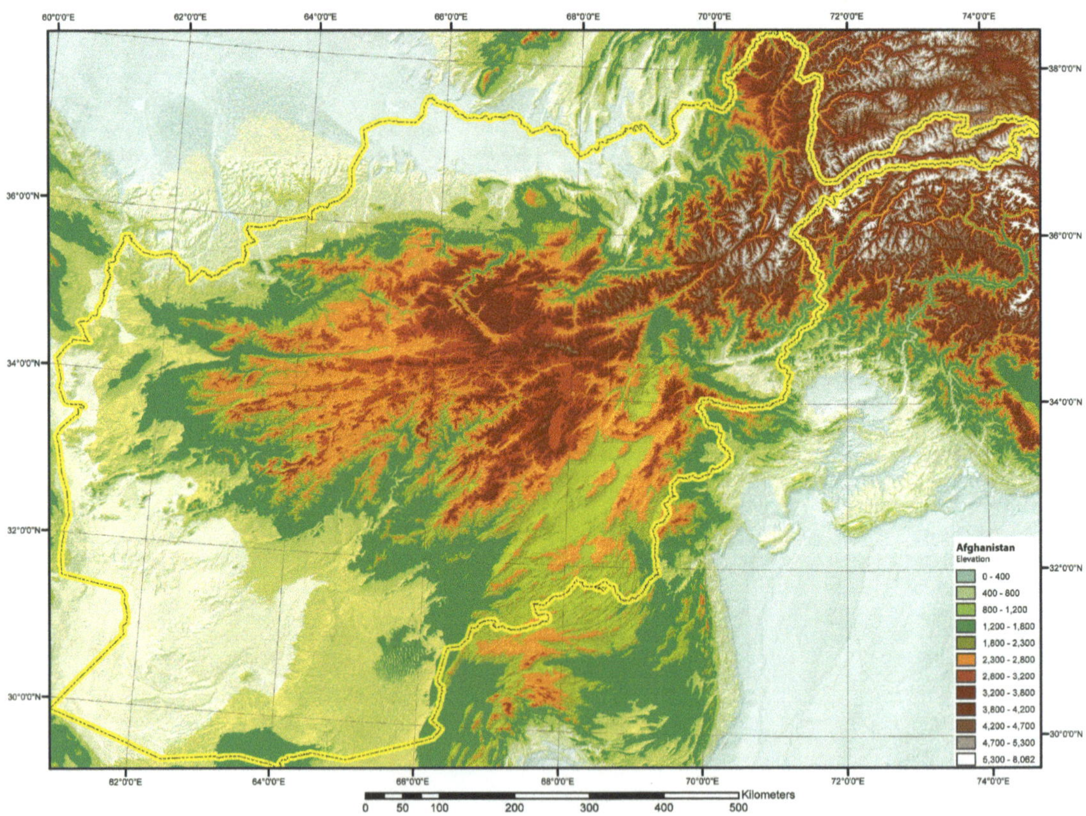

Figure 2.17 Afghanistan digital elevation model (DEM) from the ~30 m resolution, ASTER GDEM data (http://asterweb.jpl.nasa.gov/gdem.asp) **with 12 slightly irregularly spaced elevation differences (400, 500, 600 m), which show the pronounced elevation ranges of the region.** Digital representations of the complex topography now occur at many different resolution scales since the Shuttle Radar Topographic Mission (SRTM) of 2000, which generated a widespread DEM of much of the world at ~90 m resolution (compare with Figure 8.19, which is composed of classified US DOD DTED 1 and 2 data at ~90 and ~30 m (and better) resolutions). *Prepared by J. Olsenholler.*

geologists (Shroder, 1983, 1989; Shroder and Asifi, 1987); and (6) the maps of the tectonic regions of Afghanistan (Wheeler et al., 2005; Ruleman et al., 2007).

The primary means to begin subdivision of the geomorphology of any area is thus to begin with detailed maps of the geology and structure of the area to be mapped. The map of the tectonic regions of Afghanistan (Wheeler et al., 2005; Ruleman et al., 2007) is a basis for establishing the major subdivisions of Afghanistan, in that four regions of major division have been mapped and minimally described (Figure 2.8). Nonetheless, the divisions delineated in this fashion are the North Afghanistan Platform, the transpressional plate boundary, the accreted terranes, and Middle Afghanistan. Most of these four divisions are set off by major fault systems, although the Middle Afghanistan major division has been expanded somewhat from the original tectonic region zonation produced by Wheeler et al. (2005) because the bedrock faulting is either partly concealed with younger sediments or transects mountain ranges that need to be considered as one geomorphic unit, rather than being split into two or more geomorphic provinces.

(A)

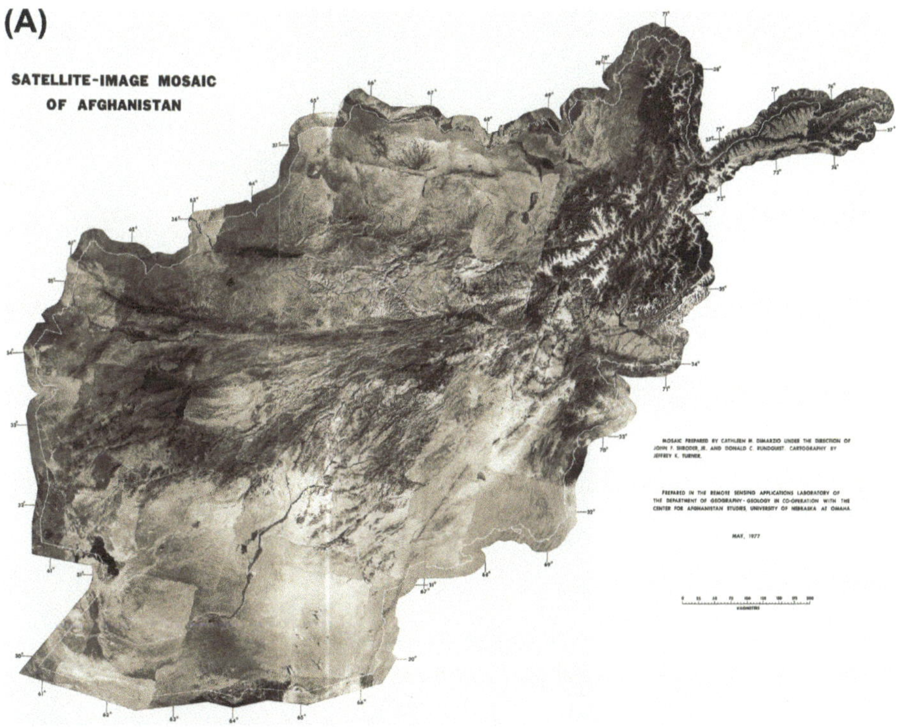

Figure 2.18 Satellite image mosaics of Afghanistan. (A) 1977 mosaic of Afghanistan made from early Landsat imagery by C. Dimarzio, J. Shroder, D. Rundquist, J. Turner. (B) 1980s mosaic of the Kabul and Panjshêr Valley made by the US CIA. (C) 2006 calibrated Landsat ETM+ nonthermal-band mosaics of Afghanistan (http://gisdata.usgs.net/website/afghan).

(B) **Afghanistan: Kabul and Panjshēr Valley**

Figure 2.18 —cont'd

Figure 2.18 —cont'd

Major Geomorphic Divisions of Afghanistan

Thus with the work of Wheeler et al. (2005) as a skeletal framework upon which to hang the main geomorphic divisions of Afghanistan, (Table 2.4) four major geomorphic *regions* exist in the country: (1) the frying-pan shaped *Transpressional Plate Boundary Geomorphic Region* extending clockwise along the eastern border with Tajikistan, China, and Pakistan, with its Wakhan Corridor Panhandle; (2) the crudely rhombohedral *Accreted Terranes Geomorphic Region* that continues clockwise to the southwest borders Pakistan and Iran; (3) the narrow *Middle Afghanistan Shear Zone Geomorphic Region* that bisects the country in a west-to-east direction from the Iranian border to north of Kabul; and (4) the ovoid- or football-shaped *North Afghanistan Platform Geomorphic Region* that extends from the borders of Iran, Tukmenistan, Uzbekistan, and Tajikistan. Each of these four major divisions has one or more geomorphic provinces associated with it.

Transpressional Plate Boundary Geomorphic Region

The ongoing collision between the northward-moving Indian subcontinent and Eurasia takes place in particular along the Chaman–Paghman–Panjshir–Central Badakhshan fault zones that run approximately southwest to northeast through the eastern quarter of Afghanistan. In the south the left-lateral, strike-slip Chaman fault is likely to have the highest slip rate of any fault in Afghanistan, whereas to the north in the Paghman, Panjshir, and Badakhshan extensions of the fault system, several other faults appear to accommodate the motion in a variety of

Table 2.4 Geomorphologic regions, provinces, and sections of Afghanistan mapped on the hierarchical basis of regional structure, lithology, and topography and described in Chapter 3

Transpressional plate boundary geomorphic region (TPB region) T-A: Eastern Pamir–Hindu Kush Mountain geomorphic province T-A1: High Pamir–Shiwa Lake geomorphic section T-A2: Greater Pamir Mountain geomorphic section T-A3: Lesser Pamir Mountain geomorphic section T-A4: Eastern Hindu Kush geomorphic section T-B: Nuristan Hindu Kush Mountains geomorphic province T-B1: Central Hindu Kush Mountain geomorphic section T-B2: Jalalabad Basin geomorphic section T-B3: Spin Ghar (Eastern Safed Koh) geomorphic section T-C: Kabul Basin geomorphic province T-D: Katawaz Mountains and Basins geomorphic province T-D1: Abi Istada geomorphic section T-D2: Cuesta-Monoclinal Ridge–Hogback geomorphic section T-D3: Khost Basin and Suleiman mountain geomorphic section
Accreted terranes geomorphic region (AT region) A-A: Farad block basin and range geomorphic province A-A1: Dashti Alishan basin and range geomorphic section A-A2: Ghor mountains geomorphic section A-A3: Khash Rud block mountains geomorphic section A-B: Helmand block geomorphic province A-B1: Helmand–Uruzgan mountains and basins geomorphic section A-B2: Arghandab mountains and basins geomorphic section A-B3: Dashti Nawar caldera geomorphic section A-B4: Spin Baldak–Qalat geomorphic section A-C: Helmand basin geomorphic province A-C1: Registan geomorphic section A-C2: Chagai Hills–Khanneshin volcano geomorphic section A-C3: Dashti Margo geomorphic section A-C4: Dashti Bakwa (Khash) geomorphic section A-C5: Frah–Harut geomorphic section A-C6: Seistan Hamoon (lakes) geomorphic section
Middle Afghanistan shear zone geomorphic region (MASZ region) M-A: Hari Rud–Ghorband geomorphic province M-A1: Herat–Hari Rud fault-zone valley geomorphic section M-A2: Chagcharan shear zone (Horst & graben) geomorphic section M-A3: Kohi Baba mountains geomorphic section M-A4: Paghman mountains geomorphic section
North Afghanistan platform geomorphic region (NAP region) N-A: Eastern Pamir–Hindu Kush mountains geomorphic province N-A1: Badakhshan Pamir mountains geomorphic section N-A2: Khwaja Mohammed (Northern) Hindu Kush mountains geomorphic section N-A3: Western Hindu Kush mountains geomorphic section N-B: North Afghanistan platform plateau geomorphic province N-B1: Takhar geomorphic section N-B2: Kunduz–Taliqan Active ranges geomorphic section N-C: Tadjik basin geomorphic province N-C1: Amu Darya Floodplain geomorphic section N-C2: Dashti Layli–Shortepa–Kunduz geomorphic section N-C3: Shiberghan–Mazar alluvial Fan geomorphic section

Table 2.4 Geomorphologic regions, provinces, and sections of Afghanistan mapped on the hierarchical basis of regional structure, lithology, and topography and described in Chapter 3—cont'd

N-D: Northern Afghanistan plateaus and Loess Hills geomorphic province
N-D1: Balkhab–Samangan plateau geomorphic section
N-D2: Faryab–Maymana Loess Hills geomorphic section
N-D3: Qalai Nau Loess Hills geomorphic section
N-E: Central Afghanistan mountains and plateaus geomorphic province
N-E1: Bandi Amir–Sari pul plateaus geomorphic section
N-E2: Feroz Koh plateaus geomorphic section
N-E3: Sefid Koh (Parapamisus) mountains geomorphic section
N-E4: Yakawlang–Bamian graben geomorphic section

transpressional and transtensional settings (Ruleman et al., 2007). Transpression, in particular, constitutes a variety of simultaneous occurrences of compression or convergence of the Earth's crust wherein rocks can be faulted upward to form positive flower structures, as well as areas of transtension, in which rocks can drop down to form negative flower structures. The result geomorphologically is a series of upthrust ranges and downwarped basins that are developed into complex terrains. The Kabul block in particular is a separate group of downdropped ancient rocks of largely Precambrian and Paleozoic age set apart by Paghman fault to the northwest and the Gardez fault to the south, and which is overlain by the much younger sediments of the Kabul Basin. Similarly the Jalalabad basin is set off between the Spin Ghar fault on the south and part of the Konar fault to the north.

Accreted Terranes Geomorphic Region

In a somewhat similar fashion, the many complex ranges of southwestern Afghanistan are of many diverse rock types and geological ages that are extensively faulted. Many of the largest northeast–southwest trending faults in the region include the Kaj Rod, Helmand-Bande Bayan, Darfashan, and Mokur fault systems between the ranges. A close geological look at the rocks of this region reveals that they have all the earmarks of being formed elsewhere and then being rafted into place on the backs of moving tectonic plates, where they ultimately docked against the Asian plate and were sutured onto it for several hundred million years. These isolated small patches of different rocks are what geologists refer to as "suspect terranes" because of their lack of similarity or real association to other nearby rocks that they are in direct contact with. The extreme southwest of this region to the borders with Iran and Pakistan has subsided sufficiently in the past few million years to serve as an entrapment basin for young sediments of the Seistan that overlie the older bedrock of the accreted terranes.

Middle Afghanistan Shear-Zone Geomorphic Region

As the rock masses of the Accreted Terranes Region and the Transpressional Plate Boundary Region were juxtaposed onto the leading edges of the Asian

Plate throughout the Phanerozoic over the past half billion years or so, the east-west-trending Hari Rud (Herat) fault zone came into existence as a suture zone and ultimately became a right-lateral, strike-slip fault zone with a number of other associated faults of various types splayed out along it. The whole zone is about 780km long east to west, 30–60km wide in the east, and is now much as 120km wide at the Iranian border where it opens out. In this region the shear-zone region splays outward with the main Hari Rud fault on the south side of the shear-zone region passing westward into Doruneh fault system at the north end of the Lut structures in Iran, and the northwest-trending part on the north side of the shear-zone region passes northwest into the right-lateral Kopeh Dagh fault system (Berberian, 1981). This designation used and defined here is in contrast to that designated to be the "Middle Afghanistan" tectonic region as first mapped by Wheeler et al. (2005), but is justified herein as a more reasonable viewpoint from the perspectives of underlying tectonics combined with geomorphology and topography.

North Afghanistan Platform Geomorphic Region

The North Afghanistan Platform region is part of the original coast and continental shelf of formerly flat-lying sediments in southern Asia before they were uplifted out of the former Tethys Ocean to become the high plateaus, mountains, and plains of today. The whole region is crossed by numerous large faults and folds of various types as a result of the collisions of other suspect terranes into it on the south and the northeast. The Bande Turkestan fault system, for example, is thought by some geologists to represent a relatively new shear zone caused by the Hari Rud fault stepping out to the northwest to relieve stress as its western extent toward Herat is at least partially locked and not moving. As a result of the change in direction from east to west of the main right-lateral Hari Rud, to the northwest of the Bande Turkestan fault, the Yakawlang pull-apart basin or graben has resulted as a prominent trench into which the Amir river flows from the lakes of Bande Amir, before breaking out to the northeast through an erosional canyon across the Dosi Mirzavalan fault and the Alburz–Marmul fault and monoclinally uplifted rocks of the North Afghanistan Platform and pouring out onto the vast plains of the Amu Darya near Mazari Sharif. Other major fault systems in the mountains of the North Afghanistan Platform rocks are the Andarab in the southeast of the region and the Darvaz and Henjvan of the northeast.

Geomorphologic Provinces of Afghanistan

Much of Afghanistan is dominated by the central and northeastern highlands that are a complex of collision tectonics and diverse rock types and sediments that are blended together, one after another, in a somewhat bewildering jumble of seemingly endless range upon range without distinction, one from another. Subdivision of these regions can thus seem somewhat arbitrary. Nevertheless, closer inspection reveals certain distinguishing rock structures, lithologies, and overlying sediments and landforms that can enable adequate geomorphologic subdivision and eventually better detailed terrain mapping. Each of the four main major geomorphic divisions of Afghanistan described above are thus further subdivided into the geomorphologic provinces discussed in more detail in Chapter 3 below.

References

Al-Absy, J., 2008. Geomorphology of Afghanistan: A Brief Overview and Basic Geomorphic Classification System. unpublished Senior Thesis; Department of Geography and Geology. University of Nebraska at Omaha. 225 p.

Anonymous, 2003. Terrain Analysis of Afghanistan. East View Cartographic, Minneapolis, MN.

Beaven, P.J., Lawrance, C.J., 1973. The application of terrain evaluation to road engineering. Conference on Road Engineering in Asia, Australasia held in Kuala Lumpur. The Institution of Engineers, Malaysia and Public Works Department, West Malaysia (obtained on-line December 9, 2008).

Berberian, M., 1981. Active faulting and tectonics of Iran. In: Gupta, H.K., Delany, F.M. (Eds.), Zagros, Hindu Kush, Himalaya Geodynamic Evolution. Geodynamics Series, vol. 3. American Geophysical Union, pp. 33–69.

Bird, P., 2003. An updated digital model of plate boundaries. G^3, Geochemistry, Geophysics, and Geosystems 4 (3), 1–52.

Chirco, P.G., Barrios, B., 2005. Void-filled SRTM digital elevation model of Afghanistan. US Geological Survey Data Series. 130. http://geology.er.usgs.gov/eespteam/terrainmodeling/ds_130.htm.

Dewey, J.W. (editor) Seismicity of Afghanistan and Vicinity. US Geological Survey Open-File Report 2006–1185, 18 p. http://pubs.usgs.gov/of/2006/1185/

Dietz, R.S., 1952. Geomorphic evolution of the continental terrace (continental shelf and slope). Am. Assoc. Pet. Geol. Bull. 36, 1802–1819.

Doebrich, J.L., Wahl, R.R., 2006. Geologic and Mineral Resource Map of Afghanistan: U.S. Geological Survey Open-File Report 2006–1038; (http://pubs.usgs.gov/of/2006/1038/).

Fenneman, N.M., 1914. Physiographic boundaries within the United States. Ann. Assoc. Am. Geogr. 4, 84–134.

Fenneman, N.M., 1917. Physiographic divisions of the United States. Ann. Assoc. Am. Geogr. 6, 19–98.

Fenneman, N.M., 1928. Physiographic divisions of the United States. Ann. Assoc. Am. Geogr. 18, 261–353.

Fenneman, N.M., 1931. Physiography of the Western United States. McGraw Hill Book Co., New York, 534 pages.

Fenneman, N.M., 1938. Physiography of the Eastern United States. McGraw-Hill Book Co., New York, 714 pages.

Geokart, 1987. Geomorphology (Map of Afghanistan). Scale 1:6,000,000. National Atlas of the Democratic Republic of Afghanistan, Warsaw, Poland.

Glicken, M., 1960. Making a map of Afghanistan. Photogrammetric Engineering 26, 743–745.

Halikov, C.H., et al., 1979. Geomorphological Map of Afghanistan, Scale 1:1,000,000. Ministry of Water and Power, Kabul, Source US/AID (Nathan Associates & Berger, International, 1992).

Hinds, N.E.A., 1925. Evolution of the California Landscape. California Department of Natural Resources, Division of Mines. Bulletin 158, 240 p.

Jade, S., 2004. Estimates of plate velocity and crustal deformation in the Indian subcontinent using GPS geodesy. Current Science 86 (10), 1443–1448.

Leith, W., February 2002. Military geology in a changing world. Geotimes. http://www.agiweb.org/geotimes/feb02/feature_military.html.

Lesley, J.P., 1856. Manual of Coal and its Topography, J.B. Lippencott and Co., Philadelphia, 224 p.
Lesley, J.P., 1869. Notes on a map intended to illustrate five types of earth-surface in the United States between Cincinnati and the Atlantic Seaboard. Am. Philos. Soc. Trans. N.S 13, 305–312.
Mitchell, C., 1991. Terrain Evaluation: An Introductory Handbook to the History, Principles, and Methods of Practical Terrain Assessment, second ed. Longman Scientific and Technical & John Wiley & Sons, NY, 441 p.
National Atlas of the Democratic Republic of Afghanistan, 1985. Geokart Poland: Organization for Surveying and Cartography. Warsaw, Poland.
Powell, C.McA., 1979. A speculative tectonic history of Pakistan and surroundings: some constraints from the Indian Ocean. In: Farah, Abdul, DeJong, K.A. (Eds.), Geodynamics of Pakistan. Geological Survey of Pakistan, Quetta, Pakistan, pp. 3–24.
Powell, J.W., 1895. Physiographic regions of the United States. Natl. Geogr. Soc. Monogr. 3, 65–100.
Rinker, J.N., 1998. Remote sensing, terrain analysis, and military operations. In: Underwood Jr., J.R., Guth, P.L. (Eds.), Military Geology in War and Peace, vol. XIII. Geological Society of America Reviews in Engineering Geology, pp. 145–150.
Ruleman, C.A., Crone, A.J., Machette, M.N., Haller, K.M., Rukstales, K.S., 2007. Map and Database of Probable and Possible Quaternary Faults in Afghanistan: U.S. Geological Survey Open-File Report 2007-1103, 39 p.
Sella, G.F., Dixon, T.H., Mao, A., 2002. REVEL—A model for recent plate velocities from space geodesy. Journal of Geophysical Research 107, 1. http://dx.doi.org/10.1029/2000JB000033.
Şengor, A.M.C., 1984. The Cimmeride Orogenic System and the Tectonics of Eurasia. Geological Soc. Am. Spec. Pap. 195.
Şengör, A.M.C., Natal'in, B.A., 1996. Paleotectonics of Asia: fragments of a synthesis. In: Yin, A., Harrison, M. (Eds.), The Tectonic Evolution of Asia. Cambridge University Press, pp. 486–640.
Shareq, A., 1981. geological and geophysical investigations carried out in Afghanistan over the period 1972–1979. In: Gupta, H.K., Delany, F.M. (Eds.), Zagros, Hindu Kush, Himalaya-Geodynamic Evolution, 3. American Geophysical Union Geodynamics Series, pp. 75–86.
Shroder Jr, J.F., 1983. The U.S.S.R. and Afghanistan mineral resources. In: Agnew, A. (Ed.), International Mineral Resources a National Perspective. American Association for Advancement of Science, pp. 115–153.
Shroder Jr, J.F., 1989. Afghanistan resources and Soviet policy in Central and South Asia. In: Hauner, M., Canfield, R. (Eds.), Afghanistan and the Soviet Union: Collision and Transformation. Westview Press, pp. 101–119.
Shroder Jr., J.F., Assifi, A. Tawab, 1987. Afghan mineral resources and Soviet exploitation. In: Klass, R. (Ed.), Afghanistan and the Soviet Union: The Great Game Revisited. Freedom House Pub, New York, pp. 97–134.
Thornbury, W.D., 1965. Regional Geomorphology of the United States. John Wiley & Sons, NY.
Underwood Jr., James R., Guth, Peter L., 1998. Military Geology in War and Peace. Geological Society of America, Boulder, CO.
U.S. Department of Agriculture, 2001. Soil Regions of Afghanistan. Original Scale 1:1,000,000. USDA National Resources Conservation Service. Hari Eswaran responsible author.
Vernant, P., Nilforoushan, F., Masson, F., Hatzfeld, D., Abbassi, M., Vigny, C., Tavakoli, F., Bayer, R., Martinod, J., Ashtiani, M., Chery, J., 2004. Contemporary crustal deformation and plate kinematics in the Middle East constrained by GPS measurements in Iran and north Oman. Geophysical Journal International 157, 381–398.
Wheeler, R.L., Bufe, C.G., Johnson, M.L., Dart, R.L., 2005. Seismotectonic map of Afghanistan, with annotated bibliography: U.S. Geological Survey Open-File Report 2005–1264, 31 p.
Wheeler, R.L., Rukstales, K.S., 2007. Seismotectonic map of Afghanistan and adjacent areas: U.S. Geological Survey Open-File Report 2007–1104.
Wilson, J.P., Gallant, J.C., 2000. Terrain Analysis: Principles and Applications. John Wiley & Sons, NY, 479 p.

In this Chapter

3

Terrains of Torment

Abstract

Geomorphologic subdivision of Afghanistan includes 13 geomorphic provinces set up on the basis of the fundamental rock structures and basic bedrock lithologies. The Transpressional Plate Boundary geomorphic region of northeast and east has the four provinces of the Eastern Pamir–Hindu Kush Mountain with four sections; the Nuristan Hindu Kush Mountains with three sections, the Kabul basin with no subsidiary sections; and the Katawaz Mountains and Basins with three sections. The Accreted Terranes geomorphic region has the three provinces of the Farah Block Basin and Range with three sections; the Helmand Block with four sections; and the Helmand Basin with its six sections that reflect its considerable geomorphic diversity. The Middle Afghanistan Shear Zone geomorphic region along the central east–west axis of the country has only the geomorphic province of the Hari Rud–Ghorband, with four different sections. The North Afghanistan Platform geomorphic region has the five geomorphic provinces of the Western Pamir–Hindu Kush Mountains with its three sections; the North Afghanistan Platform Plateau with its two sections; the Tadjik Basin with three sections; the Northern Afghanistan Plateaus and Loess Hills with three sections; and the Central Afghanistan Mountains and Plateaus with four sections. All the 4 regions, 13 provinces, and 42 sections of Afghanistan are described briefly to establish new recognition of the tremendous geomorphic diversity of the country.

Keywords: Geomorphic province; Geomorphic region; Geomorphic section.

Geomorphologic Subdivision of Afghanistan

The details of the geomorphology of Afghanistan reflect the rather tortured geologic and geomorphologic past wherein many ordinary to extremes of geomorphic process have produced such a fractured and twisted plethora of landscape elements. The chapter title, "Terrains of Torment", is a reflection of the complex configurations of landforms that have resulted, as well as incidentally, what human life can be like trying to wrest a living from such a difficult place. Earthquake-riven mountainsides, landslide-covered steep slopes and fault scarps, ice-covered highlands, bare rocky outcrops on which little will

Natural Resources in Afghanistan. http://dx.doi.org/10.1016/B978-0-12-800135-6.00003-9

grow, flood-swept lowlands, wind-blasted, hyperarid environments with dried-out lake basins; these are the geomorphic processes that have produced this striking landscape. The following is a brief explanation of some of the more salient parts of the geomorphology of the region that are based upon the basic bedrock and surficial geology (Doebrich and Wahl, 2006), which will enable a better understanding of the complex landforms of Afghanistan.

Transpressional Plate Boundary Geomorphic Region (TPB Region)

Eastern Pamir–Hindu Kush Mountain Geomorphic Province (T-A)

The Eastern Pamir–Hindu Kush (EaPaHiKu) Geomorphic Province in the extreme northeast of Afghanistan encompasses part of the high Pamir mountains of eastern Badakhshan, the Eastern Hindu Kush along the border of the Wakhan Corridor with Pakistan, and the mountains and valleys of the Greater Pamir and the Lesser Pamir mountains of the Wakhan (Figure 3.1). The province is composed dominantly of crystalline rocks of Precambrian to Phanerozoic age that have been caught up in various orogenic or mountain-building episodes to become metamorphosed and injected with igneous intrusions. Especially pronounced in the Wakhan Corridor are black argillitc rocks that are

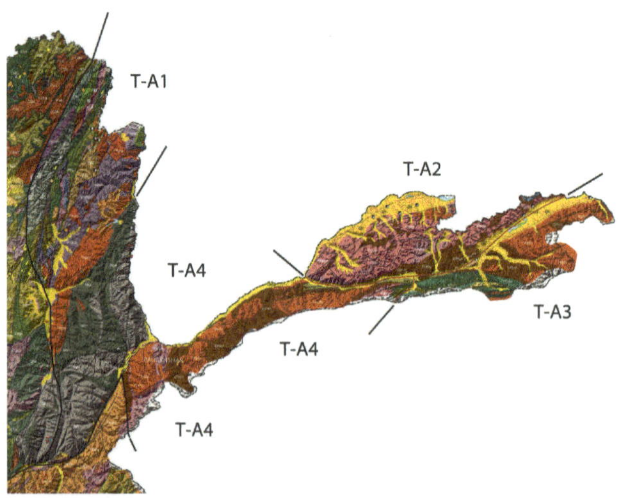

Figure 3.1 Portion of the USGS geological map of Afghanistan. This is a republication of the work done by Soviet and Afghan geologists in the 1960s and 1970s (Doebrich and Wahl, 2006). Geologic location map of the Eastern Pamir–Hindu Kush Province in eastern Badakhshan and the Wakhan Corridor showing the four sections: T-A1 High Pamir–Shiwa Lake Section; T-A2 Greater Pamir Mountains Section; T-A3 Lesser Pamir Mountains Section; and T-A4 Eastern Hindu Kush Mountains Section. The reds and pinks on the geologic map are dominantly granitoid plutonic rocks, the grays are dominantly gneisses, and the yellows are of Quaternary age that are dominantly glacial, alluvial, or mass movement in origin.

the metamorphosed equivalents of the dark marine shales and sandstones that were laid down on the marine shelf of the Eurasian plate prior to its collision and transformation as the Indian plate arrived. These black rocks are quite distinctive in appearance because of their strong color contrast with the light-colored granites that intrude them, which were produced in the heat and pressure of the collision. The Pamir Mountains of both Badakhshan and the Wakhan also have many deep mountain valleys that are filled with glacial, colluvial, and fluvial sediment deposits that are commonly covered with grasses and tundra vegetation. These broad grassy valleys are the source of the name "pamir" in the local languages (Figure 3.2).

The EaPaHiKu Geomorphic Province is subdivided into four sections, based on variations in underlying geology, as well as on topographic–geomorphologic variation, in the following fashion: (1) High Pamir–Shiwa Lake Section; (2) Greater Pamir Mountain Section; (3) Lesser Pamir Mountain Section; and (4) Eastern Hindu Kush Section. The four sections of the EaPaHiKu Geomorphic Province all are somewhat similar topographically in being high-altitude glaciated and glacierized mountain ranges, but different geologically in being dominantly either plutonic or metamorphic, as well as having greater or lesser quantities of fault shearing and fracturing disturbance, depending upon their proximity to the Central Badakhshan fault.

The EaPaHiKu Geomorphic Province is characterized throughout all its sections with landform and sediment evidence of past glaciations in Pleistocene time. Only three episodes of prior glaciations are mapped in Afghanistan and these have been designated separately as early, middle, and late Pleistocene (Bohannon and Stoesser, 2005; Lindsay, 2005; Stoesser, 2005). It should be noted, however, that with as many as 20 prior glaciations known in the deep sea records that were used to establish the standard marine isotope stages through Quaternary time, the fact that only three past glaciations have been mapped

Figure 3.2 **Broad grassy valley, or "pamir" in the local language, of the Wakhan Corridor.** *Photo by John Mott, by permission.*

here is reflective of a generally uncertain stratigraphy and a lack of modern dating applied to the sediments and glacially affected bedrock. In addition to the timing uncertainty of the glaciations, also unknown and unmapped are the many possible configurations of the valley-glacier magnitudes and directions of flow. Porter (2004) did attempt to map past glaciations in Afghanistan with Landsat imagery (Figure 3.3) but is known to have missed a number of important moraines that show glaciations far beyond his mapped extents (Desio, 1975), so new attempts to rectify this lack of modern mapping of the Quaternary must be undertaken eventually when the higher educational system of

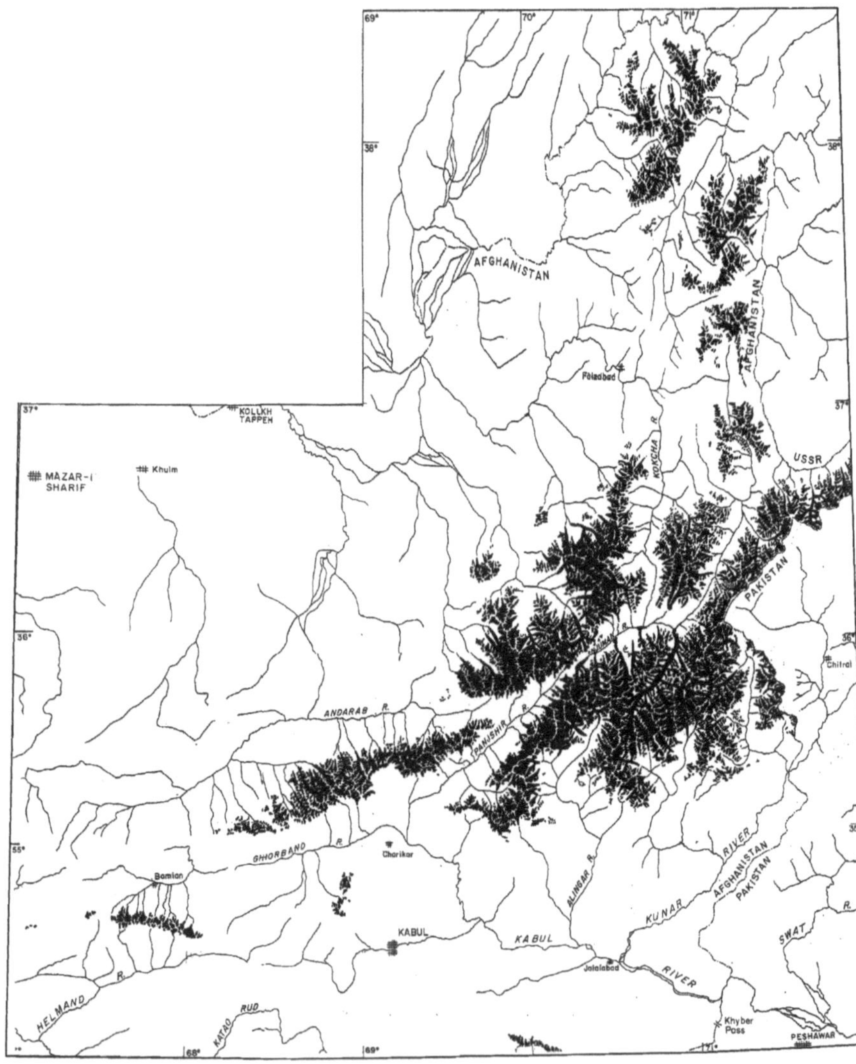

Figure 3.3 **Map of most of the Pleistocene glaciers in Afghanistan (excluding the Wakhan Corridor) by S. Porter (2004).** Porter used Landsat satellite imagery to produce this map, which enabled determination of glacial cirques and wide U-shaped valleys, but not much for end moraines, or for age dating. Consequently this map represents only minimal glaciations near the end of the late Pleistocene, rather than the true Late Glacial Maximim (LGM) when the ice was more extensive.

Afghanistan has recovered sufficiently to train students in such more advanced research possibilities, should that eventuality come to pass at some future date.

High Pamir–Shiwa Lake Geomorphic Section (T-A1)

The bedrock geology of the EaPaHiKu Geomorphic Province is made up largely of Precambrian (Archean and Proterozoic) gneisses and granites with a few volcanics, which were overlain later with Lower Paleozoic carbonate and clastic rocks, and Permian, Triassic, and Jurassic siltstones and shales. All these preexisting rocks were then intruded in Oligocene time about 30 million years ago (mya) by large igneous plutons of granitic rocks (Bohannon and Stoesser, 2005). The mountain-building forces of heat and pressure involved in the plate-tectonic collision not only metamorphosed many of the older sediments, but also forced up the Hindu Kush and Pamir ranges, and shattered the rocks extensively in the ongoing plate collision to the present day.

Old high-level deposits that predate most of the present topography exist in several places in the mountains of Badakhshan and the Wakhan that are discussed below. One deposit mapped as middle Pleistocene (Q_2a) conglomerate and sandstone (Bohannon and Stoesser, 2005) on the "thumb" of Badakhshan was mapped as shingly and detrital sediments, with sand sizes being more abundant than silt and clay (Figure 3.1). The 1:250,000-scale US DOD JOG topographic map series shows depression contours and several lakes at an altitude of ~3200 m at the site of this deposit just south and altitudinally above the town of Pajwar on the Ab-i Panj. Close inspection of the deposit with high-resolution satellite imagery from Google Earth™ actually shows the deposit to be a huge landslide (~8 × 3.4 km in area; ~1 km thick; ~27 km^3 volume) with pronounced hummocky topography and with two large lakes and one small one on it, as well as strong topographic lineations and large boulders that provide evidence of probable high-speed emplacement (Shroder et al., 2011). Other overlying and associated features developed over a long period of time, which attests to the considerable antiquity of the landslide feature.

Lake Shiwa (or Shewa), a body of water about 15 km^2 in area in northeastern Badakhshan about 10 km from the Panj River border with Tajikistan also was dammed sometime in antiquity by one or more landslides (Shroder and Weihs, 2010). A large rock-slope failure from the fault-shattered and strongly weathered Zirnokh peaks to the north moved into the valley and dammed up the river and its tributaries to a dam thickness of ~400 m, producing a 12-km-long lake that is as much as 270 m deep, leaving ~80 m of freeboard to the top of the dam. At least five separate slope failures have been mapped at the site of the landslide dam where two strike-slip faults cross each other and presumably contribute to instability (Shroder and Weihs, 2010). The uppermost part of the Lake Shewa landslide dam is a prominent rock glacier, which as a slow moving accumulation of rock debris and internal ice cement, takes centuries to millennia to form, and thus testifies to the antiquity of the dam. Water levels in Lake Shiwa fluctuate several meters annually in response to variations in inflow, as well as in outflow through the open matrix rock fragments. Progressive seepage of water through the dam face has caused several young subsidiary slump failures, which if continued long enough, or with additional seismicity from the active strike-slip faults that cross beneath the landslide dam, could threaten its integrity.

In the southwest part of the High Pamir–Shiwa Lake Section, the confluence of the Warduj River with the Zardew River, and then a few kilometers farther downstream again, the Warduj with the Kocha River, a group of different landforms was mapped that are exemplary of those in the high mountain valleys. Desio (1975) expended considerable effort mapping the detailed geology and geomorphology in this whole region to show complex interactions and geomorphic formations resulting from the actions of glaciers, landslides, rivers, and lakes. His early work on glacially streamlined hills (Figure 3.4), hummocky glacial drift, glacial erratics, large landslides, and explanatory cross-sections were generally quite perceptive for this early time.

Greater Pamir Mountain Geomorphic Section (T-A2)

The bedrock of this section in the Wakhan Corridor is predominantly Middle to Late Paleozoic metamorphosed siltstones and sandstones that are more abundant than the dark slates (Lindsay, 2005). Nonetheless, the dark slates are prominent in contrast to the light-colored granitoids that intruded them (Figure 3.5). These batholithic intrusions occurred first in Cretaceous time, and then later during the Oligocene, when the largest intrusions resulted. The central part of this section is constituted by the high Pamir mountains, which are held up by the granite core that is bounded in part to the north and south by the dark metasediments. Much of the Greater Pamir region is upfaulted highland wherein the linear east–west-trending valleys tend to be aligned along some of the faults. The faults host rising thermal waters that issues out as hot

Figure 3.4
Photograph from 1961 of glacially streamlined hill (large roche moutonnée or small flygebirge) with glacial moraine and terraces of Dashti Feraq from a hill southwest of Furmorah in the Baharak region in the High Pamir–Lake Shiwa Section. *After Desio (1975).*

springs, commonly where the water rises through talus and alluvial–colluvial cones (Buchroithner, 1978). The Greater Pamir mountains have been extensively glaciated multiple times and the previously glaciated, wide grassy *pamir* valleys are underlain by Early and Middle Pleistocene glacial tills, and other sediments (Patzelt, 1978; Lindsay, 2005). Late Pleistocene and Holocene alluvium overlies many of these deposits, commonly deposited as striking alluvial–colluvial fans, but also as linear alluvial valley fills and glacial kame terraces. As the alluvial–colluvial fans debouch from the upper mountain valleys into the lower valleys, the sediments form perfect fan cones that radiate out from the valley mouth sources (Figure 3.6).

Zori Kol lake at the extreme northeast tip of the Wakhan Corridor (Wood, 1872) is dammed by moraines that extend down from the mountains on both sides of the valley, but principally in the south from the Greater Pamir. The moraine loops show up well on the satellite imagery. The lake outlet west into the Abi Panj flows in the first 25 km of its course between high-gravelly banks that rise to extensive grasslands to the lower mountain slopes. The fresh-water lake extends E–W for ~20 km and 3–4 km wide. The glacier regions in the Greater Pamir were studied extensively by Patzelt (1978), who mapped general geomorphology and

Figure 3.5 Photograph looking northeast at the fans on both sides of the town of Rawtshun (3220 m). The cliffs above expose the light-colored Issik Granodiorite that intrudes into the dark-colored schist and quartzite of the Wakhan Formation. The light-colored granodiorite debris of the alluvial–colluvial fans is emplaced largely by rapid wet debris flows. *After Senarclens de Grancy, Kostka (1978).*

determined glacier characteristics, terminus positions, and other useful glaciological parameters (Figure 3.7). Haritashya et al. (2009) followed up with satellite-image analysis that showed pervasive glacier retreat and many new glacier lakes.

Figure 3.6 Map of Issik and Ptukh debris cones that issue from the glacierized southern valleys of the Greater Pamir. *After Patzelt (1978).*

Figure 3.7 The partly debris-covered South Issik Glacier occurring in the foreground and North Issik Glacier as they appeared in 1975. *After Patzelt (1978).*

Lesser Pamir Mountain Geomorphic Section (T-A3)

The Lesser Pamir mountains (Figure 3.1) occur in the extreme southeast portion of the Wakhan Corridor along the Pakistan border, extending eastward from the Baroghil Pass glacial col that goes into the Yarkhand Valley in Chitral Pakistan, to the border with China. The rocks of the Lesser Pamir are dominantly metasediments, with Late Paleozoic carbonates and clastics with Triassic intrusives, as well as Eocene and Oligocene clastics and volcanics, pierced by Oligocene granodiorite and granite batholiths (Lindsay, 2005).

The geomorphology of the Lesser Pamirs has not been much studied; suffice it to say that the wide *pamir* valleys that were once opened out by strong glaciations in the Pleistocene are characteristic. The lake of Chaqmatin Kol in this section occurs at an altitude of 4020 m, is about 9 km long and 2 km wide, and drains to the east into the Ak Su river. It has its source in snow and glacier ice melt to the north along the Tajikistan border and to the east and south on the Pakistan and Chinese borders (Figure 3.8).

Lake Chaqmatin Kol is a prominent remnant of past ice activities and the surrounding valleys are replete with some deranged drainage. In fact, one of these at the west end of the lake has a permanent stream that comes down it to split into distributaries in a delta into the lake (Gordon, 1876; Littledale, 1892). Some of the distributaries flow into the lake and one is the headwaters of the Abi Wakhan river system. Normally such a distribution of water high in the

Figure 3.8 Satellite image of the lake Chaqmaktin Kol in the far eastern part of Wakhan Corridor, and the Lesser Pamir portion of the Eastern Pamir–Hindu Kush Province. Note the delta with multiple distributing channels at the west (left) end of the lake where water flows to both the right into the lake and out into Tajikistan, as well as to the west and into the Wakhan Darya (Figure 5.18).

mountains is not particularly notable, unless as in this case, it causes confusion (Mirzad, 1970) in the delineation of major hydrologic-basin boundaries wherein the waters divide between the basin of the Abi Wakhan tributary in the Wakhan Corridor to the Abi Panj, and the basin of the Ak Su across Tajikistan to the Abi Panj and its downstream flow into the Amu Darya and the Aral Sea.

Eastern Hindu Kush Geomorphic Section (T-A4)

The "recumbent mushroom-shaped" Eastern Hindu Kush Section with the Wakhan Corridor as the "stem" is a mixture of Precambrian (Archean and Paleoproterozoic crystalline gneisses in the western "cap of the mushroom" in Badakhshan) and Late Paleozoic metasediments (siltstone and sandstone more abundant than slate and volcanics) and extensive Oligocene granitoid batholiths that dominate the mountains along the border with Pakistan in the eastern "stem". The black slates, however, are one of the most noticeable lithologies that make up the debris on the plentiful glaciers of the region, and the contrast to the light-colored granites is pronounced. The tallest mountain in Afghanistan, Naushak (7455 m), is a satellite peak of the huge Tirich Mir (7708 m) in Pakistan, and composed dominantly of much loose black slate that accumulates easily as profuse supraglacial debris. The result is unusual black-appearing glaciers in many places.

At the contact of the "mushroom stem" of the Eastern Hindu Kush with the "cap," a supposed middle Pleistocene Q_2a deposit occurs at an altitude of ~2800 m at the mouth of the Wakhan Corridor in the Eshkashem valley where it connects to the Abi Panj trough. In this case, the old deposit is caught up in a valley-bottom fault that apparently has protected it. High on the side walls of this Daryai Sangtec and Darrah Dora valley to the southwest from Eshkashem around Zebak also are numerous deposits of middle Pleistocene till. These mountains in Quaternary time, particularly in the Pleistocene, were extensively glaciated multiple times. Hundreds of glaciers still exist in this section, although most are now downwasting and backwasting to retreat dramatically. Light-colored, and therefore unweathered and unvegetated, glacier forefields in many places on the high peaks testify to this recent retreat.

The farthest eastern end of the Eastern Hindu Kush is a single range of high mountains, the peak summits of which constitute the border with Pakistan. The range begins at Baroghil Pass into the Yarkand Valley of Pakistan and at the base of the range there in the Wakhan Corridor in Afghanistan, the river of Abi Wakhan (Wakhan Darya) flows west to join with the Abi Panj (Panj Darya) and form the border with Tajikistan. These valleys have been the site of many trunk glaciers in Pleistocene time, the sediments of which are still preserved in lateral ablation valley, or kame terrace remnants. The Keshnikhan Glacier valley and its surroundings at the western end of the Wakhan Corridor, a short distance (17 km) northeast of Naushak, has been extensively mapped and photographed (Figures 3.9) (Braslau, 1972; Gratzl, 1972; Braslau and Bussom, 1978a,b; Shroder, 1978; Shroder and Bishop, 2010). The valley of the Abi

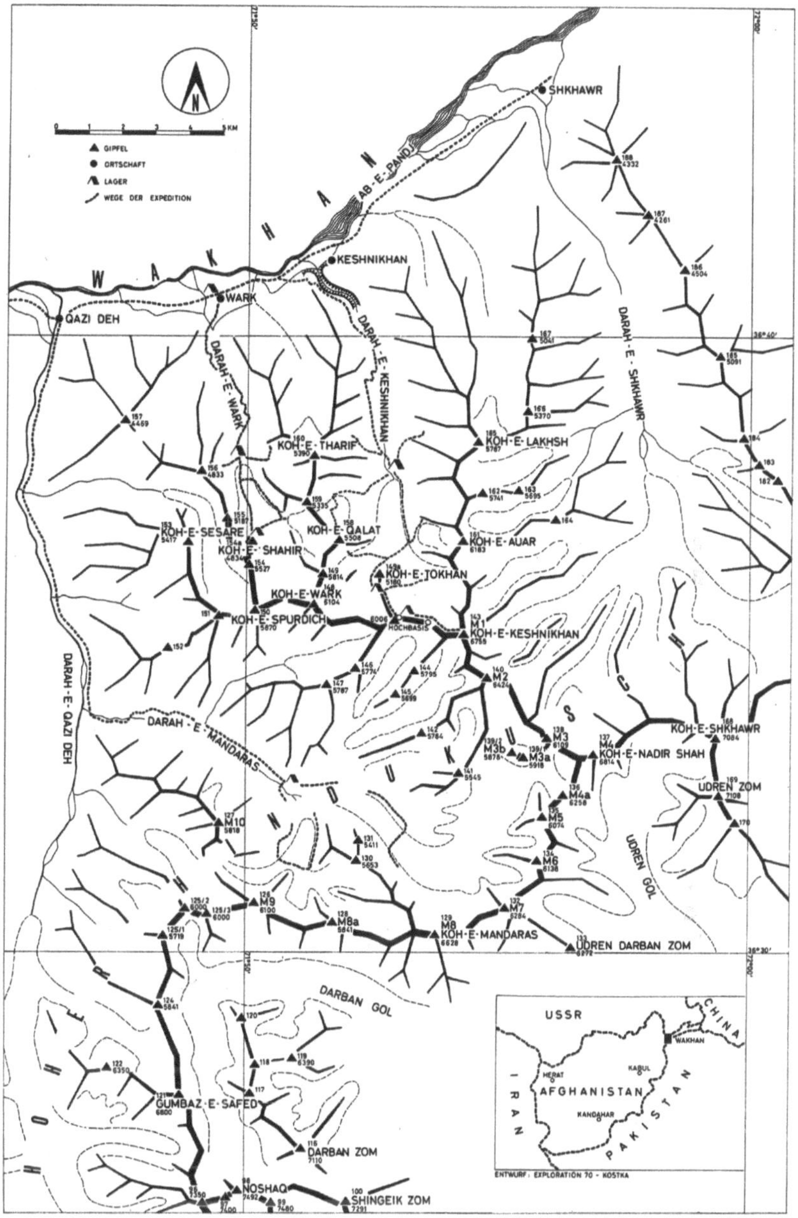

Figure 3.9 Map of the west end of the Wakhan Corridor showing peaks, glaciers, and valleys of the eastern Hindu Kush. *After Gratzl (1974).*

Panj at the confluence and downstream from the Keshnikhan valley is characterized on both sides of the river by lineated erosional ridges and depositional terraces, where one or more trunk-valley glaciers existed in Pleistocene time (Figure 3.10). Just as upstream in the valley of the Wakhan Darya south of the Greater Pamir, alluvial–colluvial fans are profuse where a combination of glacial

Figure 3.10 Village of Wark on a debris fan in the Abi Panj valley in the Wakhan just east of the mouth of the Keshnikhan Valley. The ridges in the background occur across the Panj River show glacial lineations of rock and moraine. The boulder field on the river terrace reflects entrainment and imbrication from catastrophic breakout floods upstream that are most likely glacial. *After Gratzl (1972).*

meltwater and summer cloudbursts produce rapid wet debris flows, as well as less catastrophic entrainment and deposition of sediment.

A variety of glaciers from the high Hindu Kush mountains along the border with Pakistan, as well as the mountains adjacent to the Abi Panj and the border with Tajikistan in this section have left their traces in the valleys leading north throughout the area. The Warduj valley, for example, has had glaciers in it in many places, which Desio (1975) mapped and described. His map of the moraine at Aqshira on the Warduj is accurate and shows former ice movement down the Warduj. Desio's (1975) two maps of glaciers around Zebak, however, and the supposed moraines they left are incorrectly so designated because instead the high-resolution satellite imagery from Google Earth™ shows that they are actually landslides (Figure 3.11). Thus, both the Zar Khan landslide (Figure 3.12) and the Gul Khana at Zebak are derived from unstable weathered bedrock and overlying Pliocene clastic sediments on the 3935 m mountain on Desio's map (Figure 3.12). Both landslides flowed down into the surrounding river valleys and probably blocked them for a time. The Zar Khan feature is notable in particular because it has such pronounced lateral landslide levees, which actually resemble glacial lateral moraines, but no cirque occurs on the

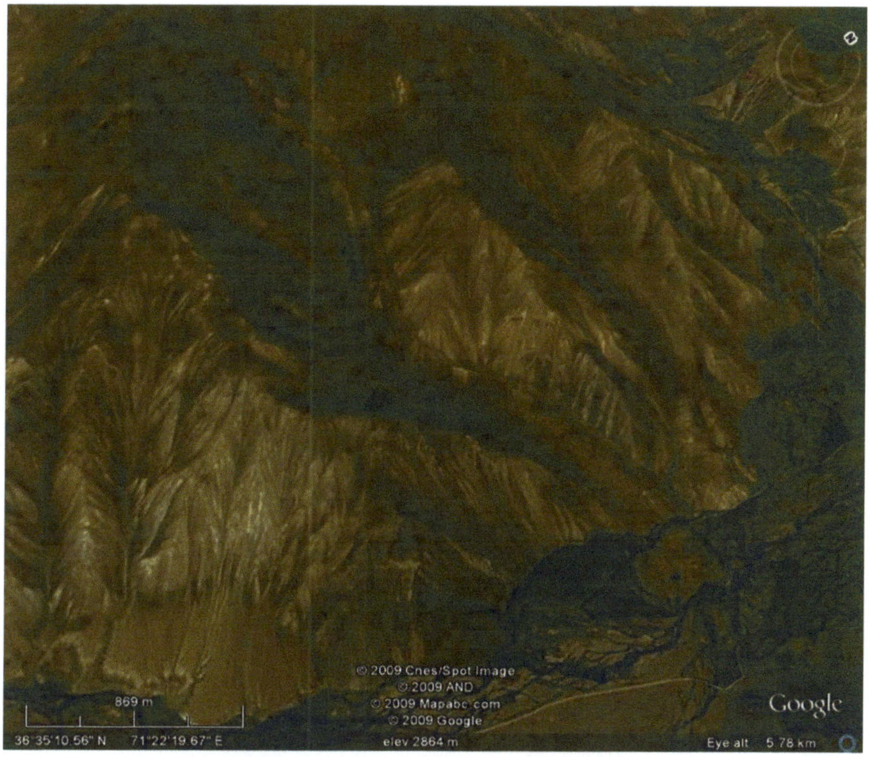

Figure 3.11 **Synthetic view looking west from an altitude of ~2864 m above the landscape at the Zar Khan landslide into the valley of Wazling north of Zebak.** Note the double landslip levees along the course of the flow-type slope failure.

mountain and a glacial origin is clearly out of the question. Farther downstream on the right bank of the Warduj, several other landslide dammed lakes also occur all along the middle altitudes of the peaks along the east banks of the river between Zebak and Aqshira.

Nuristan Hindu Kush Mountains Geomorphic Province (T-B)

The area of the Nuristan Hindu Kush Mountains Geomorphic Province (Figure 3.13) is characterized by the rugged mountains of the Central Hindu Kush Mountain Section (Stoesser, 2005; Lindsay et al., 2005), as well as the Jalalabad Basin Section to the south, and the mountains of the Spin Ghar (Safed Koh) Section on the border with Pakistan (Bohanon and Turner, 2005). The Nuristan Hindu Kush Mountains Geomorphic Province in the Transpressional Plate Boundary Region is set off largely from its neighboring provinces by major fault systems. This includes the sharp "spur" projecting to the north in contact with the Pamir–Hindu Kush Province there, as well as the fault in the Panshir Valley all along the northwest edge of this province, the Sarobi fault along the west contact of this province with the Kabul Basin Geomorphic Province,

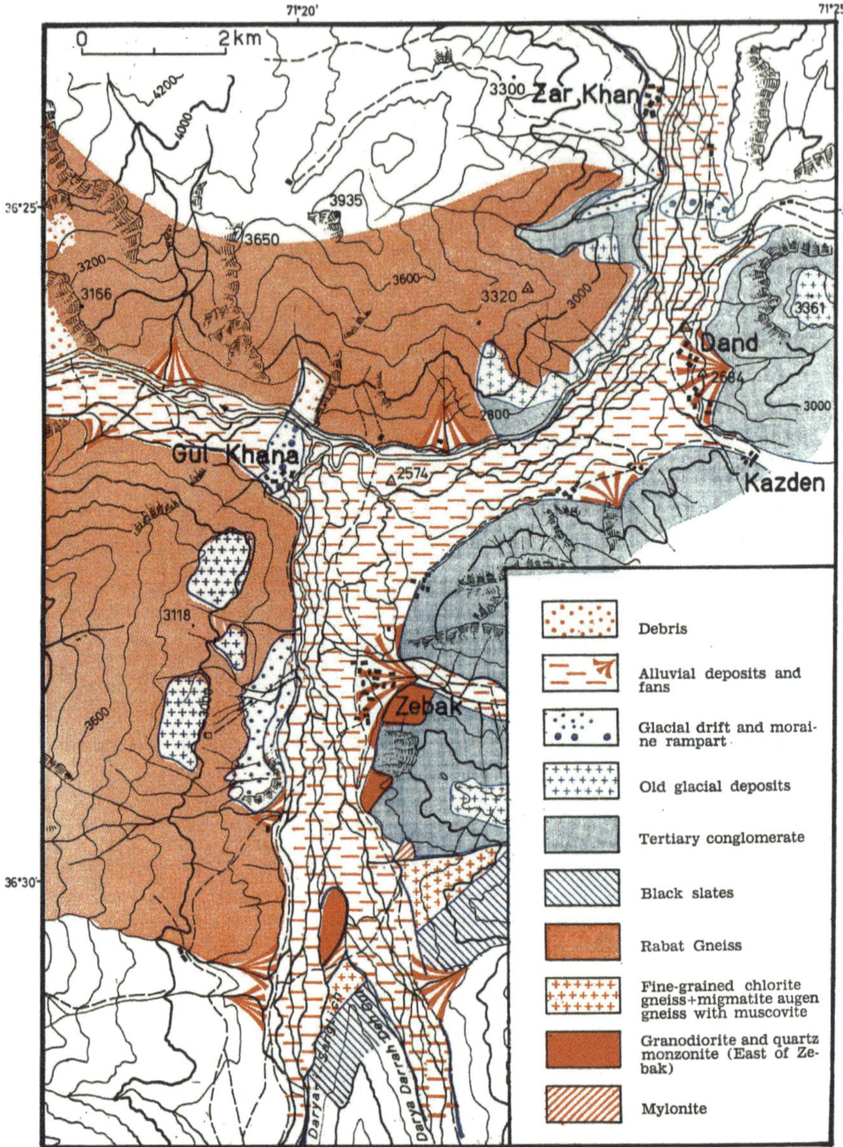

Figure 3.12 Map of the Zebak valley where the confluence of the river Daryai Sanglich from the south and the Daryai Wazling from the northeast are confluent at Gul Khana and flow off to the west as the river Darya Warduj *(after Desio, 1975)*. The river valleys are choked with sediment and commonly flooded with water because of the poor drainage. The east side of the valley has thick sediments of Pliocene age that probably once lapped up high on the 3935 m mountain to the northwest because large landslides of friable sediment have blocked the rivers at both Zar Khan, as well as Gul Khana, where Desio (1975) incorrectly mapped only glacial drift and moraine ramparts.

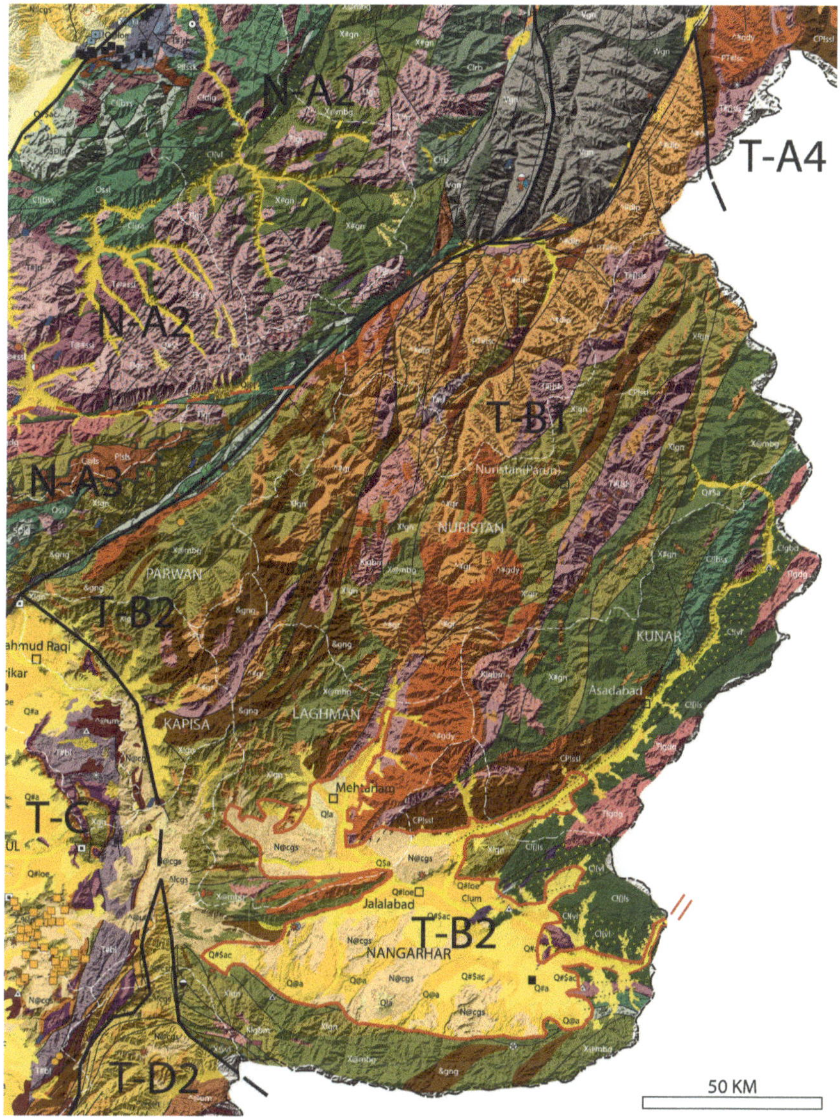

Figure 3.13 Map of Nuristan Hindu Kush Mountains Geomorphic Province (T-B) that extends from close to the mouth of the Wakhan Corridor south between the eastern border mountains with Pakistan and the Sorobi Fault that bounds the Kabul Basin to the west, south across the Jalabad Basin to the Safed Koh (Spin Ghar) mountains on the Parachinar border reentrant of Pakistan. *After Doebrich and Wahl (2006).*

and a number of other faults. The national boundary with Pakistan cuts off the south and east boundaries of the province, so that the true geologic boundary conditions are concealed by lack of further information over the border.

The area east of the Kunar River to the Pakistan border is the East Nuristan island arc, which is the westernmost continuation of the Kohistan-Ladakh island–arc complex exposed in North Pakistan and Northwest India (Treloar and Izatt, 1993). It is composed of acid to basic plutons that intruded basic volcanic and volcaniclastic sediments. The volcanic suites presumably began in Mid-Cretaceous and were sutured onto the Asian plate about 100 mya, with subduction-related volcanism continuing into the Eocene and the final closure of the Tethys Seaway by ~50 mya.

Central Hindu Kush Mountain Geomorphic Section (T-B1)

This section is the heart of the province, with range after range of crystalline rocks composed largely of diorite, plagiogranite, and granite of Oligocene age that intruded a variety of preexisting rocks, including Proterozoic gneisses and granites, as well as Late Paleozoic, Triassic, and Cretaceous metasediments and other plutons in the region. These ranges in Nuristan have been extensively glaciated, not particularly because of their altitude, but mainly because of enhanced monsoonal moisture. In summer, with altitudes in the mountains above about 4500–5000 m, the incoming monsoonal precipitation generally falls as snow, which enhanced glaciations and produced large valley glaciers that spread down most of the major big valleys.

In the northwest part of this section around a mountain massif known as Mir Samir (5809 m), a number of glaciers occur that have been studied enough to serve as examples of glacial types throughout this province (Figures 3.14). For example, the East glacier (Yakchaal-i-Sherq) and West glacier (Yakchaal-i-Charb) below the north face of Mir Samir were first assessed glaciologically by a British team in 1965 (Gilbert et al., 1969). They studied snow accumulation,

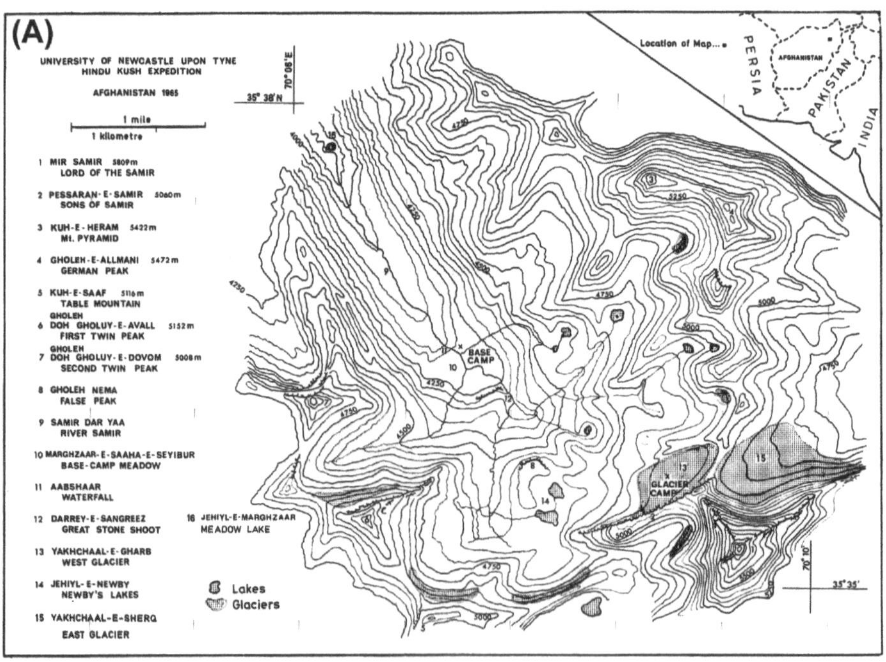

Figure 3.14
(A) Location map of glaciers at Mir Samir that were studied by Gilbert et al. (1969). The glacier numbered 13 and glacier camp is West Glacier (Yakchaal-i-Charb).

ablation, meltwater discharge, moraine lichenomtery, micrometeorological heat balance, and other glaciologic characteristics. Their simple planimetric maps of ice distribution and lichen zones on the moraine allow some comparison with modern satellite imagery for change detection (Shroder and Bishop, 2010).

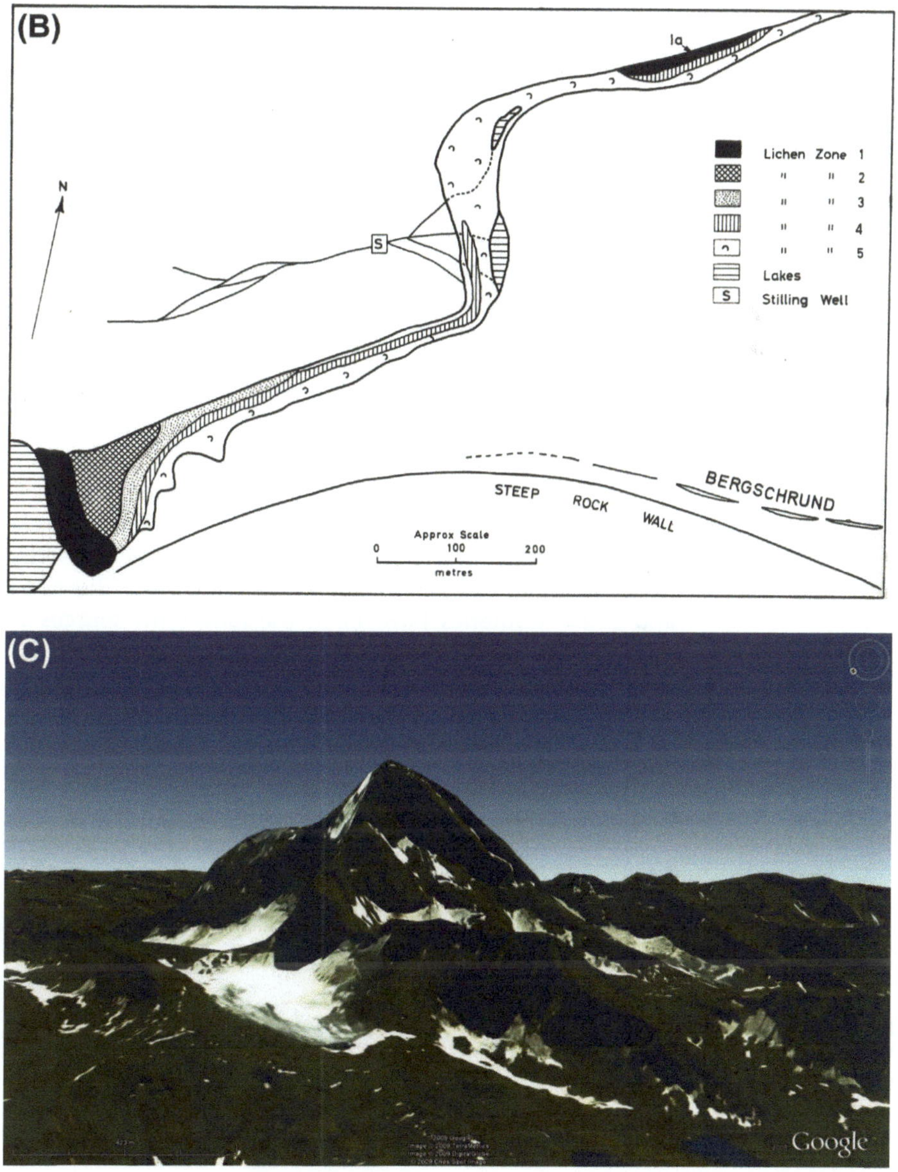

Figure 3.14 Cont'd (B) Large-scale map of West Glacier at Mir Samir that has now divided into two glaciers through major downwasting across a rock rib that has recently emerged from beneath the ice. (C) Synthetic view from Google Earth™ from the north of the mountain of Mir Samir (5809 m) which is a glacial horn that rises ~1000 m above a surrounding *gipfelflur* or generalized erosion surface of lower altitude upon which many small glaciers exist. The West Glacier in the lower center was mapped by British glaciologists in 1964 as a single mass of ice but since then the glacier has downwasted so much that it divided into two ice masses with a rock rib between them that was previously covered with ice.

Down valley from the glacier regions of today, extensive broad, U-shaped valleys extend in many parts of this section. The Munjan Valley of the northern part of the area is so well developed into an open U-shape that Porter (2004) was able to map in a large trunk valley glacier there, something he was less comfortable doing for many other less broad valleys elsewhere in Badakhshan where ice had not opened up the valleys as much. More detailed topographic assessment of this valley shows it to be entrained along a number of prominent faults that seem to have helped enable more vigorous glacial erosion there than elsewhere. At lower altitudes, these prominent river valleys tend to have developed deep V-shaped canyons with rushing torrents in their bottoms (Figure 3.15).

Jalalabad Basin Geomorphic Section (T-B2)

The basin of Jalalabad is actually two basins: the northwestern part where the Alinghar and Alishing rivers join in Laghman and flow into the Kabul River, and the main Jalalabad Basin where the Kunar River joins the Kabul as well. The two subbasins are separated by the upthrust small mountains Caper Ghar and Tor Ghar of largely crystalline rock, along the north edge of which runs the Kabul River and the modern road between Kabul and Jalalabad. A slightly lower altitude prolongation of this range protrudes up through the basin sediments past the junction of the Kunar River with the Kabul River. The Kabul River then flows across the northern part of the Jalalabad Basin before exiting Afghanistan through the low ranges north of the Khyber Pass in Pakistan. Many small braided tributaries, most of them as ephemeral stream channels that only have water and sediment part of the year, also flow north into the Kabul River from the range of the Spin Ghar to the south. The Kabul River and its large and small tributaries have been the long-term sources of sediments through the late Cenozoic Pliocene and Quaternary time that have been poured into these basins from the surrounding high mountains. The earliest conglomerates

Figure 3.15
V-shaped river valley of Landay downstream from Kamdesh village.
Photograph by author.

and sandstones of Pliocene age that began to accumulate in the Jalalabad Basin also had zones of clay, limestone, marl, gypsum, and salts, which give evidence of sporadic dry lake basins occurring at that time (Figure 3.16). These older sediments are overlain progressively by more early, middle, and late Pleistocene conglomerates and sandstones that were deposited largely as river alluvium. Late Pleistocene windblown loess dust deposits are also common. Holocene and late Pleistocene alluvial and colluvial fan deposits are also characteristic of wide-spread areas in the basin. Holocene alluvial sands and gravels predominate in the modern river valleys. The landforms of the sediments and sedimentary rocks in the Jalalabad Basin have been eroded into areas of extensive older buttes, mesas, badlands, pediment surfaces, and dry washes. These landforms are eroded from a variety of clastic sediments, coupled with extensive alluvial fans and plains, all derived from erosion of the surrounding mountain highlands. A number of braided streams flow out of the mountains of the Spin Ghar to the south and their waters are used for extensive local irrigation of the sediments of the Jalalabad Basin.

Spin Ghar (Eastern Safed Koh) Geomorphic Section (T-B3)

The Spin Ghar (Pushto) or Safed Koh (Dari) mountains on the south edge of this province along the border with Pakistan constitute a clear topographic/geomorphic division from surrounding lowlands. They are an isolated, east–west-trending, upthrust mass of largely crystalline gneisses and granites with various admixtures of gabbros, mafic metavolcanics, marbles, and even migmatites, all of Proterozoic age (Bohanon and Turner, 2005). These resistant metamorphic and igneous rocks have been subject to the forces of deep weathering for many millions of years, with the result that they have formed a variety of erosional residuals of the tor type. At higher altitudes, the vigorous freeze and thaw have tended to shatter the bedrock and produce plentiful rock rubble. In several places at the top of the Spin Ghar, Pleistocene glacial cirques occur, which in

Figure 3.16 **Photograph of part of the Laghman valley basin-fill sediments of Pliocene age showing a higher level, truncated planar erosion surface across its top.** *Photograph by author.*

some the rock rubble has accumulated over ice to form ice-cored moraine or ice-cored rock glaciers. The highest peak in the Spin Ghar near Tora Bora, where Osama bin Laden took refuge from Coalition forces before escaping to Pakistan, is only 4120m high, which is of insufficient altitude to maintain active glaciations. The Spin Ghar is high enough, however, to serve as a reasonably efficient trap for incoming monsoonal and westerly moisture sources. This results in a number of stream courses, especially on the northern side, so that these water the Jalalabad Basin and have been a partial source of many of its sediments.

Kabul Basin Geomorphic Province (T-C)

The Kabul Basin Province (Figure 3.17), or Kabul block, is one of the isolated continental fragments that was rifted from the southern landmass of Gondwana perhaps 100mya and after a long voyage north through the ancient Tethys Ocean, was later sutured onto the northern Eurasian plate (Broshears et al., 2005). Myslil et al. (1982) and Treloar and Izatt (1993) noted that the Kabul block is now bounded on the east and west by major left-lateral strike-slip faults, as well as a thrust fault at depth in the north. The rocks that outcrop in the Kabul Basin range from Precambrian Proterozoic metamorphic rocks, mainly gneisses, which were progressively overlain by various Phanerozoic (Paleozoic, Mesozoic, and Cenozoic) sedimentary rocks, as well as a few igneous rocks (Wolfhart and Wittekindt, 1980; Bohanon and Turner, 2005; Lindsay at al., 2005; Maldonado and Turner, 2005). Essentially, the Kabul block can be thought of as a continental island mass that converged on southern Asia from the south as the Paleo-Tethyan sea closed up during the Cretaceous (Şengör et al., 1988; 1996; Treloar and Izzatt, 1993; Bohannon, 2005). Ultramafic rocks in tectonic slices seem to record bits of sea-floor ophiolite sequences that record evidence of suturing of parts of the block to each other and to the southern Asian land mass.

In the course of the evolution of the Kabul block, the prime agents of formation were the fault uplifting of the surrounding crystalline rocks, and their erosion to infill the downfaulted graben basin between them. The result is a basin infilled with the Lataband Series; a variety of river alluvium, mass-movement colluvium, and fine-grained lacustrine sediment from lakes long gone (Figure 3.18). The initial development of the graben structure beneath the Kabul basin may have begun in the late Miocene when the Kabul and Panshir river valleys first came into existence (Myslil et al., 1982). As the basin filled with these sediments, the original hilly and mountainous topography of the underlying bedrock was gradually buried as much as 600m deep with Cenozoic basin-fill sediments (Figure 3.19). The progressive burial of the old bedrock hills and mountains led to the isolation of the individual masses, so that they became surrounded as kinds of "island mountains", or what are known technically as *inselbergs*.

As Bohannon (2005) has pointed out, the recent geologic history that produced the modern geomorphology of the basin is written in the courses of these two rivers. Such meandering courses (Figure 3.20) are generally developed on broad plains that are covered with soft and easily erodible sediment. But today these meandering rivers have been superimposed from the overlying cover sediments or entrenched in 500-m-deep gorges down upon the resistant bedrock beneath (Figure 3.21). An alternate viewpoint is that bedrock was

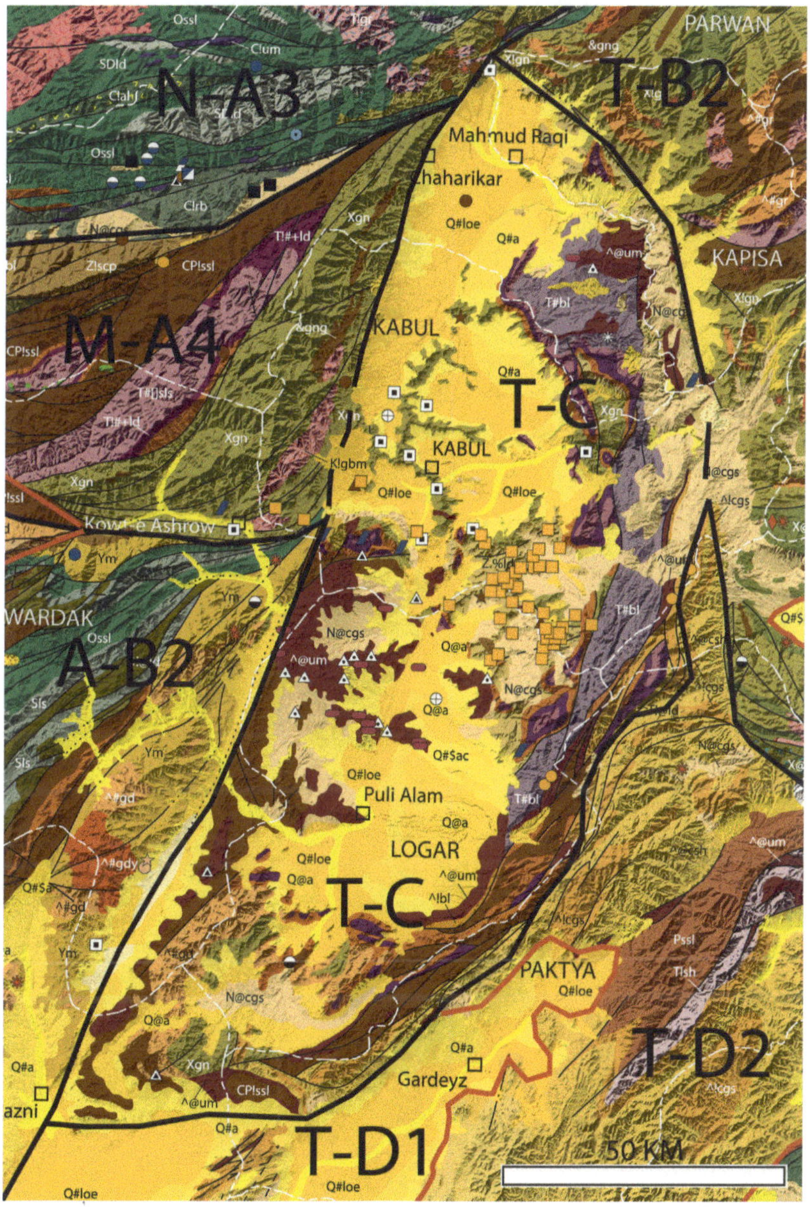

Figure 3.17 Geology map of the Kabul Basin Geomorphic Province (T-C). *After Doebrich and Wahl (2006).*

uplifted in an antecedent fashion to force the rivers to cut down through the rising rock. Bohannon (2005) presented these two ideas in an either/or fashion, but probably equally likely is that both events occurred relatively simultaneously, with some superposition of the rivers downward from the overlying cover sediments, while some uplift along the Tagab fault occurred as well. The relatively fresh looking, 500-odd-m-high fault scarp, with its deflected Tagab river at its base adds evidence to this (Figures 3.19 and 3.20). This could have

Figure 3.18 Some of the Kabul Basin infill of finer clastic sediments of the Lataband Series.

Figure 3.19 Digital elevation map of the northern part of the Kabul Basin to show in particular the topographic configuration of the Panshir River canyon, as well as the uplifted, east-facing Sarobi fault scarp.

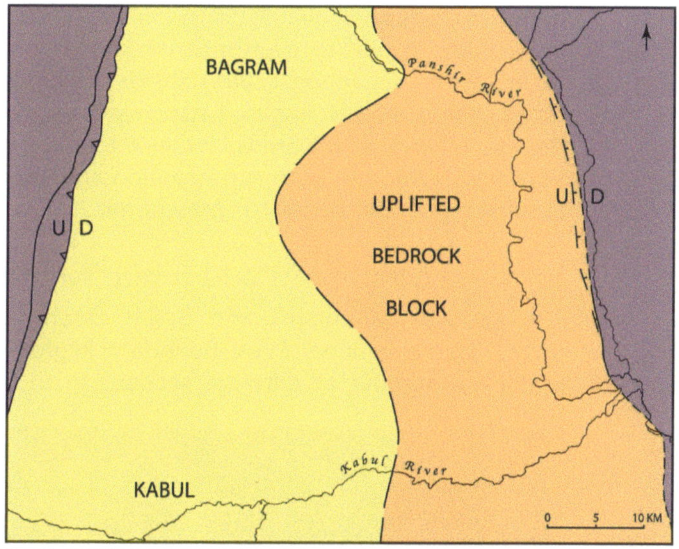

Figure 3.20 Simplified structural geologic sketch map of northern Kabul Basin with Neogene sediment infill in yellow and older bedrock in orange and purple The western Paghman fault system is thought to have been emplaced by left-lateral strike-slip motion that is now converted to reverse or thrust fault loading of the Kabul Block that has pushed down that side of the Kabul Block down. The Sorobi Fault on the east side of the Kabul Block was originally established by right-lateral motion that has recently converted to normal or vertical motion of the uplifted bedrock block (orange) with its pronounced, east-facing fault scarp so that the Panshir and Kabul rivers were forced to cut down through the overlying Neogene sedimentary cover into the bedrock beneath. *(after Bohannon, drafting by Angela Burgett).*

Figure 3.21 Photograph of part of Panshir River canyon Downcutting here is thought to be a combination of the river cutting down from overlying cover sediments, as well as differential uplift and tilting of the bedrock caused by thrust-fault loading of the Kabul Block 40–50 km to the northwest. *(after gohikeone, from Google Earth™).*

occurred through regional tilting, up-to-the-east, down-to-the-west, perhaps in response in part to overthrust loading occurring along the Paghman fault to the west. Bohannon (2005) shows a photograph of dark-colored gneiss thrust over westward dipping Lataban series rocks along the northwest edge of the Kabul basin, which lends credence to the supposition concerning the possible combination of both superposition and antecedence together to produce the drainage anomalies of the present day.

Katawaz Mountains and Basins Geomorphic Province (T-D)

The Katawaz Mountains and Basins Geomorphic Province (Figure 3.22) in the Transpressional Plate Boundary Region of Afghanistan is characterized by a number of different sections in the transpressional plate regime

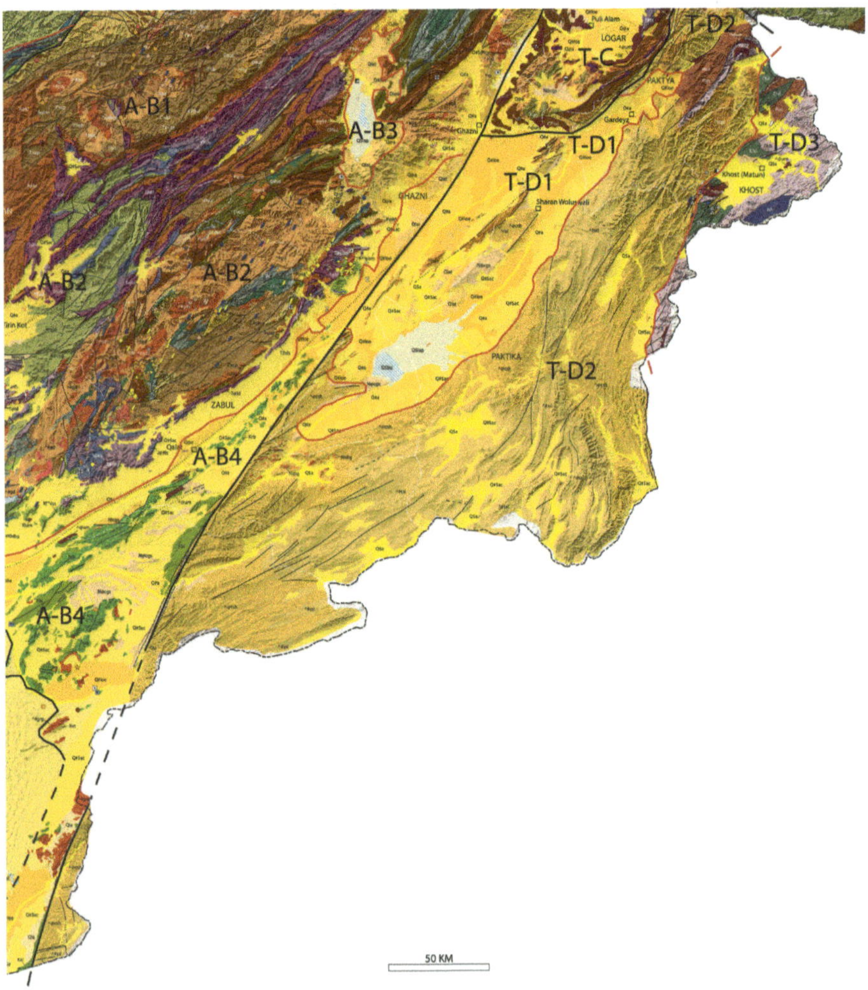

Figure 3.22 Geology map of the Katawaz Mountains and Basins Geomorphic Province. TD-1 is the Abi Istada Geomorphic Section; TD-2 is the Cuesta–Monoclinal Ride–Hogback Geomorphic Section; and T-D3 is the Khost and Sulieman Mountain Geomorphic Section. *After Doebrich and Wahl (2006).*

representing different geological occurrences that have combined to produce some topography noticeably different from much of the rest of the area. Thus three sections exist: (1) the Abi–Istada Geomorphic Section; (2) the Cuesta, Monoclinal Ridge–Hogback Geomorphic Section; and (3) Khost Basin and Suleiman Mountain Geomorphic Section, which vary geologically and topographically according to their proximity to the Chaman Fault, as well as with the past geologic histories in their assembly during the active accretionary and deformational regimes that occurred in each place in the late Cenozoic.

The Katawaz Basin exists in both eastern Afghanistan, as well as the Baluchistan Province of neighboring Pakistan. Its bedrock geology has been studied by a number of people (Wittekindt and Weippert, 1973), but the synthesis of Treloar and Izatt (1993) makes the most sense (Figure 3.23). The basin contains a thick sequence of clastics, which have been variously interpreted, but the units appear to grade upward from a series of shallow marine through estuarine and fluvial environments as a paleo-Indus fan (Brookfield, 1998) before the primary depositional river, the Indus, was captured near Nanga Parbat about 12 mya (Shroder and Bishop, 2000), diverted to more-or-less its present location in Pakistan, and extended seaward.

Abi Istada Geomorphic Section (T-D1)

This area of the Katawaz Basin is characterized by a variety of Paleocene conglomerates and sandstones, Eocene claystones and shales, and Oligocene sandstones and siltsones (Sawyer and Stoesser, 2005). A small amount of Pliocene conglomerate and sandstone occurs as well. All these sedimentary rocks are

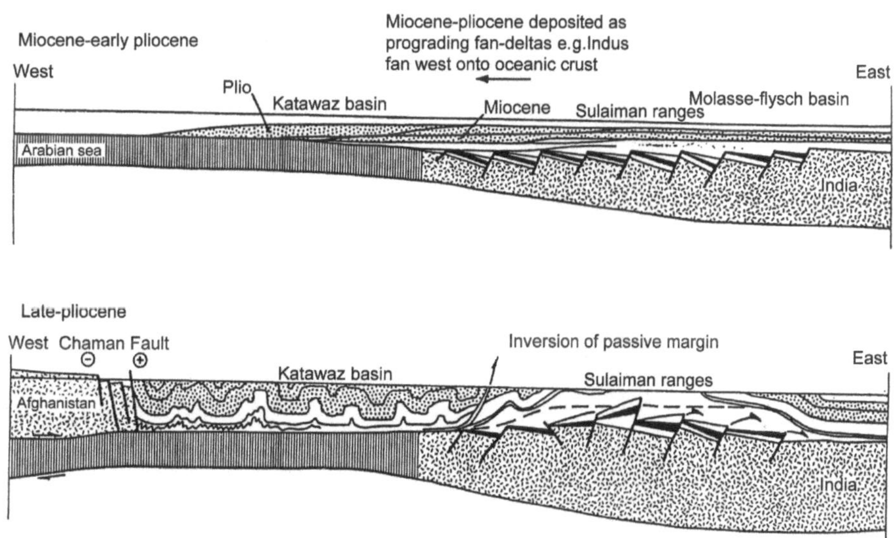

Figure 3.23 Cross-section showing the origin and structural evolution background that led to the geomorphology of the Katawaz basin. *After Treloar and Izatt (1993).*

overlain by widespread plains of wind-blown loesses, as well as fluvial conglomerates and sandstones of late Pleistocene age. Fan alluvial and colluvial deposits occur in many places where sufficient relief occurs to enable deposition of typical fan materials. Holocene playa-lake deposits occur in the direct vicinity of Abi Istada.

Abi Itsada itself (Figure 3.24), the "lake of standing water", dries up periodically, and except for its large size, resembles what might be considered a fairly typical sag pond along the strike of the Chaman Fault (see Chapter 5 for additional discussion of the lake). Probably there is some element of fault-related sagging associated with the landform, as well as the general shortening, buckling, and decollement-like nature of the adjoining Cuesta–Monoclinal Ridge–Hogback Section. This has enabled sufficient downwarping of this part of the crust to allow formation of a basin into which Abi Itsada could collect.

Cuesta–Monoclinal Ridge–Hogback Geomorphic Section (T-D2)

The Cenozoic sedimentary pile of the Katawaz Basin was shortened and inverted during the Late Tertiary collision between the Indian plate and Afghanistan (Treloar and Izatt, 1993). This section is composed predominantly of Eocene claystones and shales that are overlain by Oligocene sandstones and siltstones (Stoesser, 2005). The dominant structural elements there are tight folds characterized by relatively open synforms, and basins separated

Figure 3.24 **Lake of Abi Istada** The yellow scale bar is 5 km long. *(from Google Earth™).*

by tight antiforms and domes (Ganss, 1970; Wittekindt and Weippert, 1973). This deformation of the bedrock postdates the deposition of a majority of the sedimentary sequence, but before the deposition of the undeformed Bostam Formation of Pleistocene age. Thus it appears that much of the deformation is probably post-Miocene, with a late-Pliocene age likely for the main deformation. Significant transpressive movements are Plio-Pleistocene and younger in age (Treloar and Izatt, 1993).

Erosion throughout Quaternary time in this section for the past several million years up to the present has eroded the topography into a wide distribution of the gently dipping (<10°) cuestas, the moderately dipping (10–45°) monoclinal ridges, and the steeply dipping (>45°) hogbacks that abound throughout this section (Figure 3.25). These extensively folded rocks are overlain in almost all of the intervening synformal valleys by fan alluvium and colluvium of late Pleistocene and Holocene ages. In a few places, mostly draining out across the border with Pakistan occur Holocene gravels and sands in the alluvium of the ephemeral stream courses.

Khost Basin and Suleiman Mountain Geomorphic Section (T-D3)

This section on the border with Pakistan is composed mainly of extensively faulted shelf sediments of the Indian Plate with several prominent ophiolite sequences emplaced as well. Several cuesta-form ridges occur in these sedimentary sequences, but mainly the topography is a complexly eroded sequence of hills and small mountains of diverse characters reflecting the variable geology. The whole is partly overlain by extensive basin-fill alluvial and colluvial sequences that form broad, low-angle fan shapes in the basins.

Figure 3.25 Synthetic view to the north of the zigzag ridges of eroded and dipping sedimentary rock layers of the Katawaz Basin The two smooth-floored basins are synclines, with an anticline to the left between them. *(from Google Earth™).*

Accreted Terranes Geomorphic Region (AT Region)

Farah Block Basin and Range Geomorphic Province (A-A)

The accretion of the diverse geologic terranes of Afghanistan, as shown in Chapter 2 detailing the course of Afghan continental fragments moving across the Tethys Seaway from early Mesozoic time to their late Mesozoic impact, is the basic understanding required to discuss the geomorphic terrain of the area. Once it is understood that the sharply wedge-shaped Farah Block (Figure 3.26) is a highly sheared amalgamation of ENE-trending ranges, in which the bedrock geology is dominantly calcareous sedimentary rocks of Jurassic and Cretaceous age, then the interpretations of the landforms are much easier, even though there are vast stretches of relatively similar and rather homogeneous arid ranges to consider. The basic bedrock was rafted in and sutured onto the Asian Plate along what is now the Herat or Hari Rud Fault. The bedrock also was intensely folded and faulted along axial surfaces that tend to trend east in the north parts, and southwest in the east and south portions of the block. Overall, the block generally received a complimentary series of WSW-trending fold axes parallel to the fault shearing as well. Folding was generally upright but also tends to verge southward in the south part of the block. The block was later overlain just south of the Herat fault zone by an alkaline-basaltic volcanic and sedimentary succession of Oligocene age (Moores and Fairbridge, 1997).

The Farah Block is marked on its northeast–southwest junction with the Helmand Block by the Panjao Suture zone, which is an ophiolitic mélange where the igneous rocks are mixed up with the sedimentary types. The considerable

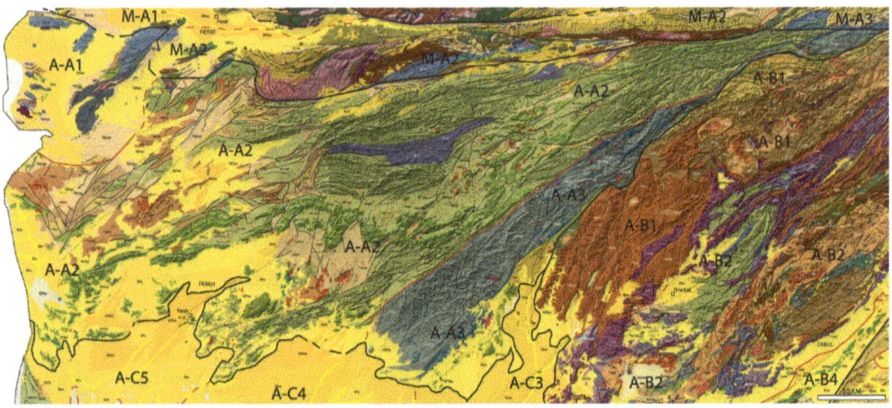

Figure 3.26 Geologic map of the Farah Block Basin and Range Province. A-A1is the Dashti Alishan Basin and Range Geomorphic Section; A-A2 is the Ghor Mountains Geomorphic Section; and A-A3 is the Kash Rud Block Mounatuins Geomorphic Section. *After Doebrich and Wahl (2006).*

extent of this Panjao Suture ophiolite indicates that a fairly sizable ocean existed between the Farah and Helmand blocks. Volcanic rocks in the ophiolites from this ocean have limestone pieces within them containing middle to late Triassic fossils, but the deformed suture rocks are unconformably overlain by sedimentary rocks of Lower Cretaceous age, which shows the approximate time of emplacement.

Because the Farah Block is so strongly and sharply wedge-shaped, with the sharp and narrow ends of the wedge being pointed to the northeast and higher altitude, with the other wide end extending out to the west and at lower elevation along the Iranian border. The result of these altitude differences is that late Cenozoic sediments of the desert basins lap up onto the bedrock ranges and surround them with numerous alluvial and colluvial fans. The desert basins themselves are characterized by plentiful ephemeral stream channels, wadis, or dry washes, and a variety of dry lakes or playas which may have a salt crust (*namaksar*) or just a layer of fine clays (*takyr*).

The Farah Block Basin and Range Geomorphic Province has been differentiated here as worthy of special geomorphologic recognition not because the desert ranges are distinctly different from those nearby, but because its rock ages and lithologies can be substantially to only slightly different.

Dashti Alishan Basin and Range Geomorphic Section (A-A1)

This section has been separated from the rest of the province because of the fact that it has broad, open desert basins. The Namaksar dry lake basin on the border with Iran has salt flats in its lowest part and measures ~25 km in length, with perhaps 125 km^2 of area in Afghanistan; possibly twice as much area in Iran. Some older Pliocene clastics exposed in this section have been deformed perhaps by either tectonics or salt movement beneath. The several desert ranges rising above the basins are dominantly Jurassic sandstones and siltstones.

Ghor Mountains Geomorphic Section (A-A2)

The region has a few smaller desert basins close to the Iranian border but is mainly composed of some late Jurassic clastics and a plentiful amount (~15,000 m) of early Cretaceous sandstone and siltstone, which is overlain in some places by less deformed Oligocene clastics that overlie their older source rocks (Figure 3.27). Floods in these desert basins can sometimes be quite extreme (Figure 3.28).

Khash Rud Block Mountains Geomorphic Section (A-A3)

The Khash Rud block is part of a right lateral shear zone or Panjaro Suture (Fernette, 2011) of clastics and sea floor volcanics in an ophiolite mélange that was welded onto the Farah Block.

Figure 3.27
Closed zigzag ridges of the Najuha syscline composed of Oligocene clastics ~75 km east of Shindand that overlie the strongly sheared clastic sedimentary rocks of early Cretaceous age in the surrounding area. The scale bar at the bottom is 3 km long.

Figure 3.28
Flood beneath the Bagh Pul Bridge on the Farah Rud in Farah City in about 2010.
Photograph by Wahidff from Google Earth™.

Helmand Block Geomorphic Province (A-B)

The Helmand Block occurs southeast of the Farah Block and is delineated along its northwest side by the Panjao suture of ophiolitic rocks marking the closure of another part of the Tethys Seaway that was complete by early Cretaceous times (Figure 3.29). Many of the rocks in this province are Paleozoic sedimentary units that are at least 10,000 m thick, with some Cretaceous sediments

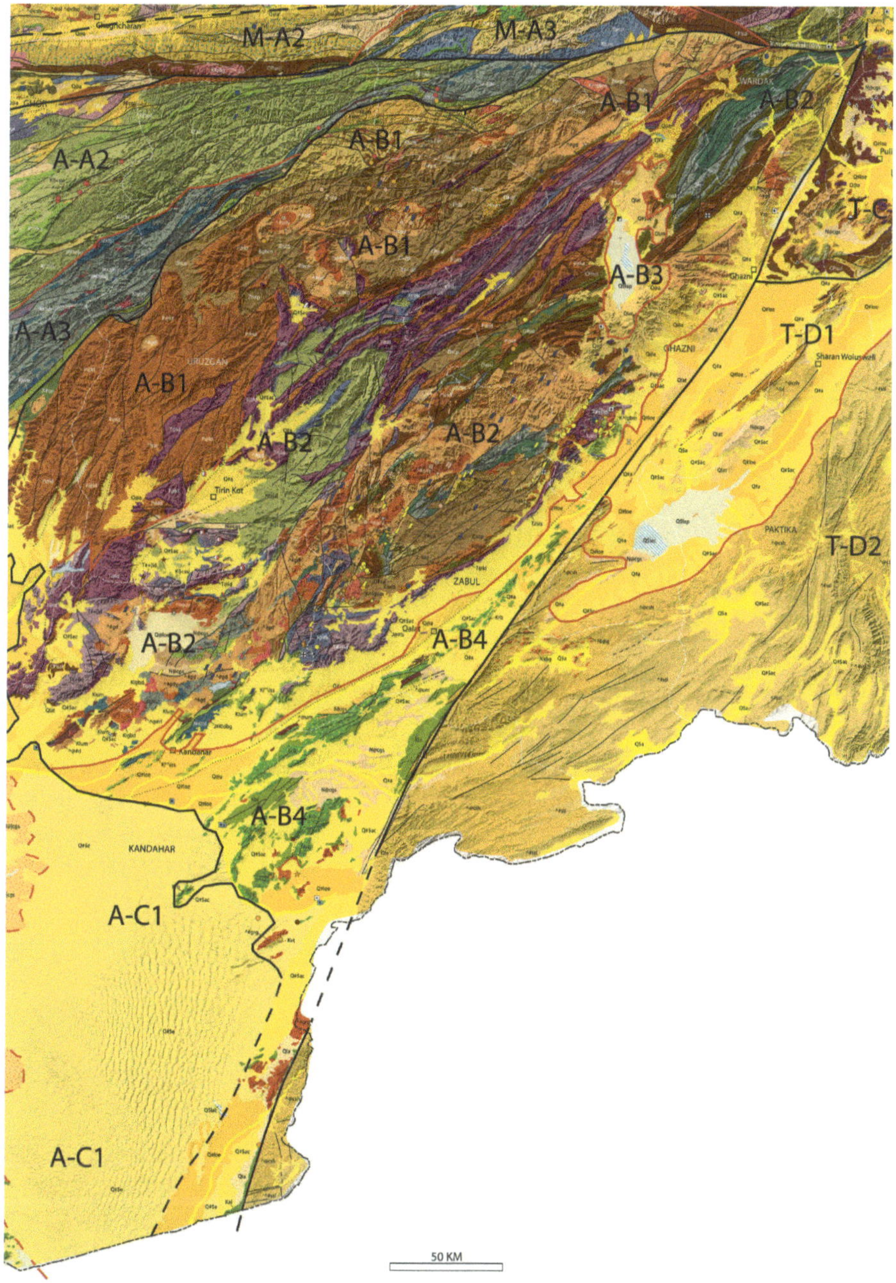

Figure 3.29 Geology map of the Helmand Block Geomorphic Province. Section A-B1 is the Helmand–Uruzghan Mountains and Basins; section A-B2 is the Arghandab Mountains and Basins; Section A-B3 is the Dashti Nawar; and Section A-B4 is the Spin Baldak–Qalat part. *After Doebrich and Wahl (2006).*

deposited across the tops of the Paleozoic units; the whole was in turn intruded by subduction-related granitoid plutons during Cretaceous to early Tertiary time. The block is transected by the Tirin and Moqur strike-slip faults as well (Moores and Fairbridge, 1997).

Helmand–Uruzgan Mountains and Basins Geomorphic Section (A-B1)

This geomorphic section of the headwaters and part of the middle reaches of the Helmand River is dominated by a complexity of bedrock composed dominantly of limestone and dolomite of Permian age that is intruded by huge Oligocene batholiths of granitoid rocks. The light-colored igneous rocks outcrop as large ovals (13–50 km) easily visible on the satellite imagery, that on the ground have characteristic weathering into corestones, tors, castle koppies, and other rugged granite landforms typical of that lithology (Figure 3.30). The region is also characterized by deep canyons cut throughout the area by the Helmand and its tributaries.

Arghandab Mountains and Basins Geomorphic Section (A-B2)

The Arghandab Section is composed of a highly sheared complex of Proterozoic metamorphic rocks overlain by Paleozoic clastics and carbonates that are also intruded by Cenozoic plutons. The landforms are a diverse group of rocky mountains and rugged canyons.

Dashti Nawar Caldera Geomorphic Section (A-B3)

The Dashti Nawar caldera is the remnants of a large composite or strato–volcano complex that developed in late Cenozoic time in south central Afghanistan about 55 km due west of what is now Ghazni. The last big eruption seems to have been about 2.2 mya, and since then the volcano complex has been largely or entirely extinct. During the final paroxysmal eruption, when the huge

Figure 3.30 Granite landforms of corestones and tors near Daikundai in Section A-B1; the Helmand–Uruzghan Mountains and Basins. *After Manaqebi from Google Earth™.*

oval caldera (~30–50 km) was created by explosion, a large mixture of rocks and mud or lahar swept SE down toward the ancestral Indus River and left the Bain Boulder Bed in Pakistan as a result (see Chapter 8). At the present time, the resulting oval caldera (15 × 45 km) is a broad grassy plain in its north, with extensive mud, peat, and salt flats at its lowest points (3127 m) further south.

Spin Baldak–Qalat Geomorphic Section (A-B4)

This section is a long narrow strip of NE–SW trending, largely Quaternary-aged sands and gravels that extend out from the mountain fronts to the Chaman fault. A number of inselbergs of eroded Cretaceous-aged igneous rocks protrude up through the cover sediments (Figure 3.31).

Helmand Basin Geomorphic Province (A-C)

The huge area of the Seistan Basin encompasses most of the whole of southwestern Afghanistan, extending from the Farah Block on the north, south ~300 km to the Baluchistan border of Pakistan (Figure 3.32). From the western

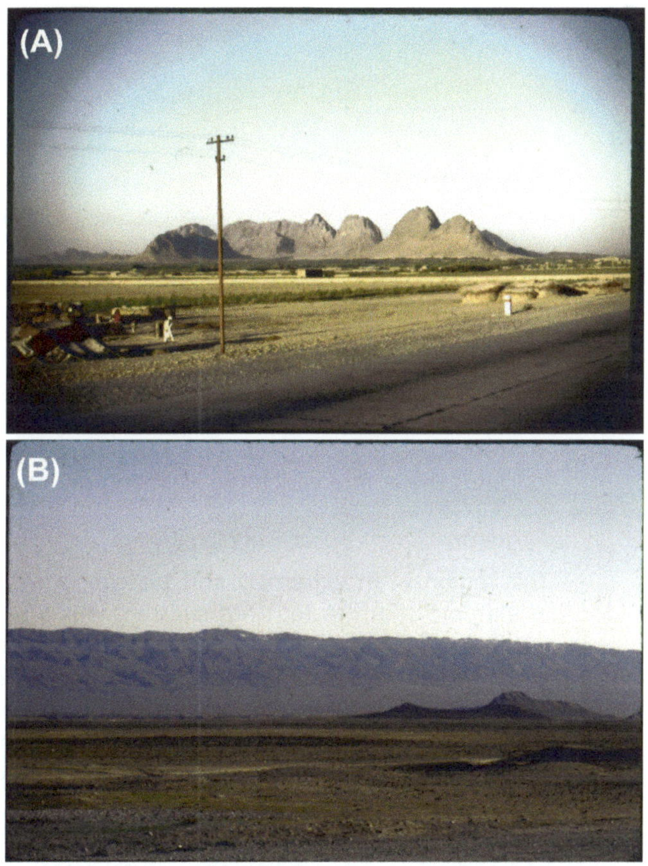

Figure 3.31
(A) Inselbergs near Kandahar in Section A-B4. (B) View of area along the Chaman fault northeast of Kandahar in Section AB4.

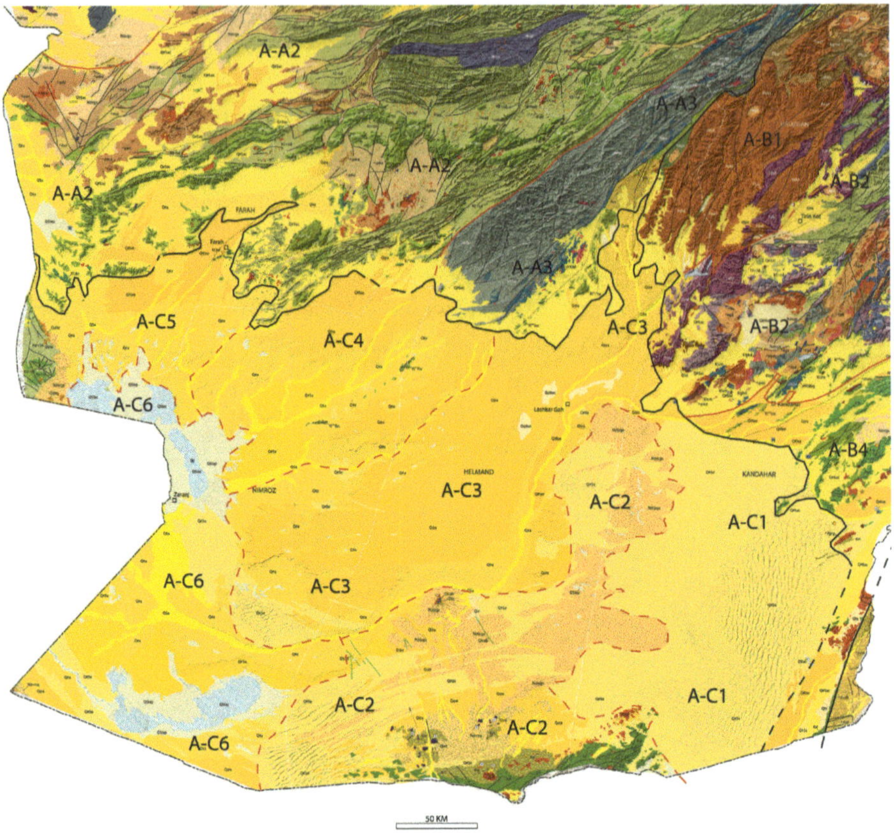

Figure 3.32 Geology map of the Helmand Basin Province. *After Doebrich and Wahl (2006).*

border with Iran, it extends about 450 km to the east as well. From east to west, the six sections of the province include Registan, the Helmand Valley, the Kohi Khanneshin carbonatite volcano and the Chagai Hills, the Dashti Margo and other desert expanses extending to the Iranian border, and the hamun lowlands there, into which the Helmand drains.

Registan Geomorphic Section (A-C1)

This region is an extremely arid plateau of semistabilized sand dunes located in extreme southwestern Afghanistan. It is a sandy desert consisting of small, reddish sand hills about 15–30 m high, sand-covered plains, and open rocky and clay-covered areas that extends 140 km E–W and 200 km N–S. The underlying base of the hills consists of partly stabilized ancient transverse dunes, reflecting the west-to-east wind of 120 days and nearly unlimited sand supply, which has been blown in for a very long time by western wind deflation of the Seistan basin. The reddening is a reflection of long-term weathering of minerals in the sand (Figure 3.33). During recent droughts, the desert has gradually encroached on surrounding agricultural areas.

Figure 3.33 Oblique-angle aerial view across the north end of the Registan sand dunes where they come up to the Arghandab River south of Kandahar. *After Millemaci Antonio in Google Earth™.*

Chagai Hills–Khanneshin Volcano Geomorphic Section (A-C2)

This section is on the left bank of the Helmand and is dominated by sand blown in particularly from the Helmand delta in the west and across strong desert pavement lag gravels. The extinct and eroded Kohi Khanneshin carbonate volcano remnant of Quaternary age deflects the west-flowing Helmand River around it in this section, and the volcanic Chagai Hills to the south have a variety of Cretaceous tuffs and andesitic lavas, with later Oligocene plutonic granites and granodiorites that now exist as inselbergs in the desert.

Dasht-i Margo Geomorphic Section (A-C3)

The "Desert of Death" is the main right-bank area of the Helmand River and is an extensively wind-deflated area with only an extensive lag gravel on most of its surface (Figure 3.34). Inside the north curve of the Helmand delta, however, an extensive area of active sand dunes occurs that is known as Jehanoum (Hell). The lower altitude parts of the Dashti Margo are pocked with small deflation basins that have fine-grained sediments in their bottoms where minimal winter rains accumulate occasionally; toward the northeast

Figure 3.34 Lag gravel in the Dashti Margo.

Figure 3.35 One of the numerous ruined structures in the Seistan, most of which were built when Lake Seistan was much larger than the present.

and closer to the mountain fronts water-washed alluvial gravels that are evidence of rare flash floods and rapid wet debris flows that have been the long-term source of much of the gravel lags of the desert area. In the western parts of the section where towns sprang up in antiquity around a much larger Lake Seistan that was once there (Figure 3.35), their ancient graveyards have deflated in many places so that carpets of bones occur as lag gravels as well, which may have given rise to the name for the desert area overall (Figure 3.36).

Dasht-i Bakwa (Khash) Geomorphic Section (A-C4)

This desert region is dominated by the gravelly remnants of the rapid, wet debris flows that have swept out of the mountains for millennia, as well as the extensive alluvium from rivers through the region (Figure 3.37).

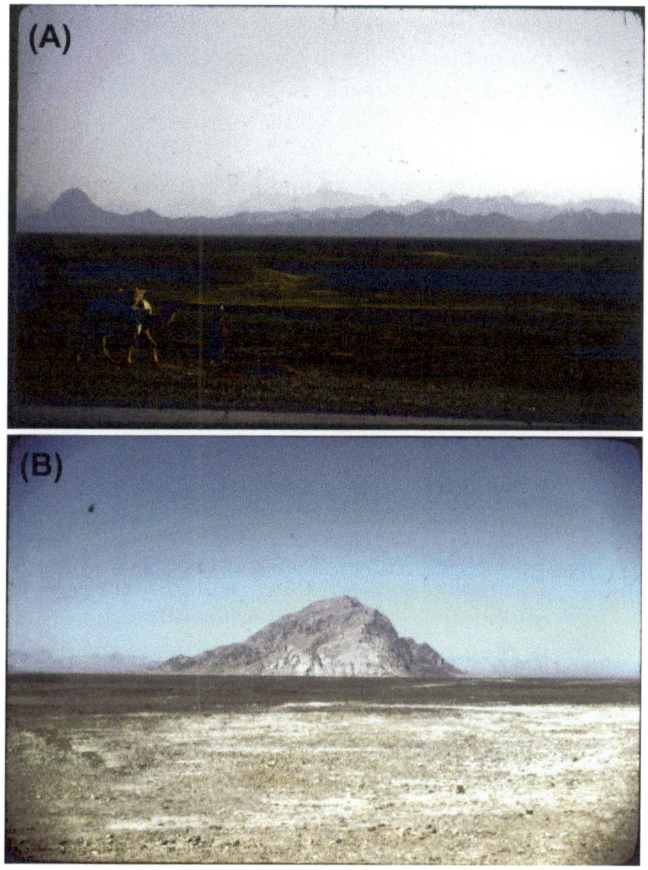

Figure 3.36
(A) Desert scene near Dileram.
(B) Inselberg in the desert near Dileram.

Figure 3.37 Part of the remnant Helmand delta in Lake Seistan.

Frah–Harut Geomorphic Section (A-C5)

In a fashion similar to the preceding section, a dominance of alluvial and debris flow deposits prevails here.

Seistan Hamun (Lakes) Geomorphic Section (A-C6)

The hamun lakes are intermittent and depend upon the spring discharge of the Helmand River to fill them. As explained in Chapter 5, these lake basins are a series of intermittent water bodies subject to complete desiccation in some seasons, or filling that allows spillover flow all the way through them into Iran and then back into Afghanistan into the final sink of Gaudi Zirreh, which is sometimes the last salty remnant water from an exceptionally wet year by the Helmand. In mid-Holocene time around 5000 years ago, Lake Seistan was then much larger and deeper, leaving behind the Helmand delta sediments when it dried up (Figure 3.37).

Middle Afghanistan Shear Zone Geomorphic Region (MASZ Region)

This region does not differ geomorphically in any substantial fashion from most of the other regions around it, other than the fact that most of the rocks and landforms tend to be oriented E–W in the direction of the pronounced Hari Rud or Herat shear zone that marks the south edge of the Asian plate onto which these rocks were sutured (Figure 3.38). As the Farah Block was rafted into position against Asia in early Mesozoic time, it would have crushed into the preexisting sediments and rocks and caused the lineations that now exist in the region. Further pressures from the east as the whole of the Indian subcontinent continued to push into Asia following initial docking collision from early Cenozoic on in time caused continued shear in this region that further fractured the rocks to cause isolated zones of quite different lithologies.

Hari Rud–Ghorband Geomorphic Province (M-A)

This province extends over 730 km from central Afghanistan near Kabul to the border with Iran (Figure 3.38) and includes four sections that delineate various crustal fragments sutured together along the shear zone.

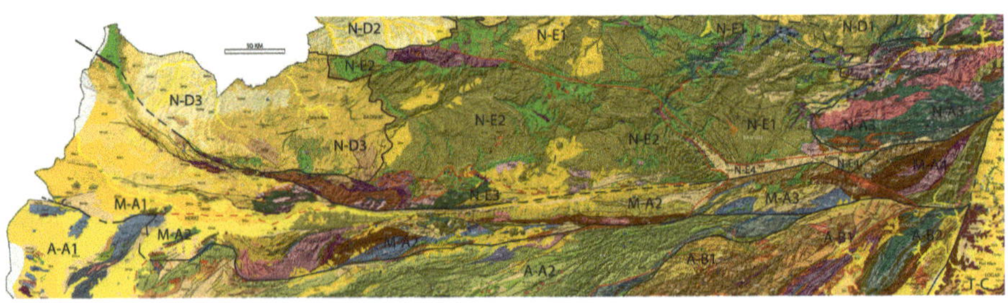

Figure 3.38 **Geology map of the Hari Rud–Ghorband Geomorphic Province.** *After Doebrich and Wahl (2006).*

Herat–Hari Rud Fault-Zone Valley Geomorphic Section (M-A1)

This western-most of the sections is a typical open desert environment with the broad braidplain of the Hari Rud and long gently sloping alluvial fans spreading in all directions from the surrounding low mountains. A few hills of older rocks such as gneiss and granite, as well as middle Cenozoic sedimentary rocks occur.

Chagcharan Shear Zone (Horst & Graben) Geomorphic Section (M-A2)

The main central and western portions of the region extended some 450 km E–W from near Herat nearly to Bamiyan (Figure 3.39). The section has a mélange of sheared Precambrian crystallines, as well as Phanerozoic sedimentaries.

Koh-i Baba Mountains Geomorphic Section (M-A3)

The high glaciated mountains of granite in the southern part of this region are surrounded on the north, and to the east and west by other rock types, including Precambrian crystalline rocks, as well as Mesozoic and Cenozoic sedimentary rocks and some lava (Figure 3.40).

Paghman Mountains Geomorphic Section (M-A4)

The previously glaciated mountains of Paghman that rise above Kabul are composed of a variety of Precambrian crystalline rocks, as well as late Paleozoic and lower Mesozoic clastic sedimentary rocks (Figure 3.41).

Figure 3.39 Oblique aerial photograph of the upper Hari Rud in the Chagcharan Shear Zone Geomorphic Section (M-A2).

Figure 3.40 **Photograph looking south at the Kohi Baba range near Bamiyan.**

Figure 3.41 **Oblique aerial photograph of the Sari Vorki (~4200 m) erosional residual in the Paghman Mountains.** The mountain consists of nearly flat-lying Cretaceous clastics overlying Triassic clastics in an angular unconformity. *Photograph by James Springer, by permission.*

North Afghanistan Platform Geomorphic Region (NAP Region)

The northern platform of Afghanistan is constituted of sequences of rocks and structures that represent a continental shelf of mainly clastic and carbonate sediments of Mesozoic age that were caught up in the ongoing collision of the Indian subcontinent as it docked with Asia in early Cenozoic time. In the east, these rocks were extensively sheared, metamorphosed, intruded, and fault-sutured onto other similar rocks of the Eastern Pamir–Hindu Kush Mountain Geomorphic Province to become the Western Pamir–Hindu Kush Mountains Geomorphic Province. The other four geomorphic provinces—the North Afghanistan Platform Plateau, Tajik

Basin, Northern Afghanistan Plateaus and Loess Hills, and the Central Afghanistan Mountains and Plateaus—all constitute the only moderately folded and faulted sedimentary rocks and sediments of the northern region. Active ongoing neotectonism in the region is superimposing new faulting and folding across the whole region, commonly with new preferred directions of motion and lineation that are at considerable variance to other directions not very much like those of a few million years ago. Across this vast region, plentiful quantities of variously sized clastic debris were spread by glacier river break-out floods, finer caliber alluvial sands and gravels, and wind-borne sand dunes and loess dust. Numerous occurrences of anthropogenic geomorphology were formed back in antiquity and on up to modern times as humans modified the landscape through construction of canals, huge burial mounds, and saline soil build-ups through long-term, unremediated irrigation.

Western Pamir–Hindu Kush Mountains Geomorphic Province (N-A)

This province, like its neighbor directly to the east, is mountainous terrain composed mainly of resistant crystalline rocks of metamorphic and igneous origins (Figure 3.42). The province is characterized by overlying cover sediments that occur in many places as either higher altitude blankets of Pliocene clastics, or deeper river valley alluvium, colluvium, and glacial deposits of Quaternary age.

Badakhshan Pamir Mountains Geomorphic Section (N-A1)

This section is characterized by a highly sheared terrain of Precambrian gneisses and Palozoic-aged metasediments with extensive cover blankets of Pliocene conglomerates and sandstones that were dominantly derived from erosion of local source rocks and spread west over this province and the adjacent North Afghanistan Platform Plateau Geomorphic Province.

Khwaja Mohammed (Northern) Hindu Kush Mountains Geomorphic Section (N-A2)

These mountains are dominated by Proterozoic gneisses and Paleozoic-aged metasedimentary clastics and a few similarly aged igneous rocks that were intruded by large resistant granitic plutons of Triassic age. Deeply carved river valleys have been eroded across and into these mountains to provide thin bands of alluvium in valley bottoms onto which people manage some limited agriculture. In the western part of the region, thin loesses have blown over and been deposited on the crystalline rocks to slightly mute the underlying crystalline ruggedness (Figure 3.43).

Western Hindu Kush Mountains Geomorphic Section (N-A3)

In this section, the Paleozoic clastic and carbonate metasedimentary rocks have been extensively intruded by granite batholiths of Triassic age. The

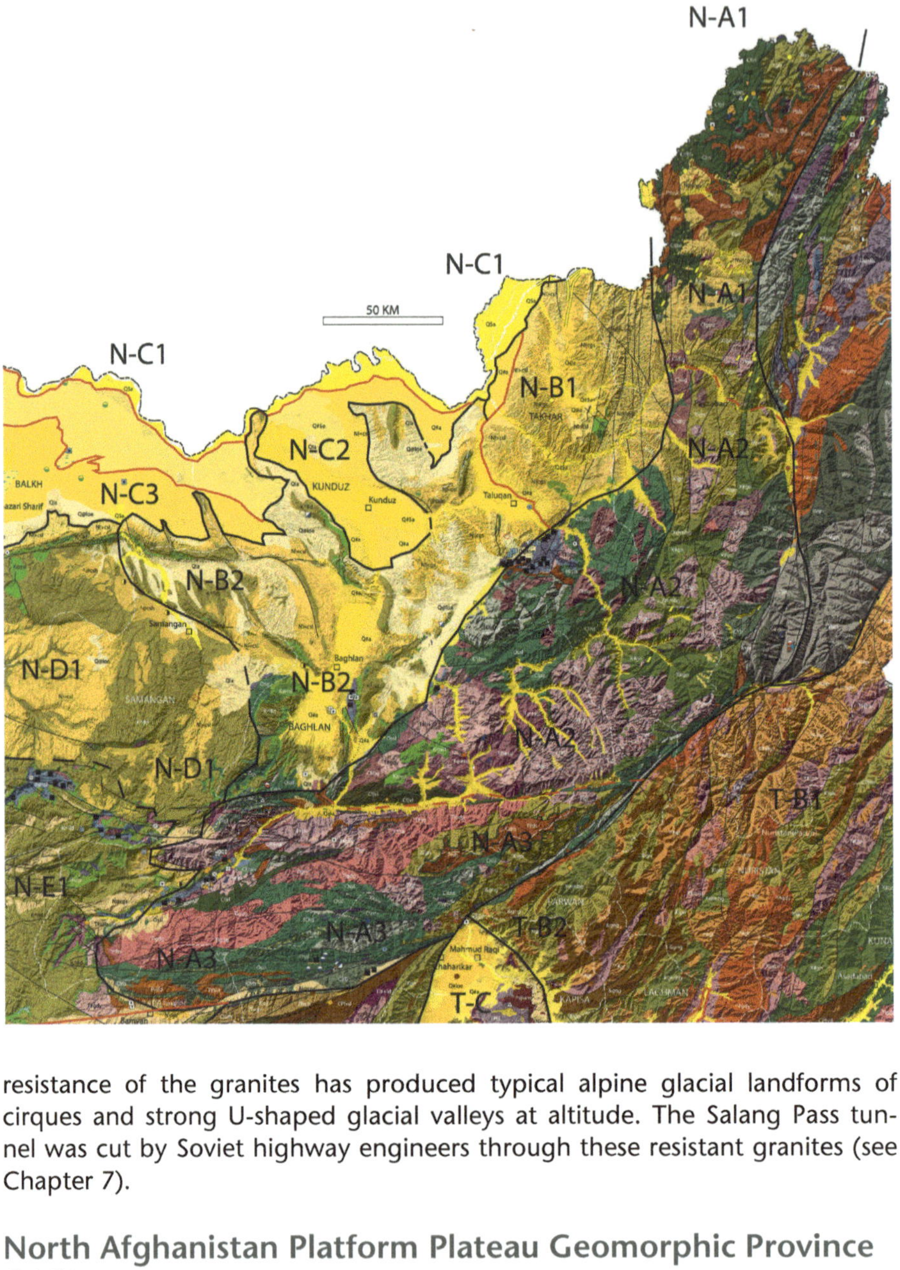

Figure 3.42 Geologic map of the Badakhshan Mountains Geomorphic Province. Badakhshan Pamir Mountains Section (N-A1); Khwaja Mohammad (Northern) Hindu Kush Mounatins Section (N-A2); and Western Hindu Kush Mountains Section (N-A3). *After Doebrich and Wahl (2006).*

resistance of the granites has produced typical alpine glacial landforms of cirques and strong U-shaped glacial valleys at altitude. The Salang Pass tunnel was cut by Soviet highway engineers through these resistant granites (see Chapter 7).

North Afghanistan Platform Plateau Geomorphic Province (N-B)

This region of Afghanistan just west of the mountains is characterized by the general suite of shelf sediments that have begun to be deformed by active tectonism, the folding and faulting of which have accomplished a number of rock and landform effects (Figure 3.44).

Figure 3.43 View of steeply dipping gneissic foliation along the Fayzabad River truncated by a river strath terrace. These two-mica gneisses are of early Paleoproterozoic age. *Photograph taken by JWZCSH from Google Earth™.*

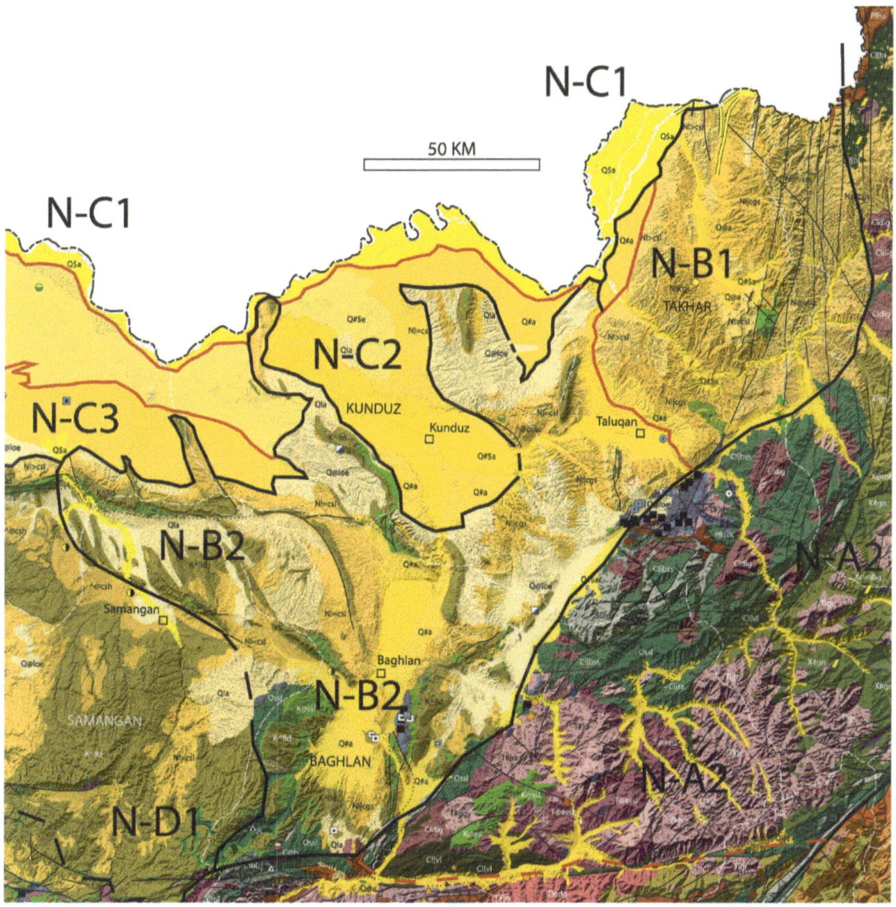

Figure 3.44 Geologic map of the North Afghanistan Platform Plateau Geomorphic Province. *After Doebrich and Wahl (2006).*

Takhar Geomorphic Section (N-B1)

This section at the far northeast in Takhar Province is dominated by rolling hills of Miocene and Pliocene conglomerates, sandstones, and finer clastics overlying late Mesozoic shelf sedimentary rocks, all of which are transected by numerous N–S faults. Lapparent (1970) described a number of river captures in this region that were forced by active differential uplift on various anticlines. For example, the Neogene anticline of Ghani Darra that is situated directly west of the town of Cha Ab, caused the beheading and diversion of a tributary to join the Rustak River, whereas previously it had flowed north to join the Panj River just above where that river now exits the mountains. Similarly, farther south at Taliqan, the river there had once flowed north to join the Kokcha and then the Amu Darya, but movements on the Archi anticline allowed the Khanabad tributary to the Kunduz to capture and divert the river to the west to become a new tributary in that section.

Kunduz–Taliqan Active Ranges Geomorphic Section (N-B2)

The active ranges are a reflection of the fact that the Tajik basin is actively closing as a result of ongoing mountain building rather than being displaced and transported in a westerly direction (Ruleman et al., 2007). For example, several reaches of the Panj and Amu Darya are incised though rising anticlines. South of there from Poli Khumri north to Aybak (Samangan) and Baghlan, the area is marked by a network of northwest-trending, right-lateral strike-slip faults, which where left steps occur in the fault system, pressure ridges have developed with thrust faults beneath (Figure 3.45).

Tajik Basin Geomorphic Province (N-C)

This northernmost of the geomorphic provinces of Afghanistan is based on the southern rim of the greater structural Tajik Basin that occurs on further north

Figure 3.45 Part of the NW–SE-trending pressure-ridge system between Baghlan and Samangan associated with right-lateral strike-slip faulting.

beneath the countries of Turkmenistan, Uzbekistan, Tajikistan, and related areas (Figure 3.46). The southernmost extension of these rocks and sediments constitutes the sedimentary overlapping of these materials on the underlying rocks of the rising Hindu Kush and Badakhshan Pamir that on the eastern edge of this province are actively rising as antecedent structures. Some of these structures are active salt domes and anticlines, which in places have pierced the overlying rocks and extruded on the surface. The province itself involves all the Quaternary-aged sediment and sedimentary rocks that dominate the floodplain and associated alluvial plains of the Panj and Amu Darya river valleys constituting the northern border areas with the neighboring countries named above. Three geomorphic sections can be easily delineated on the basis of type and age of the Quaternary materials: (N-C1) Amu Darya Floodplain Geomorphic Section; (N-C2) Dashti Layli–Shortepa–Kunduz Geomorphic Section; and (N-C3) Shiberghan–Mazar Alluvial Fan Geomorphic Section (Figure 3.46).

Amu Darya Floodplain Geomorphic Section (N-C1)

This section is dominantly composed of the (Q_4a) alluvial sediment that is shingly and detrital, with gravel and sand more abundant than silt and clay (Doebrich and Wahl, 2006). The sediment as a unit was apparently mapped originally by Soviet-era geologists as conglomerate and sandstone of Holocene age, which is an adequate descriptor of the material until more advanced work is done.

One of the more interesting facets of the sedimentologic–geomorphologic nature of this region is the northeast edge of the section where the Panj River tributary to the Amu Darya exits from the Badakhshan Pamir and disgorges the copious sediment (Figure 3.47) it has gained from glacial and mass-movement sources upstream. The resulting Panj Fan (Figure 3.48) has been spread out into a valley that appears to be actively structural in nature. The northeastern-most part of the fan, for example, has the active Hoja Mumin (also Kohi Namak) piercement salt dome at its edge which may have been involved in the removal of salt from beneath by the differential loading of the Panj Fan, allowing subsequent subsidence

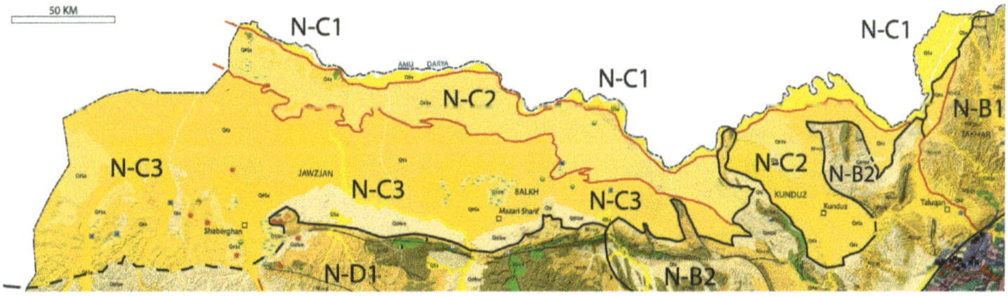

Figure 3.46 Geologic map of the Tadjik Basin Geomorphic Province. N-C1, Amu Darya Floodplain Geomorphic Section; N-C2, Dashti Layli–Shortepa–Kunduz Geomorphic Section; N-C3, Shiberghan–Mazar Alluvial Fan Geomorphic Section. *After Doebrich and Wahl (2006).*

Figure 3.47 View to the west of the delta of the Panj River as it exits the mountains and spreads out on both sides of the border between Tajikistan from where this picture was taken *(Poroshenko from Google Earth™)*, and Afghanistan on the left (south). The light-colored far mountain running N–S across most of the horizon is the active anticline of Nurek Ai Khanoum, with the Kizil Su coming from the north in Tajikistan alongside the anticline and running into the Panj here as well.

in the area. The fan measures some 65 km wide north-to-south, and extends out from the mountain front some 35 km until it butts up against an active anticline on its west side, which has controlled the diversion of the river and trained it to the south and west. The fan is large and extensive, which suggests that it has had a history of break-out flooding from glacier- and landslide-dammed lakes in the past. Indeed, extensive photographs upstream in the mountains (Gratzl, 1972) show imbricated boulders that attest to past such floods (Figure 3.10). Photographs of the fan in the drier seasons show widespread bare sand and gravel flats with several braided stream channels incised into it that are many km wide, but in full spring flood coming out of the mountains, the same areas are completely inundated.

The Panj River itself runs almost due west as several braided channels ~15 km long from the mountains before swinging abruptly SSW (Figures 3.47 and 3.48) and running ~40 km to its confluence with the Kokcha River coming in from the southeast on the left back at Ai-Khanoum where the French archaeology teams (Barnard, 1994) have identified and dug an extensive Greco-Bactrian-aged city from several thousand years ago, that was finally destroyed about 145 BCE. Overall, the winding course of the Panj and Amu Darya around the rising anticlines reflects multiple diversions of the rivers as a result of the ongoing tectonism.

The constriction of the Panj channel at Ai-Khanoum apparently also has caused meander loops with amplitudes of ~3 km to develop as much as 9–10 km upstream from the constriction. Downstream ~35 km from the constriction

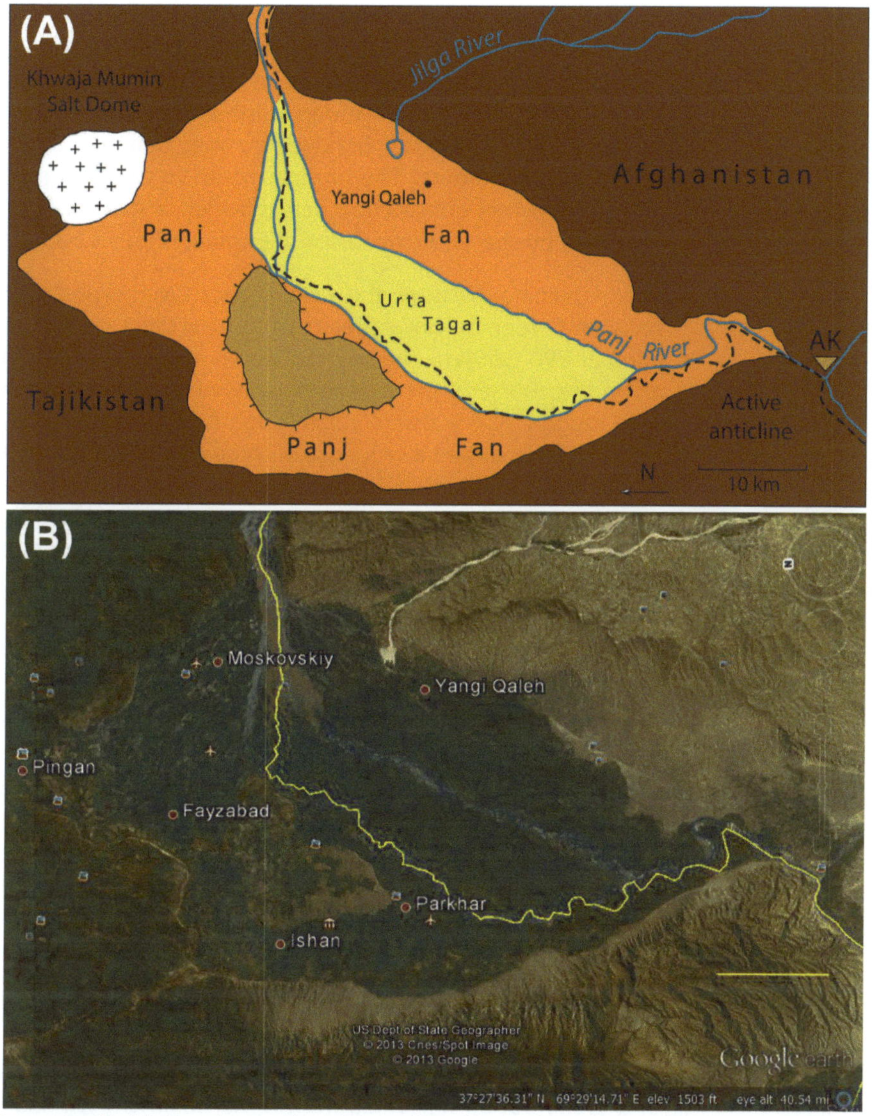

Figure 3.48 (A) Sketch map (north is to the left); (B) from Google Earth™ image of Panj Fan area where it exits the mountains and spreads out in a basin that may be in part the result of subsidence where rock salt at depth have migrated north and south into piercement salt domes, as well as the rise of the Nurek Ai Khanoum anticline in front of it. The Jilga River on the upper right is characterized by plentiful recent flood debris of eroded loess and related sediment eroded from degraded catchments upstream. AK–Ai Khanoum is ancient Greek-aged archaeological site. The Kokcha River is tributary to the Panj at Ai Khanoum and the combined waters of both pass over a saddle in the Archi to Nurek Ai Khanoum anticlines at this location.

occurs a prominent set of about four meander loops north of Imam Saheb in Afghanistan with amplitudes averaging about 7 km that occurred when the border of Afghanistan was first delineated by the Russians in the 1880s. At the present time, the old meander loops are almost entirely gone, having been replaced by a straighter braided channel that has cut through the necks of the former meanders. The original meanders had formed about 18 km above another channel constriction across which a new bridge was built recently to Tajikistan by the US Army Corps of Engineers (see Chapters 7 and 9).

The remaining parts of the Amu Darya Geomorphic Section continue as the national border on downstream without a great deal of significant variation for another 290–300 km. The channel way is restricted in a number of other locations along the course of the river, especially where rising anticlines or other structures with roughly N–S axes are cut through by the course of the river. Fouache et al. (2012) have mapped a sequence of three alluvial terraces directly north and northwest of Mazari Sharif at the edge of the active floodplain that attests to downcutting of the river away from the south shoreline in this location. The highest T3 terrace at ~285 m is not utilized for agriculture, no doubt because of a dearth of irrigation water, but the two lower terraces, T2 at ~280 m and T1 at ~275 m, are covered with agricultural fields. The T3 terrace, on the basis of surficial Neolithic artifacts, as well as optically stimulated luminescence dates from sediments beneath the terrace, was judged to be ~9000 years old (Fouache et al., 2012).

In a number of other areas along the Amu Darya, the southern embankment of the channel on the Afghanistan side has been encroached upon periodically by river-bank erosion, which the Afghans have attempted to stop by piling in wrecked armor from the Soviet invasion of the 1980s (Figure 9.2). Elsewhere on the north banks in Tajikistan and Uzbekistan, the wealthier residents there have armored their banks against erosion by proper stone riprap and earthen levees that are maintained.

Dashti Layli–Shortepa–Kunduz Geomorphic Section (N-C2)

The western two-thirds of this region is underlain by sediments that are mapped as dominantly eolian sand deposits of Holocene and late Pleistocene ages (Doebrich and Wahl, 2006). Reworked from fluvial sediment washed west out of the Pamir ranges by the Amu Darya and its tributaries, the clastic sediments are in turn blown back eastward and south-eastward as migrating sand dunes of blowout, transverse, and barchanoid types. In a few cases, trains of longitudinal dune complexes are composed dominantly of nested barchanoid, or overlapping blowout dunes like shingles on a roof. In many interdune depressions, dry lake basins of fine-grained silts and clays occur as what are called "takyrs" in the local Turkic-based languages. As the dunes encroach downwind back onto the floodplains of the Amu Darya and Panj rivers, in many places, the Afghan farmers have erected long, mud-walls of earth in an attempt to keep the sand from overwhelming their fields.

In fact, a long history of anthropogenic manipulation of the environment has occurred on the Balkh fan through numerous irrigation changes and salt build-up over the centuries (Fouache et al., 2012). For example, over 5000 years ago in the Bronze Age, irrigation development was concentrated mainly to the northwest of Balkh, which itself is located near the fan head. Then in two subsequent time periods, irrigation locations rotated first northward from Balk, and then northeastwards of Balkh, before in the aftermath of the Islamic conquest about 1200 years ago, irrigation switched to the far northwest around Aqcha (Figure 3.49). In fact, over this length of time since human manipulation and reshifting of the main channel of the Balkh Ab, the channel way has had sufficient time to develop its own, apparently natural-appearing meandering patterns. Nonetheless, only the rather unnatural, almost orthogonal direction that the channel takes on the Balkh fan once the river leaves the mountains gives evidence of the anthropogenic channel change. The new archaeological evidence

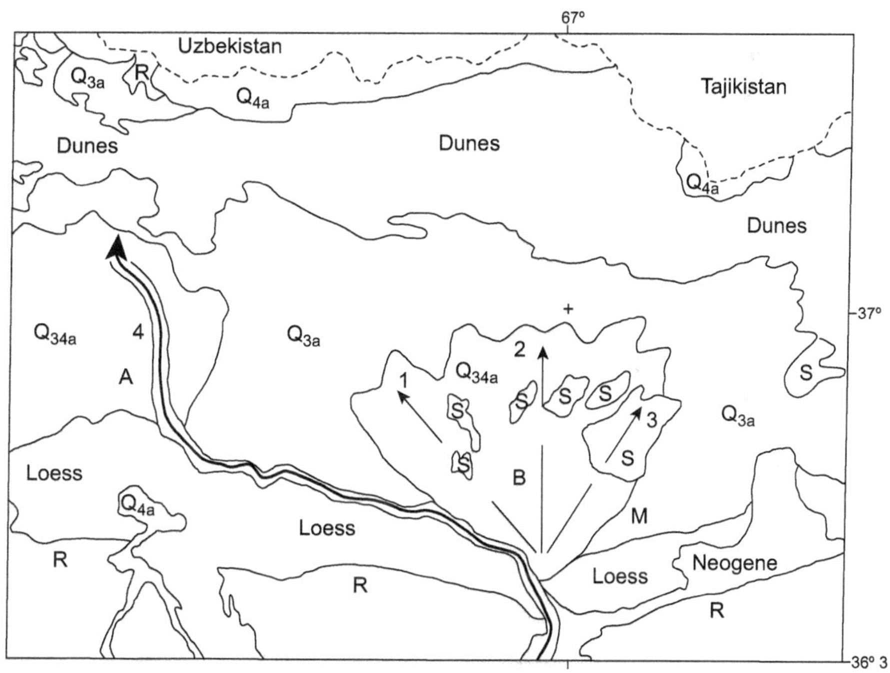

Figure 3.49 Sketch map of Balkh Fan showing progressive movement in a clockwise direction across the fan of the main irrigation network over the past five millennia, probably in response to progressive salinization and reduced crop yields until around the time of the Islamic conquest around 1200 years ago, at which time the main stem of the Balkh Ab was diverted abruptly back west to irrigate a new agricultural area around Aqcha. (1) Bronze Age (~5000–4000 years ago); (2) Bactro-Achaemenid (~2000–5000 years ago); (3) Kushan-Sasanid (~1500 years ago); (4) post-Islamic diversion to Aqcha (compare with Figure 5.10 satellite image). *After Fouache et al. (2012).*

presented by Fouache et al. (2012) now confirm geomorphological speculation made by the author in manuscript map form several decades prior to their work.

Further, during the Bactro–Achaemenid Period (sixth to fourth century BCE), an aqueduct for the Balkh Ab irrigation system of the time was constructed some 45–50 km north of Mazari Sharif, which now exists beneath what is currently only extensive dune fields (Fouache et al., 2012). Although other direct evidence is lacking, it is tempting to speculate that at this time about 2500 years ago that the dunes were not as active then or few people would have attempted to construct such a structure through what is now an active dune field.

In a few places, particularly toward the east of the geomorphic province, rising anticlines protrude through the eolian sediments as low hills and dissected structural remnants. Shebaqqli Mountain, for example, 36 km east of Kholm (Tashkurghan) is an asymmetric anticline (gentle eastward dips–steep western dips) that has been deeply eroded and truncated on its western, windward side (Figure 3.50), probably mostly by eolian abrasion and deflation.

The Kunduz Basin at the eastern margins of this section is composed predominantly of infilled sandy gravels of early to late Pleistocene, and Holocene age that have eroded from the surrounding hills and mountains of older rocks. The geomorphic configuration of most of these features is that they are composed of alluvial fan gravels and their stony, clast-covered surfaces leave no doubt about their origin. At the southeast end of the basin where the Kunduz airport is situated, a flat-topped Dashti Eshantop plateau exists that was also mapped originally by the Soviet geologists, probably because of its location on the basin margins, as gravely and sandy alluvium (Q_3a). Apparently on this basis, a number of buildings and mortared walls for a new Afghanistan Army Base were built without regard to the substrate so that water drainage from them was not

Figure 3.50 Photograph of Shebaqqli mountain near Kholm (Tashkughan) from the west. It is a breached, asymmetric anticline that appears to have been heavily eroded by wind abrasion and deflation on its western side. *After Mike (1971) from Google Earth™.*

channeled safely away. Unfortunately, the material was actually loess, which in fact emerged when the author was hired in 2009 to investigate the matter (Shroder, 2010). This wind-blown dust is quite common in Afghanistan but generally forms steep-sided hills and valleys that are accentuated by repeated deposition during each successive dust storm event. Here the upper surface of the older loess (middle to late Pleistocene?) had been eroded to a smooth plain, dissected only by gullies in some places (Figures 3.51 and 3.52). An east–west trending fault scarp that is ~25–30 m high and covered with burial tumuli on the top of the up-thrown block occurs as well.

Shiberghan–Mazar Alluvial Fan Geomorphic Section (N-C3)

This section of the Tadjik Basin Province constitutes the higher altitude margins of the basin on its southern edges. It is a heterogeneous amalgamation of alluvial fans that show unmistakable archaeological signs of human manipulation for millennia (Fouache et al., 2012), coupled with diverse loess hills and loess blankets, large and small deflation basins or blowouts, small playa salt flats, and various flat-floored river valleys cut though areas of thick loesses. Its margins are dominantly lithological and partly structural, especially at its southern-most edge where strong east–west oriented, faulting, and monoclonal upwarping of indurated Cretaceous-aged and other bedrock of the Alburz–Marmol Fault occur directly south of Mazari Sharif and Kholm (Tashkurghan).

Northern Afghanistan Plateaus and Loess Hills Geomorphic Province (N-D)

This province, which extends all across the northwest of Afghanistan, south of Mazari Sharif, and farther to the west, constitutes part of the border with

Figure 3.51 Flat-floored Shoraw Channel eroded into loess on the Dashti Eshantop plateau south of the Kunduz airport. *After Thiarmen from Google Earth™.*

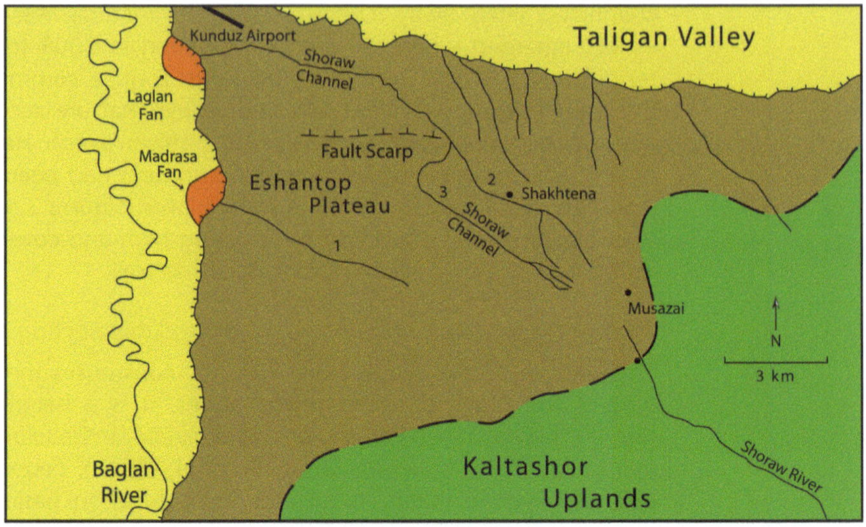

Figure 3.52 Sketch map of geomorphology of Dashti Eshantop Plateau (north is to the top). The ephemeral Shoraw River that rises in the Kaltashor Uplands on the lower right (southeast), has had a history of anthropogenic diversion of its lower channels, possibly in the 1, 2, 3 sequence indicated on the map. At some point in the course of these diversions, the rise of the 25–30 m fault scarp may have affected the incision as well.

Turkmenistan as well (Figure 3.53). It is dominated by shelf sedimentary rocks of Cretaceous age that are partially to completely covered with various thicknesses of loess.

Balkhab–Samangan Plateau Geomorphic Section (N-D1)

This section just south of the Tadjik Basin Province is marked in particular by the Alburz–Marmol fault, which is especially pronounced by the exposed and eroded drag folds that run roughly E–W out of the Kunduz–Taliqan Active Ranges Geomorphic Section (N-B2) (Figure 3.54) and into this section (Figures 3.55 and 3.56).

Faryab–Maymana Loess Hills Geomorphic Section (N-D2)

In this section, thick loesses have been blown out of the Karakumsky Desert of Turkmenistan to the west and northwest and accumulated over the Cretaceous bedrock. In such regions where loess is thick, it forms characteristic steep-sided hills that have complex depositional and subsequent incisional histories (Figures 3.57 and 3.58).

Qala-i Nau Loess Hills Geomorphic Section (N-D3)

This area in extreme northwest Afghanistan is dominated by extensive and complexly eroded hills of loess (Figure 3.59).

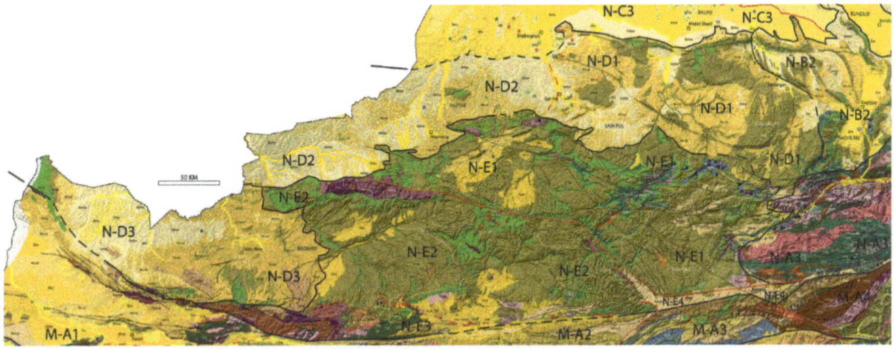

Figure 3.53 Geologic map of the Northern Afghanistan Plateaus and Loess Hills Geomorphic Province (N-D). (N-D1), Balkhab–Samangan Plateau Geomorphic Section; (N-D2), Faryab–Maymana Loess Hills Geomorphic Section; (N-D3), Qala-i Nau Loess Hills Geomorphic Section. *After Doebrich and Wahl (2006).*

Figure 3.54 Photograph of the strong hogback ridge rising above and to the southeast of Kholm (Tashkurghan), which is a manifestation of drag folds associated with the Alburz–Marmol fault in this region at the west edge of the Kunduz–Taliqan Active Ranges Geomorphic Section (N-B2).

Central Afghanistan Mountains and Plateaus Geomorphic Province (N-E)

The province is dominated by largely flat-lying or gently dipping Cretaceous shelf sediments that are cut through by sheaves of NW-trending faults; the far southwest part of the province, however, is characterized by mountains composed of late Paleozoic and early Mesozoic clastics, carbonates, and a few flow rocks that are intruded by plutons of granitoids (Figure 3.60).

Figure 3.55 Aerial view by Chris Siggs (Google Earth™) of the hogback ridges along the Alburz–Marmol fault south of Mazari Sharif.

Figure 3.56 View of the strong folding associated with the Alburz–Marmol fault south of Kholm (Tashkurghan) in the Balkhab–Samangan Plateau Geomorphic Section (N-D1).

Figure 3.57 Loess hills near Shurin Tagao in the Faryab–Maymana Loess Hills Geomorphic section (N-D2).

Figure 3.58 Photograph of loess hills on the ring road (A76) near Qaisar, taken by author in 1978. This part of the ring road is being reconstructed but it has taken considerable time because of the fragile and collapsible nature of unstable loess that must be compacted adequately to support transient vehicular loads.

Bandi Amir–Sari Pul Plateaus Geomorphic Section (N-E1)

The famous self-damming lakes of Bandi Amir (Chapter 5) have formed in valleys incised into the shelf sedimentary rocks of Cretaceous age that dominate the region (Figure 3.61). The valleys are deep because of the lower base level of the Balk Ab River that drains from the high plateau into the adjacent Yakawlang–Bamian Graben Geomorphic Section (N-E4) (Figure 3.62).

Feroz Koh Plateaus Geomorphic Section (N-E2)

This rugged and remote region is characterized by deep canyons cut into the bedrock.

Figure 3.59 **Loess hills south of Qala Nau.** *After Frank Pamar, Google Earth™.*

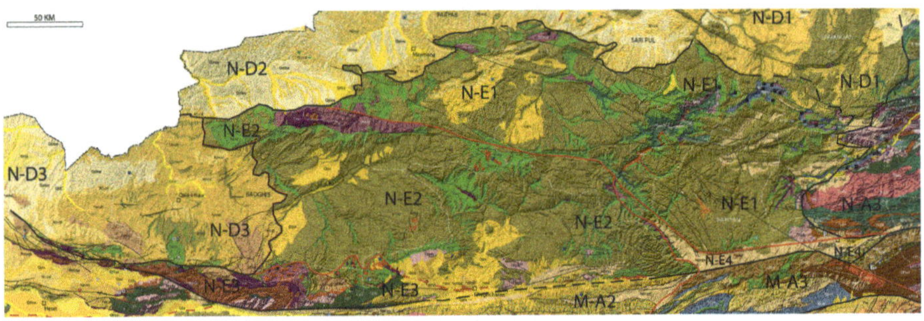

Figure 3.60 Geologic map of the Central Afghanistan Mountains and Plateaus Geomorphic Province. (N-E1), Bandi Amir–Sari Pul Plateaus Geomorphic Section; (N-E2), Feroz Koh Plateaus Geomorphic Section; (N-E3), Sefid Koh (Parapamisus) Mountains Geomorphic Section; (N-E4), Yakawlang–Bamian Graben Geomorphic Section. *After Doebrich and Wahl (2006).*

Sefid Koh (Parapamisus) Mountains Geomorphic Section (N-E3)

These mountains that have formed at the southeastern margins of this province, no doubt owe their origin to stress differentials set up as a result of the great strains produced directly adjacent to the great right-lateral Hari Rud fault on its southern margins. Apparently, the mountains of late Paleozoic and early Mesozoic sedimentary rocks (Figure 3.63) intruded by granitoid plutons were forced upward as a result.

Yakawlang–Bamian Graben Geomorphic Section (N-E4)

The Yakawlang–Bamiyan graben system is a result of major crustal cracking wherein the right-lateral Hari Rud fault at some time in the mid-Cenozoic, perhaps about Miocene time, appears to have locked in its western segments. This

Figure 3.61
Deep canyons cut into the nearly flat-lying Cretaceous shelf sediments of the Bandi Amir–Sari Pul Plateaus Geomorphic Province.

Figure 3.62
Shrine to Ali at Bandi Amir. Note the surrounding shales, sandstones, and limestones of the shelf sediments in the background that are characteristic of the region (Figure 5.8).

Figure 3.63 Aerial view of Sabzak Pass over the Safid Koh. *After Xli, Google Earth™.*

Figure 3.64 The **Bamiyan Valley in 1978 with the Buddha statues largely intact.** They were destroyed with explosives by the Taliban in 2001.

forced the accumulating stresses to divert to the northwest for a distance of about 80 km, opening a pull-apart basin or rhombochasm before they once again diverted back west again. The resulting pull-apart basin became the Yakawlang graben, that is marked by a variety of Cenozoic volcanism, which is quite understandable, given the 5–10-km-wide rift in the crust that developed there (Figures 13.3(A) and (B)). Early-Middle Cenozoic andesitic lava poured out into the graben structure near Bamiyan, as did pryroclastics south into the adjacent Chagcharan Section of the Middle Afghanistan Shear Zone geomorphic region. Conglomeratic sandstones of Miocene age that infilled the eastern part of the rift zone and lithified were carved anthropogenically in historic times into the famous Buddahas of Bamiyan (Figure 3.64). The crustal depression of the Yakawlang graben allowed the development of drainage from the Bandi Amir lakes to flow southwest and then curve around northwest down into the graben as the Bandi Amir River (Rudi

Bandi Amir). Upon the river's exit through a canyon from the Yakawlang graben and its flow to the northeast toward Mazari Sharif, it is named the Balkh Ab.

References

Barnard, P., 1994. The Greek Kingdoms of Central Asia. In: Harmatta, J. (Ed.), History of Civilizations of Central Asia. V. II. The Development of the Sedentary and Nomadic Civilizations: 700 B.C. to A.D. 250. UNESCO Publishing, Paris. ISBN: 92-3-102846-4.

Bohannon, R.G., 2005. Geologic map of quadrangles 3168 and 3268, Yahya-Wona (703), Wersek (704), Khayr-kot (521), and Urgon (522) quadrangles, Afghanistan. USGS Open-File Report 2005-1117-A.

Bohannon, R.G., 2005. Geologic Map of Quadrangles 3168 and 3268, Yahya-Wona (703), Wersek (704), Khayr-Kot (521), and Urgon (522) Quadrangles, Afghanistan. USGS Open-File Report 2005-1117-A; AGS Open-File Report (703/704/521/522) 2005-1117-A;USGS Afghanistan Project Product No. 020.

Bohannon, R.G., Stoesser, D.B., 2005. Geologic map of Quadrangles 3870 and 3770, Maymayk (211), Jamarj-i-Bala (212), Faydzabad (217), and Parkhaw (218) Quadrangles, Afghanistan. US Geologic Survey Open-File Report 2005-1095A.

Bohanon, R.G., Turner, K.J., 2005. Geologic map of quadrangle 3468, Chak Wardak – Syahjerd (509) and Kabul (510) quadrangles, Afghanistan. US Geological Survey Open-File Report 2005-1107A.

Braslau, D., 1972. The glaciers of Keshnikhan. In: Gratzl, K. (Ed.), Hindukusch-Osterreichische Forschungsexpedition in den Wakhan 1970. Akademische Druck- u. Verlagsanstalt, Graz, Austria, pp. 112–116.

Braslau, D., Bussom, D.E., 1978a. A glacier inventory method using Landsat MSS CCT. In: Rundquist, D.C. (Ed.), The Use of Landsat Digital Information for Assessing Glacier Inventory Parameters; an Evaluation: Final Report. Project Sponsored by Temporary Technical Secretariat for World Glacier Inventory, International Commission of Snow and Ice, and United Nations Educational, Scientific and Cultural Organization. pp. 23–40.

Braslau, D., Bussom, D.E., 1978b. Landsat sensing of glaciers with application to mass-balance and runoff. In: Colbeck, S.C., Ray, M. (Eds.), Proceedings of a Meeting on Modeling of Snow Cover Runoff 26–28 September 1978, Hanover New Hampshire. U.S. Army Cold Regions Research and Engineering Laboratory, CRREL SR 79–36, pp. 77–82.

Brookfield, M.E., 1998. The evolution of the great river systems of southern Asia during the Cenozoic India-Asia collision: rivers draining southward. Geomorphology 22, 285–312.

Broshears, R.E., Akbari, M.A., Chornack, M.P., Mueller, D.K., Ruddy, B.C., 2005. Inventory of Ground-Water Resources in the Kabul Basin, Afghanistan. U.S. Geological Survey; Scientific Investigations Report 2005–5090.

Buchroithner, M.F., 1978. Ein Beitrag zu den Thermen des Östlichen Wakhan. In: Gratzl, K. (Ed.), Hindukusch-Osterreichische Forschungsexpedition in den Wakhan 1970. Akademische Druck- u. Verlagsanstalt, Graz, Austria, pp. 151–165.

Desio, A., 1975. Geology of Central Badakhshan (North-East Afghanistan) and Surrounding Countries. Italian Expeditions to the Karakoram (K2) and Hindu Kush. E.J. Brill, Leiden, 628 pp.

Doebrich, J.L., Wahl, R.R., 2006. Geologic and Mineral Resource Map of Afghanistan. Version 2 (1: 850,000)/U.S. Geological Survey & Afghan Geological Survey; USGS Open-File Report 2006-1038; AGS Open-File Report 2006-1038; USGS Afghanistan Project Product No. 071.

Fernette, G., 2011. Summary of the Kharnak-Kanjar mercury area of interest. In: Peters, S.G., King, T.V.V., Mack, T.J., Chornak, M.P. (Eds.), Summaries of Important Areas of Mineral Investment and Production Opportunities of Nonfuel Minerals in Afghanistan. pp. 667–692. http://pubs.usgs.gov/of/2011/1204/.

Fouache, E., Besenval, R., Cosandey, C., Coussot, C., Ghilardi, M., Huot, S., Lamothe, M., 2012. Palaeochannels of the Balkh river (northern Afghanistan) and human occupation since the Bronze Age period. Journal of Archaeological Science 39, 3415–3427.

Ganss, O., 1970. Zur geologi von Sudost Afghanistan. Geologische Jahrbruch Beihefte 84 pp.

Gilbert, O., Jamieson, D., Lister, H., Pendlington, A., 1969. Regime of an Afghan glacier. Journal of Glaciology 8 (52), 51–65.

Gordon, T.E., 1876. The Roof of the World: Being the Narrative of a Journey over the High Plateau of Tibet to the Russian Frontier and the Oxus sources on Pamir. Edmonston and Douglas, Edinburgh, Reprint by Ch'eng Wen Publishing Company, Taipei. 1971.

Gratzl, K. (Ed.), 1972. Hindukusch-Osterreichische Forschungs Expedition in den Wakhan 1970. Akademische Druck- u. Verlagsanstalt, Graz, Austria.

Gratzl, K., 1974. Hindukusch: Österreichisches Forschungsexpedition in den Wakhan 1970. Akademishe Druck – u. Verlagsanstalt, Graz, Austria, 131 p.

Haritashya, U.K., Bishop, M.P., Shroder, J.F., Bush, A.B.G., Bulley, H.N.N., 2009. Space-Based Assessment of Glacier Fluctuations in the Wakhan Pamir, Afghanistan. Climatic Change 94, 5–18.

Lapparent, de, A.F., 1970. Quelques exemple de néotectonique dans le Turkestan afghan. Compte Rendu Academie Science Paris. Série D 271, 896–899.

Lindsay, C.R., 2005. Geologic map of Quadrangles 3772, 3774, 3672, and 3674, Gaz-Khan (313), Sarhad (314), Kil-i-Chaqmaktin (315), Khandud (319), Deh-Ghulamah (320), and Ertfah (321) Quadrangles, Afghanistan. US Geologic Survey Open-File Report 2005-1097A.

Lindsay, C.R., Snee, L.W., Bohannon, R.G., Wahl, R.R., Sawyer, D.A., 2005. Geologic map of quadrangle 3568, PoleKhromri (5030) AND Charikar (504) quadrangles, Afghanistan. US Geological Survey Open-File Report 2005-1101A.

Littledale, St G.R., 1892. A journey across the Pamir from North to South. Proceedings Royal Geographical Society XIV (1), 1–35.

Maldonado, F., Turner, K.J., 2005. Geologic map of quadrangle 3368 and part of quadrangle 3370, Ghazni (515), Gardez (516), and part of Jaji-Maydan (517) quadrangles, Afghanistan. US Geological Survey Open-File Report 2005 Mirzad, 1970.

Mirzad, A.G., 1970. Nekotoryye ozera Afghanistana i ikh proiskhozhdeniye. (Lakes of Afghanistan and their origin). Vestnik Moskovskogo Universiteta, Seriya V, Geografiya 25 (5), 106–110.

Moores, E.M., Fairbridge, R.W., 1997. Encyclopedia of European and Asia regional geology. Springer.

Myslil, V., Eqrar, N.N., Hafisi, M., 1982. Hydrogeology of Kabul Basin (in Russian). Sponsored by United Nations Children's Fund and the Ministry of Water and Power, Democratic Republic of Afghanistan.

Patzelt, G., 1978. Gletscherkundliche Untersuchlungen im 'Grossen Pamir'. In: Grancy, R., Kostka, R. (Eds.), Grosser Pamir. Akademische Druck-u, Graz, pp. 131–149.

Porter, S.C., 2004. Late Pleistocene glaciation of the Hindu Kush, Afghanistan. In: J.Ehlers, Gibbard, P.I. (Eds.), Quaternary Glaciations – Extent and Chronology, Part III. Developments in Quaternary Science 2; Series editor: Jim Rose: 1–2.

Ruleman, C.A., Crone, A.J., Machette, M.N., Haller, K.M., Rukstales, K.S., 2007. Map and Database of Probable and Possible Quaternary Faults in Afghanistan. Open-File Report 2007–1103; USGS Afghanistan Project Product No. 150; 45 p.

Sawyer, D.A., Stoeser, D.B., 2005. Geologic map of Quadrangle 3266, Ouruzgan (519) and Moqur (520) Quadrangles, Afghanistan. U.S. Geological Survey 1, 250,000.

Senarclens de Grancy, R., Kostka, R., 1978. Grosser Pamir: Österreichisches Forschungsunternehmen 1975 in den Wakhan-Pamir/Afghanistan. Akademishe Druck – u. Verlagsanstalt, Graz, Austria, 400 pp.

Şengör, A.M.C., Altiner, D., Cin, A., Ustaömer, T., Hsü, K.J., 1988. Origin and assembly of the Tethyside orogenic collage at the expense of Gondwanaland. Geological Society London, Special Publication 37, 110–181.

Şengör, A.M.C., Natal'in, B.A., 1996. Paleotectonics of Asia: fragments of a synthesis. In: Yin, A., Harrison, M. (Eds.), The Tectonic Evolution of Asia. Cambridge University Press, pp. 486–640.

Shroder Jr, J.F., 1978. Remote sensing of Afghanistan. Afghanistan Journal, vol. 5, 123–128.
Shroder, J.F., 2010. Geological character and history of Dashti Eshantop region, Kunduz, Afghanistan. Consulting Geological Report. 04/30/2010. Mueser Rutledge Consulting Engineers, New York. Report by Shroder on physical collapse of a new Afghan Army base because of reliance by US Geological Survey scientists on flawed Soviet reports and maps.
Shroder Jr., J.F., Bishop, M.P., 2000. Unroofing the Nanga Parbat Himalaya. In: Asif Khan, M., Treloar, P.J., Searle, M.P., Qasim Jan, M. (Eds.), Tectonics of the Western Himalaya and Karakoram. Geological Society of London Special Publication, vol. 170. pp. 163–179.
Shroder Jr., J.F., Bishop, M.P., 2010. Satellite glacier inventory of Afghanistan. In: Williams Jr., R.S., Ferrigno, J.G. (Eds.), Satellite Image Atlas of Glaciers. U.S. Geological Survey Professional Paper 1386-F.
Shroder Jr., J.F., Weihs, B.J., 2010. Geomorphology of the Lake Shewa landslide dam, Badakhshan, Afghanistan, using remote sensing data. Geografiska Annaler 92A (4), 471–486.
Shroder Jr., J.F., Weihs, B., Schettler, M., 2011. Mass movement in northeast Afghanistan. Journal of Physics and Chemistry of the Earth 36, 1267–1286. http://dx.doi.org/10.1016/j.pce.2011.03.003.
Stoesser, D.B., 2005. Geologic map of Quadrangle 3670, Jam-Keshem (223) and Zebak (224) Quadrangles, Afghanistan. US Geologic Survey Open-File Report 2005-1096A.
Treloar, P.J., Izatt, C.N., 1993. Tectonics of the Himalayan collision between the Indian Plate and the Afghan Block – a synthesis. In: Treloar, P.J., Searle (Eds.), Himalayan Tectonics. Geological Society, London, Special Publications, vol. 74. , pp. 69–87.
Wittekindt, H., Weippert, D., 1973. Geologisches karte von Zentral und Sud-Afghanistan. 1:500, 000, 4 sheets. Bundesanst für Bodenforsch, Hannover.
Wolfhart, R., Wittekindt, H., 1980. Geologie von Afghanistan. Gebrüder Borntraeger, Berlin.
Wood, J., 1872. A Journey to the Source of the River Oxus; with an Essay on the Geography of the Valley of the Oxus. Oxford University Press, Karachi, reprint of John Murray (London) edition.

In this Chapter

4

Soils and Vegetation *In extremis*

Abstract

In Afghanistan both the vegetation and the soil that it grows on are formed and controlled by a strong climate that ranges in extremes from high altitude, cold alpine ice and tundra, down through midaltitude/latitude forests and steppes to the low-altitude and lower-latitude, hot deserts. Actual potential vegetation has been mapped effectively on the land, even while extensive deforestation, dearbification (deshrubification), and overgrazing have added to soil salinization and soil erosion to alter the landscape considerably. Climate classification has been analytically well done in recent years and is thereby not greatly unknown or controversial, whereas soil types have been mapped in a number of different ways and along with the considerable soil degradation in recent centuries, strong indicators show a need for considerably greater attention to this resource.

Keywords: Biogeography; Climate change; Climate classification; Climate; Dearbification; Debris flows; Deforestation; ITCZ; Natural potential vegetation; Precipitation; Salinization; Soil classification types; Soil erosion; Soil regions; Temperature; Wind of 120 days.

The long changing climate and environmental stresses in Afghanistan have been extreme for millennia; both natural and human induced, with the result that shows soils and vegetation that stress more than in many other places in the world (See, Chapter 17 on Environmental Degradation). The overall results described in the pages below are that in the generally arid climate of Afghanistan, the soils are weakly developed and highly eroded or truncated so that the topsoils are commonly gone, whereas the vegetation also has been stripped away through overgrazing and deforestation or dearbification (deshrubification). Long-term climate change has added its stress to the surficial environment as well, so that in some regions, conditions have become rather harsh for soils and plants.

Climate

Afghanistan occurs in the subtropical, dry zone of Southwest Asia and has an extreme continental arid climate that is characterized by desert, steppe, and highland temperature and precipitation regimes (Humlum, 1959). The climate

Natural Resources in Afghanistan. http://dx.doi.org/10.1016/B978-0-12-800135-6.00004-0

of the country is characterized by strong solar radiation and copious sunshine hours (>3000 h annually in many places), low relative humidity, and high evaporation. The winter weather is driven dominantly by westerly airflow, although cold air drainage from the high-pressure, high-altitude Tibetan Plateau area to the east can also occur. In the summer season, however, complex monsoonal airflows associated with the Intertropical Convergence Zone (ITCZ) take over. During these times every year, the average main monsoonal trough lies approximately along the Durand Line border between Afghanistan and Pakistan (Figure 4.1). Low-altitude monsoonal circulation advances from the Indian Ocean through Pakistan and into the Jalalabad Basin (Figure 4.2), but is separated from upper monsoon circulation air masses by a boundary layer temperature inversion (Figure 4.3), which sometimes allows the upper monsoonal air masses to pass over the subsiding continental air masses crossing the Hindu Kush in the westerly airflow. When this happens torrential rains can occur in the interior or northern regions of Afghanistan, with resulting floods and catastrophic masses

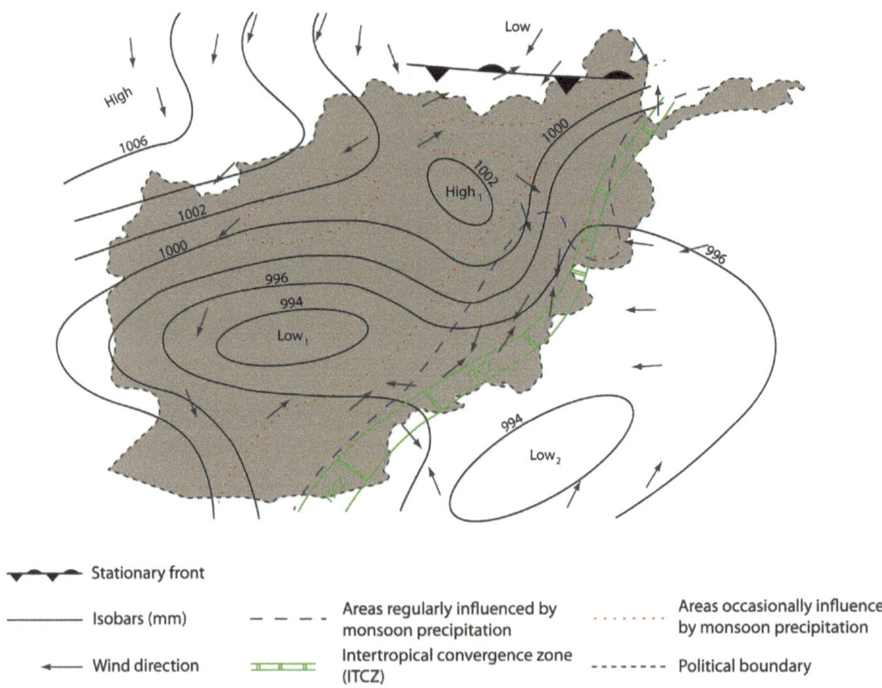

Figure 4.1 Map of summer weather conditions in Afghanistan showing air pressure relations, winds, and the approximate position of the Intertropical convergence zone (ITCZ) that affects the southeast part of the country. Monsoon conditions aloft can periodically carry strong monsoon moisture over the Hindu Kush (dots), which can result in catastrophic rapid wet debris flows (*sailab, duab*) in central and northern Afghanistan. *After Sival (1977), Shroder (1989), and Breckle et al. (2010). Drafting by Brandon Weihs.*

of mud, boulders, and debris that are eroded from the overgrazed and deforested mountain slopes. The result is that these *sailab* or *duab* (rapid, wet debris flows) sweep down into the lowlands as masses of mud and rock, destroying everything in their paths. Also because of the low-pressure zones of the ITCZ, air masses are drawn south from Central Asia, around the edges of the blocking Hindu Kush highlands, and then down approximately along the Iran—Afghanistan border lowlands as the *badi sado bist roz* (wind of 120 days; Dupree (1973)) (Figure 4.1), which can reach sustained velocities of 100 knots per hour. These fierce, hot summer winds are most debilitating and are responsible for tremendous dust and sand storms, massive soil erosion, and a variety of respiratory and ocular diseases. In addition, northwesterly katabatic winds (cold air drainage down from the highlands) also can stir up much dust in the Kabul Basin in summer as well (Fischer, 1968).

Temperature

High annual fluctuations amounting to 25–28 °C are characteristic with a sudden transition from summer to winter and vice versa (Fischer, 1968). Maps of winter (January) isotherms indicate temperatures >6 °C for the Jalalabad Basin (where oranges are grown), as well as several other low-lying areas along the southeastern border of the country with Pakistan (Figure 4.4). The same warm temperature regime prevails in the southwestern part of Afghanistan up to a zone extending approximately from Kandahar to Herat and west to the Iranian border. The Turkestan plains of the north average 0–3 °C in winter, and elsewhere in the country temperatures decrease regularly with altitude so that minima of less than −15 °C

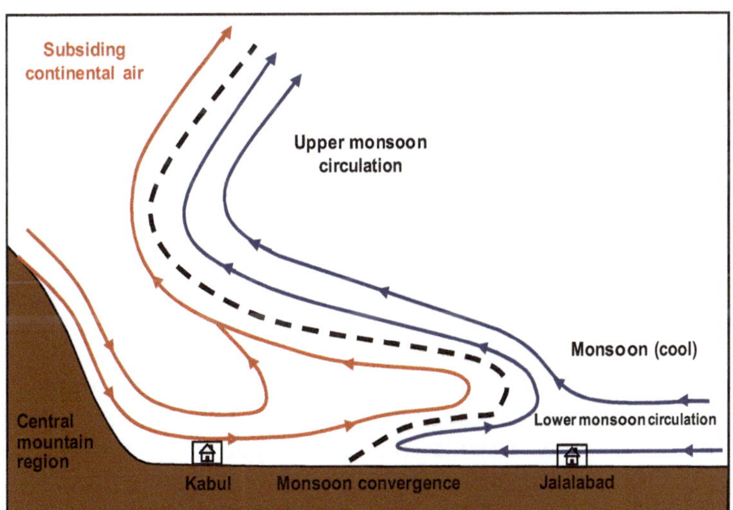

Figure 4.2 Circulation of the monsoon over eastern Afghanistan during the peak flow on July 15, 1970 in which the lower altitude monsoon circulation in the east is overlain by the main monsoon circulation aloft.
By permission, after Breckle et al. (2010) and Sival (1977).

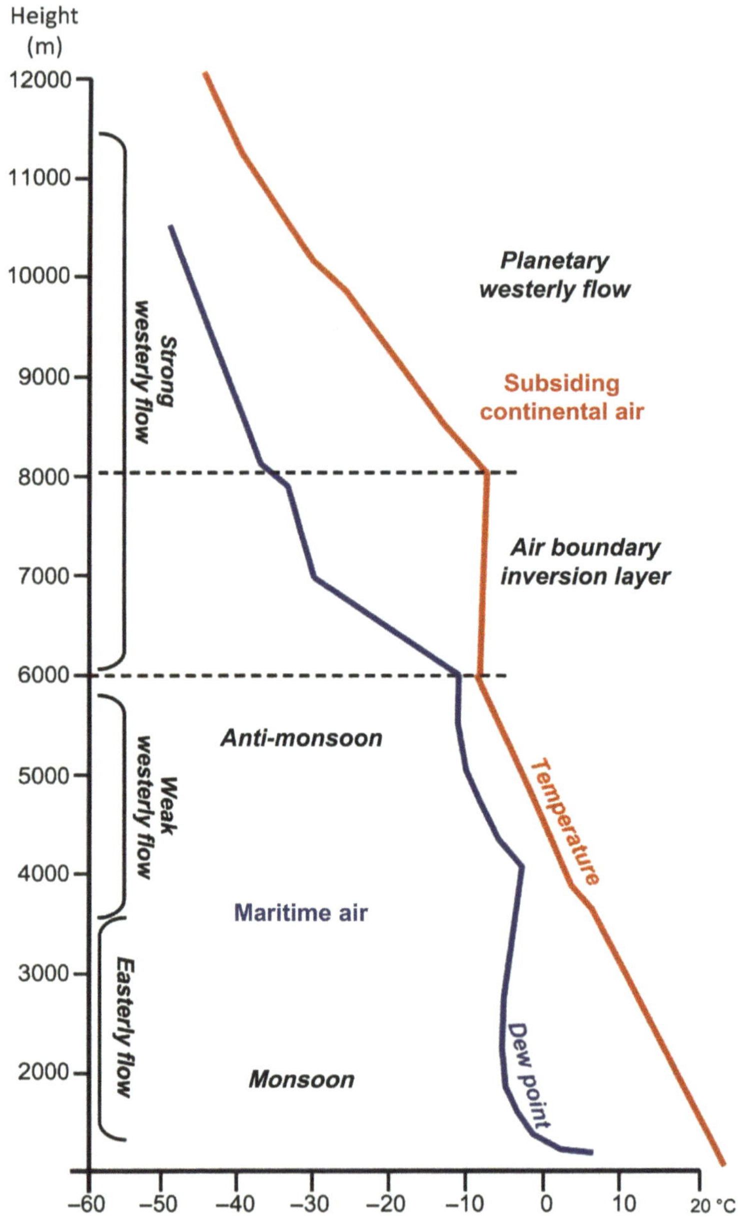

Figure 4.3 Synoptic graph of monsoon circulation above Kabul on July 26, 1965 showing a 2000-m thick, isothermal inversion layer of air from 6000–8000 m altitude that separates the lower, cooler maritime air from the upper dry and subsiding air that is warming aloft. *By permission, after Breckle et al. (2010) and Sival (1977).*

are attained in the central high mountain areas, the glacierized northeast, and in the Wakhan Corridor and the Pamir Knot area. The July isotherms show a similar distribution pattern (Figure 4.5), with the highest of 35 °C or greater in the Seistan lowlands around Zaranj in the extreme southwest. Most of the Seistan lowlands and other lowlands northeast to the Jalalabad

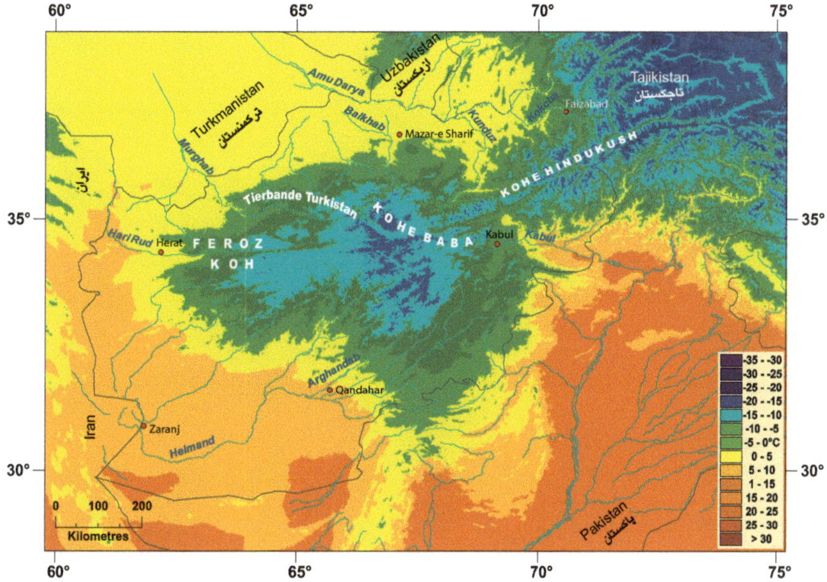

Figure 4.4 Mean winter temperature across Southwest Asia around Afghanistan in the month of January. *By permission, after Breckle et al. (2010).*

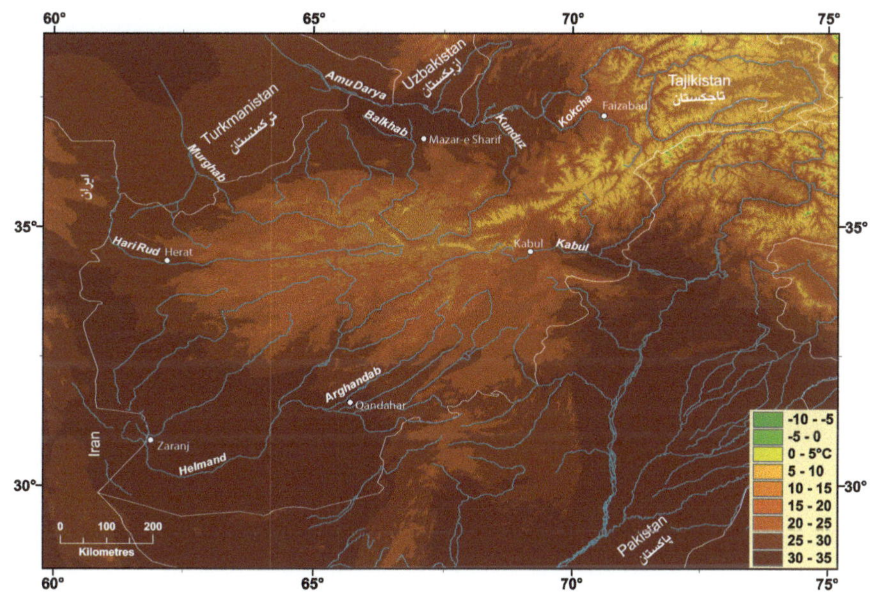

Figure 4.5 Mean temperatures for July over Afghanistan and adjacent areas. *By permission, after Breckle et al. (2010).*

Basin average temperatures between 29 and 32 °C. Elsewhere in the country, the temperatures decrease regularly with altitude down to a minimum of less than 10 °C in some highland areas. Mean annual temperatures (Figure 4.6) show generally warm lowlands surrounding the central highland core of cooler temperatures.

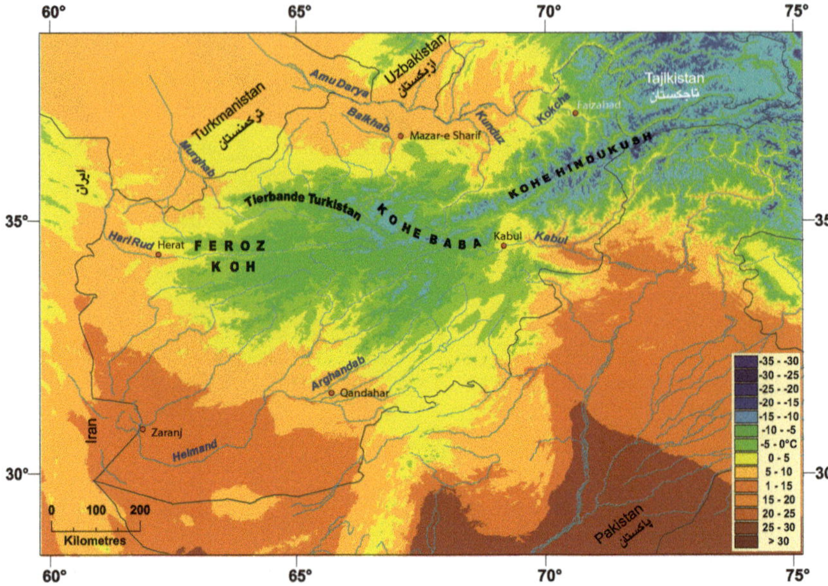

Figure 4.6 Mean annual temperature for the southwestern Asian area of Afghanistan. *By permission, after Breckle et al. (2010).*

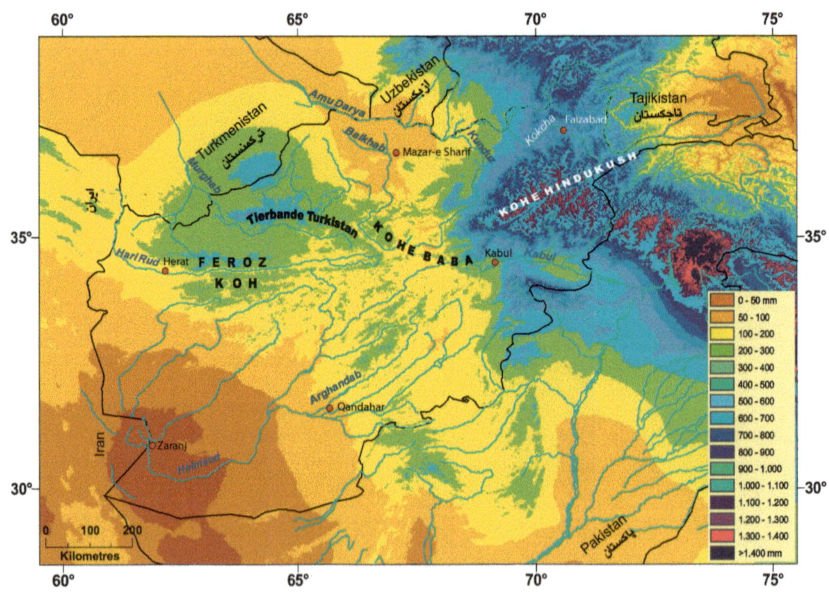

Figure 4.7 Mean annual rainfall in millimeter across Southwest Asia centered on Afghanistan showing the strong gradients from the wetter northeast to the drier southwest. *After Hijmans et al. (2005). By permission, after Breckle et al. (2010).*

Precipitation

Afghanistan is a mostly quite arid country with extreme minima of 1.3–1.4 mm in the Seistan Basin to the southwest, and 50–100 mm in the Wakhan to the far northeast (Figure 4.7). Elsewhere precipitation increases with altitude, especially in the northeastern mountains, but also to a certain extent more precipitation into the Turkmenistan regions. The highly variable summer precipitation tends to be convectional everywhere except in the southeastern monsoonal orographic zones. Winter precipitation tends to be largely cyclonic throughout the country, but is also highly variable. Blizzards are common in the highlands, where also strong snow avalanches unrestrained by trees that have long been cut down have swept away travelers in large numbers.

Climate Patterns

The long-term effects of the many individual weather events in variable temperature (T) and precipitation (P) give us the climate of Afghanistan, which can be presented in a T–P matrix of 19 different levels of temperature and humidity (Figures 4.8 and 4.9) to produce an ecoclimate map favored by some plant geographers (Breckle et al., 2010), or the T–P data, together with some precise calendric timing of weather events, can be mapped in a more conventional Köppen–Geiger fashion, (de Bie et al., 2007) wherein 10 well-known (desert, steppe, humid continental, etc.) climatic zones can be designated to compare with other localities in the world (Figure 4.10). The dry climates are divided into four subgroups, warm and cold deserts (Bwh, Bwk), and warm and cold semiarid steppes (Bsh, Bsk). The warmer deserts occur in the Seistan Basin of the southwest, with warm steppes in the Jalalabad Basin and the east part of Registan, whereas the cooler deserts and steppes occur in the north. The humid subtropical (Csa) and Mediterranean climates (Csb) with mild winters and dry

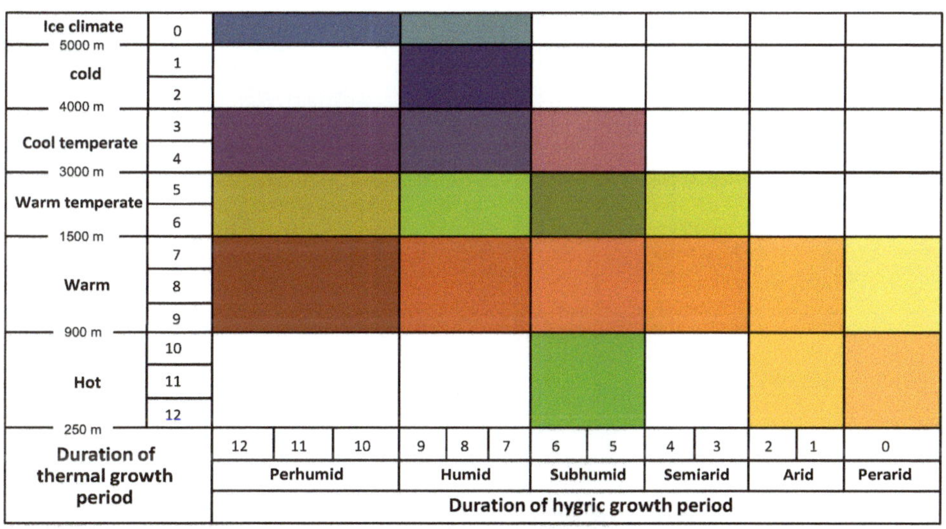

Figure 4.8 Afghanistan's 19 climate types (colored boxes) are schematically arranged by altitude (temperature) on the vertical axis and moisture on the horizontal. The white boxes indicate that a corresponding climate type does not occur in Afghanistan. Compare with Figure 4.9, the ecoclimatic classification of Afghanistan to determine what climate types occur and where they are located. *By permission, after Breckle et al. (2010).*

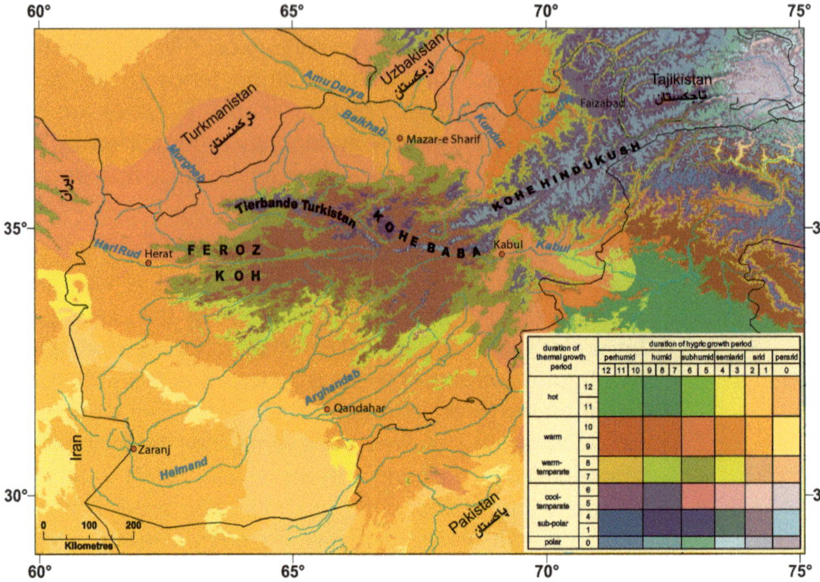

Figure 4.9 **Ecoclimate classification of Afghanistan and surrounding regions.** *By permission, after Breckle et al. (2010).*

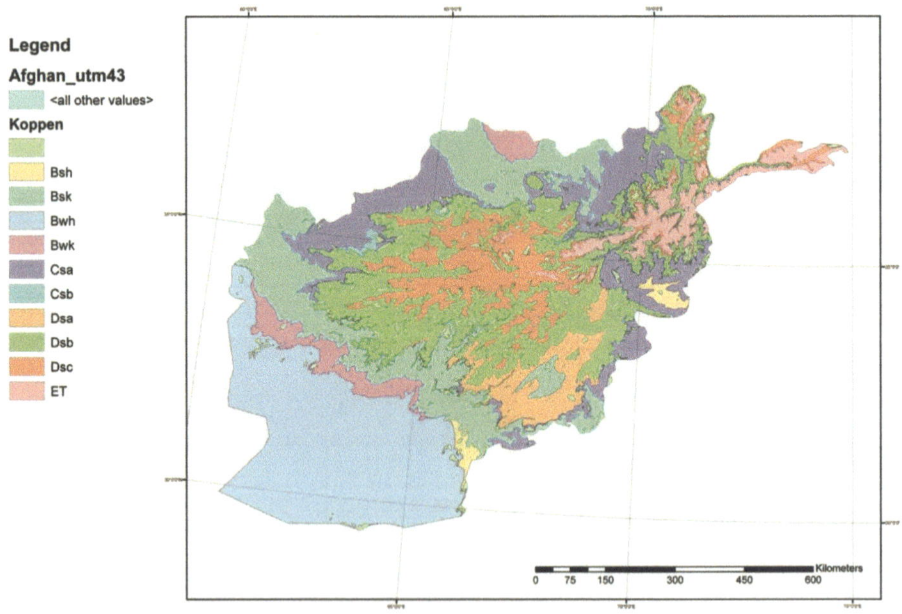

Figure 4.10 **Köppen–Geiger climate of Afghanistan.** Bsh subtropical steppe; Bsk midlatitude steppe; Bwh subtropical desert; Bwk midlatitude desert; Csa humid, subtropical; Csb Mediterranean, mild with dry, warm summer; Dsa humid continental with severe dry winter; Dsb humid continental with severe warm, dry summers; Dsc humid continental with severe cool summers; ET tundra with no true summer (also includes some ice). *After de Bie (2007) and Shroder (2012).*

summers occur in both the north and the south at lower elevations, whereas the humid continental (D) climates of the warmer, to cooler, and colder variations are altitude controlled from lower to higher. The cold extreme tundra (ET) and ice climates are at the highest altitudes in the mountains. Afghanistan has been more extensively glaciated in the past (Porter, 1985, 2004), and is still glacierized, especially in the higher mountains of the northeast (Shroder and Bishop, 2010).

Soils

Soils as features on the landscape have a history of centuries of classification that is not as long as their use by humans for agriculture, but is lengthy nonetheless. This intensive and long-lived human effort at classification is also a measure of the highly variable possibilities that can be used to classify soils, with legions of points of view and prior attempts as soil scientists have sought to deal with this prime resource so essential to human well-being. Thus some of the ways to classify soils have included: (1) depth of unconsolidated material above bedrock; (2) granulometry, or percentages of different grain sizes (gravel, sand, silt, clay); (3) composition (organic matter versus different mineralogies); (4) soil genesis; (5) soil moisture regime (SMR); (6) soil temperature regimes (STR); or other classification styles.

These different classification schemes are partly controlled by the different academic disciplines and the purpose to which the classification is being put. For example, an engineer would be interested in the sediment above bedrock, no matter how thick, as well as what was its granulometry, whereas an agricultural specialist or a climatologist might only be interested in SMR and STR. A geoscientist, on the other hand, might only be interested in the upper few meters with different layers that express the soil-forming factors and the genesis of the soil.

The soil-forming factors in any area include five main features that control the resulting soil type: (1) parent material; (2) climate; (3) topography; (4) biological factors; and (5) time. Parent material is the weathered rock or sediment from which the soil was formed and consequently can have a highly varied composition, depending upon what the original material was made up of. Soils vary in character, depending upon the temperature and moisture of a soil, because weathering types and amounts are so controlled by climate. The slope and aspect of a soil control the moisture and temperature of a soil, with the result that topography controls soil genesis as well. This produces localized soil variation that is called a soil catena. Biological factors strongly influence soils through different kinds of plant roots or organisms that burrow in soil and mix its components. Finally, the amount of time over which the above soil-forming factors act is a continuous control over their effects. The longer the time, the greater is the change.

The most effort at general classification of soils has been expended in attempting to understand their genesis, with the resulting soil taxonomic

Table 4.1 Characteristics of the main soil orders of Afghanistan

Alfisols are mineral soils that are relatively low in organic matter (not much rich topsoil) with relatively high saturation in base elements (calcium, sodium, etc.). Many have illuvial clay particles that have been moved and concentrated by water in the soil profile and the soils generally have sufficient soil moisture to mature an agricultural crop.
Aridosols are mineral soils also low in organic matter that can have some developed soil horizons but not enough soil moisture to mature a crop.
Entisols are mineral soils with weak or no developed soil horizons and variable moisture content and deep wide cracks in most years.
Inceptisols are mineral soils that contain some developed soil horizons other than of illuvial clays, and moisture is available to mature a crop.
Mollisols are mineral soils with thick dark topsoils relatively rich in organic matter and with a high saturation of base elements that are commonly calcium-rich.
Vertisols are clay-rich soils with deep wide cracks at some time in most years depending upon variable moisture content.

definition of the 12 great soil orders of the world. On the next level down in the hierarchy, 64 suborders are recognized; then some 300 great soil groups, >2400 subgroups, and myriads of local soil series are also recognized. Out of the 12 possible soil orders, because of its actively eroding landscape over such long time periods, Afghanistan is unfortunate in having mostly only weakly developed Alfisols, Aridosols, Entisols, Inceptosols, Mollisols, and Vertisols (Table 4.1).

Soil Classification in Afghanistan

Although some thinking about soil types had been done for Afghanistan by various soil scientists in the past, the work was always rather limited and without much detailed field work. For example, Geokart™ (1985) presented only a rather simplistic division of soils (Figure 4.11) in Afghanistan based on a very limited use of a few Russian classification terms such as *serozems* that are generally alkaline desert soils, *brown forest soils* that commonly form under deciduous vegetation, moderately rich in humus, and rich in calcium carbonate, *saline soils*, and *takyrs*, or dry lake basins, and a few other soil forms.

Only Salem and Hole (1969) and the US Department of Agriculture (Eswaran, 2001) have ever invested much effort to describe and map the soils of the country using more modern analysis and terminology, although the United Nations Food and Agriculture Organization (FAO) (Subramanian et al., 1962) earlier had attempted a very small-scale map, and with only 12

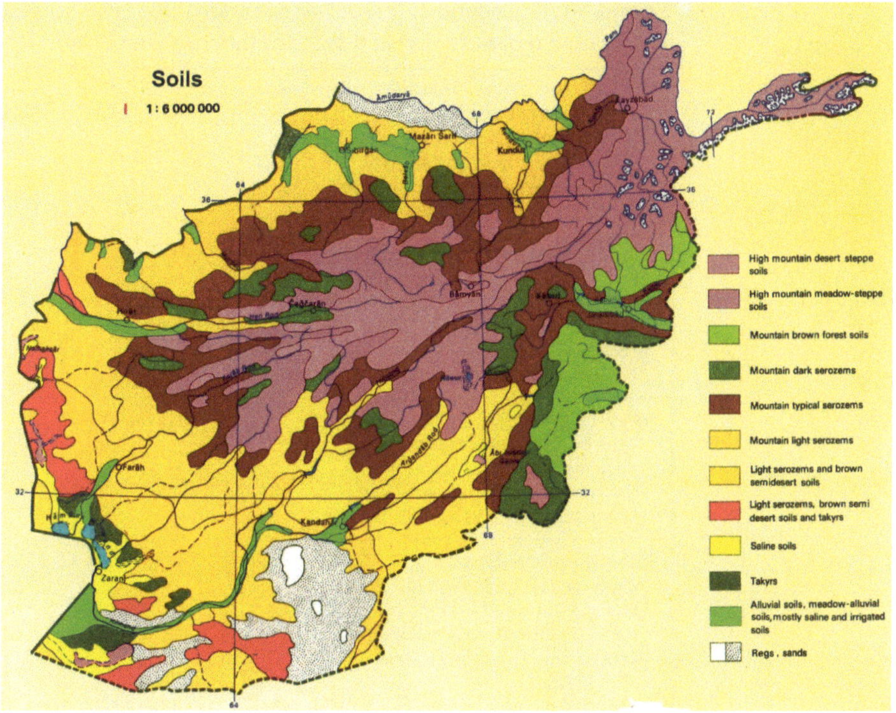

Figure 4.11 Soils of Afghanistan according to the Atlas of the Democratic Republic of Afghanistan (c. 1984). Serozemic soils are typical gray soils of semideserts with low (1–3 percent) humus content and an upper A horizon-rich in calcium carbonate, whereas the B horizon abounds in clay minerals. Takyrs are dry lake basins.

soil regions. The work of Salem and Hole (1969) itself, however, involved detailed assessment of only eight local series soil profiles, of which five were mostly Aridosols in the basins of Kabul, Helmand, and Baghlan River valleys, two were alluvial Entisols, and one was a special area of seepage. This limited field sample did not allow detailed mapping, although they did present a small-scale map at a scale of about 1:12,500,000 that was after Subramanian et al. (1962).

The soil regions of Afghanistan mapped by United States Department of Agriculture (USDA) (Eswaran, 2001), because of an obvious necessity for strategic national planning at that time, used an earlier classification by the Central Intelligence Agency that was not based on field work but on the most probable regional temperatures based on altitude. Thus the new USDA work (Eswaran, 2001) was a major classification by temperature (STR), which is of most direct relevance to agriculture (Figure 3.2), rather than a classification by soil orders. The five STR delineations were, however, further divided into a number of suborders (Table 4.3, Figure 4.11) for a total of 25 soil regions. As this map was not based upon

Table 4.2 Legend of the USDA (Eswaran, 2001) that was based primarily on soil temperature regimes (STR) and soil moisture regimes (SMR)

Soils with cryic/frigid or colder STR (mean annual soil temperature <8 °C)
Rocky land with ice-capped bare rock
Rocky land with lithic cryorthents
Rocky land with lithic haplocryids
Rocky land with lithic haplocambids
Torrifluvents with torripsamments
Soils with mesic STR (mean annual soil temperature 8–15 °C)
Rocky land with lithic haplocambids
Haplocambids with torriorthents
Torriorthents with torrifluvents
Torrifluvents with torripsamments
Torripsamments with dunes
Soils with a xerix SMR and mesic STR
Xerorthents with xerosamments
Xerochrepts with xeroorthents
Calcixeralfs with xerochrepts
Natrixeralfs with halaquepts
Soils with aridic SMR and thermic STR (mean annual soil temperature 15–22 °C)
Rocky land with torriorthents
Haplocalcids with torriorthents
Haplocamids with torripsamments
Torrifluvents
Haplosalids
Torripsamments with dunes
Soils with hyperthermic STR (mean annual soil temperature >22 °C)
Haplocambids with torripsamments
Torripsamments with torriorthents
Torrifluvents with haplogypsids
Torrifluvents with haplosalids
Aquisalids with torrifluvents
Dunes
Salt flats
Lakes
Rivers

field assessments but rather upon topographic or altitudinal gradients (Breckle et al., 2010), it is very theoretical and has a great similarity with the prior maps of the hypsometrically arranged geology, climate, and even vegetation of the country, which is considerable over simplification, although of course, the soil-forming factors, as noted above, do partly control the resulting soils. Breckle et al. (2010) have affirmed that the pedological conditions in Afghanistan are much more complicated than any of the prior maps made and therefore that the latest USDA map (Eswaran, 2001) is rather inaccurate (Figure 4.12). In part this is likely also a scale issue in that the details

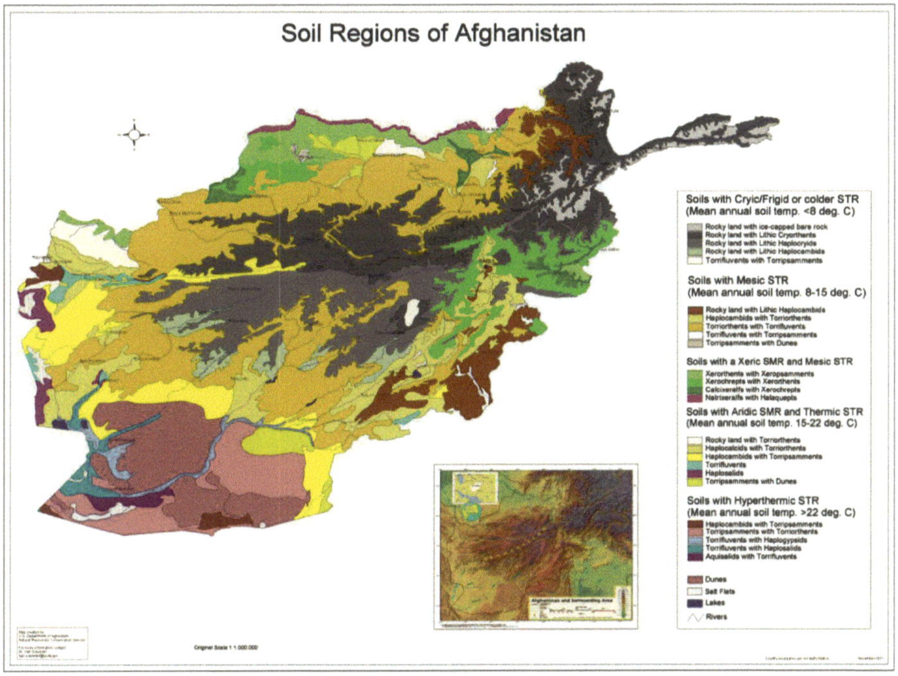

Figure 4.12 Soil regions of Afghanistan according to the U.S. Department of Agriculture. *Eswaran (2001).*

of soils in any particular area need to be considered in a context of finer scales, or even down to facies variations of the soil catenas that vary as a result of local topography. In fact the difficult terminology of the USDA classification (Tables 4.2 and 4.3) which tends to be an unfortunate mixture of Greek and Latin roots, has to be dealt with for the duration until new terminology can be developed in country that is locally specific to Afghanistan. In fact, this should be done by new scientifically trained, Afghan personnel who can bring in local sensibilities and terminology, while following rigorous scientific procedures to produce better terminology to help develop their country agriculturally according to the best modern procedures.

Breckle et al. (2010) have determined that the rather extreme contradictions in soil quality in Afghanistan are the result of the varied topography, the underlying petrological or lithological basis of the parent material, the climatic differentiation in its hypsometric arrangement, and the vegetation cover, as one could expect. Thus the four following major soil regions can be identified:

1. The higher altitude regions of central Afghanistan and the Hindu Kush are characterized by weak soil formation without any clear differentiation in the soil profile because they are developed on newer frost-riven debris and constitute Cryorthents developed on rock fragments that can also be considered Entisols.

Table 4.3 Explanation of soil nomenclature as used by USDA for Afghanistan (USDA, 2001). This difficult terminology, which tends to be an unfortunate mixture of Greek and Latin roots, has to be dealt with for the duration until new terminology can be developed that is scientifically robust but more locally specific to Afghanistan. Older Russian-derived terminology is also outdated

Aquisalids	aqui—wet	sal—salt accumulation	id—aridosol (arid soil)
Calixeralfs	cal—calcium carbonate-rich	xer—moist in winter, dry in summer	alf—mineral soil (alfisol)
Cryorthents	cryo—cold	orth—right (most representative)	ent—entisol (undeveloped or recent soil)
Halaquepts	hal—salt-rich	aqu—wet	ept—inceptisol (just beginning soil)
Haplocalcid	haplo—simple horizons	calc—calcium carbonate-rich	id—aridosol (arid soil)
Haplocambid	haplo—simple horizons	camb—only slightly altered	id—aridosol (arid soil)
Haplocryids	haplo—simple horizons	cry—cold	id—aridosol (arid soil)
Haplogypsids	haplo—simple horizons	gyp—gypsum accumulation	id—aridosol (arid soil)
Haplosalids	haplo—simple horizons	sal—salt accumulation	id—aridosol (arid soil)
Natrixeralfs	natr—sodium salt-rich	xer—moist in winter, dry in summer	alf—mineral soil (alfisol)
Torrifluvents	torri—too dry for crops w/o irrigation	fluv—recent alluvium	ent—entisol (undeveloped or recent soil)
Torriorthent	torri—too dry for crops w/o irrigation	orth—right (most representative)	ent—entisol (undeveloped or recent soil)
Torripsamment	torri—too dry for crops w/o irrigation	psamm—sandy	ent—entisol (undeveloped or recent soil)
Xerochrepts	xer—moist in winter, dry in summer	chr—high chroma (color)	ept—inceptisol (just beginning soil)
Xeropsamments	xer—moist in winter, dry in summer	psamm—sandy	ent—entisol (undeveloped or recent soil)
Xerorthents	xer—moist in winter, dry in summer	orth—right (most representative)	ent—entisol (undeveloped or recent soil)

2. The hot sandy plains of northern Afghanistan occur as rolling countryside and alluvial lowlands, commonly overlain by wind-blown loess dust, all of which results in mineral-rich soils from the Xeropsamments or sandy

Arenosols (FAO classification). Along the south banks of the Amu Darya and its tributaries, extensive sand dunes are also developed.

3. In the deserts of the southwest, soils have been developed from the group of Torripsamments or sandy Arenosols (FAO classification) in varying ways, typical of arid regions with water deficits resulting in the aridic or torric moisture regimes. The extensive sand dunes of Registan are characteristic, as are saline soils, and much clay as in the Dashti Margo (Desert of Death).
4. In all of the major river valleys of the country, fertile soils have evolved from the Torrifluvents (Fluvisols of the FAO classification) or alluvial soils that generally have an arid (torric) moisture regime and must be irrigated to maintain their characteristic extensive agriculture.

Only in the wetter eastern areas of Afghanistan where monsoonal rains and warmer temperatures predominate, the soils become more developed (brown mountain forest soils) (Breckle et al., 2010). Otherwise throughout the country the soils tend to be thin, dry, humus-poor, and rather pedogenically undeveloped.

Biogeography (Vegetation)

Because the precipitation in Afghanistan is low and the mountains are high, the plants of Afghanistan have apparently always been under considerable stress. This biogeographic factor is compounded by the extreme desertification that the country has been observed by the author to have been undergoing for >40 years. Comparative historical descriptions beginning some 500years ago (Thackston, 2002) also show that much of Afghanistan was once forested but vegetation all over Afghanistan has been severely influenced by man and only a few very dry desert areas and very high mountains retain a quasi-natural vegetation cover (Breckle et al., 2010). Natural desertification through climate change, coupled with anthropogenic desertification from human depredation, has thus put the flora of Afghanistan into a rather delicate state. Such past desertification has had a strong effect on surface runoff, with the result that soil erosion is known to have been extreme in almost all parts of the country. On a less harmful basis, however, the presence of any rotting vegetation in the soils adds organic humus to the soils, which in turn adds carbon dioxide (CO_2) to the soil moisture (H_2O), which together react to produce carbonic acid (H_2CO_3). This carbonic acid, in spite of being weak, is more than sufficient over time to produce any number of weathering changes in the soils, and ultimately to generate a host of weathering-related landforms.

The natural potential vegetation of any area is the vegetation that would result from the existing climate, providing no disturbance by human activities. That such has not been the case in Afghanistan for many centuries is obvious, with the result that maps of potential vegetation can be instructive, even where some variation may result, partly as a result of real changes between studies, as well as some differences in interpretation between German (Freitag, 1971; Breckle, 1973, 1981, 1983, 2004, 2007; Breckle et al., 2010) and Russian (Nedjalkov, 1983) or Polish (c. 1984) sources (Figures 4.13 and 4.14). The two sets of maps are very similar but some of the differences between the maps may result from real changes, such as wherein many of the almond (*Amygdalus*) and pistachio (*Pistacia*) trees mapped in the north were later cut down. Insofar as the actual or potential vegetation types on these maps relate to surface processes, strong relations exist in Afghanistan between vegetation type and wind-induced deflation of sand and dust, with attendant sand dunes, loess hills and loess blankets in many places, saline desert environments, alpine ecosystems, as well as various riverine (fluvial), and lacustrine (lake) associations and interactions. Cross sections of altitude vs vegetation across Afghanistan from southeast to north (Figure 4.15) and southwest to north (Figure 4.16) show interesting topographic and ecologic or edaphic (water content, organic content, texture, and pH) relations.

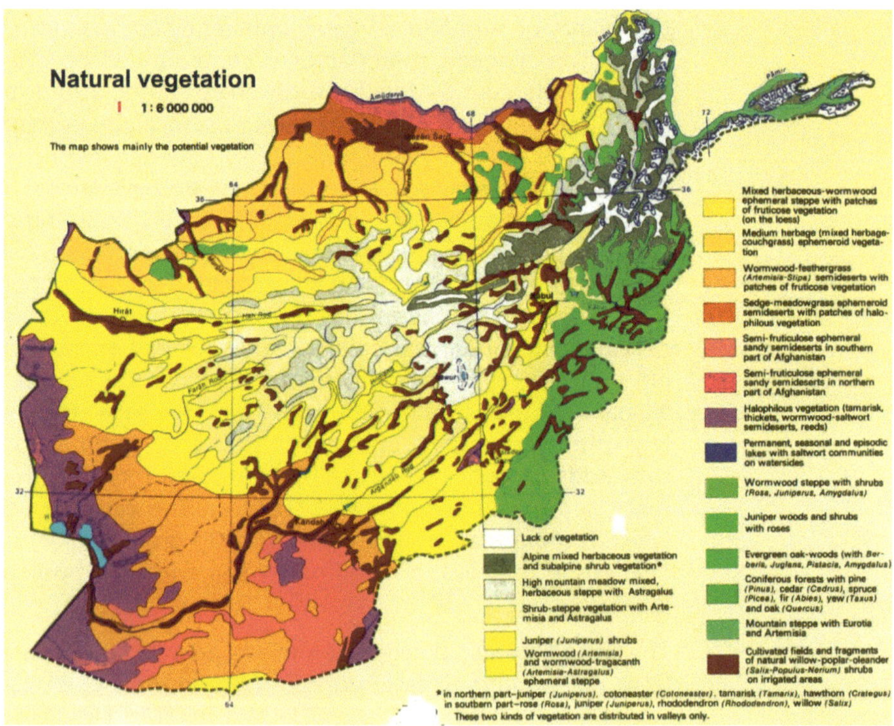

Figure 4.13 Natural vegetation of Afghanistan according to the Polish Geokart Company who produced the original Atlas of the Democratic Republic of Afghanistan (c. 1984).

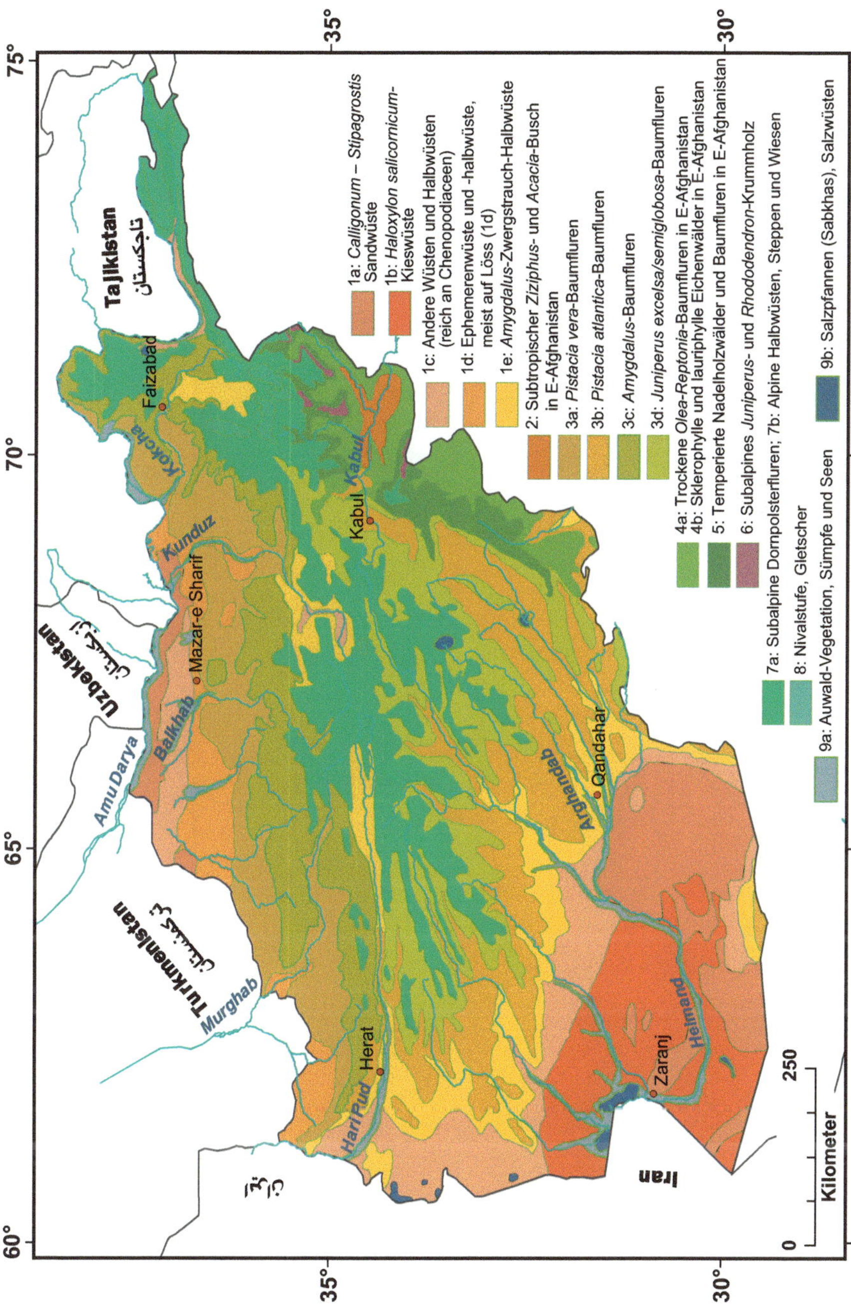

Figure 4.14 Natural potential vegetation zonation of Afghanistan. *By permission, after Breckle et al. (2010).*

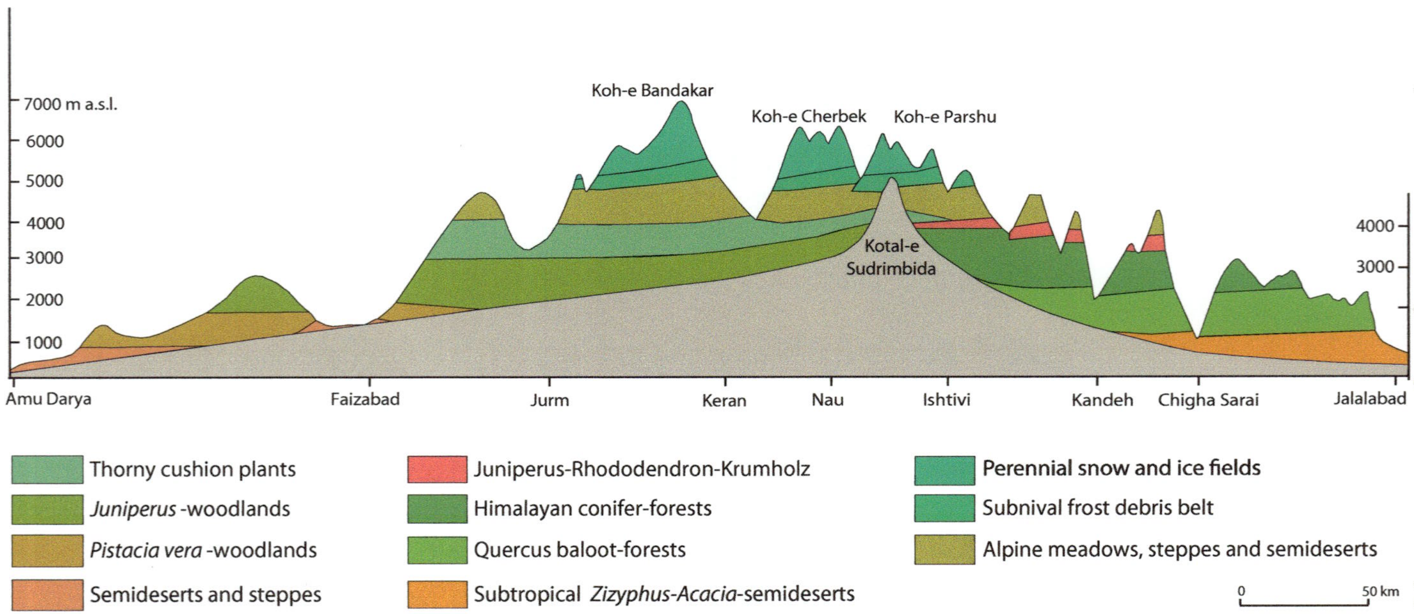

Figure 4.15 Cross section of the altitudinal belts of vegetation from Jalalabad in the southeast to the Amu Darya in the north of Afghanistan. *By permission, after Breckle et al. (2010).*

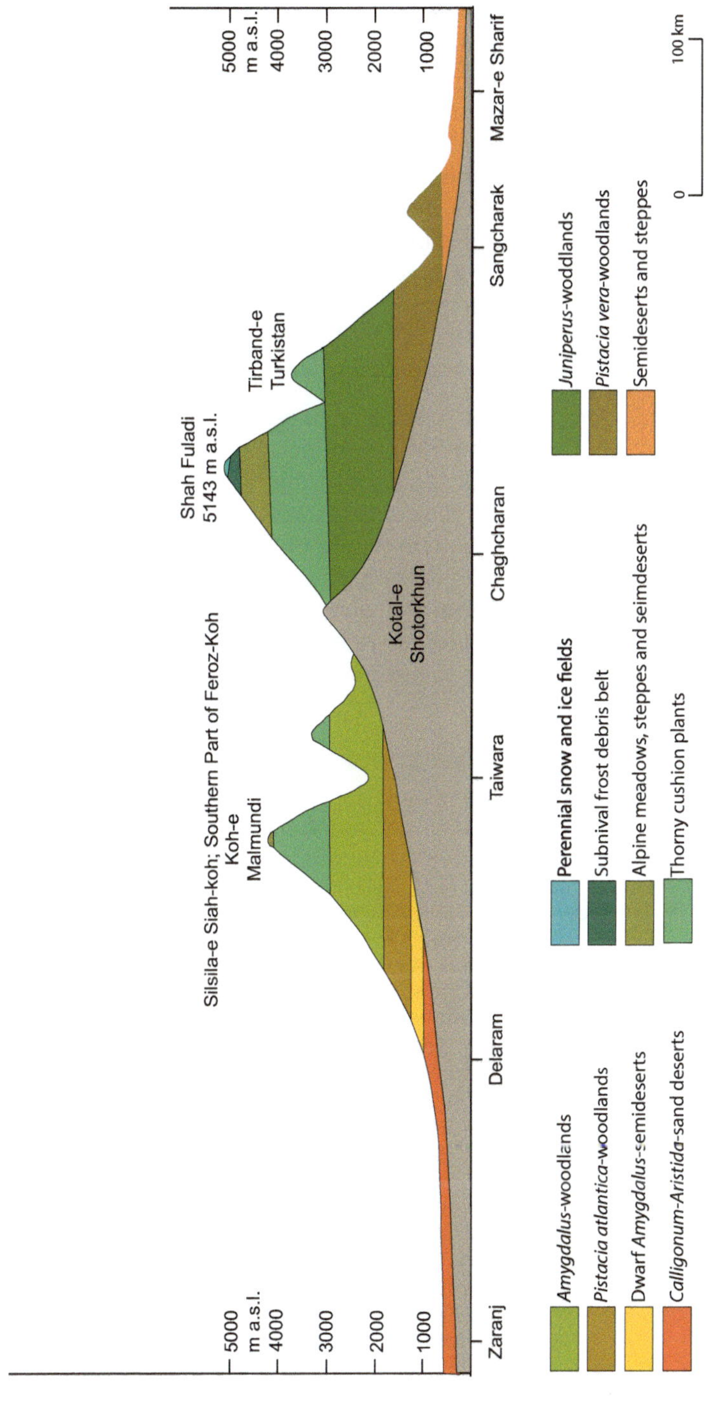

Figure 4.16 Cross section of the altitudinal belts of vegetation from Zaranj in the southeast to the Amu Darya in the north of Afghanistan. *By permission, after Breckle et al. (2010).*

Climate Change

With global climate change obvious to everyone, various quantitative interpretations have been made (IPCC, 2007; Johanson and Fu, 2009) that show the changes in the north to south enlargement of the atmospheric Hadley cells and east to west Walker circulation. These changes have serious implications for changes in the monsoon and for general precipitation in Afghanistan. In some projections for the region, greater dryness is predicted for southern Afghanistan as a result of global warming, and perhaps an increase of precipitation in the north. Such a precipitation increase and glacier expansion has been observed in the higher elevation areas of the Western Himalaya adjacent to Afghanistan (Copland et al., 2009), but not in the Pamir of the lower altitude Wakhan Corridor or the Hindu Kush (Haritashya et al., 2009).

The future of climate change in Afghanistan is therefore obviously unknown, but given the long-term increases in aridity that have allowed the progressive diminution of the lakes in the Seistan Basin, in prehistorical as well as historical time, the implications for the future of the country are most severe. A possible partial solution would be a renegotiation of international water-rights treaties with neighboring countries so that Afghanistan can utilize greater water impoundments and diversions to augment its dwindling surface supplies and recharge its groundwater in certain places, but this will be difficult and contentious, adding to further instability in the already unstable region.

References

Breckle, S.-W., 1973. Mikroklimatische Messungen und ökologische Beobachtungen in der alpine Stufe des afghanistchen Hindukusch. Botanische Jahrbuche System 93, 25–55.

Breckle, S.-W., 1981. Zum Stand der Erforschung von Flora und Vegetation Afghanistans. In: Rathens, C. (Ed.), Neue Forschungen in Afghanistan. Leske, Opladen, pp. 87–104.

Breckle, S.-W., 1983. Temperate deserts and semideserts of Afghanistan and Iran. In: West, N.E. (Ed.), Temperate Deserts and Semideserts. In: Goodall, D.W. (Ed.), Ecosystems of the World, vol. 5. Elsevier, Amsterdam, pp. 271–319.

Breckle, S.-W., 2004. Flora, Vegetation und Ökologie der alpin-nivalen Stufe des Hindukusch (Afghanistan). In: Breckle, S.-W., Schweizer, B., Fangmeier, A. (Eds.), Proceed. 2nd Symposium A.F.W. Schimper-Foundation: Results of Worldwide Ecological Studies. Hohenheim, Stuttgart, pp. 97–117.

Breckle, A.-W., 2007. Flora and vegetation of Afghanistan. Basic and Applied Dryland Research 1 (2), 155–194. ISSN 1864–3191 (BADR on-line journal).

Breckle, S.-W., Dittmann, A., Rafiqpoor, M.D. (Eds.), 2010. Field Guide Afghanistan: Flora and Vegetation. Scientia Bonnensis, Bonn, Germany, p. 862.

Copland, L., Pope, S., Bishop, M., Shroder, J., Clendon, P., Bush, A., Kamp, U., 2009. Glacier velocities across the central Karakoram. Annals of Glaciology 50 (52), 41–49.

de Bie, C.A.J.M., Skidmore, A.K., Toxopeus, B., Venus, V., 2007. An Updated Köppen-Geiger Climate Classification of the World Using Very High Resolution Interpolated Climate Surfaces of Monthly P and T Data from 1950 to 2000. ITC, Enschede, The Netherlands. http://www.itc.nl/personal/debie/An%20updated%20Köppen.htm.

Dupree, L., 1973. Afghanistan. Princeton University Press, Princeton, NJ, 760 pages.

Eswaran, H., 2001. Soil Regions of Afghanistan (Original Scale 1:1,000,000). U.S. Department of Agriculture. http://soils.usda.gov/use/worldsoils/mapindex/afghanistan-soil.html.

Fischer, L., 1968. Afghanistan, a Geomedical Monograph (with 10 Maps), Berlin.
Freitag, H., 1971a. Die natürliche Vegetation Afghanistans. Beitrage zur Flora und Vegetation Afghanistans I. Vegetatio 22, 286–344.
Freitag, H., 1971b. Studies on the natural vegetation of Afghanistan. In: Davies, P.H., Harpur, P.C., Hedge, I.C. (Eds.), Plant Life of South-West Asia, pp. 89–106. Edinburgh.
Geokart™, 1985. Atlas of the Democratic Republic of Afghanistan. Poland, Warsaw.
Haritashya, U.K., Bishop, M.P., Shroder, J.F., Bush, A.B.G., Bulley, H.N.N., 2009. Space-based assessment of glacier fluctuations in the Wakhan Pamir, Afghanistan. Climatic Change 94, 5–18.
Hijmans, R.J., Cameron, S.E., Parra, J.L., Jones, P.G., Jarvis, A., 2005. Very high resolution interpolated climate surfaces for global land areas. International Journal of Climatology 25, 1965–1978.
Humlum, J., 1959. La Geographie de l'Afghanistan: Etude d'un pays aride. Copenhagen, Denmark.
IPCC, 2007. Climate Change 2007: Working Group I: The Physical Science Basis. http://www.ipcc.ch/publications_and_data/ar4/wg1/en/ch3s3-7.html.
Johanson, C.M., Fu, Q., 2009. Hadley cell widening: model simulations versus observations. Journal of Climate 22, 2713–2725.
Nedjalkov, S.T., 1983. Ecological classification of vegetation cover of Afghanistan. In: Sokolov, V.E. (Ed.), Ecology and Biogeography in Afghanistan. USSR MAB, Moscow, pp. 30–53 (In Russian).
Porter, S.C., 1985. Extent of late Pleistocene glaciers in Afghanistan based on interpretation of LANDSAT imagery. In: Agrawal, D.P., Kusumgar, S., Krishnamurthy, E.V. (Eds.), Climate and Geomology of Kashmir and Central Asia: The Last Four Million Years. Today and Tomorrow's Printers and Publishers, New Delhi, pp. 191–195. Current Trends in Geomorphology.
Porter, S.C., 2004. Late Pleistocene glaciation of the Hindu Kush, Afghanistan. In: Ehlers, J., Gibbard, P.L. (Eds.), Quaternary Glaciations—Extent and Chronology, Part III. Developments in Quaternary Science, 2. Elsevier, Amsterdam, pp. 1–2. Series Editor, Rose, J.
Salem, M.Z., Hole, F.D., 1969. Soil geography and factors of soil formation in Afghanistan. Soil Science 107 (4), 289–295.
Shroder, J.F., Jr., 1989. Slope failure: extent and economic significance in Afghanistan and Pakistan. In: Landslides: extent and economic significance in the world. In: Brabb, E.E., Harrod, B.L., Balkema, A.A. (Eds.), Rotterdam, Netherlands, pp. 325–341.
Shroder Jr, J.F., 2012. Afghanistan: rich resource base and existing environmental despoliation. Environmental Earth Scienceshttp://dx.doi.org/10.1007/s12665-012-1638-7.
Shroder Jr., J.F., Bishop, M.P. 2010. Satellite glacier inventory of Afghanistan. In: Williams Jr., R.S., Ferrigno, J.G. (Eds), Satellite Image Atlas of Glaciers. U.S. Geological Survey Professional Paper 1386-F. in press.
Sival, T.R., 1977. Synoptic-climatologic study of the Asian summer monsoon in Afghanistan. Geografiska Annaler 59A, 67–87.
Subranamian, V., Nasirov, V., Salem, M.Z., 1962. Generalized soil map of Afghanistan. First soil correlation seminar for South and Central Asia. Soil map of the World FAO/UNESCO project World Soil Resources Reports 4, 56–59.
Thackston, W.M. (Ed.), 2002. The Baburnama: Memoirs of Babur, Prince and Emperor. Zahir-ud-din Mohammad Babur. Modern Library Palang-faack Edition, New York.
Westfall, A.O., Latkovich, V.J., 1966. Surface Water Resources Investigation Plan for Afghanistan. U.S. Geological Survey Open-File Report. 62 pages.

In this Chapter

5

Watersheds of Want

Abstract

The hydrological cycle in Afghanistan is one of the high altitude snow and rain, commonly torrential, which produces catastrophic downstream effects such as avalanches and floods. Most of the precipitation that drives the river-flow lifeblood of the country outward from the tops of the watersheds in Afghanistan diminishes toward the borders from its highs in the northeast of the nation. The main river systems, listed in a counterclockwise direction around Afghanistan. include the Amu Darya, Hari Rud–Murgab, Helmand–Arghandab, and the Kabul, each of which is discussed in more detail herein. Lakes in Afghanistan include glacial, landslide-dammed, carbonate-precipitate types, diastrophic (tectonic and volcanic) lakes, tectonic—mixed types, and multiple deflation-basin sorts of lakes, many of which are intermittently dry. Underground water in Afghanistan occurs in aquifer basins throughout the country, with the basin beneath Kabul City undergoing severe drawdown.

Keywords: Hydrological cycle; Lakes; Precipitation; Rivers; Underground water; Water sources.

The natural water supplies of any country are generally that which comes from the sky, either on an annual basis to produce river runoff, or the portion that is stored underground, perhaps derived from precipitation that has been received and held underground for a very long time. Thus in order to understand the watersheds and their importance in an arid country such as Afghanistan, we must first reconsider the climate of the region, with its variations in precipitation, as we focus on the all-important hydrology of Afghanistan, mainly in terms of the effects of glacier ice, the rivers, and the lakes.

As noted in Chapter 4 on soils and vegetation, where controlling climate factors were discussed briefly, the climates of the Earth are controlled largely by the amount of precipitation and temperature delivered at different times and places as a function of latitude, altitude, and position on a continent relative to ocean water masses. A mountainous and desertic country such as Afghanistan, far from oceanic moisture sources and subject to the vicissitudes of variable winds and erratic supplies of precipitation, is characterized also by extremes of continentally wherein winters can be quite cold and summers excessively hot and dry so that any water derived from snowmelt and rainstorm precipitation is essential but evaporates quickly (Breckle, Rafiqpoor, 2010).

Natural Resources in Afghanistan. http://dx.doi.org/10.1016/B978-0-12-800135-6.00005-2

In general, Afghanistan is dominantly an arid to semiarid country, except in the mountains where orographic precipitation is caused when air masses are forced to rise and the attendant cooling causes condensation of moisture out of the air. The amount of precipitation increases to the northeast in Afghanistan in response to the higher altitudes there. Average annual precipitation is commonly <210 mm in many areas, declining to <110 mm in the southwestern deserts, and increasing to >1000 mm in the high mountains. In all, over 80 percent of the country's water resources have their origin in the mountains >2000 m in altitude, which function as a natural storage of snow and ice that supports perennial flow in all the major rivers in summer (ICARDA USAID, 2002).

The main period of precipitation is borne on the westerly winds and extends from November to May, but is shortened in the south to December through April. About 50 percent of the precipitation occurs in winter (January to March), mostly as snow. A further 30 percent falls in spring (April to June). In the summer season some monsoonal precipitation makes it into extreme southeastern Afghanistan along the border with Pakistan, although occasionally monsoon moisture sources go well into the Hindu Kush in Central and Northern Afghanistan (Figures 4.1 and 4.2 in the previous chapter).

Hydrologic Cycle in Afghanistan

The water cycle is absolutely critical to modern Afghanistan, just as it has been nearly totally responsible for so much of the geomorphology and a large share of the natural hazards of the country. As an overall semiarid to arid nation, a number of places in the north and south have no surficial drainage. Elsewhere, however, the drainage net is well developed (Figures 5.1 and 5.2) and provides the basic water source to the people, even as it is also responsible for the general character of most of the landscapes.

The watershed boundaries in Afghanistan begin high on ridge and mountain tops, generally but not entirely, in the middle and northeast of the country and thereafter the rivers flow radially outward over the borders into neighboring countries. This position at the top of the watersheds places Afghanistan geomorphically as a source of snow and glacier ice storage and meltwater river delivery downstream that combine to produce most of the geomorphology of the country. Below the land surface of the country the fractures in the bedrock, and the natural porosity in the sediment-filled basins throughout the nation (Kabul, Jalalabad, Seistan, and northern basins), have considerable storage of underground water that is obtained through gently inclined karez tunnels, as well as boreholes and pumps.

Snow and Ice

The accumulation of snow in the highlands in winter both melts and runs off into the rivers in spring and summer, or it remains throughout the year and becomes in a few years, a glacial river of ice. About 3000 separate glaciers exist

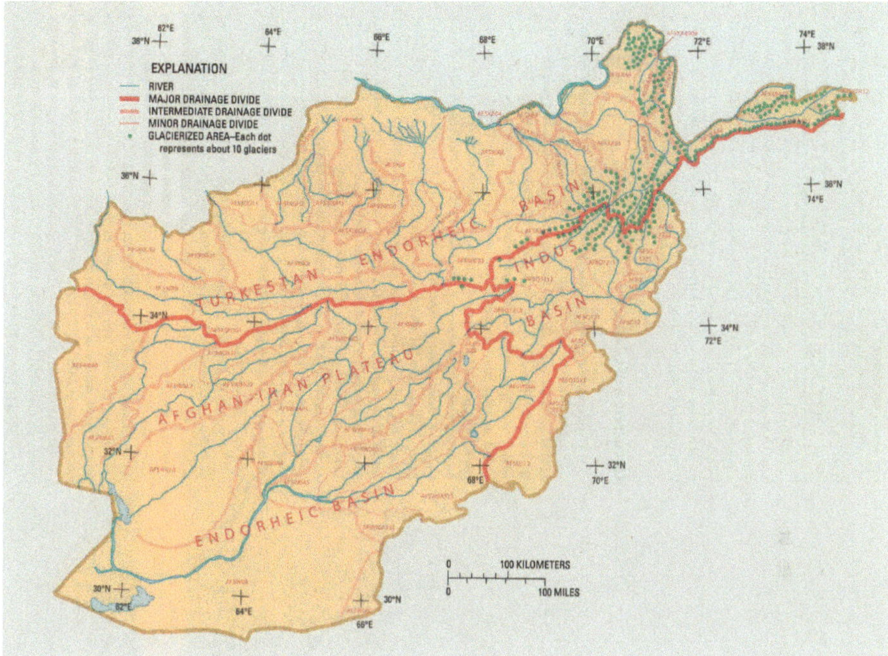

Figure 5.1 Basic hydrologic basins and drainage network of Afghanistan. *After Shroder and Bishop (2010); Figure 2.*

in Afghanistan with an estimated area of ~2700 km^2 (Shroder, 1980, 1989; Shroder and Bishop, 2010) (Figure 5.1), but they were once far more numerous than now and they were also much larger back >10,000 years ago in the Pleistocene (Ice Age) time (Porter, 1985, 2004). At present the glaciers of Afghanistan are concentrated in the higher northeastern drainages of the country (Amu Darya, Panjshēr—Kunar—Kabul). High seismicity, frequent avalanches of snow and ice as well as rocks, coupled with widespread stagnation, retreat, and downwasting (Bishop et al., 2014), have produced numerous debris-covered ice and rock glaciers, many of which have been essential sources of meltwater in past late summers. Some of these important sources have probably melted out and dried up in recent years. The largest remaining glaciers in the Wakhan Pamir have been mapped in detail in past years (Patzelt, 1978), and assessed with comprehensive remote sensing (Haritashya et al., 2009), with the result that they can be seen through their pervasive retreat and downwasting to have mainly negative mass balances in recent decades, unlike some nearby higher-altitude glaciers in the Karakoram Himalaya of Pakistan (Copland et al., 2009, 2012; Bishop et al., 2014).

Glaciers are markedly efficient erosion agents and high in the mountains of Afghanistan they produced many spectacular landforms that are mapped and discussed elsewhere in the volume. All of the stored ice is also a vital source of meltwater downstream, especially in late summer and early fall when second cropping can provide important additional food to a hungry people. If this

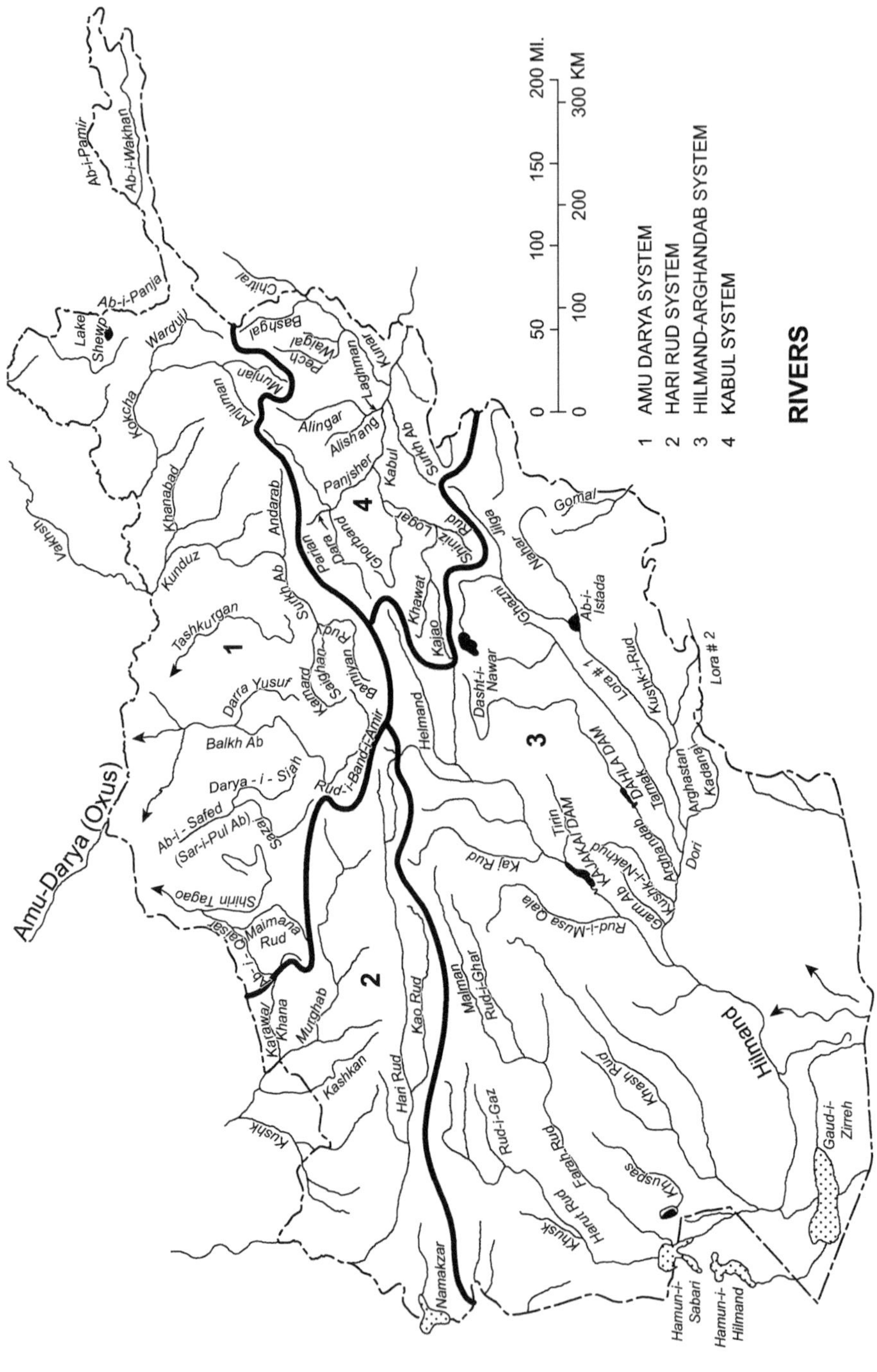

Figure 5.2 Rivers of Afghanistan showing names in English and generalized hydrologic basins. *After Dupree (1976).*

observed trend of glacier and ground-ice diminution in Afghanistan continues, the implications to the arid country are obviously most serious.

Rivers

The rivers of Afghanistan are remarkable for their diversity, even as their activity is such a strong agent in the formation of fluvial landforms, while their waters are the critical lifeblood for the Afghan people. The larger rivers of arid Afghanistan rise in the Hindu Kush and Pamir mountains, mainly from snow and glacier melt, and flow on fairly steep gradients to the lowlands. Maximal flows in spring and summer meltwater river discharges are followed by fall and winter minima, when some large rivers even dry up completely. Most runoff is delivered from melting of seasonal winter snow or older glacier ice; high silt loads are characteristic (Westfall and Latkovich, 1966). Ephemeral stream channels abound in the lower elevations and are commonly filled by torrential summer thunderstorms so typical of many desert areas, especially where monsoon precipitation has penetrated from the south.

Rivers in Afghanistan commonly reflect three major spatial controls by location of drainage basin (Figures 5.1 and 5.2): (1) the north and northwestern flow of the Nile-sized Amu Darya and several other rivers into the central Asian depressions of the Turkestan endorheic basin (See figures of the border rivers in Chapter 3); (2) the strong west and southwest flow of the Afghan–Iran Plateau endorheic basin, largely structurally controlled, with the main Helmand River into basins such as the Seistan on the Iranian border (Figure 5.3); and (3) the southeastward flow by the Kabul River (Figure 5.4) and its tributaries (Figures 5.5 and 5.6) into the Indus basin system in Pakistan. Most of these master drainages rise in the higher central and eastern part of Afghanistan where the altitude is great enough to trap moisture arriving in the westerly winds, as well as Indo–Pakistani monsoonal moisture coming from the south or southeast. Of all the major rivers that rise in the highlands of Afghanistan, only the waters of the exorheic Kabul River reach the sea because they join the Indus at Attock in Pakistan. The other major endorheic rivers that exit the country's borders include the Amu Darya and tributaries that form the northern border of the country and drain to the Turkestan endorheic basin with the drastically declining Aral Sea as the final sink (Micklin, 1988, 2007), the Murghab and the Hari Rud in the northwest that also go to the Turkestan endorheic basin, and the Helmand in the southwest, which drains into closed depressions in the Seistan Basin.

These rivers all rise in the highlands from meltwaters coming from snow, or glacier and rock-glacier ice, before passing out through mountain gorges and into wider valleys where flights of abandoned stream terraces and lower altitude floodplains start to occur. In these regions a number of flat-floored mountain valleys (Figure 5.7) occur, many of which are the remnants of infilled prior lake basins that were formerly impounded behind a variety of natural moraine,

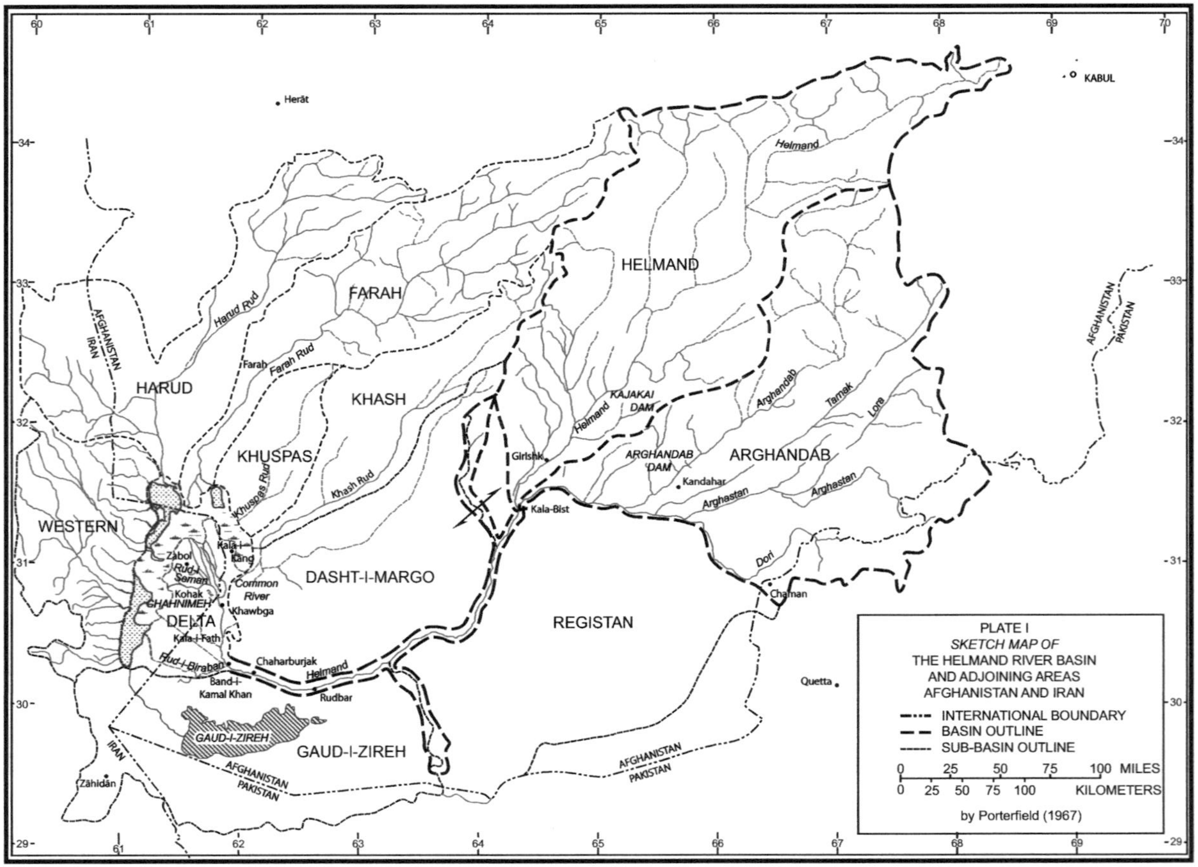

Figure 5.3 Map of lower Helmand River flow into the *hamuns* and playas of the Iran–Afghanistan border. *After Porterfield (1967).*

Figure 5.4 (A) The Kabul River in winter 1977 *(Photo by the author)*. This river has run well for centuries but began to dry up in Kabul in recent years during the height of droughts in the past decade. (B) Drier and more polluted Kabul River about 30 years later. *Photograph by T. Gusinov, by permission.*

landslide, and biohermal travertine dams (Figure 5.8), some of which have since been breached and drained, or filled to the brim with sediment. Smaller drainages, generally ephemeral stream valleys, commonly flow out of the mountains where sediment discharges of varying types (stream flow, rapid, and wet debris flows) at the mountain fronts result in the common alluvial fans and of dramatically different gradients and sizes (Figures 5.9 and 5.10). Many of the alluvial fan regions have networks of channels across them that are partly natural, and partly diversions made by human inhabitants seeking to divert seasonal water for irrigation. These alluvial fans and river floodplains grade out further into the lowlands as vast, gently sloping alluvial plains, before ending in large ephemeral lake basins or swamps, many of which are at least partly covered with dune field regions (corruption of Arabic *erg* to *reg*, as in Registan) that were mobilized out of the clastic alluvial sediments that had been brought into the basins originally by river flow into the lakes before they dried out. In prior centuries, historical records

Figure 5.5 The Panjshēr River in 1977 above its confluence with the Kabul River *(Photo by the author).* The Panjshēr has its sources in many glaciers and rock glaciers high in the Hindu Kush.

Figure 5.6 The Kunar River in 1978 above Its confluence with the Kabul River in the Jalalabad basin. *Photo by the author.*

(Beveridge, 1922; Jentsch, 1972) indicate that much of the lower altitude uplands in antiquity and later were well mantled with soil and extensively forested or grass and shrub covered, which would have had good effect on the infiltration of precipitation waters and retarded runoff. In addition to the dramatically increased deforestation in Afghanistan in the past three decades or so (Shroder, 2008), and

Figure 5.7 Flat-floored mountain valley near Bamiyan. *Photo by the author.*

Figure 5.8 Bandi Amir self-damming lake of biochemical travertine. *Photo by the author.*

other forms of desertification (see Part IV), flooding of several kinds also seems to have increased in Afghanistan, which is discussed in Chapter 8.

The hydrologic regimes of Afghanistan vary because of different runoff controls throughout the year (Humlum, 1959). Thus some of the rivers have a pluvial regime, in which rain and snowmelt produce a late winter–spring maximum (Figure 5.11), and a late summer—early fall minimum when many are entirely dry. Most of the small watercourses carry no water at all from April, May to November, December. Torrential pluvial regimes occur in the many dry

Figure 5.9 **Small alluvial fan with snow cover near Ghazni.** *Photo by the author taken in 1977 from airplane of US Ambassador, T. Eliot.*

Figure 5.10 **Huge alluvial fan of Balkh with scale bar of 20 km in left lower part below Aqcha.** The fan of Aqcha on the left was produced in post-Islamic times by the diversion of the Balkh Ab (river) for irrigation purposes over a millennium ago. *See Figure 3.49, for comparison.*

watercourses (*darrah*, *wadi*, and *juie or 'jōy*) in Afghanistan, especially at times of rapid snowmelt or summer rainstorms. Rivers with a nival regime in the northeast high mountains of Afghanistan tend to have a minimum discharge in winter when they freeze strongly, and a summer maximum when the snow and ice melt. Those with a monsoon regime in the southeast also have a summer maximum, but from increasing precipitation at that time.

The estimates of the actual availability of water and its use in Afghanistan are fraught with misinformation and a high variability by as much as one-third in the numbers, so much so that one suspects varieties of estimation apparently

Figure 5.11 Khash Rud in full flow, spring 1977. *Photo by the author.*

Table 5.1 Mean annual surface water flow in Afghanistan (in billion cubic meters (bcm)).

River Basin	Surface Area (%)	Mean Annual Water Flow (bcm)	Total Annual Flow (%)
Amu Darya	14	48	57
Kabul	12	22	26
Helmand	43	9.3	11
Hari Rud–Murghab	18	3	4
Northern Basin	13	1.8	2

Favre and Kamal (2004).

based upon what appear to be rather unreliable assumptions, or weak estimation methods, or both. In part this is because lack of reliable data collecting over the past 30 years of war, and in part perhaps also, because of changes in hydrological forecasting methods. In any case, in reference to the mean annual surface water flow of Afghanistan, or total river discharge in the country, the older estimates of the United Nations FAO (1997) and Azimi and McCauley (2002) were values of 55–57 billion cubic meters (bcm) for the whole country, whereas the values estimated at a slightly later time by Favre and Kamal (2004) indicated 84 bcm (Table 5.1). This apparent increase of 28 bcm is apparently derived from unknown citational sources but is likely not real, inasmuch as the country has been undergoing a severe drought cycle in recent years and one would expect such numbers to decline, not increase.

Table 5.2 Groundwater characteristics in Afghanistan (in million cubic meters per year (mcm))

River Basin	Recharge (mcm)	Usage (mcm)
Kabul	1920	530
Helmand	2480	1500
Hari Rud–Murghab	1140	460
Northern basins	2140	210
Amu Darya	2970	100
Total	10,650	2800

After Uhl, Tahiri, (2003).

Amu Darya River System

The Amu Darya (Oxus of antiquity) originates in the Wakhan Corridor of Afghanistan as three separate drainages, the Aq Su, the Abi Pamir, and Abi Wakhan that are confluent downstream (Figure 5.12). For example, the latter two join in the Wakhan to form the Panj River of Badakhshan, which constitutes most of the national border of Afghanistan with Tajikistan. Where the Panj River is confluent with the Kokcha River from southern Badakhshan, the Amu Darya begins and goes on the form the rest of the entire southern border of Tajikistan and Uzbekistan with Afghanistan, as well as the eastern portion of the south border of Turkestan with Afghanistan.

Estimates vary because of the lack of reliable measurements or preservation of past flow records, but most of the Nile-sized flow of the Amu Darya originates in Tajikistan at about 61 percent and in Afghanistan at ~30 percent (Fipps, 2007). The downstream republics of Uzbekistan and Turkmenistan combined use about 52 percent of the total flow of the Amu Darya but contribute only 9 percent of the total volume. In comparison Afghanistan currently uses less than 10 percent of its contribution (1.5–2 km^3) and only about 2 percent of the total discharge. The average discharge of the Amu Darya appears to be around 2000 m^3/s per year. The Kokcha itself, one of the major rivers of northern Afghanistan, has measured discharge rates of 108–163 m^3/s per year, whereas the Kunduz River tributary to the Amu Darya also measures similar amounts with a mean discharge of 108 m^3/s per year. All other rivers in this northernmost drainage basin of Afghanistan have waters in insufficient quantities to travel from their mountain sources to be confluent with the Amu Darya; they are diverted off for irrigation and what is left soaks into the desert sands that are peripheral to the Amu Darya.

Hari Rud–Murghab River System

This fluvial system represents about 12 percent of the water resources of Afghanistan and is dominant in the heavily irrigated regions around Herat

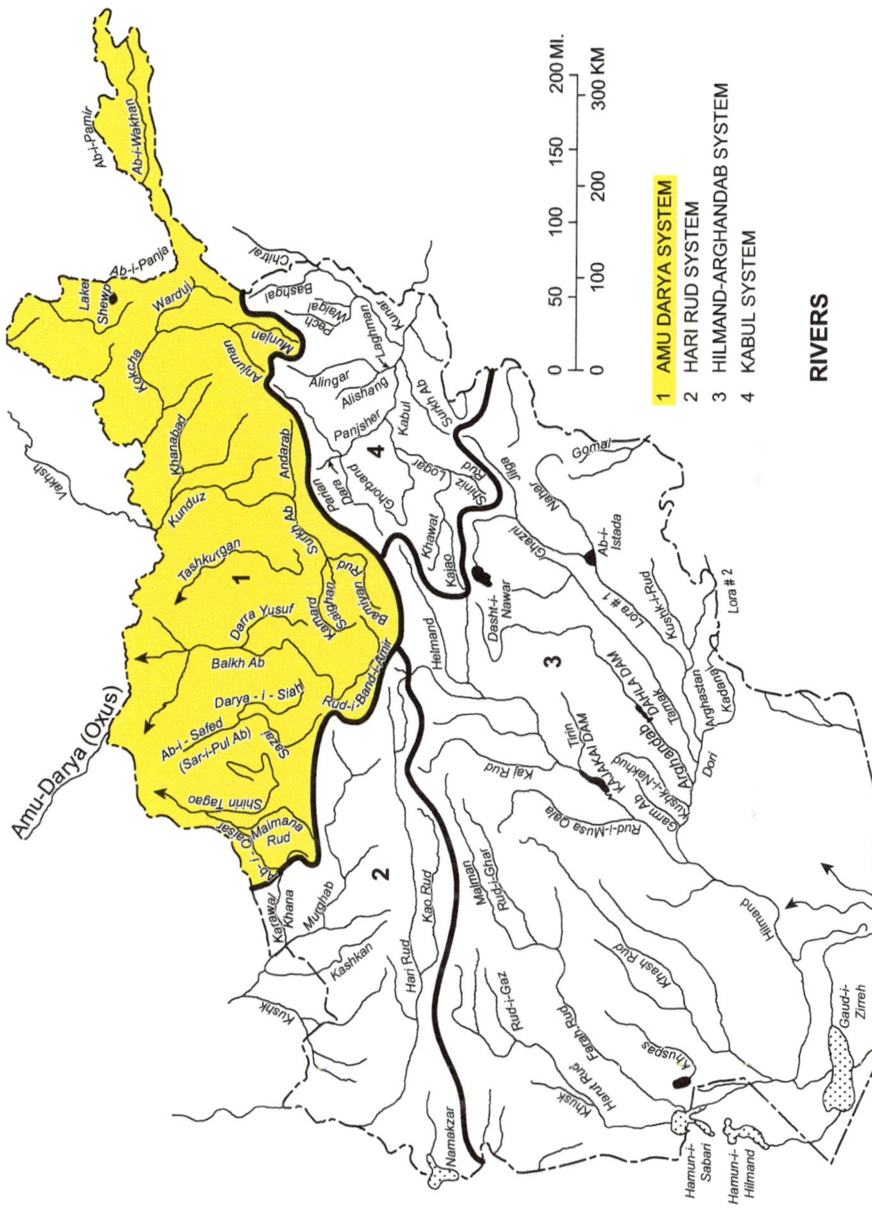

Figure 5.12 Map of Amu Darya River system (#1). *Map after Dupree (1973) and Clausen (1987).*

(Figure 5.13). The waters of the Hari Rud are mainly sourced in central Afghanistan and flow in a nearly straight westerly direction down a fault-guided valley trace toward the Iranian border, before looping north to form Afghanistan's far northwest border with Iran, and then between Iran and Turkmenistan, before its last remnants soak into the sands of the Karakumsky Desert. The Murghab River in this basin also flows out of Afghanistan into the Karakumsky Desert as well, at first in Afghanistan through a strong canyon whose narrow width does not allow much room for irrigation along its banks (Fipps, 2007; King and Sturtewagen, 2010; Clausen, 1987). The average annual discharge of the Hari Rud is about 55 m^3/s, but during a spring flood in 1939 the discharge went up to 1090 m^3/s (Westfall, 1969). The mean annual discharge of the Murghab was about 41 m^3/s but during a flood in 1886 a discharge rate of 367 m^3/s was measured (Westfall, 1969).

Helmand–Arghandab River System

This important river system drains a wide region extending from the Bamiyan region in central Afghanistan, southwest into the Seistan depression and the Iranian border to encompass about 43 percent of the total land area of the nation (Figures 5.14 and 5.15). The Arghandab River that flows through Kandahar, joins the Helmand near Lashkargah. Two major dams in this system were built in the 1950s; the Kajaki Dam on the Helmand about 70 km upstream from Girishk, and the Dahla Dam on the Arghandab 50 km northeast of Kandahar. Total flow in nondrought years was estimated at about 7 km^3 (Fipps, 2007), with mean discharges of 90–4000 m^3/s and maxima of 18,000–20,000 m^3/s (Clausen, 1987). The main branch of the Helmand in its delta in the Seistan depression forms the Iranian border where the waters drain off in *hamun* (*hamoon*) wetlands and lakes that are discussed further below. The location of the Khash Rud, shown in Figure 5.11 in full spring flow, is also portrayed in Figure 5.15 where it empties into the Seistan depression near Chakansur.

Kabul River System

Collectively, the Kabul River shown in Figure 5.4 rises in the mountains of central Afghanistan west of the city, whereas its tributary, the Panjshēr comes from the Hindu Kush to the northeast, and its other important feeder, the Kunar River enters Afghanistan from northwest Pakistan, make up about 12 percent of the land area of Afghanistan, with about 26 percent of the annual flow of the country (Figure 5.16). The river supports over 300,000 ha of intensively irrigated areas and high-value crops in Afghanistan. The river goes on to flow back into Pakistan just north of the Khyber Pass, where it supports some 50,000 ha of agricultural crops before its eventual confluence with the Indus River at Attock (Fipps, 2007; King and Sturtewagen, 2010; Clausen, 1987). The river has had discharges of 33–460 m^3/s, but in recent drought years has run nearly completely dry several times in the city of Kabul with its population of ~3 million people, which is a most serious shortfall. Because of mountain snow and glacier

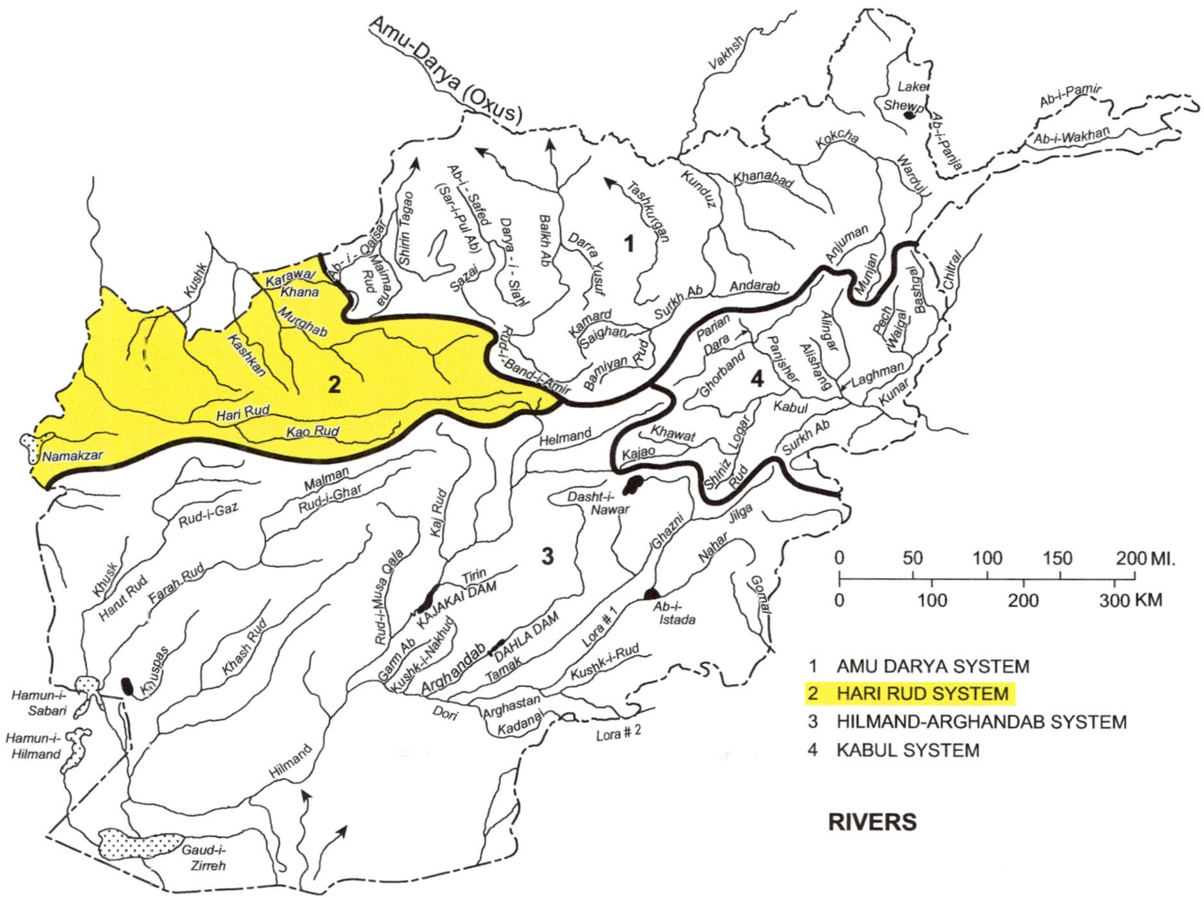

Figure 5.13 Map of Hari Rud–Murghab River system (#2). *Map after Dupree (1973) and Clausen (1987).*

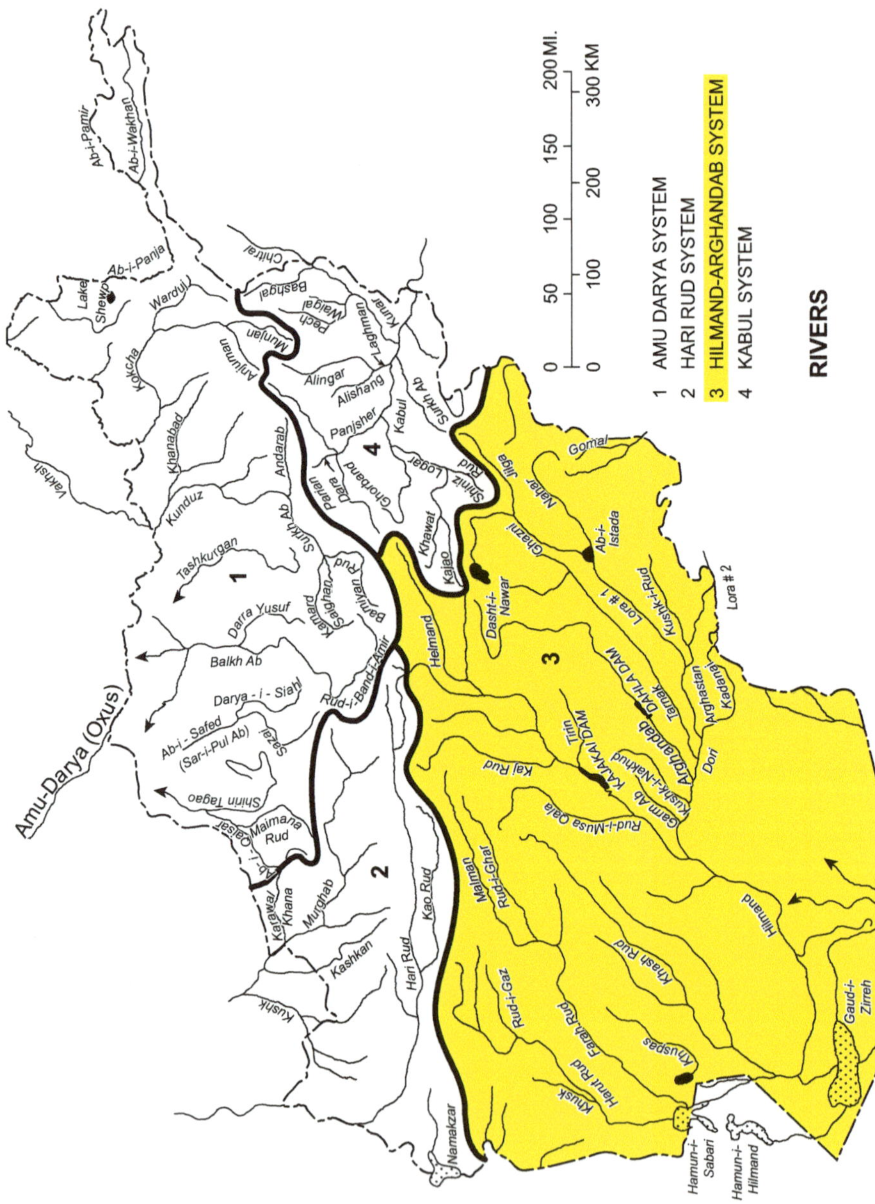

Figure 5.14 Map of the Helmand–Arghandab River system (#3). *Map after Dupree (1973) and Clausen (1987).*

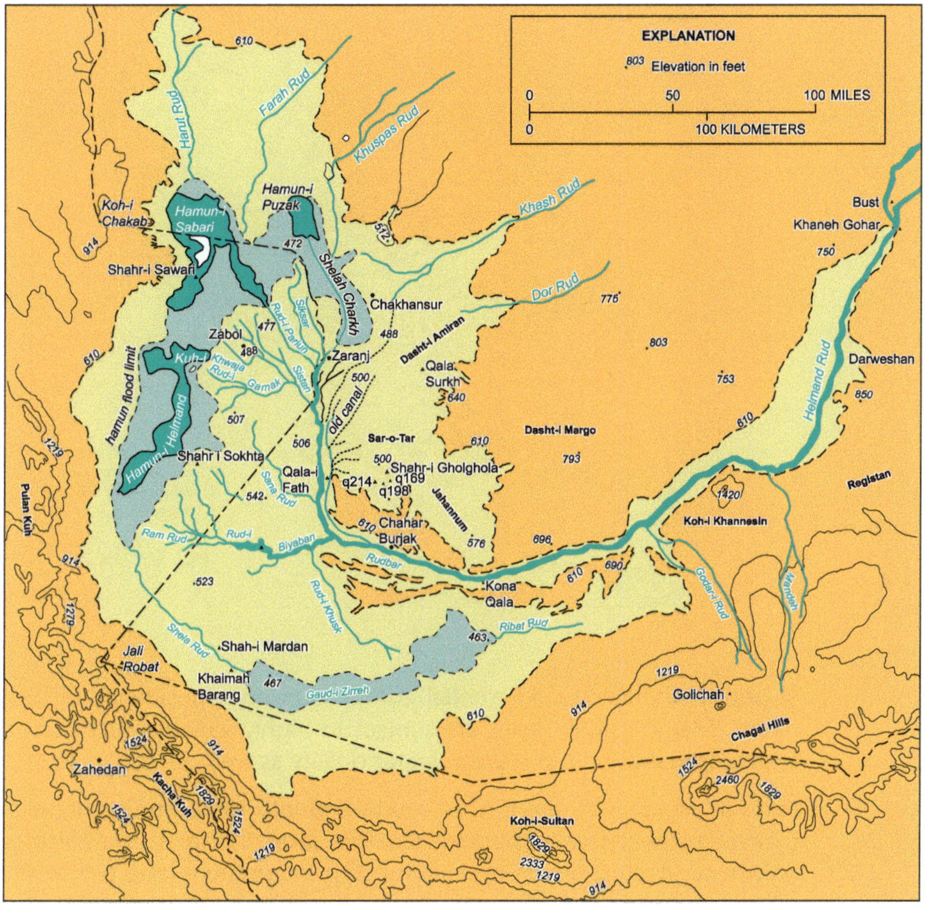

Figure 5.15 Map of the Helmand Delta region. *After Figure 2 in Whitney (2006).*

meltwaters, the Panjshēr and Kunar tributaries provide the bulk of flow in the lower Kabul where it can flow with a mean annual discharge of 700 m^3/s before it enters Pakistan (Westfall and Latkovich, 1966).

Lakes

Hundreds to thousands of lakes exist in mostly dry, arid Afghanistan, but few are large enough, of sufficient water quality, or close enough to population centers to be of direct use to the people of the country. The main causes of lakes in Afghanistan (in part after Mirzad, 1970) are: (1) supraglacial lakes and glacially eroded rock basins and moraine dams; (2) slope-failure dams; (3) carbonate-precipitate dams; (4) deflation basin *takyrs* (playas); (5) tectonic basins; (6) volcanic depressions and extrusive dams; (7) tectonic–fluvial depressions; and (8) tectonic-deflation depressions. A few may be some combination

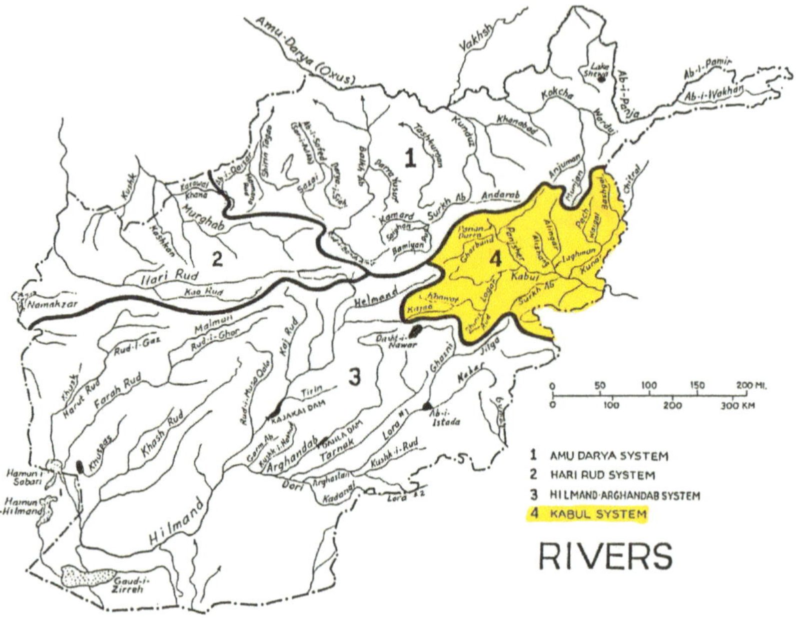

Figure 5.16 Map of the Kabul River system. *Map after Dupree (1973) and Clausen (1987).*

of poorly understood differential compaction and subsidence of sediments in basins. Many of these lakes are subject to highly variable fluctuations in water level seasonally and therefore in water quality as well.

Lakes in Afghanistan are further subdivided into three groups depending upon their permanence or impermanence. The lakes that exist all year around have only minor fluctuations of level and quality. Intermittent lakes (*hamuns*) are either full most of the time, or contain a considerable quantity of water at least once a year, and the quality is not excessively saline. The *takyrs* (known as *playas* in the West) or *namaksar* (salt flats) are dry most of the year, or for many years in a row, and are commonly highly saline. Gradations between types exist, depending upon amplitude and periodicity of the various climatic determinants.

Glacial Lakes

Lakes of glacial origin in Afghanistan number in the thousands and range from small ephemeral ponds to large bodies of water many square kilometers in size. They are most plentiful from high in the eastern Pamirs of the Wakhan (Figure 5.17) and the eastern Hindu Kush, to the western-most range of the Kohi Baba mountains south of Bamiyan. Two main types occur, the tarns in high glacial cirques, and the larger rock basin and moraine–dammed types in lower valleys. Tarns and other such cirque lakes in Afghanistan (James and Hubbick, 1969) tend to be fairly small, shallow (a few meters), neutral pH to only slightly alkaline (only a few milligram per liter of Na, Ca, SO_4, Cl, and

Figure 5.17 Glacial lake in Wakhan Pamir. *Photograph by Arthis, Google Earth™.*

Figure 5.18 Lake Chaqmatin Kol in the Wakhan Corridor. *Photograph by Arthis, Google Earth™. See Figure 3.8, for comparison.*

CO_3), cold (3–4 °C), and ultraoligotrophic (exceptionally few nutrients). Direct entries by glacial streams commonly deposit large quantities of silt and the lakes generally freeze to the bottom in winter. These factors indicate that these lakes tend to be rather sterile and unproductive of biomass. The larger rock basin and moraine-dammed types include such well-known large lakes as Zori Kol and Chaqmatin Kol in the far Wakhan Corridor (Figure 5.18).

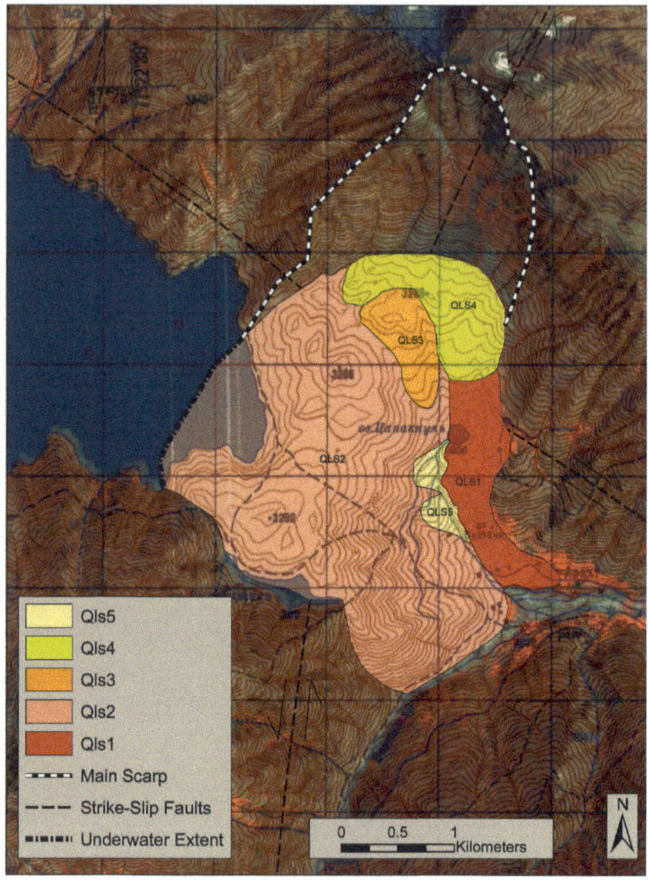

Figure 5.19 **Map of Lake Shewa landslide dam in eastern Badakhshan.** At least five different mass movement events have occurred at the location as a result of the crosscutting strike-slip faults that cross through the area.

Landslide-Dammed Lakes

A number of these occur in Afghanistan but the largest and best known is Lake Shewa in northern Badakhshan (Figure 5.19). A number of prior visitors (Lindberg, 1961; Sawata, 1962; Desio, 1975) thought that the dam was glacial moraine, but Ruleman et al. (2007) and Shroder and Weihs (2010) first recognized the slope-failure origin. Lake Shewa was dammed sometime in antiquity when a large rock avalanche (sturzstrom) from the fault-shattered and strongly weathered Archean gneisses of the Zirnokh peaks to the north moved down into the Arakht River valley. This rock avalanche dammed up the river and its tributaries to a thickness of ~400 m, producing a 12-km-long lake that is as much as 270 m deep, leaving ~80 m of freeboard to the top of the dam. At least four separate instances of slope failure have been mapped at the location of the emplaced landslide dam, as well as a rock glacier on its top, using remotely sensed data, historical maps, and Google Earth™. Spring seepage through the dam face has caused several recent subsidiary debris slides, which if continued at a large enough scale for long enough, or with additional seismicity from the active strike-slip faults that cross beneath the landslide dam, could threaten its integrity. Otherwise the clean water that emerges from the dam face could be the source of an unvarying mini-hydroelectric power source, in addition to the agricultural irrigation that it provides at the present time.

Carbonate-Precipitate Lakes

Numerous small and large, active or dry lakes and lake basins occur in the central mountains of Afghanistan where precipitation of carbonates across water courses has formed these self-damming features (Lapparent, 1966; Boutière, 1969; Lang, 1970; Bouyx and Pias, 1971; Albul et al., 1975a; Linder, 1979, 1982). The carbonate is derived from the extensive marine limestones of the central highlands where the rock becomes more easily dissolved in the cold ground waters there. Physical factors of formation of the travertine dams that cause the initiation of precipitation involve some apparently random agitation of flowing surface water that is supersaturated with dissolved bicarbonate ions, whereby as the partial pressure of the solution changes in the agitated water, a small portion of carbonate travertine is precipitated. This is followed thereafter by biochemically enhanced precipitation to build up wall-like, travertine masses such as the famous lakes of Bandi Amir (Figure 5.8), or Bandi Hajar and several other places. The fate of many of these lakes is to fail and drain, sometimes it is thought because of the general high seismicity, or in other cases with the help of man, which can leave behind only an irregular valley. In some cases, however, the lakes ultimately fill in with sediment and become agriculturally useful, flat-floored valley bottoms (Figure 5.7).

Diastrophic (Tectonic and Volcanic) Lakes

Basins that retain water as a result of faulting or warping occur in a number of places in Afghanistan. For example, the Sari Hawz lake in Faryab Province in northwestern Afghanistan is the headwaters of the Maimaneh River that flows northeast to Andkhoy. The lake was formed by right-lateral, strike-slip faulting that strikes approximately directly across the river so that higher ground on the downstream block was effectively drawn directly across the river from northwest to southeast in past seismic events to impound the drainage upstream on the southern block. Past seismic events have resulted in multiple outflow channels to the northeast that are cut into the northern block at successively declining altitudes of ~1835 m, ~1815 m, and the present-day channel at ~1714 m (Figure 5.20). The lake, now with human-made dam to reduce outflow, is of variable size, depending upon weather-controlled inflow so that it can range up to several kilometers to tenths of a kilometer long, depending upon the season of the year.

Other lakes such as Abi Istada 125 km south of Ghazni, occur in complex sagged or rifted ground along the left-lateral, Chaman strike-slip fault system (Figure 5.21). Abi Istada, or the "lake of standing water" was first noted by

Figure 5.20 Google Earth™ oblique view to the northeast showing the 0.5 × 0.4 km Sari Hawz lake in the center that is entrapped behind a right-lateral strike-slip fault that occurs at the base of the steep bluffs in the middle ground. At present the lake is also impounded by a man-made dam but in the past active faulting entrapped the original lake naturally. Several former channels cut through the uplifted block on the right to drain the Maimaneh River into the large open valley beyond. Each of the former spillways is several hundred meters lower from right (northeast) to left, which no doubt reflects offsetting by the faulting followed by additional downcutting. Area (N-E1) occurs in the Central Afghanistan mountains and plateaus geomorphic province.

Emperor Babur in 1504 CE to be quite shallow and possessed of a rich avifauna, whereas its inflowing streams were variously in flood after spring rains or quite dry at other times so that the lake surface varies greatly in extent over time as it is brimful or empty. The lake basin went completely dry in 2000. The lake was reported on by many British explorers in the nineteenth century and various other nationalities thereafter. Abi Istada (Abi Istadeh-ye Moqor) is roughly half-moon shaped and when full it measures some 11 × 15 km in size. It has no surface outlet and is known to freeze in winter. The water is highly alkaline, with the salt content varying by season, with the result that freshwater fish brought down by the Ghazni River inflow can be killed en masse at various times. The associated rich insect and plant life, as well as the algae, plankton, and crustaceans and mollusks enable a temporally variable but sometime profuse bird population of greater flamingos (*Phoenicopterus ruber roseus*), as well as Siberians cranes (*Grus leucogeranus*), and other migratory types.

Lang (1971) noted three sets of beach terraces around the lake, with one set 2–3 m above normal water levels, the second at 6–7 m up, and the third at 9–10 m above, at which point the lake probably drained into the Lora, which is a tributary to the Arghastan (Figure 5.2). On the contour maps the 1975-m contour corresponds approximately with the second terrace, which would

Figure 5.21 Vertical Google Earth™ view of Abi Istada lake 125 southwest of Ghazni city with 5 km scale bar at the bottom.

double the size of the lake, as it exists today. Two channels, the Akasi Mandh and the Sekva Mandeh lead out of the south end of the lake basin and into marshy areas that are the headwaters of the Lora. At the present time, only some groundwater connections between the lake and the Lora drainage are hypothesized.

The only lake of volcanic origin in Afghanistan is that of Dashti Nawar (Figures 5.2 and 5.22), an extinct caldera 55 km west of Ghazni whose last major eruptive cycle was about 2–2.2 million years ago when the top of the major peak exploded and sent a lahar (volcanic mudflow) downhill at least 340 km all the way to the Indus River in what is now Pakistan (Khan et al., 1985). The Bain diamictite, which outcrops there, is full of boulders of explosive volcanic pumice, as well as many other diverse lithologies picked up along the way by the liquid mass of wet mud and rocks, which is the remnant of the world's largest known lahar deposit at $30 \times 10^9 m^3$ in size. The deposit is some 15–20 m thick in its observed distal parts close to the Indus River, so the original deposit would have extended even farther east past the present Indus from where it now outcrops at that thickness. The 40 × 15-km-wide basin of Dashti Nawar left behind is now an elliptic-shaped depression rimmed by remnant peaks up to 4800 m in altitude that are the remnant stubs of a once much more lofty stratovolcano that may have resembled a mountain something like Kilimanjaro with glaciers on top when it was fully developed back in Plio-Pleistocene time before its final paroxysmal phases.

At the present time the Nawar plain and ephemeral brackish lake exist as an inverse remnant of the once gigantic mountain that now varies only as a lake in size from intermittent dryness in some winters to nearly 20 million m^3 of water during spring snowmelt. This is delivered by 6–10 ephemeral stream channels around the basin, along with three partially thermal springs, two in the south

Figure 5.22 Vertical Google Earth™ view of Dashti Nawar 57 km west of Ghazni city with 5 km scale bar at the top. The darkest colors at the lakes edge are peat, whereas the lighter colors are salts. The area occurs as a geomorphic section (A-B3) of the Helmand Block geomorphic province.

and one in the north. About 40 islands varying in size from 35–500 m^2 are situated at various places in the lake as well. As a basin with no outlet, various salts have accumulated at its bottom, particularly sodium carbonate that can be used in glass manufacture, as well as lithium salts that are the raw materials for modern battery production. Parts of the basin also hold peat deposits that have not yet been exploited as a fuel.

Tectonic–Mixed Origin (Fluvial Depression, Differential Compaction, Levee and Backswamp Configuration, and Subsidence)

Although back in antiquity, Lake Hashmat Khan was once much larger, it is now a small lake just southeast of the Bala Hissar citadel in Kabul that varies from 0.5 to 0.7 km in size up to perhaps 1 × 2 km when extra full (Figure 5.23). The origin of the depression, once on the outskirts, but now surrounded by Kabul City, is enigmatic in that it may be the result of some combination of tectonism, differential compaction, or even digging of mud for house building so far back in antiquity that its human origin is lost. Alternatively because the depression occurs against the mountain front about 3.5–4 m from the

Figure 5.23 Lake Hashmat Khan at the edges of Kabul City. *Photograph from Google Earth™ by Shuhada Salaheen.*

Kabul River, its origin may also be some cause related to natural levee construction and groundwater elevation in a backswamp location. The water level was always quite shallow, not more than 1.5 m deep, and reeds and waterfowl have abounded in the past. Until the first half of the twentieth century the lake was much bigger than the present, and was used to irrigate agricultural land in the Logar Valley. In fact at that time, water used to cover what is now the Karte Nau and Chaman Uzari sections of Kabul, where now only houses abound. Now, however, a combination of dry conditions and the diversion of water for irrigation, water table drawdown through household use, and urban encroachment have largely dried up this wetland.

Deflation Basin Lakes

The effects of wind erosion to produce deflation basins of varying sizes are most common in Afghanistan, and where the groundwater table or river inflow rise and fall periodically, an intermittent lake or later dry playa can result. Deflation basins or blowouts commonly have angular erosional remnant hills or yardangs associated with them, which provide other evidence of such aeolian activity. Also because the wind is most effective in very dry areas, and it cannot erode below the water table because the water there holds the particles together, with the result that dry lake playas commonly have a groundwater table of salty brine directly beneath the surface. This water table can rise during wetter times of the year to flood the surface and produce ephemeral lakes. Such deflation basin playas (*takyr* in central Asian languages or *namaksar* if salt-covered) occur by the thousands in the more easily erodible, basin-fill sediments that occur all around to the north, west, and south of the central highlands of Afghanistan. Small deflation basins tend to be linear and oriented with the prevailing or dominant winds; Namaksar Taskurghan is such a feature. Such larger basins may be either aligned with the wind, or transverse to it, and may have their location controlled by eddy currents tens of kilometers upwind in front of mountain massifs, as Smith (1973) thought of Namaksar Andkhoy in the northwest of Afghanistan, and Gaudi Zirreh in the extreme southwest and the distal end of the Helmand River delta and *hamun* ephemeral lake system.

Namaksar Andkhoy (Khwaja Mod)

This playa occurs ~38 km south of Andkhoy and ~60 km west of Shiberghan. It is located in a large closed depression that is ~12 km long normal to the prevailing westerly winds and ~4 km wide, with white salt layers prominent on its surface (Figure 5.24(A,B)). The rim of the deflation basin is 150–250 m above the surface of the salt, which itself is at an altitude of about 257 m above sea level (asl). The basin floor consists of six broad depressions elongated parallel to the strong winds that formed the basin, and the depressions are divided by swampy areas or meter-high ridges. The two middlemost salt pans are connected, and the third to the north is separate. The depression was deflated out by the wind and redeposited downwind toward Shiberghan as a loess blanket that is perhaps at least 250 m thick. The typically compact internal structure of loess allows it to stand as steep bluffs and complexly dissected hills with steep slopes where it is eroded by rainfall and snowmelt after its original deposition by fallout from the wind. When a United Nations team visited the area of Namaksar Andkhoy in the mid-1970s they became lost in a dust storm that cut visibility down to 20 m (Neilson and Gannon, 1976). Such storms both erode and deposit the loessic dust of the region, as well as importing the salts deflated also from the other takyrs of the region, with the result that the groundwater here tends to be excessively salty.

Namaksar Tashkurghan (Sari Namak)

This playa is about 28 km north-northwest of Tashkurghan (Kohlm) in northern Afghanistan and is about 10 km long and 3 km wide. It has an irregular or blotchy appearance on the satellite imagery, which is likely the result of variable substrate from a combination of fluvial, lacustrine, and aeolian deposition of clastic sediments. Aeolian barchan dunes occur on both sides. When Smith (1973) visited the area in October 1971, the groundwater brine was 60 cm below the surface and from it about 1100 metric tons of salts were produced annually. Most of this was NaCl but considerable Mg and SO_4 were also present, and the presence of this unpalatable epsomite ($MgSO_4.7H_2O$) and related salts make it useful only to help control intestinal parasites in sheep.

Namaksar Herat (Koli Namaksar)

This playa (Figure 5.25) occurs on the border with Iran just 145 km west-southwest of Herat at an altitude of about 600 m asl, with the lowest point

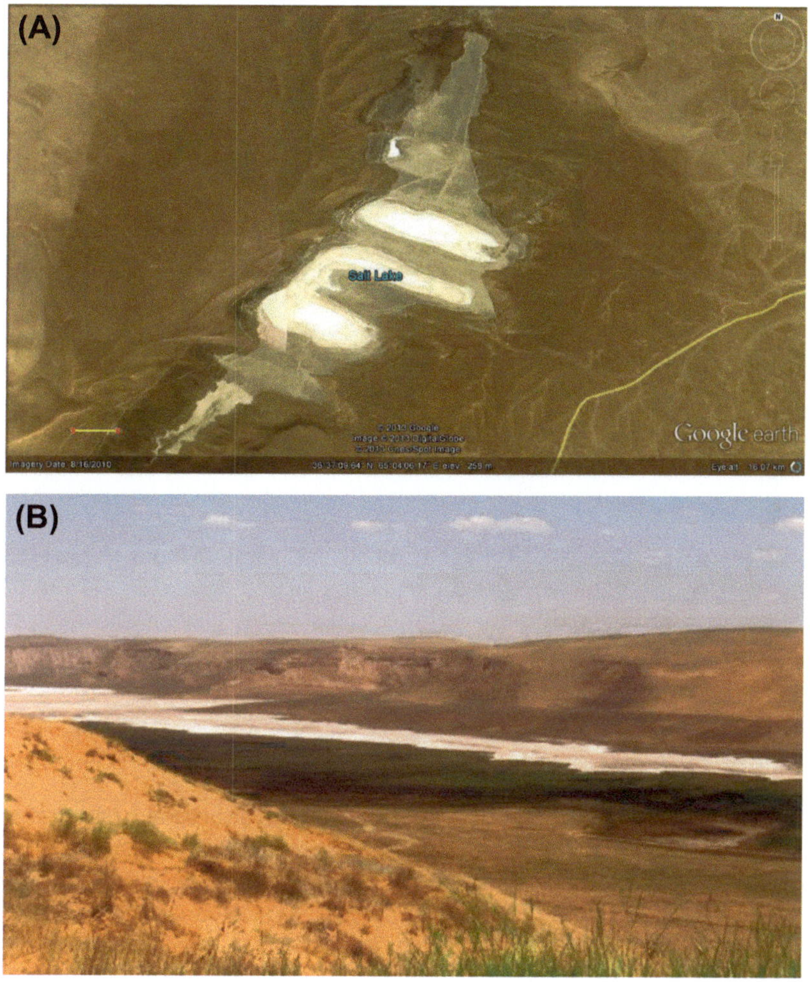

Figure 5.24 (A) Namaksar Andkhoy dry lake from vertically overhead. *(Photograph from Google Earth™).* (B) Namaksar Andkhoy dry lake. *(Photograph from Google Earth™ by Zmurik79).* The area occurs as part of the Shiberghan–Mazar Alluvial Fan Geomorphic Section (N-C3) of the Tajik Basin geomorphic province.

(~594 m asl) in Afghanistan where the most valuable lithium and other salts accumulate in a basin about 25 km long eastwest and 10 km wide northsouth. The closed desert basin itself that contains the playa appears to have resulted from tectonic down faulting and warping (Smith, 1973). Several higher lake levels occurred in Quaternary time as old lacustrine deposits have been uplifted and folded to dip 5°–25°, but whether the origin of this is tectonics or salt related is unknown. The youngest deposits 5–10 km east of the present playa surface are flat lying, up to 10 m thick and seem to record original lake levels that were up to 10-m deep. Smith (1973) interpreted gravel beach bars present there as late Pleistocene in age.

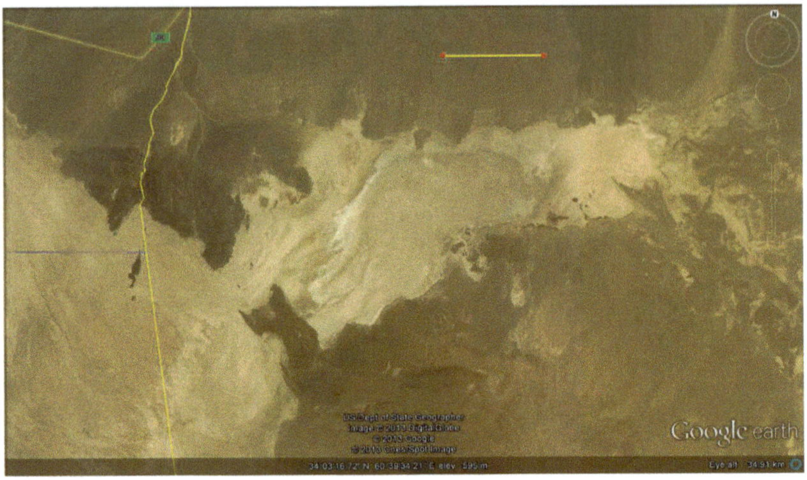

Figure 5.25 Namaksar Herat with 5 km scale bar. The yellow line on the left is the border with Iran and the playa over the border there drains east into Namaksar Herat. This area is part of the Dashti Alishan Geomorphic Section (A-A1) of the Farah Block Basin and Range Province.

The playa surface is mostly tan silt and the area covered by salt is about 30 km^2. Water can flood part of the lake to depths of ~1 m during winter but evaporates by midsummer. The salt crust of the playa is known to crack into polygonal forms with small ridges 1–5 cm high. Slabs of salt about 15 cm thick are mined from the surface, with a brine depth at 15 cm (Smith, 1973). The salt is mostly NaCl, but moderately large Mg, SO_4, and K are also reported, as well as Li in recent years.

Lakes of the Seistan

The Seistan depression is a broad structural basin that is several hundred kilometers in size in extreme southwestern Afghanistan. The basin is surrounded on all sides by mountains, with the lowest point on the rim of the basin at the Hamuni Lora at the east end of the Chagai Hills on the border with Baluchistan, Pakistan, where an ancient outlet to the sea may have occurred many millions of years before. The lakes of the Seistan are filled and shaped primarily by the waters and the sediment of the Helmand River (Figures 5.3 and 5.15), but the most salient point is that they are but diminutive remnants of a much larger lacustrine system that prevailed in the area for much of the last several million years of the Quaternary time period (Figure 5.26). Although a gross oversimplification of what was undoubtedly a complex and ever-changing series of waxing and waning lakes, at least three major lake systems are thought to have occurred here. Each of the three systems had numerous different water levels within its own time span, but because a comprehensive study of the lake histories has yet to be performed, these three lacustrine systems from oldest to the modern, are herein called Lakes Seistan I, II, and III. This avoids cumbersome special nomenclature for the lakes, the spatial and temporal extents of which are known only in a reconnaissance fashion (Huntington, 1905a,b; Tate, 1910–1912; Lang, 1971; Pias, 1972; Anderson, 1974; Smith, 1974; Whitney, 2006).

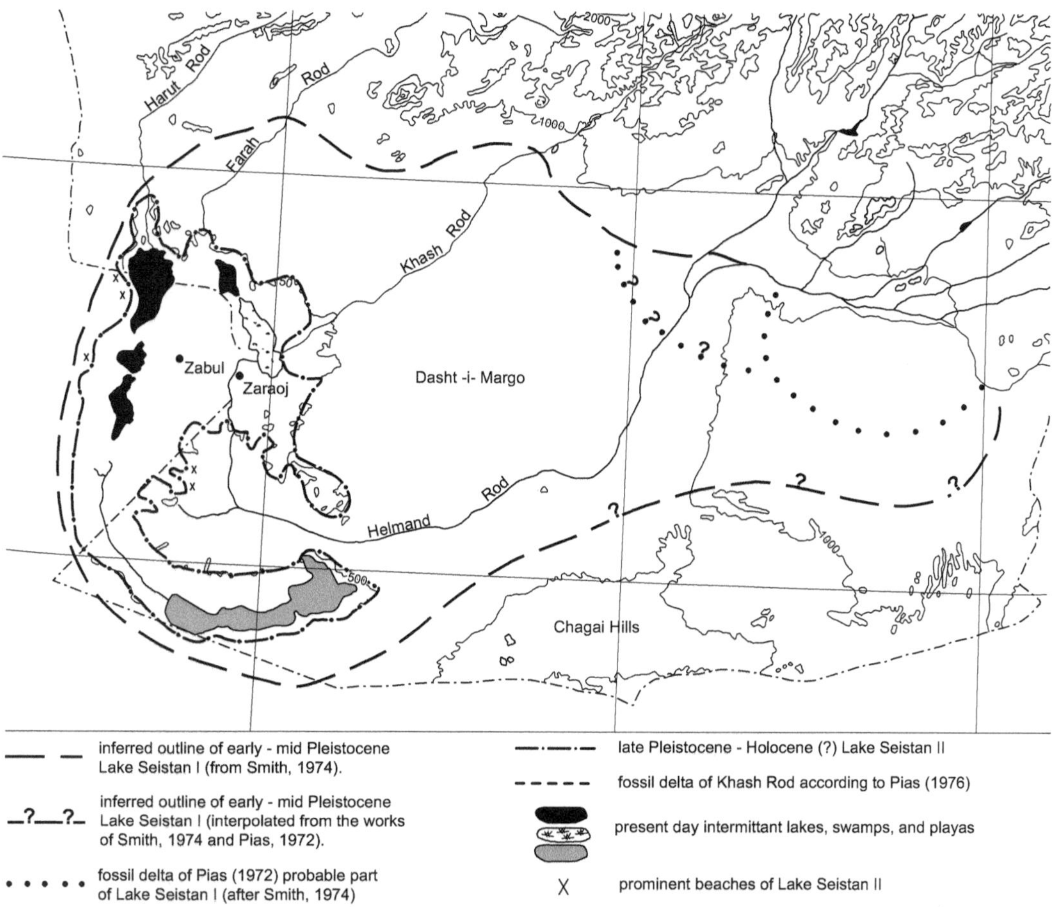

Figure 5.26 Outline map of possible prior lacustrine limits in the Seistan depression of southwest Afghanistan. The large size of the lake toward the east is unlikely but has been proposed in the past.

Lake Seistan I was hypothesized to occupy a large part of the Seistan depression during early to middle Pleistocene time, although Whitney (2006) disputed this large size proposed by Smith (1974). A hundred meters or more of flat-lying, fine-grained clays, silts, and sands with uniformly flat bedding, laminar layering, fine clastic sizes, disseminated calcitic and gypsiferous content, and well-sorted nature, with a few channel sands would seem to indicate some shallow lake context like this. The most common tan to grayish yellow of the units was thought to be characteristic of sedimentation into intermittent or permanent lakes that allowed oxidation of the bottom sediments, whereas the rarer greenish gray colors are the product of oxygen-deficient bottom conditions. To the northeast towards Kandahar the lacustrine facies grade into an alluvial facies system that seems to represent the original delta of the Paleo-Helmand–Argandab River system during that time.

The paleohydrology of Lake Seistan I (Smith, 1974) indicates that an inflow at least 13 times greater than modern averages would have been required to have maintained a lake of the size retrodicted here. The present-day evaporation rate of ~2 m per year would have been much less then, but even if reduced by half, and inflow of 7–8 times higher is still required. This implies a dramatically increased precipitation in the central Hindu Kush headwaters, which is commensurate with the far larger areas of glaciation present there that occurred at that time.

If the inferred boundaries of the earliest lake are even marginally correct, an approximate 300 m difference in altitude between the sediments to the north and south compared to those in the center and west of the basin must be accounted for in some fashion. This disparity can be reconciled by the eruption in Middle Pleistocene of a number of basalt volcanoes at Chakab in extreme western Afghanistan as well as eastern Iran at the west edge of the lake where magma withdrawal at depth and overlying subsidence might account for the altitude differences.

Directly overlying the fine-grained clastics of the first lake is a vast sheet, 2–10 m thick of alluvial gravels, washed down out of the mountains to the east in a series of debris flows that emplaced well-rounded cobbles in a sandy matrix, which became the present surface of the Dashti Margo (Desert of Death; Figure 5.27). Downwarping and deflation erosion of this surface seems to have produced the depression into which Lake Seistan II developed in late Pleistocene and Holocene

Figure 5.27 Dashti Margo (Desert of Death) in southwest Afghanistan showing extensive stony pavement caused largely by deflation of fine clastics (clay, silt, and sand) by wind, leaving gravel lag behind. Note round camera lens cap for scale in lower center left.

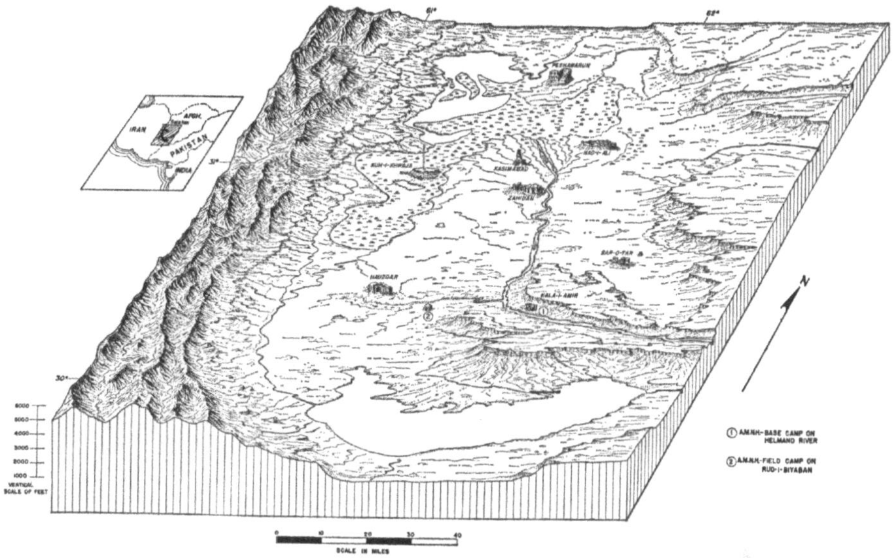

Figure 5.28 Relief diagram of topography surrounding lake basins of Seistan. The pronounced relief of the Helmand delta into Lake Seistan II is obvious in the lower center at the circled 1. *After Fairservis (1961).*

time (Figure 5.26). Into this lake poured a large delta of sediment from the Helmand River that measured some 75–80 km in diameter that filled in the center of the depression to leave a distorted horseshoe or U-shape of deeper basins beyond and around it on both sides (Figure 5.28). Prominent beaches, bars, and spits can be observed as remnants in many places in both Afghanistan and Iran around the delta, as well as the ancient shorelines beyond. In the middle to late Holocene, this lake attracted profuse human settlement, the archaeological remnants of which abound around the edges of the ancient water body.

Continual annual deflation of the Seistan (Figure 5.29) is impressive, with nearly the whole of Lake Seistan II surrounded by steep bluffs up to 100 m above the present floor (Figure 5.28) and composed of the older deposits of Lake Seistan II (Anderson, 1973). The overall Seistan Basin surrounding the present Helmand delta is about 28,000 km^2, or about one-third bigger than the well-known Qattara depression of Egypt, which is the world's largest known deflation basin. The wind intensity of the Seistan is legendary and the *badi sado bist roz*, or wind of 120 days that flows into the lows of the monsoon off to the east, can sustain wind gusts at 100 knots. Tate (1909) noted sustained velocities of 88 mph, with gusts up to 120 mph (141–192 kmph). It is an intense, hot summer blast that mobilizes much sediment. Frontal storms can occur in winter as well that produce plentiful sand and dust storms (Figure 5.30).

The high velocity winds of this environment move a great deal of sediment and produce numerous erosional remnants or yardangs up to 25 m high. Gaudi Zirreh, the final lowest sink of the Helmand in the extreme southwest corner of Afghanistan (Figures 5.3 and 5.15), has had some 200 km^3 of sediment

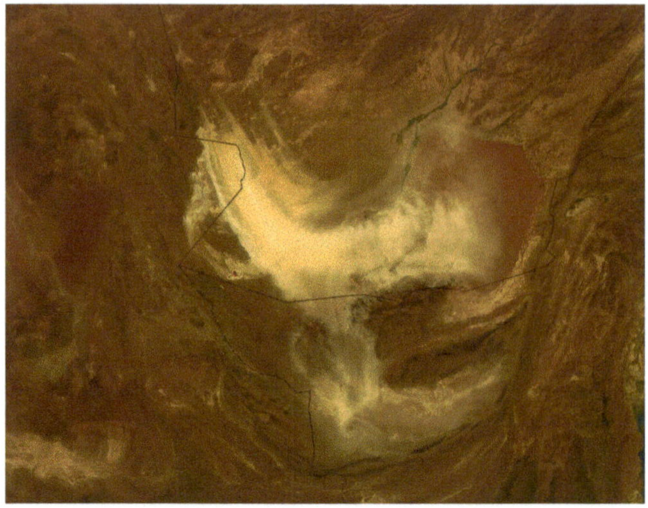

Figure 5.29 Modis weather satellite scene of June 14, 2004 showing deflation effects from strong winds during severe drought.

Figure 5.30 Photograph by the author of the leading edge of a frontal westerly, sand, and dust storm coming over the Dashti Margo near Dilaram, southwestern Afghanistan in spring 1978.

deflated from it to produce the lowest basin in the region (Smith, 1974). The downwind result of all the deflation is a profusion of dunes that have streamed out from the entrained Helmand River delta sediment in a retrograde fashion mainly because the sediment has been moved in part back upstream. Three main sand dune regions occur here: (1) barchanoid types blown relatively recently from the right bank of the Helmand delta into the Jahannum (Hell) depression (Figures 5.15 and 5.31) and back into the Helmand upstream above Chahar Burjak; (2) longitudinal dunes blown from Gaudi Zirreh east between the Kohi Khaneshin volcano and the Chagai Hills; and (3) the great transverse dunes of Registan (Figure 5.32) where the sands from the Helmand delta had first wrapped around both sides of the Chagai Hills and made their way 400 km downwind to this location where many dunes have later become marginally

Figure 5.31
Sand Dunes in the Jahannum depression north of the Helmand delta that were deflated from the bottom of Lake Seistan II.
Photograph from Whitney (2006).

Figure 5.32
Photograph of dunes of Registan taken from desert location south of Lashkargah by the author in spring 1978.

stabilized by shrubby vegetation. The finer grained dust from all this sand movement has moved even farther afield into the loessic blankets that occur in many valleys farther northeast in Afghanistan and east into Pakistan.

As the original Lake Seistan II continued to fluctuate, but overall diminish down to the nineteenth, twentieth, and twenty-first centuries, the present series of *hamun* (intermittent lakes and playas) developed, with a complex of distributary channels on the Helmand delta that flowed, dried up, or were diverted by humans at various times. The normal sequence of water infilling of the hamuns by the Helmand River in the Seistan Basin is arranged in counterclockwise order first with the Hamuni Puzak in Afghanistan, next the small connecting wetlands of Chone Sorkh and Baringak in Iran that allow flow into the Hamuni Saberi that straddles the border, then the large Hamun–Hirmand Basin in Iran to the south, and then the Sheila Rod ephemeral water course that in years of exceptional flow is the overflow channel back into Gaudi Zirreh in Afghanistan (Figures 5.3 and 5.15). Over the past 28 years since 1985, the sequence of filling and desiccation events have been: (1) low water 1985–1988; (2) high water 1989–1993; (3) medium water 1994–1999; (4) dry 2000–2004; (5) mostly dry 2004–2013 (Vekerdy and Dost, 2005).

The Kamal Khan Flood Control Project in Afghanistan was conceived originally to divert water from the lower Helmand through a new canal and directly into Gaudi Zirreh without the excess water going into the hamuns as usual during wet years to relieve flooding. The idea was a combination of flood control as well as new irrigation in both Afghanistan and Iran but the excess salts in Gaudi Zirreh make it doubtful that such a reservoir would provide much usable irrigation (Vekerdy and Dost, 2006).

Underground Water

Shallow groundwater at the water table in Afghanistan has been exploited traditionally by shallow, hand-dug wells and long, gently sloping karez tunnels for many thousands of years (Figure 5.33). More recently drilled wells have been being used, although in many cases local overexploitation, especially with diesel pumps, the water table can be drawn down enough that the water tables

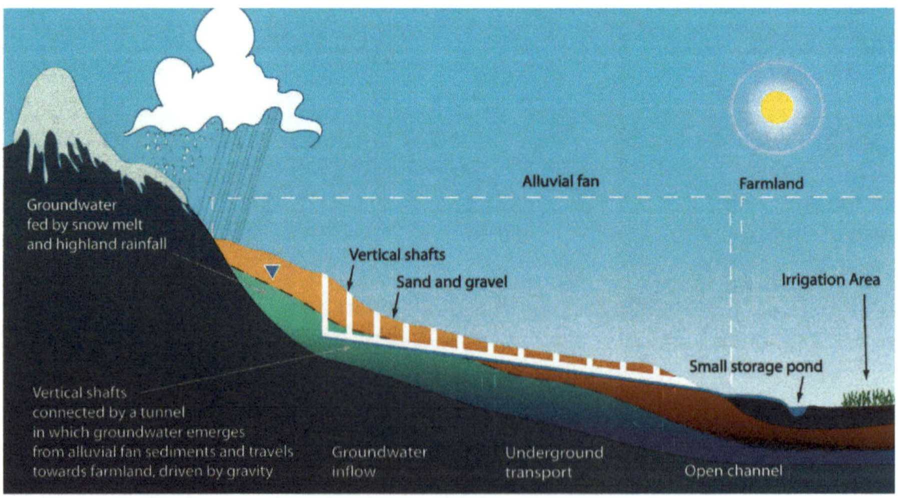

Figure 5.33 Diagram of typical karez irrigation systems that abound in Afghanistan. *By permission from Earth magazine.*

drop and the karez dry up. Deep-drilled wells have become more common only in a few valleys, such as the Ghazni, Kabul, Logar, and Tarnak areas. Artesian waters from groundwater aquifers buried deeply enough to have beds of water under pressure that will rise some distance up in the well have been known about and exploited only in a limited number of cases.

Annual groundwater recharge in normal years is about 18 bcm, whereas total groundwater extraction is some 3.7 bcm, which indicates that a good deal more can be extracted (Azimi and McCauley, 2004). Nonetheless, in spite of the groundwater being underutilized in some places in the country, it is overused in a few places, such as beneath several of the cities, especially Kabul (Uhl and Tahir, 2003; Mack et al., 2009). Groundwater drawdown (Uhl and Tahir, 2003) has been a matter of considerable discussion because of water table declines of 4–6 m and a few to ten meters in the Kabul Basin, 5–8 m in Kandahar, 2–4 m in the Herat Basin, and 2 m in Zabul, with many wells going dry. Natural and artificial recharge of aquifers in these urban areas is only in the infancy of discussion and analysis seeking solutions.

The result has been that the United Nations and other agencies that work in Afghanistan have become somewhat cautious about groundwater exploitation, given the well understood proclivities of far too many poor Afghans to overtax whatever resource that they can, once they learn how to do it. Norwegian Church Aid Afghanistan (Kirkens Nødhjelp), one of many nongovernmental organizations that work in Afghanistan, for example, have developed a comprehensive plan for sustainable use of groundwater in the country (Banks, 2002). Already, however, the water table beneath Kabul City has undergone considerable exploitation and drawdown, so much so that discussions concerning new canals or piping water from the Panjshēr River tributary have occurred as a possible future groundwater recharge or direct flow measure to augment existing supplies.

In spite of some attempts in the 1970s by Soviet geologists to better understand the groundwater of Afghanistan from its basic bedrock (Albul et al., 1975b; Shareq et al., 1977) (Figure 5.34), the country has never had a dedicated and comprehensive groundwater survey of a modern kind, with test drilling, pump tests, and other techniques to more robustly define the quantity, quality, and availability of the resource using modern techniques. The map of Shareq et al. (1977) has basic bedrock aquifer types, sparse water table contours, and limited water chemistry and flow rates. An example of a modern study was that done recently by the US Geological Survey in the Kabul Basin (Mack et al., 2009), with a full hydrogeological assessment (Figure 5.35). Additional such studies in many other areas of the country were scheduled but had to be canceled because of the considerable danger from the Taliban insurgency against the hydrogeologists who customarily undertake a good deal of exposed field work in order to obtain the requisite information.

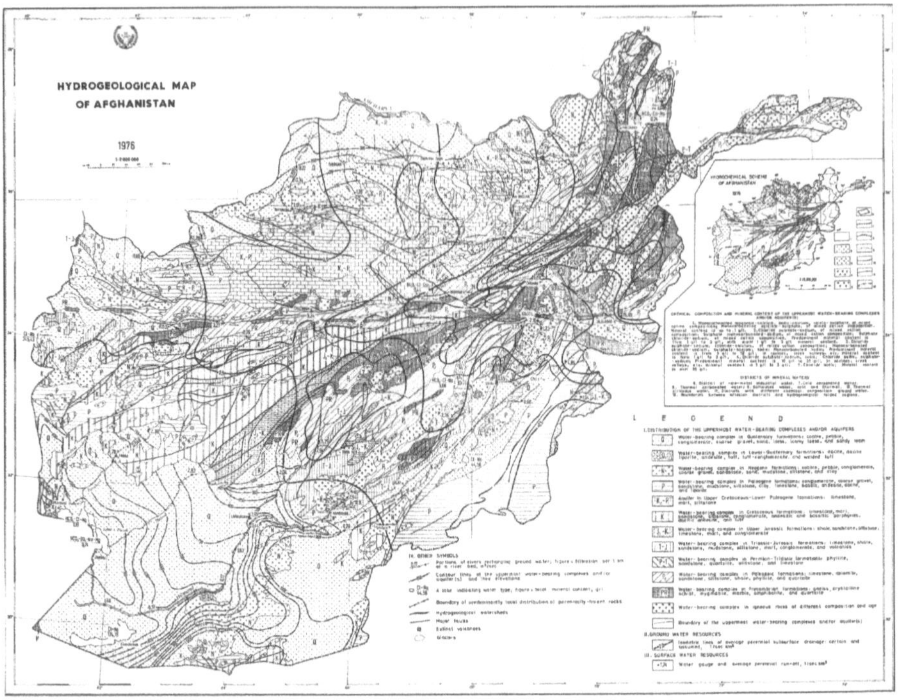

Figure 5.34 Hydrogeological map of Afghanistan done largely by Soviet hydrogeologists and first published by Shareq et al. (1977).

Malyarov et al. (1977a,b) (although often cited as Malyarov and Chmyriov [1976] internally with no other details of the reference) did the original hydrogeological zoning of Afghanistan that was first published in English in Shareq et al. (1977) (Figure 5.34; Table 5.3). This work of subdivision of the hydrogeology of Afghanistan was done by the Soviet hydrogeological community that worked in Afghanistan once they had the basic bedrock geology mapped out fairly thoroughly so that they could do the general overview required. The hydrogeological subdivisions of Afghanistan were based upon geological–structural analysis and were established as conceptual tools primarily upon designation of: (1) artesian-water basins with predominantly layered water aquifers; (2) hydrogeological massifs of crystalline rock with water mainly restricted to fissures and fractures in the bedrock; (3) subartesian basins with both layered and fissure water that are generally restricted to the large troughs that parallel the other aquifers. The hydrogeological massifs are large-sized zones of crystalline rock that are predominantly igneous, metamorphic and strongly metamorphosed sedimentary rock so that the predominant groundwater is only in joints, faults, foliation planes, and other fissures in the otherwise rather impermeable rock. The subartesian basins are those in

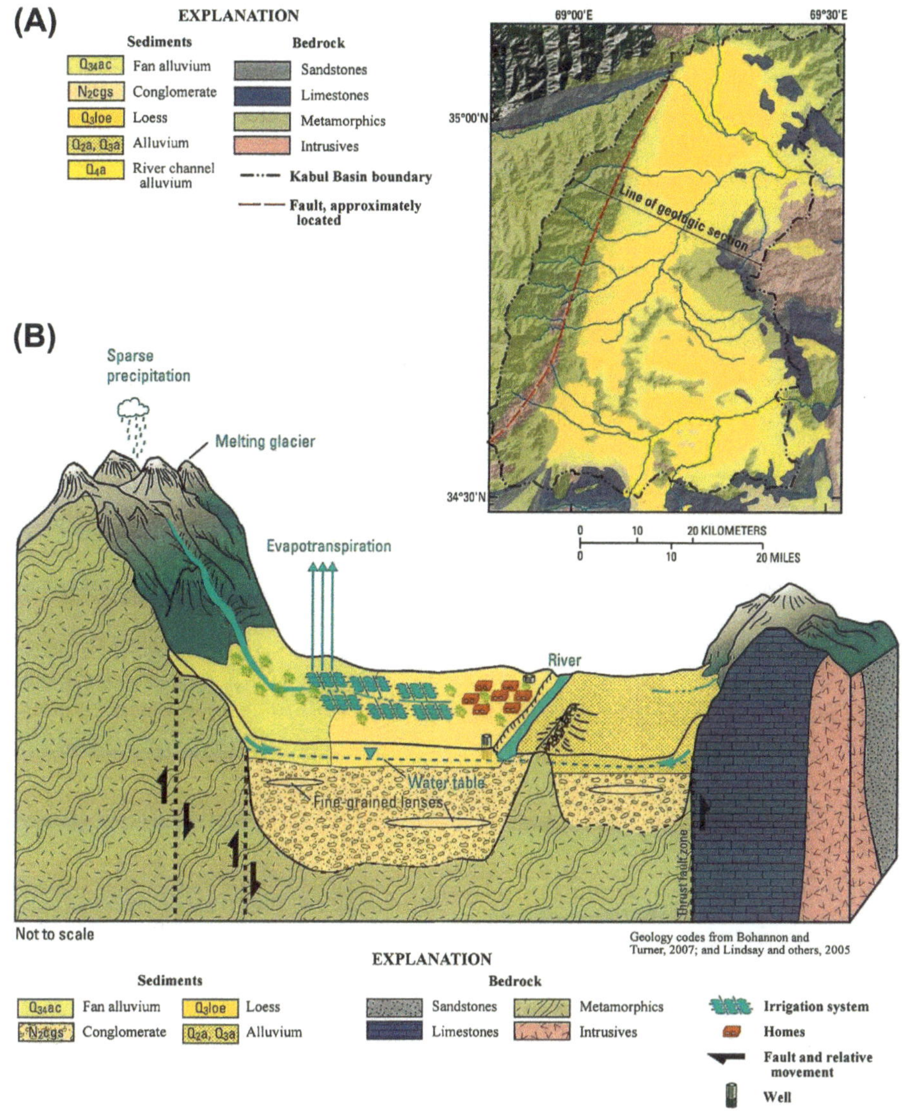

Figure 5.35 (A) Map and (B) Cross section of bedrock and sediments in Kabul Basin showing modern conceptions of water table in the basin. *Diagram from Mack et al. (2009).*

which the groundwater does not have enough artesian pressure head to rise above the aquifer that contains it.

In each of the three main regions discussed, a listing of the main aquifers was enumerated, described, and defined in some detail. In some cases detailed water information was described where sufficient detail was possible to obtain the thickness of different aquifers, along with the depth to the water table, the

Table 5.3 Scheme of hydrogeological zonation of Afghanistan. This was based on the structural and stratigraphic-lithologic fundamentals of geological outcrops and presented in small-scale (~1:10.5 million) map form

I. North Afghanistan artesian area					
Artesian Basins					
I_1 Amu Darya		I_2 Kunduz	I_3 Kulab–Kokcha	I_4 Sheberghan	I_6 Kushka
Basins with Fracture-Karst Water					
I_7 Murghab		I_8 Maymana	I_9 Shashan		
Hydrogeological Massifs					
I_5 Bande–Turkestan (dissected)		I_{10} Surkhab (strongly dissected)			
II. Central Afghanistan hydrogeological folded area					
Hydrogeological Massifs					
II_1 Hindu Kush—Badakshan (partly cryogenic, strongly dissected)		II_2 Hari Rud (dissected)	II_3 Upper Helmand—Sarobi (dissected)	II_{11} Arghandab (dissected)	
Basins with Fracture-Karst Water					
II_6 Middle Helmand		II_7 Tyrin			
Subartesian Basins					
II_5 Harutrod		II_{10} Uruzgan			
Basins in Superimposed Formations					
II_4 Adreskan		II_8 Dashti Nawar	II_9 Farah Rod		
III. Southern Afghanistan artesian area					
Artesian Basins					
III_1 Katawaz		III_2 Lower Helmand	III_3 Registan		
Basins with Fracture-Karst Water					
III_4 Tarnak		III_5 Chaghay			
Basins in Superimposed Troughs					
A. Kakhan	B. Wakhan	C. Balkh	D. Baghlan	E. Herat	F. Middle Afghanistan
G. Kabul	H. Jalalabad	I. Aynak	J. Sehdaraht	K. Dakki-Tundy	L. Abi Istada

After Malyrov et al. (1977a,b) and Shareq et al. (1977).

specific discharge in liters per second (l/s) from drill holes, discharge (l/s) from wells, springs, and karezes, and mineral content in grams per liter.

The best of the principal aquifer systems of Afghanistan (Uhl and Tahir, 2003) include Quaternary and Neogene-aged sediments that occur along major river systems, and intermontane basins, both of which are the most prolific aquifers in the country. Most are coarser clastics deposited along mountain areas, commonly as porous and permeable alluvial fan materials that can produce excellent aquifers. The intermontane river basins and structural river basins of tectonic (trough) or erosional (valley) origin are infilled with alluvial, slope wash (proluvial), colluvial, and local glacial deposits that provide essential water. The main aquifers of these types include the Hari Rud, Tarnak Trough, Ghazni River valley, Logar Basin tributary to Kabul River, the Khost–Yaqubi Basin, Jalalabad Basin, and Parwan Basin with the Panjshēr River, as well as the Ghor and Bamiyan intermontane basins.

Carbonate rock systems with interbedded sandstone and conglomerate provide aquifers on northern and southern flanks of the Hindu Kush; upper Cretaceous in the north, Permian and Jurassic in the south. Consolidated (crystalline) bedrock aquifer systems in the central highlands have been little studied, as also is the case with the so-called Great Southern Plain, which includes the dunes of Registan, the Dashti Margo, and the Piedmont areas. The Northern Plain and Amu Darya Basin have freshwater aquifers close to the mountains but a good deal of salty water throughout the main lower altitude and some deeper parts of the basins.

References

Albul, S.P., Makhorin, A.A., Chmyrev, V.M., 1975a. Mineral'nyye uglekislyye vody Tsentral'nogo Afganistana. (Mineral carbon dioxide waters in central Afghanistan). Izvestiya Vysshikh Uchebnykh Zavedeniy. Geologiya i Razvedka 4, 103–107.

Albul, S.P., Makhorin, A.A., Nadzhibulla, Safdari, 1975b. Podzemnyye vody allyuvial'nykh otlozheniy rayona g. Kabula, (Ground water in alluvial deposits of the Kabul region). Izvestiya Vysshikh Uchebnykh Zavedeniy. Geologiya i Razvedka 1975 (11), 121–125.

Anderson, A.O., 1973. Pleistocene geology and archeological problems of the Seistan Basin, southwest Afghanistan. unpublished Master's thesis. University of New Mexico, Albuquerque, 105 pages.

Anderson, O.J., 1974. Pleistocene Geology and Archeological Problems of the Seistan Basin, Southwest Afghanistan (Masters thesis). University of New Mexico, Albuquerque, NM, USA, 105 p.

Azimi, A., McCauley, D., 2002. Afghanistan Environment in Transition. Asian Development Bank, South Asia Department. http://www.amazon.com/Afghanistans-Environment-Transition-Ali-Azimi/dp/9715614787.

Azimi, A., McCauley, D.S., 2004. Afghanistan's environment in transition. Asian Development Bank. 40 pages.

Banks, D., 2002. Guidelines for Sustainable Use of Ground Water in Afghanistan. Norwegian Church Aid—Afghanistan Program (NCAAP), 63 p. http://www.holymoor.co.uk/Poldok.PDF.

Beveridge, A.S., 1922. (Reprinted 1975) Babur-Nama (Memoirs of Babur), vols I & II, Sange-Meel Publications, Chowk Urdu Bazar, Lahore, Pakistan, 962 p.

Bishop, M.P., Shroder, J.F., Ghazanfar Ali, Bush, A.B.G., Haritashya, U., Rakshan Roohi, Sarikaya, M.A., Weihs, B.J., 2014. Remote sensing of glaciers in Afghanistan and Pakistan. Chapter 23 In: J.S., Kargel, G.J., Leonard, M.P., Bisop, A., Kääb, and B.H., Raup, (editors); Global Land Ice Measurements from Space. Praxis Publishing, Springer Verlag, 509–548.

Boutière, A., 1969. Formations récentes (volcanisme, travertins) observées dans la region du Dashte-Nawar (Afghanistan central). Etudes françaises sur le Quaternaire, Bulletin AFEQ, Supplement, VIII Congress International INQUA 1969, 241–245. 5 Abb., Paris.

Bouyx, E., Pias, J., 1971. Signification géologique, pédologique et paléoclimatique des travertins d'Awpar (valée de Ghandak, Afghanistan central).

Breckle, S.-W., Rafiqpoor, M.D., 2010. Field Guide Afghanistan: Flora and Vegetation. Scientia Bonnensis, Bonn, Germany, 862 p.

Clausen, J.L., 1987. Hydrogeology of Afghanistan. Independent Research in Geomorphology; Unpublished Senior Thesis in Geology, University of Nebraska at Omaha, 58 p.

Copland, L., Pope, S., Bishop, M., Shroder, J., Clendon, P., Bush, A., Kamp, U., 2009. Glacier velocities across the central Karakoram. Annals of Glaciology 50 (52), 41–49.

Copland, L., Sylvestre, T., Bishop, M.P., Shroder, J.F., Seong, Y.B., Owen, L.A., Bush, A., Kamp, U., 2012. Expanded and recently increased glacier surging in the Karakoram. Arctic, Antarctic, and Alpine Research 43 (4), 503–516. 2011. doi: http://dx.doi.org/10.1657/1938-4246-43.4.503.

Desio, A., 1975. Geology of Central Badakhshan (North-East Afghanistan) and surrounding countries. Italian expeditions to the Karakoram and Hindu Kush, scientific reports III. Geology–Petrology, E.J. Brill, Leiden, 628 p.

Dupree, I., 1973. Afghanistan. Princeton University Press, Princeton, NJ, USA, 760 p.

Fairservis, W.A., 1961. Archaeological studies in the Seistan Basin of southwestern Afghanistan and eastern Iran. 48(1). Anthroplogical papers of the American Museum of Natural History, NY.

FAO, 1997. Afghanistan Agriculture Strategy, Rome.

Favre, R., Kamal, G.M., 2004. Watershed Atlas of Afghanistan: 1st Edition—Working Document for Planners. FAO and Afghanistan Information Management Agency (AIMS).

Fipps, G., 2007. Appendix to the transboundary water policy of Afghanistan: transboundary water issues. Draft, 10 p. http://gfipps.tamu.edu/Afghanistan/Transboundary%20Water%20Issues26-04-07.pdf.

Haritashya, U.K., Bishop, M.P., Shroder, J.F., Bush, A.B.G., Bulley, H.N.N., 2009. Space-Based Assessment of Glacier Fluctuations in the Wakhan Pamir, Afghanistan. Climatic Change 94, 5–18.

Humlum, J., 1959. La géographie de l'Afghanistan. Etude d'un pays Aride. Scandinavian University Books, Copenhagen.

Huntington, E., 1905a. A geologic and physiographic reconnaissance in Central Turkestan. In: Pumpelly, R., Davis and, W.M., Huntington, E. (Eds.), Explorations in Turkestan, with an account of the basin of eastern Persia and Sistan, pp. 159–218. Expedition of 1903. Pages..

Huntington, E., 1905b. The basin of eastern Persia and Sistan. In: Pumpelly, R., Davis, W.M., Huntington, E. (Eds.), Explorations in Turkestan, with an account of the basin of eastern Persia and Sistan, pp. 219–317. Expedition of 1903. Pages..

ICARDA, USAID (International Center for Agricultural Research in the Dry Areas, US Agency for International Development), 2002. Needs assessment on soil and water in Afghanistan.

James, A., Hubbick, J., 1969. Studies on some high altitude lakes in the Hindu Kush. Verhandlungen des Internationalen Verein Limnologie 17, 362–372.

Jentsch, C., 1972. Afghanistan: Naturräumliche Möglichkeiten der Weiderwirschaffte. Map a scale of 1:3 500 000.

Khan, M.J., Opdyke, N.D., Shroder Jr, J.F., 1985. Bain Diamctite: Lithology, Age Origin. Geological Bulletin University of Peshawar, Pakistan, 18 (53–64).

King, M., Sturtewagen, B., 2010. Making the most of Afghanistan's river basins. EastWest Institute; Gerda Henkel Stiftung, 27 pages http://www.isn.ethz.ch/Digital-Library/Publications/Detail/?ots591=0c54e3b3-1e9c-be1e-2c24-a6a8c7060233&lng=en&id=113057.

Lang, H.D., 1971. Ueber das Jungtertiaer und Quartaer in Sued–Afghanistan. (The upper tertiary and quaternary in southern Afghanistan). Geologisches Jahrbuch, Beihefte 96, 167–201.

Lang, J., 1970. Contribution à des lacs de Band-e-Amir (Afghanistan central). Bulletin Societie géologique de France XII (5), 834–842 (7).

Lapparent, A.F. de, 1966. Les depots de travertins des montagnes Afghanes a l'ouest de Kabul. Revue de Géographie Physique et de Géologie Dynamique VIII (5), 351–357 (2).

Linder, L., 1979. Zarys stratygrafii mlodszego czwartorzedu srodkowego Afganistu w swietle badan grobli trawertynowych jezior zaporowych w dolinie Bande Amir. (Stratigraphy of the upper Quaternary of central Afghanistan; studies on travertine dams, Bande Amir Valley). Przeglad Geologiczny 27 (5), 286–291.

Lindner, L., 1982. The age of the Bande Amir travertines in Afganistan. Acta Geologica Polonica 32 (3–4), 289–296.

Lindberg, K., 1961. Le lac de Chiva (Afghanistan). Acta Geographica Société de Géographie 37, 35–45.

Mack, T.J., Akbari, M.A., Ashoor, N.H., Chornak, M.P., Coplen, T.B., Emerson, D.G., Hubbard, B.E., Litke, D.W., Michel, R.L., Plummer, L.N., Rezai, M.T., Senay, G.B., Verdin, J.P., Verstraeten, I.M., 2009. Conceptual model of water resources in the Kabul Basin. U.S. Geological Survey Scientific Investigations Report 2009–5262, 81 p.

Malyarov, E.P., Albul, S.P., Chmyrev, V.M., Azimi, N.A., Makhorin, A.A., 1977a. Osnovnyye cherty gidrogeologii Afganistana, (The principal features of the hydrogeology of Afghanistan). Izvestiya Vysshikh Uchebnykh Zavedeniy, Geologiya i Razvedka 1977 (10), 112–121.

Malyarov, E.P., Chmyrev, V.M., Albul, S.P., Azimi, N.A., Makhorin, A.A., 1977b. Gidrogeologicheskoye rayonirovvaniya territorii Afghanistana. (Hydrolgeologic zoning of the territory of Afghanistan). Izvestiya Vysshikh Uchebnykh Zavedeniy, Geologiya i Razvedka 1977 (3), 79–84.

Micklin, P.P., 1988. Desiccation of the Aral Sea: a water management disaster in the Soviet Union. Science 241, 1170–1176.

Micklin, P., 2007. The Aral Sea disaster. Annual Review of Earth and Planetary Sciences 35, 47–72.

Mirzad, A.G., 1970. Nekotoryye ozera Afghanistana i ikh proiskhozhdeniye. (Lakes of Afghanistan and their origin). Vestnik Moskovskogo Universiteta, Seriya V, Geografiya 25 (5), 106–110.

Neilson, J.M., Gannon, P.J., 1976. Mineral evaluation project Afghanistan. United Nations Development Program unpublished Report AFG/74/OOZ; 2 vols, UN Library, Kabul.

Patzelt, G., 1978. Gletscherkundliche Untersuchlungen im "Grossen Pamir" [glacier research in the "Greater Pamir"]. In: Senarclens de Grancy, R., Kostka, R. (Eds.), Grosser Pamir. Akademiche Druck-u. Verlagsanstalt, Graz, Austria, pp. 131–149.

Pias, J., 1972. Signification geologique pedologique et paleoclimatique de formations paleolacustres et detaiques au Seistan (Afghanistan meridional). Comptes Rendus de l'Academie des Sciences, 274. Paris.

Porter, S.C., 1985. Extent of late Pleistocene glaciers in Afghanistan based on interpretation of LANDSAT imagery. In: Agrawal, D.P., Kusumgar, S., Krishnamurthy (Eds.), Climate and Geology of Kashmir and Central Asia: The Last Four Million Years. Current Trends in Geology, 4. Today and Tomorrow's Printers and Publishers, New Delhi, pp. 191–195.

Porter, S.C., 2004. Late Pleistocene glaciation of the Hindu Kush, Afghanistan. Quaternary glaciations extent and chronology—part III: South America, Asia, Africa, Australia, Antarctica. Developments in Quaternary Sciences 2 (C), 1–2.

Porterfield, G., 1967. Reconnaissance sediment survey of Kajaki and Arghandab Reservoirs, Helmand River Basin, Afghanistan. US Geological Survey Open-File Report, 8 p.

Ruleman, C.A., Crone, A.J., Machette, M.N., Haller, K.M., Rukstales, K.S., 2007. Map and database of probable and possible Quaternary faults in Afghanistan. US Geological Survey Open-File Report 2007–1103, 39 p.

Sawata, H., 1962. Preliminary geologic and geographic notes of the trip to Mt. Noshaq—Lake Shignon (L. Shewa) region, Western Pamirs. Journal of Geography (Chigaku Zasshi), Toyko, Japan 71 (3 and 4), 119–135. and 167–185.

Shareq, A., Chhmyriov, W.M., Stazhilo-Alekseev, K.F., Dronov, V.I., Gannon, P.J., Lubemov, B.K., Kafarskiy, A.Kh., Malarov, E.P., Rossovskiy, L.N., 1977. Mineral Resources of Afghanistan, second ed. United Nations Development Program Project AFG/74/012. 419 p.

Shroder Jr., J.F., 1980. Special problems of glacial inventory in Afghanistan. World Glacier Inventory Proceedings Reideralp Workshop, September 1978 (IAHS-AISH) Publication No. 126, Hydrological Sciences Bulletin, pp. 142–147.

Shroder Jr, J.F., 1989. Glacierized areas of Afghanistan. In: Haeberli, W., Bosch, H., Scherler, K., Ostrem, G., Wallen, C.C. (Eds.), World Glacier Inventory, Status 1988. IAHS (ICSI)-UNEP-UNESCO, Teufen, Switzerland, pp. C39–C40. C346-C353.

Shroder Jr, J.F., 2008. Afghanistan's development and functionality: renewing a collapsed state. Geojournal 70 (2–3), 91–107.

Shroder Jr., J.F., Bishop, M.P., 2010. Glaciers of Afghanistan. In: Williams Jr., R.S., Ferrigno, J.G. (Eds.), Satellite Image Atlas of Glaciers of the World—Asia. U.S. Geological Survey Professional Paper 1386–F3, pp. 167–199.

Shroder, J.F., Weihs, B.J., 2010. Geomorphology of the Lake Shewa landslide dam, Badakhshan, Afghanistan, using remote sensing data. Geografiska Annaler 92 A (4), 471–486.

Smith, G.I., 1973. Potash and other evaporite resources of Afghanistan, U.S. Geological Survey Project Report, Afghanistan Investigations, 62 p. (Listed as: Smith, G.I., 1975. Potash and other evaporite resources of Afghanistan, U.S. Geological Survey Open–File Report 75–89, 63 p. http://pubs.er.usgs.gov/usgspubs/ofr/ofr7589 but no longer available online or in print).

Smith, G.I., 1974. Quaternary deposits in southwestern Afghanistan. Quaternary Research 4 (1), 39–52.

Tate, G.P., 1909. The Frontiers of Baluchistan. Witherby, London, 261 p.

Tate, G.P., 1910–1912. Seistan: A memoir on the History, Topography, Ruins and People, 2 vols, Government Printing Office, Calcutta.

Uhl, V.W., Tahir, M.Q., 2003. Afghanistan: An Overview of Groundwater Resources and Challenges. http://www.vuawater.com/pages/Afghanistan_GW_Study.pdf.

Verkerdy, Z., Dost, R., 2006. History of Environmental Change in the Sistan Basin: Based on Satellite Image Analysis: 1976–2005. UNEP Post-Conflict Branch, Geneva, 56 p.

Westfall, A.O., 1969. Surface water investigations in Afghanistan: a summary of activities from 1952 to 1969. US Geological Survey US Agency for International Development. 50 p.

Westfall, A.O., Latkovich, V.J., 1966. Surface Water Resources Investigations Plan for Afghanistan: Administrative Report; United States Geological Survey and the Water and Soil Survey Department. Ministry of Agriculture and Irrigation, Royal Government of Afghanistan. 62 p.

Whitney, J.W., 2006. Geology, water, and wind in the lower Helmand Basin, southern Afghanistan. US Geological Survey. Scientific Investigations Report 2006–5182.

In this Chapter

6

Ethnic Patchworks

Abstract

The ethnic patchwork of Afghanistan is notable for its diversity, with 49 different languages spoken by its multiple ethnic groups. The largest groups, all Moslem, include the Pashtuns, Tajiks, Hazaras, Uzbeks, Turkmen, and Aimaqs; the smaller groups are the Nuristanis, Pashai, Qizilbash, Baluchi, Arabs, and Pamiri; and the very smallest groups are Jugis, Jats, Brahuies, and Kirghiz. Non-Moslems include a few Sikhs and Hindus, with perhaps a very few Christians. War crimes genocides in Afghanistan, as a result of these ethnic and religious differences, occurred mainly in the 1990s, especially when Sunni Pashtuns tried to exterminate many Shia Hazara, which has been feared will be a resurgent practice again following the withdrawal of the Western coalition forces in 2014.

Keywords: Aimaq; Arab; Baluchi; Hazara; Nuristani; Pamiri; Pashai; Qizilbash; Tajik; Turkmen; Uzbek.

The country of Afghanistan, particularly the rural part, provides an outstanding example of a place where ethnic and tribal groupings take primacy above the individual. These ethnic–tribal divisions have played key roles in the history of Afghanistan (Barfield, 2010). The product of many thousands of years of human migration from multiple directions, the geographic position of Afghanistan, with its mountain barriers and hidden fastnesses, has repeatedly diverted people off into remote locations where they felt safe enough to settle down and establish a tribal sanctuary. The result is the ethnic patchwork of Afghanistan, more so a patchwork amalgamation than most places. In more modern times, rulers in Kabul have even exiled some of their most fractious subjects into regions populated by other hostile ethnicities in order to subdue some of their more difficult people. The Iron Amir, Abdur Rahman Khan, for example, did this in the nineteenth century when he sent some of his more problematic Ghilzai Pashtuns north to live in isolated groups amongst the Tajiks, Turkmen, and Uzbeks, where these Pashtuns then became far more loyal to him as a result of then being surrounded by their traditional competitors and enemies.

Barfield (2010) noted that unlike many other parts of the world, no ethnic group in Afghanistan makes a claim of always having been there. Instead the various groups note how they came from somewhere else and either

Natural Resources in Afghanistan. http://dx.doi.org/10.1016/B978-0-12-800135-6.00006-4

conquered the land in antiquity, or moved into the mountains to live as they pleased, or as nomads they moved about at will and settled where they liked if they so chose. Ethnic identity in Afghanistan can be a problem in identification, especially in urban environments where mixed marriages are fairly common. Barth (1969) defined ethnic variation as: (1) largely biologically replicating; (2) a sharing of fundamental cultural values; (3) constituting a field of mutual interaction and communication; and (4) being constituted through self-identification as well as being identified as such by others. The practical rule of thumb is that if a group of people define themselves as such and so, and their neighbors agree, then that definition is considered a fact (Barfield, 2010). Political scientists and anthropologists are given to arguing about the true meanings of ethnicity, but for our purposes here, these arcane arguments are not useful. Afghans themselves, often assert that ethnicity produces distinct physical appearances, that which is called "profiling" when the police are accused of such practice in the West. In Afghanistan, this can be true to a certain extent where some stereotypical variance in things such as eye shape, skin color, nose length, and other factors occur, but it is commonly in error in the cities where much mixing has occurred. Certain types of dress, skull cap types, beard cut, turban differences, jewelry, and other such features also distinguish certain ethnicities and religion in the countryside where traditional tribal dress is more common.

Approximately 49 different languages are spoken by the people of Afghanistan (Emadi, 2005), most of which are mutually unintelligible, which is a good measure of the high variance of the ethnic patchwork. Demographic statistics are really nonexistent and only crude estimates are possible because they are politically highly charged, with each ethnic group tending to inflate its own numbers to make itself seem more important. Any real census information is unavailable because home owners do not discuss the number of people behind closed doors and high walls through fear of prying strangers, exclusion of women from society, vulnerabilities if the number of warriors inside is exposed, or even the bad luck of counting children who so often die at a young age. About the most that can be done with such limited data sources is to make reasonable estimates of population sizes, which means that a three-tiered approach is about the best useful approximation that can be made: (1) large groups, such as the Pashtuns, Tajiks, Hazaras, Uzbeks, Turkmen, and Aimaqs; (2) smaller groups of Nuristanis, Pashai, Qizilbash, Baluch, Arabs, and Pamiris; and (3) the very small groups such as the Jugis and Jats, Brahuies, Kirghiz, and non-Muslims such as Hindus, Sikhs, and a few others. The Central Intelligence Agency (CIA) World Factbook demographic estimates, revised downward by several million for Afghanistan a few years ago, now estimates that Afghanistan presently has a population of about 30 and a half million with 42 percent Pashtuns, 27 percent Tajiks, 9 percent Hazaras and Uzbeks, 4 percent Aimaqs, 3 percent Turkmen, 2 percent Baluchs, and the remainder of the others constituting about 4 percent in total (Figure 6.1).

(A)

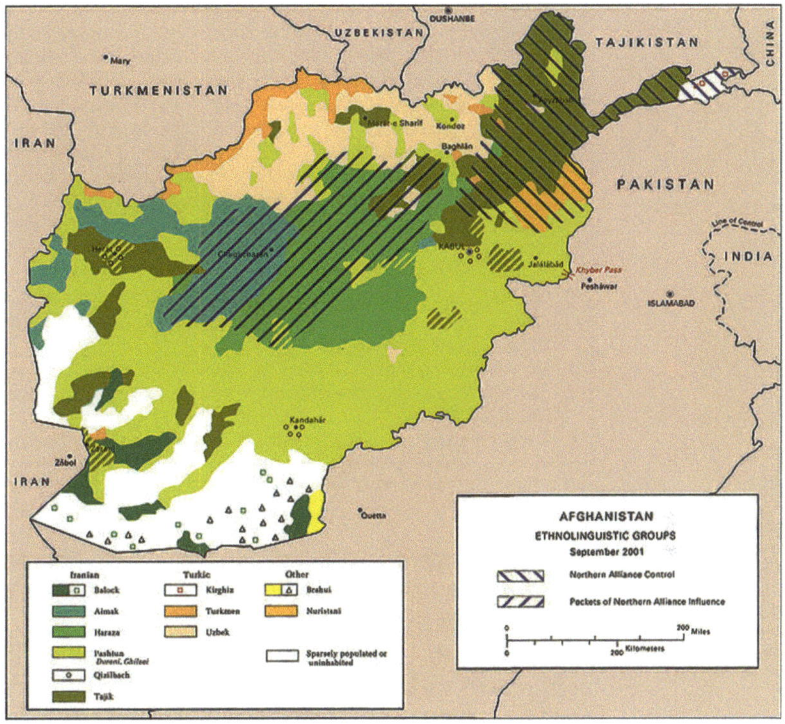

(B)

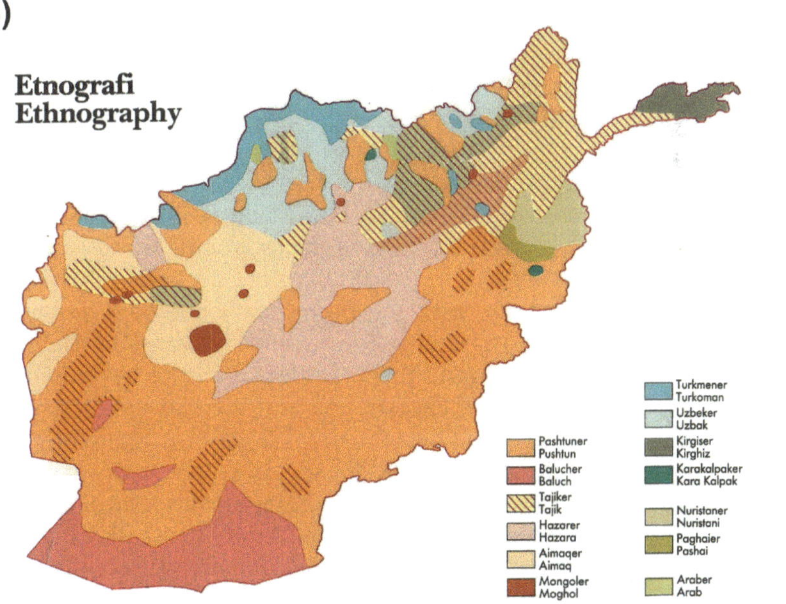

Figure 6.1 (A) Ethnolinguistic map of Afghanistan showing Iranian, Turkic, and other ethnic divisions (Anonymous, 2001). This map is that used at small scale by the US Army, and is quite close to similar versions utilized by both the CIA and the National Geographic Society. (B) Ethnographic map by the Swedish Committee (1986, second edition, 1991) during the 1980s Soviet–Afghan War.

Table 6.1 Map "families" of ethnolinguistic groupings in Afghanistan

1. CIA–US Army–National Geographic Society–Institute for the Study of War map family—1972; 1982; 1997; 2001; 2001; 2003; 2009. Most authoritative, with Pashtun, Tajik, Hazara, Uzbek, Aimaq, Baluchi, Kyrgyz, Turkmen, Nuristani, Pamiri, and Other. Some versions (CIA 1997; US Army 2001) include overprint of Qizilbash and Brahui. Institute for the Study of War includes ethnic breakdown list by city (Figure 6.2). The Swedish Committee map of 1986–1991 resembles these.
2. Dupree–Barfield map family—1973; 2010. Slightly quirky with missing groups in certain known locations, but includes all ethnicities of the #1 map family, plus Arabs and Pamiris (Ismiali Shiites) added.
3. CNN map version indicates sources are merged Encyclopedia Britannica and CIA World Factbook but will not print out from the Internet so appears not meant to be definitive. Maps only Pashtun at 38%; Tajik 25%; Hazara 19%; Uzbek 8%; and other 10%. Other ethnicities missing are presumably included in other category.
4. PBS and Maps of World map family—2006 and unknown date. Only eight ethnicities (Pashtun, Baluchi, Tajik, Kyrgyz, Hazara, Nuristani, Turkmen, and Uzbek) common to both, and one religious affiliation (Ismiali) exchanged with Aimaq.
5. Map of ethnic groups in Afghanistan by political district—old Wikipedia file map shows only dominant ethnicity in each district so only Pashtun, Tajik, Hazara, Uzbek, Turkmen, Baluch, Nuristani, and Pashai are shown.
6. Soviet ethnicity map in 1980s with 21 different ethnicities are listed and mapped.

Comparisons of the lineages of different cartographic groupings of the different ethnic maps through time.

As a further attempt at understanding where people live in Afghanistan, numerous attempts have been made by different sets of people to map the distribution of ethnic groups in Afghanistan (Table 6.1). Five different "families" of maps seem to exist, all apparently based on various levels of definition and presumable variable accuracy in what seems to have been a group of different levels of rather hit-and-miss estimates. The apparent best "guesses" are the #1 CIA–USA Army–National Geographic Society suite of at least five maps (Anonymous, 1972, 2001, 2003, 2009a,b), all showing approximately the same boundaries of the ethnic groups. Probably the best version is that of the US Army (2001) that uses the 1997 CIA mapped version of ethnolinguistic groupings, with an overlay of Northern Alliance Control and Pockets of Northern Alliance Influence (Figure 6.1). The second, or #2 map family of Table 6.1 are those by the scholars Louis Dupree (1973) and Thomas Barfield (2010) who both traveled around Afghanistan interacting with the local people and deducing ethnicity in various ways. Dupree drew heavy, dark lines around Pashtun areas ("fox with the bushy tail") but only dashes or dots around the other groupings; otherwise he wrote the names of the many different ethnic groups where he found them. Barfield seems to have used Dupree's many names as a base to draw lines around and then covered the whole map with an overprint symbol to indicate the dominant ethnic group. Importantly, Barfield has singled out the Ismaili Shiites for special emphasis as what he calls "Pamiris" as different from the Sunni Tajiks. The #3 groupings favored by the CNN group (Anonymous, 2009c) seem to have some of the same basis as the others, with some odd variance and a good deal left out or dramatically oversimplified. The #4 group of U.S. Public Broadcasting Service PBS (Wolly, 2006) and the maps of the world are even more oversimplified, with (PBS) leaving out the Ismailis entirely, but the maps of

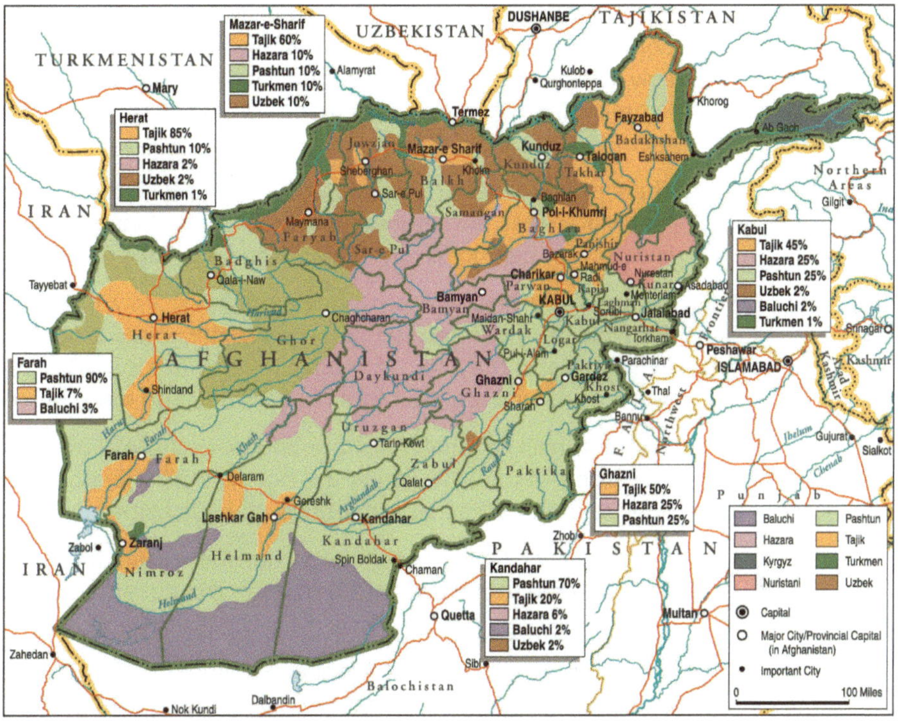

Figure 6.2 Ethnolinguistic map of Afghanistan from the Institute for the Study of War (Anonymous, 2009e). Mapped information is similar to prior CIA efforts but is more generalized, except where composition of ethnicity is presented in the cities of Kabul, Ghazni, Kandahar, Farah, Herat, and Mazar-i-Sharif as well. Notably however, the important Ismiali Shiite Pamiris in northeastern Badakshan and the Wakhan are not identified on the legend but are mapped in a mistaken shade of green the same as that of the Turkmen, which they are decidedly not.

the world includes them. The old Wikipedia attempt of #5 (Anonymous, 2009d, 2013) is hopeless in that it is quite misleading with its single ethnic group per political subdivision, and is therefore hugely and misleadingly oversimplified.

Ethnic Groups: The Afghan people are not "Afghanis", as the media would commonly have it, because that is actually the term for the money of the country, but "Afghan" with a guttural or throat-clearing type of pronunciation. The term "Afghan" properly applies to the Pashtun people only but has come to mean any of the residents who live inside the national borders, no matter their ethnicity. *Pashtuns*, or Pathans to the Indians and British Raj, constitute about 38–40 percent of the total population of Afghanistan, and have dominated the politics of the country for centuries. The Pashtun heartland straddles the Durand Line border on the southeast to northeast of the country, because of course, the British policy of divide and rule in the nineteenth century created that boundary deliberately to partition the bellicose Pashtuns to make it easier for the British Raj to attempt to control them. Other centers of Pashtun population, however, do occur in the north and northwest of Afghanistan, particularly where groups of Ghilzai families were forcibly sent there by Amir Abdur Rahman Khan in the late nineteenth century to make it easier for him to control them.

The Pashtun tribes are composed of lineages of descent (the Pashto *zai* means "sons of") that are aggregated into larger clans referred to as "khel". These lines in turn belong to four major descent groups (Caroe, 1958; Barfield, 2010):

Durrani (Abdali several centuries ago) live mainly in the south and southwest, with major tribal components being the Zirak (Popalzai, Alikozai, Barakzai, Achakzai) and the Panjpao (Nurzai, Alizai, Isaqzai). The Shinwari, Yusefzai, and Mohmand, who are prominent in Pakistan, also claim descent through this line.

Ghilzai (Khalji or Ghalji) are the largest group of Pashtuns in Afghanistan and live mainly in the east, where the Hotaki, Tokhi, Kharoti, Nasiri, Taraki, Suliaman Khel, and Ahmadzai, among others, claim membership.

Gurghusht include tribes such as the Kakar and Musa Khel, who border the Baluch in the southwest, and the Safi in the Kunar region of the northeast.

Karlanri straddle the Durand Line, with most in the Khyber Paktunkhwa Province (formerly Northwest Frontier Province) of Pakistan. Their tribal constituents are the Wardak, Orakzai, Afridi, Wazir, Jaji, Tani, Khattak, Zadran, Mangal, Mahsud, and Khugiani.

The overwhelming proportions of Pashtuns are Sunnis of the Hanifi School of jurisprudence in Islam, but a small number are also Shia. These Shias belong to theTuri tribe, whereas the Bangash tribe is about 50/50 Shia and the rest Sunni, who live mainly in Pakistan where it is very dangerous for them because they are targeted as apostates. Most rural Pashtuns in Afghanistan are subsistence farmers, but a minority are nomads. Most of the *kuchi* or *maldar* nomads are from the Nurzai of the Durrani, or the Kharoti or Nasiri of the Ghilzais.

Tajiks are generally nontribal, Persian (Dari) speakers of dominantly Sunni Moslem persuasion, who constitute about 27–30 percent of the population. They live all over Afghanistan, with a dominance in the northeast part of the country, and especially in the cities where they are known for being bureaucrats, merchants, and educated clergy. They are quite literate in Dari (Afghan) Persian, and because of that language of government administration, have long had important roles in government. Many Tajiks are known by their geographic location, as for example, the Panshiris from the Panshir River valley where Ahmad Shah Masood, leader of the Northern Alliance, maintained his home base.

Pamiris from the Pamir ranges in Badakhshan, Afghanistan, or the Wakhi from the Wakhan are considered Tajiks misleadingly by Russian ethnographers who were relating to things as they are in the own former provinces, but in Afghanistan these people are Ismiali Shiites who are looked down upon and not regarded as the "normal" Tajiks who are dominantly Sunni. In the nearby upland Pakistan where these Ismialis also occur they are commonly referred to as "Aga Khanis", because that is their religious allegiance. They are widely known as aspirants for education, development, modernization, and fair treatment of women, more than many others of their countrymen, which is yet another reason for ethnic hatreds that in the past have targeted them for ethnic cleansing.

Hazaras constitute about 9–15 percent of the population and tend to live in the central, mountainous part of the country where they can practice their Shia Moslem faith without interference, although they have traditionally also been a dominant manual labor force in the large cities where they are looked down upon as lower-class descendants of Ghengis Khan's invading Mongol hordes, in part because of a common Asiatic cast of facial features. Because they ranked at the bottom of the ethnic hierarchy in Afghanistan they were systematically excluded from almost all government positions and educational opportunities by the dominant Pashtun-controlled governments, and were the major targets for ethnic cleansing by the Taliban. At the present time, however, because of their important status in the Northern Alliance when Kabul was wrested from Taliban control in 2001, the Hazaras have risen greatly in stature and jobs in the government in Kabul and elsewhere.

Uzbek people are a main Turkic ethnicity in the country, are Sunni Muslims, and reside mostly in the north of the country, especially where their ethnicity overlaps the border into Uzbekistan in Central Asia. They descend from nomadic tribal confederations of conquers that moved in a series of waves and settled into these areas over a thousand years ago. They settled down as sedentary farmers in the sixteenth century on the loess steppes and irrigated valleys in northern Afghanistan, and many more came over the border to escape the Stalinist USSR later. They are generally bilingual, between the Uzbek language and the Dari variety of Persian and constitute some 9 percent of the population. Many Uzbeks are prominent in politics, with Abdul Rashid Dostum, leader of the Junbishi Milli Afghanistan party, Abdul Rauf Ibrahimi, speaker of the Afghan Parliament, and Muhammad Yunus Nawandish, mayor of Kabul, being among the more noted.

Aimaq ("tribe" in Turkic languages) are generally Sunni Muslims and are not actually a separate ethnic group but a group of tribally oriented, seminomadic herders and agricultural tribal groups of various origins that formed only about 500 years ago. They make up about 4 percent of the population.

Turkmen are a smaller Turkic group of Sunni Muslims who make up about 3 percent of the population, with an origin similar to the Uzbeks, although they tend to be more nomadic. They are known for breeding karakul sheep and doing the highest quality of carpet weaving.

Baluch are a separate ethnic group south of the Pashtuns in southwestern Afghanistan and over the border into Baluchistan Province in Pakistan. They are thought to be an offshoot of the Kurds, reaching Afghanistan perhaps 3000years ago. They are Sunni Muslim, making up about 2 percent of the population, and are generally pastoral nomads, who are well-known for being smugglers from Afghanistan into Iran and India.

Nuristani people are an isolated group of tribes that inhabit some of the most mountainous and isolated valleys in northeastern Afghanistan and northwestern Pakistan who are thought to most likely represent isolated pockets of early Aryan invaders, with about one-third recessive blonde hair. In Afghanistan, they were forcibly converted to Islam by Amir Abdur Rahman Khan in the mid-nineteenth century, and are actually a group of small tribes who speak languages separate from almost all others so they grouped together as a people because of the geographic proximity only.

Pashai people are culturally similar to Nuristani people but speak Pashtu, maintain their own identities, and adopted Islam sooner than the Nuristanis. Pashais live in Kapisa, Laghman, and Kunar provinces adjacent to Nuristan, and although most are Sunnis, a few are Shias of the Ismiali type.

Tatars live mainly in the northern areas of Bamiyan Province and Samangan Province, as well as north of the national border in Uzbekistan. They are Sunnis and claim descent from Mongols.

Arab people in Afghanistan claim descent from the conquering Arabian armies that swept through 1300-odd years ago in the initial Muslim conquest, but they do not retain that language, now speaking Dari, and in the farther north they speak Uzbek as well.

Qizilbash were originally Shiite Turkish military units that helped to find Afghanistan in the mid-1700s and tend to have remained in urban areas where they speak Dari and are prominent in the government, military, trade, medicine, and education.

Brahui people are Sunnis of dark-skinned Dravidian descent who are mainly peasant farmers in extreme southwest Afghanistan.

Karakalpak people or "black hats" are Sunni Muslims who speak a Turkic language similar to Uzbek.

Khyrghiz people live as herders in the extreme far northeastern Wakhan Corridor.

Hindus and Sikhs live in a few places as well, in addition to a few Christians. Most all of the Jews of Afghanistan have long since migrated to Israel to escape persecution.

Jugis and *Jats* are terms much like "gypsy" in the West, where itinerant crafts people (barbers, knife, or sieve makers) or peddlers of special services (snake charming, etc.) come across borders and travel around in small numbers to provide their services.

War Crimes in Afghanistan: In the investigation of war crimes in Afghanistan, many of the crimes were related to the so-called "ethnic cleansing" during the Second Civil War. It is clear that a strong asymmetry exists in war crimes that are investigated and prosecuted successfully. This procedure has been applied mainly against criminal members of the Western coalition troops, even though more than three-fourths of the more outrageous war crimes in the past decade of war have been perpetrated by the Taliban. In fact, however, Afghanistan has been at war since 1978 in a series of phases involving a number of different actors, and numerous incidents of war crimes have occurred, some of which are documented in various ways, and many of which are little more than massacre grave sites, or that have been entirely lost to history because no one survives to testify.

The five phases that produced war crimes were: (1) the Saur revolution of April 1978; the factionalized Marxist–Leninist People's Democratic Party of Afghanistan (PDPA), with its radical reform measures and brutal measures against those who resisted; (2) the Soviet invasion in December 1979 and the following decade of occupation and counterinsurgency war; (3) the Soviet withdrawal and the following civil war period of the early 1990s; (4) the Taliban period of repressive rule and clear ethnic cleansing of the late 1990s; and (5) the following decade of the 2000s during the U.S.-led coalition intervention against the Taliban. During these five phases over the past more than 30 years of war, different armed factions, both Afghan and foreign, committed serious war crimes and crimes against humanity. These crimes included large-scale massacres, deliberate targeting of civilians in warfare, executions of tens of thousands of people, indiscriminant rocketing and bombing that has killed hundreds of thousands of civilians, and torture and rape of thousands. In the 35 years since the wars began, no serious effort has been undertaken either internationally or domestically to document or account for these crimes.

To address at least some of this oversight, in late 2001 the independent and nonpartisan Afghanistan Justice Project (AJP) was set up as an unconstrained research and advocacy operation to discover and document serious war crimes or crimes against humanity throughout all the five phases of the wars in Afghanistan. The personnel of the AJP have set out to interview and document the facts concerning specific incidents in order to establish a historical record that can be provided to the government of Afghanistan, the international donor communities, and international policy makers to pursue accountability for these war crimes. Of prime concern to them are the following: (1) exclude from political office all those who have been responsible for war crimes or crimes against humanity; (2) institutionalize vetting procedures for politicians seeking office; (3) establish mechanisms at local, regional, and/or national level for documentation, investigation, and whatever other truth-seeking measures about war crimes can be set up; and (4) recognize the sufferings of many war victims and acknowledge their right to the truth about what happened to their family who disappeared or who were known to be killed in the course of the wars. The first two of these apparent mandates concerning politicians have been routinely violated repeatedly in recent years.

Although there are hosts of grave war crimes that have been perpetrated over the years, and whether or not they conform to ethnic cleansing may not be known, only a few of the most egregious from the AJP report are listed here:

- People's Democratic Party of Afghanistan (PDPA) period, April 1978 to December 1979: Arrests, disappearances, and summary executions; Kerala massacre by PDPA and Soviets, 1979; Herat uprising, March 1979; bombings, disappearances, and resistance, Hazarajat, 1979; crackdown on uprisings, Kabul; and torture in mujahideen prisons.
- PDPA period and Soviet Occupation 1980–1988: Arrests, detention, and torture; indiscriminate bombardments and reprisals against civilians in countryside.
- Period of Najibullah government after the Soviet withdrawal and resistance: Continuing bombardments; abuses by government-backed militias; attacks on Afghans in Pakistan; and torture in mujahidin prisons, particularly the Lejdey facility operated by Shurai-Nazar faction in northeast Afghanistan.
- Period of Second Civil War,1992–1996, and United Front period, 1996–1998: Indiscriminate bombardment and rocketing of Kabul by all parties, 1992–1995; Afshar massacre and mass rape in Kabul by Abdul Rasul Sayaf's Ittihadi Islami and Ahmad Shah Massoud's Jamiat/Shurai-Nazar forces, 1993; torture, rape, and summary executions by Ittihad, Hizbi Wahdat, and Shurai-Nazar forces in Kabul; torture, rape, and summary executions by Junbish forces under command of General Abdul Malik Pahlawan in Kabul and the north; massacre of Taliban prisoners in Mazari Sharif, 1997.
- Taliban era: Massacre in Mazari Sharif, 1998; massacres in Sari Pul and Gosfandi, 1999–2000; massacre in Rabatak, 2000; massacre in Yakawlang, 2001; burnings and deliberate destruction: Shomali campaign of mass burning and destruction of means of livelihood, 1999–2000; summary executions by the Taliban in districts of Bagram, Kalakan, Qarabagh, and Mir Bachakot, 1999; Yakawlang and Bamyan, 2001.
- U.S./coalition forces era: Bagram torture and prisoner abuse, 2001–2012; Maiwand District killings, 2010; CIA and U.S. Army torture and abuse cases; Stryker team murders.

This list is far from exhaustive but does provide some foundation to establish the nature and dimensions of the problem.

Under considerable pressure from human rights groups, in 2005 the government of Afghanistan adopted a plan to investigate war crimes and human rights violations committed from 1978 until the ending of rule by the Taliban in 2001. The legislation allowed for the possibility of either a truth and reconciliation commission or a criminal court. The plan was seen as a major advance, especially because it ruled out granting immunity to anyone for past crimes. This was not to last, however, because of the great pressure from some of the war crime criminals in the government, the National Stability and Reconciliation Law was passed in 2007 that granted amnesty for war crimes and human rights abuses committed prior to December 2001, but did not take effect until it was published in the official gazette in 2010. The law also appeared to grant amnesty to anyone involved in current fighting in Afghanistan if they agreed to reconciliation with the government. Criminal claims may still be brought against individuals but the rampant corruption, lack of functioning courts, and absence of any form of witness protection makes the use of that provision unlikely.

Both the United Nations Human Rights Council and Amnesty International issued calls for the Taliban and other insurgent groups in Afghanistan to be tried for war crimes for targeting civilians, as well as investigations into illegal actions by American and British military forces. Human Rights Watch called for the Afghanistan government to retract the amnesty law (Anonymous (AJP), 2004; AIHRC, 2005; Arce, 2010; Brunner, 2005; Schimizzi, 2010; Stemple, 2010).

At the present time as the coalition troops are withdrawing or getting ready to withdraw, many people in Afghanistan are preparing for a return to the extreme ethnic-related violence of Taliban time. For example, in the northern, Mazari Sharif area, the area now has only a small number of Pashtun inhabitants left as many fled due to reprisals after the collapse of the Taliban regime in 2001. Nowadays the population there is mainly Tajik, Uzbek, and Hazara. An increase in Taliban activity there in 2011 has led to ethnic tension and a rearming of Tajik and other militia groups (Australian Government, 2011). Much of the recent violence was thought to be the result of power plays along ethnic lines of allegiance as tribal leaders, elders, ex-mujahideen commanders, and warlords jostle about for money and power in the uncertainty of departures of coalition troops. For example, a total of 70 tribal leaders were killed since 2001, with 30 in 2010 alone.

Postcoalition—Withdrawal Civil War? A number of cogent articles have been written about the likelihood or not of a renewed civil war with the withdrawal of the coalition forces by 2014 (Chiovenda, 2012; Khan, 2012; McCaffrey, 2013; Rafiq, 2012; Reichmann, 2012; Rubin and Rahimi, 2012; Shanker, 2013a,b; Shalizi, 2013). For example, the talk of partition roughly along roughly east–west lines, and probably though Kabul as a partitioned city similar to Jerusalem, is taboo for diplomats and political suicide for Afghan politicians because the idea is so unpopular amongst Afghans in general. Nevertheless, many inside the country do feel that a division into Dari speakers in the north and Pashto speakers in the south would be a solution, although others note that this would be a sure route to renewed civil war. This would be because the actual partition plan would be strikingly contentious with the spoils of resources, irrigation waters, and complexities of ethnic distributions impossible to partition fairly in any sort of mapped plan. The partition of India and Pakistan in 1947, with its huge death toll of hundreds of thousands and displacement of people in the millions, might pale in comparison. Such a partition in Afghanistan would benefit Pakistan in the sense that they might aspire to graft on the Pashtun part of Afghanistan as a new province, but this would greatly threaten the Punjabi power structure, as well as encourage India to ally with the Northern Alliance in the north of Afghanistan, which collectively would be a net loss to Pakistan.

The problem, of course, is that in the jostling for power against the imminent withdrawal of coalition forces, the ethnic killings are increasing in such places where Hazaras interact with Pashtuns, as for example, along the Ghorband Road to Bamiyan where Taliban roadblocks are now targeting people for execution or kidnapping, or in Oruzgan where executions of Hazaras have prompted revenge reprisal killings against Pashtuns, as well as in Wardak Province where several thousand Pashtun Kuchi nomads have attacked and burned Hazara households in land disputes (Chiovenda, 2012; Rubin and Rahimi, 2012; Anonymous, 2012).

Shalizi (2013) has noted that if all the coalition forces are withdrawn too precipitously, which has been suggested by Washington and reinforced by Kabul, the odds of another civil war greatly increase. Full withdrawal before the country is ready could spell disaster. Recently, Washington officials indicated that a post-2014 presence of 3000–9000 troops would be most likely, which is significantly lower than the number (6000–15,000) given by top US Commander, General John Allen before his unexpected retirement in March 2013. The overall North Atlantic Treaty Organization (NATO) mission post-2014 was estimated at 8000–12,000 troops (Shanker, 2013b).

Most recently, NATO is considering a proposal to sustain the Afghanistan military security forces at 352,000 troops through the next five years to 2018, although the cost would be $6.5 billion per year, of which Afghanistan would pay only $500 million, international partners $300 million, and the US would provide the remaining $5.7 billion per year (Shanker, 2013a). This would increase the Afghanistan National Security Forces (ANSF) by 20,000 from their present number of 332,000 (Afghan Army 184,000; Afghan Police 148,000), but they already hampered by huge attrition and desertion rates (McCaffrey, 2013). Furthermore, the competence and reliability of the ANSF is not high and has been unraveling, with 20 percent of the US combat deaths caused by treachery from these security forces. In addition, the intensity of fighting against the Taliban has increased rapidly to 110 attacks a day in June 2012, which is the highest since the war began. Meanwhile, however, the US has effectively targeted the Taliban leadership with the drone strikes and other means, the Taliban reportedly want to come home from Pakistan, because they recognize that it would likely take at least 10 years to win the war, and the Pakistani armed forces have been taking action against the Taliban in border regions. The USA has also tripled nonmilitary aid to Pakistan to $1.5 billion for five years, with a total of $12.3 billion since 2002. The result has been that a number of senior Taliban officials have left for third countries without interference from Pakistani officials, which suggests that the postwithdrawal political situation will not be as easy for the Taliban as some of the advocates of the predicted coming civil war may think.

Outgoing US Ambassador Ryan Crocker noted that a civil war would be unlikely because of the high pressure on the Taliban, and that the increase in northern militias actually indicated more of an increase in criminal activity, as well as jockeying for political position as the coalition troops withdraw, rather than being an indicator of a looming civil war. He did say that the withdrawal will likely drop the gross national product from roughly 11 percent of the economy to about 5 percent, but which he felt would not be too bad for an economy like Afghanistan's (Reichmann, 2012).

It would seem obvious that between the optimistic and pessimistic scenarios of whatever combination occurs of renewed ethnic violence, partition, accommodation to existing realities, and/or rigorous military control of the situation, the future has any number of possibilities that will play out with consequences yet to be seen. In any case it would seem that so much ethnic violence has occurred in the past, little likelihood exists that it will not continue on into the future. If the general peace is managed to be kept, however, in the end, if the standard of living is raised enough, as the external neighbors are able to extract the resources and pay in the appropriate fees without too much overt corruption, then the situation may remain in control and Afghanistan might have a chance.

References

Afghanistan Independent Human Rights Commission (AIHRC), January 25, 2005. A Call for Justice – A National Consultation on Past Human Rights Violations in Afghanistan. http://www.unhcr.org/refworld/docid/47fdfad50.html.

Anonymous, 1972. Afghanistan Ethnolinguistic Groups (Map). Central Intelligence Agency (CIA).

Anonymous, December 2001. A Patchwork of Ethnic Minorities (Map). Supplement to National Geographic. National Geographic Society.

Anonymous, 2003. Afghanistan Ethnic Map. National Geographic Society. http://ngm.nationalgeographic.com/ngm/0311/feature2/images/mp_download.2.pdf.

Anonymous (AJP), 2004. Amnesty would jeopardize national reconciliation and security. The Afghanistan Justice Project. http://www.afghanistanjusticeproject.org.

Anonymous, 2009a. Afghanistan's Ethno-linguistic Groups (Map). Institute for the Study of War. http://www.understandingwar.org/map/afghanistans-ethno-linguistic-groups.

Anonymous, 2009b. Afghanistan Ethnic Map: Afghanistan: Ethnic Group Map. Maps of the World. http://www.mapsofworld.com/afghanistan/afghanistan-ethnic-groups-map.html.

Anonymous, 2009c. Afghanistan's Ethnic Diversity: Ethnic Groups (Map). CNN.com/asia. http://www.cnn.com/2009/WORLD/asiapcf/08/17/afghanistan.ethnic.groups/index.html.

Anonymous, 2009d. File: Map of Ethnic Groups in Afghanistan. by District.svg Wikipedia. http://ss.wikipedia.org/wiki/File:Map_of_Ethnic_Groups_in_Afghanistan,_by_district.svg.

Anonymous, 2009e. Afghanistan's Ethno-Lingistic Groups. Institute for the Study of War. http://www.understandingwar.org/sites/default/files/Afghanistan%20Page%20Size%20Ethno%20Linguistic.jpg.

Anonymous, August 6, 2012. Afghan ethnic violence shows signs of resurgence. Dawn. http://dawn.com/2012/08/06/afghan-ethnic-violence-shows-signs-or-resurgence/.

Anonymous, 2013. Ethnic Groups in Afghanistan (with 2001 Map by US Army). http://en.wikipedia.org/wiki/Ethnic_groups_in_Afghanistan.

Arce, D., August 10, 2010. Taliban Should Face War Crimes Charges over Civilian Casualties: AI. http://jurist.org/paperchase/2010/08/taliban-should-face-war-crimes-charges-ai.php.

Australian Government, June 2, 2011. Country Advice: Afghanistan; Refugee Review Tribunal; Afghanistan – AFG38731 – Tajiks in Kabul and Mazar-e-Sharif – Taliban. http://www.mrt-rrt.gov.au/CMSPages/GetFile.aspx?guid=44a15bfc-81fe-4218-b6bd-d50d2b32a2f7.

Barfield, T., 2010. Afghanistan: A Cultural and Political History. Princeton University Press, Princeton, NJ.

Barth, F., 1969. Ethnic Groups and Boundaries: The Social Organization of Cultural Differences. Universitetsforlaget, Oslo, Norway.

Brunner, L., December 12, 2005. Afghanistan approves plan to investigate past war crimes. Jurist, Legal News and Research. http://jurist.org/paperchase/2005/12/afghanistan-approves-plan-to.php.

Caroe, O., 1958, reprint 1973. The Pathans 550 B.C. – A.D. 1957. Macmillan Co Ltd, London and Union Book Stall, Karachi, Pakistan.

Chiovenda, M.K., 2012. Afghanistan: Road Dangers and the Rising Threat of Renewed Ethnic Violence. Eurasianet.org. http://www.eurasianet.org/node/66147.

Dupree, L., 1973. Afghanistan. Princeton University Press, Princeton, NJ.

Emadi, H., 2005. Culture and Customs of Afghanistan. Greenwood Press, Westport, CN, 252 pages.

Khan, A.R., 2012. Afghanistan and the Coming Civil War. Macleans.ca. http://www2.macleans.ca/2012/07/03/the-coming-civil-war-in-afghanistan/.

McCaffrey, B.R., 2013. Withdrawal under Pressure: Afghanistan 2013–2014: The Coming Civil War. http://www.washingtonspeakers.com/images/pdfs/McCaffreyBarry.WithdrawalUnderPressure.2.13.pdf.

Rafiq, A., 2012. The Coming Civil War in Afghanistan. Foreign Policy, 3 February; and JK Alternatives Viewpoint; 22 February. http://www.foreignpolicy.com/articles/2012/02/03/the_coming_civil_war_in_afghanistan.

Reichmann, D., 2012. Afghanistan Civil War Unlikely, U.S. Official Claims. Huffington Post World.

Rubin, A.J., Rahimi, S., August 3, 2012. Afghan officials cite revenge killings in latest outbreak of ethnic hatred. New York Times. http://www.nytimes.com/2012/08/04/world/asia/9-afghans-killed-in-latest-ethnic-violence.html?_r=0.

Schimizzi, C., September 27, 2010. Former UN rights expert urges Afghanistan war crimes investigation. Jurist, Legal News and Research. http://jurist.org/paperchase/2010/09/former-un-rights-expert-urges-afghanistan-war-crimes-investigation.php.

Shalizi, H., January 9, 2013. Afghanistan War: 'Zero Option' Could Lead to Civil War, Taliban Takeover, Afghan Officials Say. Huffington Post. http://www.huffingtonpost.com/2013/01/09/afghanistan-zero-option-civil-war-taliban-disaster_n_2439920.html.

Shanker, T., February 22, 2013a. NATO weighs financing larger Afghan security force. New York Times. http://www.nytimes.com/2013/02/22/us/philip-m-breedlove-is-likely-candidate-for-top-nato-post.html?_r=0.
Shanker, T., February 23, 2013b. NATO plans aid to Afghanistan after 2014. New York Times. Actual online headline differs from print version (NATO in talks on scale of Afghan role after 2014 deadline). http://www.nytimes.com/2013/02/23/world/asia/nato-plans-assistance-to-afghanistan-after-combat-role-ends.html?_r=0.
Stemple, H., March 10, 2010. Rights group urges Afghanistan to repeal war crimes amnesty law. Jurist, Legal News and Research. http://jurist.org/paperchase/2010/03/rights-group-urges-afghanistan-to.php.
Swedish Committee for Afghanistan, 1991. Afghanistan (Map at 1:1,500,000 Scale) Ethnography.
Wolly, B., October 3, 2006. Afghanistan and the war on terror (map). PBS NewsHour. http://www.pbs.org/newshour/indepth_coverage/asia/afghanistan/map_flash.html.

In this Chapter

7

Silk Road Nexus

Abstract

The Silk Road across the Middle East to China has been established and interconnected for millennia, with Afghanistan as a focal-point nexus of multiple routes. Roads and highways that had descended from the rough caravan tracks of antiquity were established in the twentieth century, paved in the midcentury, and rebuilt recently following extensive destruction in the prior decades of war. Railroads were deliberately excluded during nineteenth and twentieth centuries because of long-held fears of their use in foreign invasions. Now however, a new railroad has been built south to Mazari Sharif, with plans to extend it on south to Kabul and elsewhere throughout the country. Small oil and gas pipelines have been built several times in the past and resurgent plans are to transport gas by pipeline from Turkmenistan across Afghanistan to Pakistan, and perhaps India. The electrical power grid, air transport, airports, and critical Afghan Air Force are all in extensive development.

Keywords: Afghanistan Air Force; Air transport; Chinese silk; Coalition military supply routes; Electric power grid; Highways; New seaport access; Pipelines; Railroads; Roads; Silk Road.

The Silk Road is a modern name that refers to the historical network of interlinking trade routes across the Afro–Eurasian landmass, with many routes through Afghanistan, which connected eastern, southern, and western Asia with the Mediterranean and European world, as well as parts of northern and eastern Africa (Figure 7.1). Extending over more than 6500 km, the Silk Road got its name from the lucrative Chinese silk trade along it, which began over two thousand years ago. The Roman Empire, for example, greatly admired Chinese silk. The central Asian sections of the trade routes were expanded around 2100 years ago by the Han Dynasty of China, but earlier trade routes across the regions had already existed. Trade on the Silk Road was a significant factor in the development of the civilizations of Arabia, China, Europe, India, and Persia. Though silk was certainly the major trade item from China, many other goods were also traded, and various technologies, religions, and philosophies, as well as conquering armies and the bubonic plague or "Black Death", also traveled these routes. The main traders during antiquity were the Indian and Bactrian traders, but then from the fifth to the eighth century the Persian-speaking

Natural Resources in Afghanistan. http://dx.doi.org/10.1016/B978-0-12-800135-6.00007-6

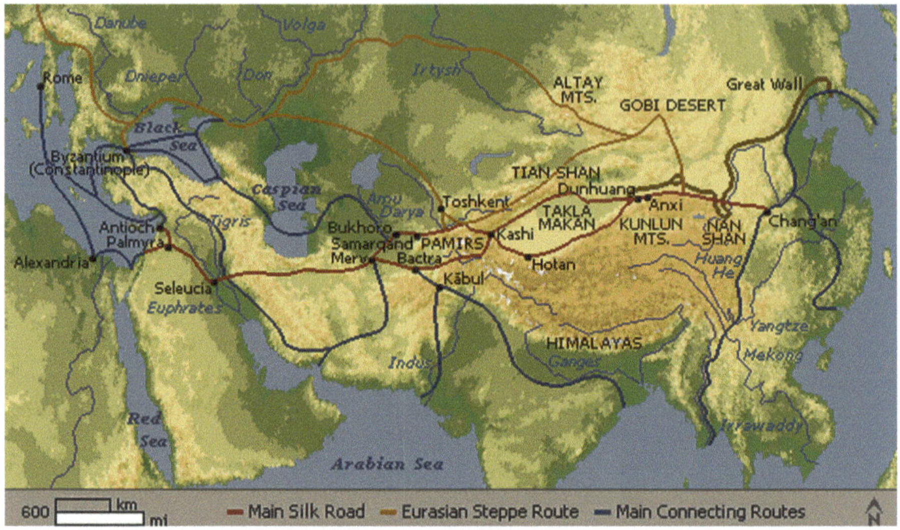

Figure 7.1 The so-called **"Silk Road"**, a complexity of routes of overland travel across South and Central Asia to and from China to the Middle East, North Africa, and the Mediterranean in antiquity.

Sogdian traders came along it from what is now Uzbekistan and Tajikistan, and then afterward the Arab and Persian traders used it too. Marco Polo himself even went through Afghanistan in the late thirteenth century on his way from Venice to China. The hundreds of trails, passes, and ancient caravanserais, or overnight rest houses that dot the landscape of Afghanistan are silent testimonials to the millennia of Silk Road traffic through the country (Isphani, 1989).

The economic redevelopment of Afghanistan, after its three decades of war and total economic devastation, is absolutely critical to move away from its near total dependence on foreign assistance, and instead to move from aid to trade. This can only be accomplished as the first of the four pillars of job creation, which also includes provision of basic services, infrastructure construction, and development of fiscal sustainability. The only approach that will succeed by directly and manifestly improving the lives of Afghans and their neighbors, while simultaneously pursuing economic and military strategies together, giving the Government of the Islamic Republic of Afghanistan (GIRoA) a sustainable revenue stream, and quickly enough to meet the goal of transferring full sovereignty to the GIRoA by 2014, is to focus on reestablishing Afghanistan as a hub of transport and trade. This Silk Road revisited, to link Europe and the Middle East with the Indian subcontinent and all of South and Southeast Asia, is absolutely critical to Afghanistan's future. The vast mineral wealth will be useless and of little commercial value without this transportation-based strategy. Impediments to long-distance road and rail transport, as well as transmission of hydrocarbon and hydroelectric energy within the country must also be removed (Starr and Kuchins, 2010; Kuchins, 2011).

Roads and Highways

The many centuries of prior traffic through Afghanistan have left behind a mixed geography museum of remnant ethnicities, languages, religions, archeologies, and profuse warring ideas and allegiances. What this traffic did not do was to ever lead to any coherent or organized transport system of highways, railroads, or the like. One could walk behind a camel into the country, but vehicular traffic on roads was something else altogether. In fact as a result of the many invading armies back and forth, from Alexander the Great onward, a certain distaste for making it easy for the conquerors developed. The result was many routes in and out of Afghanistan, but none was a walk in the park, so to speak. In fact the very construction of a drivable road has been contentious for many tribes who prefer their safer isolation and inaccessibility from and to the world. In the Korngal Valley of the remote Nuristan region, for example, a place where incessant fighting against Coalition troops has been a near constant threat for the past decade, a new road was bulldozed into the valley to make travel easier for the military and the local people, and to get goods to market; in short order thereafter, however, the road was destroyed about 2010 by the locals (2012 email from embedded journalist, Wesley Morgan).

Other road projects all across the tribal territories on both sides of the Durand Line border between Afghanistan and Pakistan have been successfully opposed repeatedly for decades. More than 200 passes occur through the mountains of this border region, and 90 can accommodate vehicles, but only two are paved. The first is the famous Khyber Pass, and the second is a border crossing point from Spin Baldak to Chaman and the Khojak Pass in Pakistan. The lack of pavement in much of Afghanistan is not only because of poverty, but also because some of the tribes do not want significant vehicular penetration of any of their strongholds because they want to exclude outsiders; this has been a ruling paradigm of theirs for many years. Tribesmen commonly take down bridges and tear out road posts just to assert themselves and make their presence felt, as well as to make sure that the government continues its subsidies to them and recognizes their independent nature. They want to remain isolated. Nonetheless, by the 1920s it was increasingly possible to carefully drive many of the vehicles of the day into Afghanistan along a few carefully chosen routes, albeit with plenty of spare parts, extra fuel cans, and no little risk to life and limb (Gregorian, 1969).

The caravan traffic through Afghanistan from time immemorial was especially increased during the British Raj of the nineteenth century when successive expeditions for exploration and military conquest moved into the country. Early British explorers were always looking for animal fodder and water sources as they openly and clandestinely mapped Afghanistan for their own benefit. Later generations looked for other things. As the country became moderately opened up in the later nineteenth century, the Iron Amir, Abdur Rahman Khan attempted to improve the condition and security of the roads. He saw to the repair of the old roads and bridges, as he ordered new ones to be built as well. The major trade routes were vastly improved, especially the Kabul to Peshawar road, the northwestern extent of the old Grand Trunk Highway of India that had been a major trading route for thousands of years. In addition the roads were improved: Kabul–Balkh, Kabul–Ghazni–Herat, Herat–Kandahar, and Jalalabad–Kafiristan (now Nuristan). The Iron Amir held the inhabitants of the villages and towns along the way to be responsible for highway maintenance and landscaping, as well as for the safety of the traveling public and merchants. In an unusual move to boost his authority and augment his income, as well as to maintain tight control, the Amir also required people to obtain permission and purchase passes (*rahdari*) to travel more than 10 km outside Kabul (Gregorian, 1969).

In the very earliest twentieth century, when the new Amir Habibullah took over the government after the death of his father, the Iron Amir, additional efforts were made to improve and extend the highways. The new Amir made no concessions to the conservative Moslem clergy either; in some instances the roads were put right through graveyards if they were in the way. The first iron bridge was put up in this period as well, where previously only arched spans of brick or rock had been constructed. The old policies of road upkeep by the local villagers were maintained as well, with forced labor and heavy fines for noncompliance or neglect. New roads were constructed by labor forces of some 5000–8000 men, mostly Hazaras, in what amounted to a first public works program. When low pay and unpleasant working conditions caused some to run away, others were forcibly conscripted with no recourse (Gregorian, 1969).

The most important development occurred in 1905 when the Amir brought the first motor vehicle into the country. Periodically he drove it from Kabul to Jalalabad as a demonstration of its usefulness. He founded the Afghan Motor Transport Co. and appealed to wealthy families to invest in it, but by 1918 still only about 30 cars were in the country. It was set up to haul Afghan exports out of the country and bring freight in. The greater share of Afghan trade still moved by pack animals, however, with elephants doing the heavy work of carrying industrial machinery. The Khyber Pass was kept open by the British Raj only twice a week so that under normal circumstances, the trip between Kabul and Peshawar took about 10 days.

During King Amanullah's reign in the 1920s he had the old roads repaired and made plans for new roads southeast to Peshawar, southwest to Kandahar, and north to Afghan Turkestan. He had hotels for travelers built in Kabul, Jalalabad, Kandahar, and Herat. A postal bus service began operating in 1925, between Kabul, Jalalabad, and the Khyber Pass, as well as to Kandahar and Herat. By 1933 Afghanistan possessed some 4608–5120 km (2880–3200 miles) of road, very few of good quality, however. That same year a new, all year round connection was built through the Hindu Kush for the first time, the Great

North Road, the building of which was supervised by a Hungarian engineer (Gregorian, 1969).

Through World War II (WWII) and the 1940s, during which nothing much of significance happened with road transport in Afghanistan, road travel there continued to languish while the rest of the world was speeding along. Finally in the 1950s, in part to increase their trade as well as to look south toward other opportunities, the Soviets began an ambitious road-building program in the country. In fact, the transport infrastructure was one of the top two components of the Soviet economic aid to Afghanistan. As with their British masters before them, the Pakistanis and their American friends became immediately suspicious of Soviet designs on access to warm water ports on the Indian Ocean, which would involve taking over part of Pakistan to do it because Afghanistan is landlocked, with no access to the sea. The nineteenth century Great Game between Russia and England was afoot again, with much worry on all sides about Soviet intentions. To counter this worry in Afghanistan, the country turned repeatedly to the Americans, who seemed far enough away to be quite safe, and did not seem either to be much interested in colonialism. In the USA aid program that resulted, transport was one of the top three priorities. Accordingly, Afghanistan road construction was divided between American and Soviet engineers.

The road-building effort in Afghanistan in the late 1940s through the 1960s was a concerted effort to upgrade all the rude tracks and dirt roads that prevailed throughout the country, crossing the vast desert plains and high mountain passes through the Hindu Kush and adjacent ranges on the border with Pakistan. At the start of the Royal Government of Afghanistan's (RGA) Five Year Plan in 1956, some 2080 km of dirt and gravel main roads began to be modernized by asphalting. In the late 1950s, the International Cooperation Administration (ICA) of the US Department of State decided to join with the RGA with its own programs to counter the growing Soviet aid assistance programs in road building. In short order however, the ICA was discovered by the US Embassy to be deeply in over its head in road building as it had contracted with various American firms. Road building in such an unknown environment, with unsampled and untested problem soils, old and obsolete equipment, some of WWII vintage, faulty engineering, and myriads of logistical and materiel problems caused considerable pressure for change. The ICA was found to have written its contracts based on faulty assumptions, incomplete specifications, and with contractors who could not perform. US Ambassador Henry A. Byroade, a West Point graduate and former US Army officer, was instrumental in seeing that the US Army Corps of Engineers (USACE) was brought in to take over most of the American road building in Afghanistan (Figure 7.2; Table 7.1).

The American effort to counter the work of the Soviets was mainly in the south of Afghanistan, especially at first to construct or reconstruct the roads from

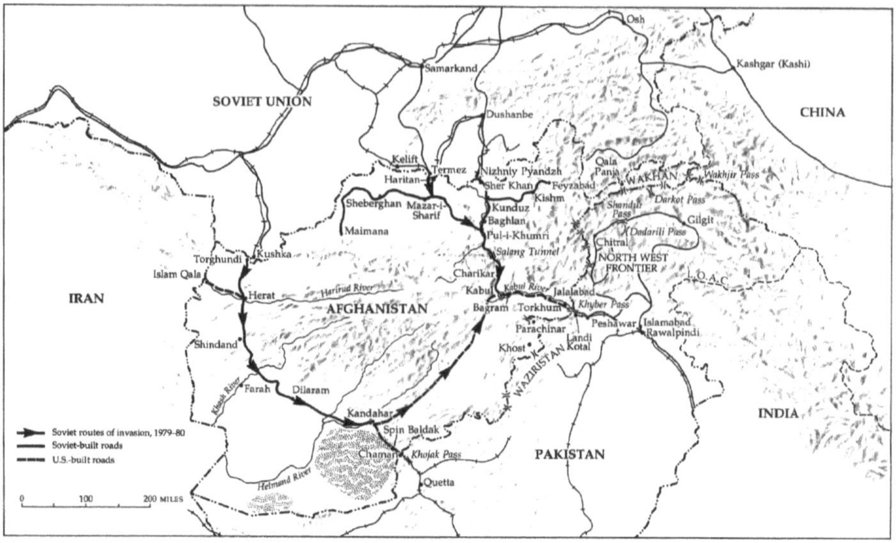

Figure 7.2 Map of road building in Afghanistan in the 1950s and 1960s as well as invasion routes by the USSR in their war in Afghanistan in the 1980s. *After Isphani, 1989.*

Table 7.1 Original road construction (subgrade, bridges, and fords, concrete, asphalt) in Afghanistan

Highway Location	Distance (km)	Dates	Contactor/Country
Kabul–Torkam	232	1959–1961	RGA (Kabul Gorge section)
Kabul–Kandahar	515	1961–1966	USACE
Kandahar–Spin Baldak	100	1948–1949	Morrison–Knudson (gravel)
		1966–1967	USACE
Herat–Islam Qala	120	1966–1967	USACE
Kandahar–Girisk	113		USSR
Girisk–Lashkargah–Darweshan	~100		Morrison–Knudson
Herat–Girisk	640	1959	USSR
Kabul–Jabalus Seraj	76		USSR
Jabalus-Seraj–Doshi (Salang)	109	1958–1964	USSR
Doshi–Sher Khan–Bandar	214		USSR
Mazar-i-Sharif–Khelif	120		USSR

ICA—International Cooperation Administration in USA; USACE—US Army Corps of Engineers; RGA—Royal Government of Afghanistan.

Pakistan into the country. In the farthest south, the road from Quetta to the border at Spin Baldak needed to be upgraded after the Morrison and Knudsen Company in 1948–1949 had put down a rather anemic, approximately 150-mm subbase, overlain by another thin 150-mm layer of processed gravel from the border to Kandahar. The USACE built bridges and fords, and paved the route in 1960–1961 (Table 7.1). In the northeast where the RGA had been working on the difficult job of blasting and tunneling its way through the Kabul River gorge to produce a new route between the capital, Kabul, and the Jalalabad Basin and its Torkam gate to the Khyber Pass and Pakistan (Figures 7.3 and 7.4). This route was taken over by the USACE and brought to completion in 1961 (Table 7.1). The next big effort was to take on the all-important Kabul-to-Kandahar highway to link the essential cities with an all-weather road. Work got underway in late summer 1961 and proceeded slowly as border frictions between Afghanistan and Pakistan limited the movement of supplies, equipment, and materials to the work sites (Cronin, 2010). The project took 5 years but eventually allowed rapid driving between the two cities so that what used to take days now was only a 6-hour drive. Aside from miscellaneous road-building efforts by the Morrison Knudsen Company around Girishk, Lashkargah, and Darweshan in the Helmand Valley Project, the other major road building effort by the USACE was to connect Herat west to the Iranian border at Islam Qala (Table 7.1). That project was completed in 1967.

The well-known British historian, Arnold Toynbee, at the time the roads were being constructed, offered his opinion that the many new roads being built promised to reinstate Afghanistan back into position as a transit point and economic bonus recipient of big power competition, much the same as in

Figure 7.3 Photograph of the torturous highway through the Kabul River gorge that was built by the government of Afghanistan in the 1950s before being assisted by the US Army Corps of Engineers in the 1960s to complete paved access between Jalalabad and Kabul. *Photograph by Dr William F. Podlich, Jr., 1968; by permission of the family.*

Figure 7.4 American road construction on the highway between Kabul and Khyber in 1963. *Photograph by James Burke from Life Magazine; Courtesy of the Embassy of the Islamic Republic of Afghanistan, Washington, DC.*

the Great Game of the nineteenth century (Toynbee, 1961a, b). The Cold War competition between the USSR and USA in Afghanistan led to bonuses that were obviously of considerable benefit to the Afghans, but the risk was high. As he said, "Roundabouts [such as Afghanistan] are strategic as well as economic assets, and strategic assets are tempting political prizes." Prescient words indeed in light of later events!

The Soviet road-building effort was designed to enhance development of access from their country into Afghanistan at several locations across their northern borders. In 1959 the Soviets began planning for their road building in the west of Afghanistan, where the 680-km-long Kushka–Torghundi–Herat–Kandahar road was contracted and started in 1960. The road cost about $140 million, of which only $11 million had to be paid by the Afghanistan government because the rest was an outright gift by the Soviets. On the way by, they also built the Shindand military airfield, perhaps as a gratuitous gift that they may have possibly intended to use at some later date when the time was ripe, which of course was, and only two decades later when they invaded the country.

The Kushka–Kandahar road was unusual also in having a reinforced concrete-slab surface, where unlike American road building in which the cement was poured in place along the route by hand or by special machines built for the purpose, instead these huge, approximately 10–12 cm thick, slabs were in some cases poured off at the road's edge and then hauled and wrestled onto their final place on the roadbed with much effort (Figure 7.5). The roadbeds in many cases were not well prepared and so the slabs were commonly tipping in irregular fashions (Schubert, 1991). The concrete-slab construction was commonly also overcome by the grotesquely overloaded trucks (Figure 7.6), because no weight limit laws or vehicle scales have ever been used in Afghanistan. On the other hand, the military attachés of the US Embassy took special note of the fact that the bridges built by the Soviets along the route had carrying capacities far in excess of the heaviest Afghan civilian traffic; in fact, the heaviest load limit was precisely the weight of the heaviest Soviet battle tank.

Figure 7.5 Photograph of the Soviet-built highway of concrete slabs between Herat and Dileram in 1977. Many of the slabs were poured off-site and later dragged into position. The western part of this ring road has been repaved with asphalt in recent years. *Photograph by J. Shroder in September, 1977.*

Figure 7.6 Severely overloaded truck in Afghanistan; the lack of weight restriction on Afghanistan highways leads to severe destruction of pavement. *Photograph from Cronin, 2010.*

In the late 1950s, the Soviet road-building effort north of Kabul to their border went to great and sometimes dangerous lengths to improve the highways. The most impressive undertaking was the Salang Pass bored 2.5 km through the heights of the Hindu Kush at an altitude of over 3380 m and protected at both ends against rock falls and snow avalanches by many kilometers of concrete sheds over the roadway (Table 7.1). The tunnel was about 5–6 m wide and 7–8 m high, with a vertical ventilator shaft about halfway along that exits on the mountain slopes above. The highest tunnel in the world at the time, the Salang became a major conduit for transport and trade

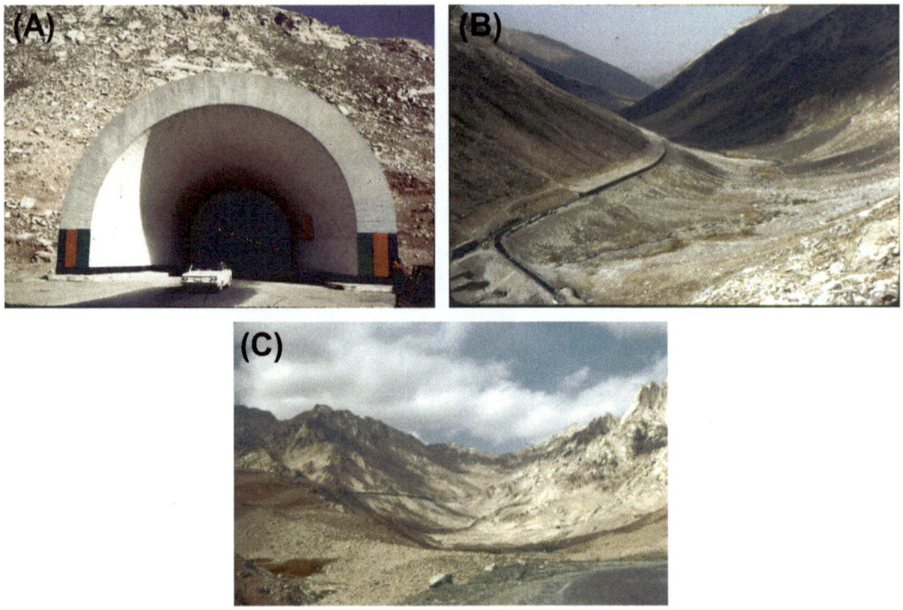

Figure 7.7 (A) The north portal of the Soviet-built tunnel through Salang Pass as it was in 1968 *(Photograph by Dr William F. Podlich, Jr., 1968; by permission of the family.)* (B) View of the north portal from the slopes above in 1978 with a North Korean convoy passing through to welcome the new communist regime headed by Taraki in Kabul. *(Photograph by J. Shroder in September, 1978.)* (C) South side of Salang Pass showing extensive snow sheds. *Photograph by J. Shroder in September, 1978.*

(Figure 7.7(A–C)), although it has also been subject to considerable mortality through internal vehicle fires, overwhelming exhaust fumes, and external tragedy in the form of snow avalanches that have swept off many cars and trucks. The Salang project took 6 years to complete, from 1958 until 1964, at a cost of about $638 million.

Then in 1965 the Soviets also worked on reconstruction of the road from Poli Khumri to Mazari Sharif and Sheberghan, as well as in the 1970s from Shebergan southwest to Maimana, and eastward from Kunduz–Khanabad–Kishim toward Feyzabad, provincial capital of mountainous Badakhshan. Some of these highways were built with newer asphalt technology, but again a number of Afghans were worried that they had been built wide enough and sufficiently strong to accommodate troop transporters and tank traffic. In 1978 a new $50 million project was begun to put a road and rail bridge across the Amu Darya from Termez in Uzbekistan to Hairatan on the south bank in Afghanistan. The Friendship Bridge was opened in the spring of 1982 by the Soviet invaders and the new communist regime in Kabul (Chapter 20; Figure 7.3). Soviet road building in Afghanistan after their invasion in 1979 continued at a rapid rate, especially in the north closer to their border so much so that considerable concern existed at the time that they might try to annex the northern provinces of Afghanistan together with their own "-stans", thereby consolidating a number of their, sometimes problematic ethnicities on both sides of the border (Turkmen, Uzbeks, Tajiks, Kirghiz). In the nineteenth century, the Russians had noted as Afghanistan's border delineations were first taking place under British auspices, that rivers made bad borders and that they actually preferred more impenetrable, natural landmarks, such as mountain ridge lines (Shroder, 1983, 1989; Chapter 20). But the feared annexations never took place, the Soviets departed in 1988–1989 along the roads they had built to invasion standards, and the Second Afghanistan Civil War erupted in the 1990s.

During the Soviet–Afghan War of the 1980s, disruption of highway transport was a preferred mode of harassment of the communist regime by the guerilla mujahideen of native Afghan forces (Figure 7.8). Planting bombs (improvised explosive devices—IEDs) beneath roadways was a favorite tactic of the mujahideen. In one case related by a mujahideen commander to me in Peshawar right after it happened in 1983, several large aerial bombs, which had not exploded upon impact after they were dropped by the Soviets, were carefully inserted into a hand-dug tunnel into the side of an embankment beneath a roadway frequented by Soviet convoys. When the next convoy came by, the IED was detonated by wire, with the result that many of the Soviet troops "were reduced to small pieces". For this reason, in recent years it has been noted that paved highways are harder to plant IED bombs beneath than unpaved roads that can be dig into at will, but it is still possible where an embankment lends ease of access and the possibility of concealed digging for the guerilla fighters.

In any case, the two decades of combat between the Soviets and the Afghans, and between the Afghans themselves in their own civil war after the Soviets

Figure 7.8 Afghan mujahideen fighter sighting down on a road in Afghanistan during the Soviet–Afghanistan War in the 1980s. *Photograph by Joseph Polleross, JB Pictures; from book jacket of Roads and Rivals by M.Z. Ispahani, I.B. Tauris, London.*

went home, resulted in the devastation of the much vaunted road network across the country. Travel along the much mined roads and blown-up bridges took many days again, just like the old days before the 1960s, with endless careful deviations around explosion craters, detours down off road embankments, past blown-up bridges, and across river bottoms before going back up the other side to rejoin the road. All the while the fears of exploding mines, random deadly roadblocks, armed hijackers and extortionists, and other similar perils awaited the traveler. With the arrival of the Coalition forces in the country after the 2001 invasion, it was time to refurbish the old and bring in the new as soon as possible. In fact, trying to bring back roads where there were no legal truck load limits to weight, where a totally overloaded truck with enough other heavy vehicles loaded on it was sufficient to crack through virtually any pavement (Figure 7.9); these conditions made it difficult to bring back a war-scourged road at all (GAO, 2008).

The primary road network in Afghanistan was about 4500 km long in total, but in 2002 about half of that total was in exceptionally poor condition (very poor—47%; poor—15%; fair—24%; good—14%). Rehabilitation of the half-destroyed roads was judged at that time to be about $1.7 billions, which was a hefty price tag for which GIRoA had no funding, so outside foreign donors had to pay the price. The USA and Japan picked up much of the tab for the essential Kabul–Kandahar refurbishment, but overall after the USA at 42 percent of the total costs nationwide, the Asian Development Bank (ADB) was the second-most robust contributor at 24 percent (Figure 7.10). The most important parts of the ring road, which encircles the country, were taken care of predominantly by the USA, Japan, the World Bank, and the ADB. Only the northwestern portion of the ring road around Badghis, which was stopped for 5 years because of ongoing security concerns, is still delayed and so far unfinished (Figure 7.11). Part of the problem also is the substrate of the countryside in the northwest is wind-blown loess dust, which if not precompacted properly prior to paving, will collapse internally and destroy any new road surface put over it. ADB provided $477 million in grants and technical assistance to finish the approximately 233-km job.

Figure 7.9 **The destroyed Kabul-to-Kandahar highway near Ghazni, February 27, 2003.** *Photograph by Joe Pecht, from Cronin, 2010.*

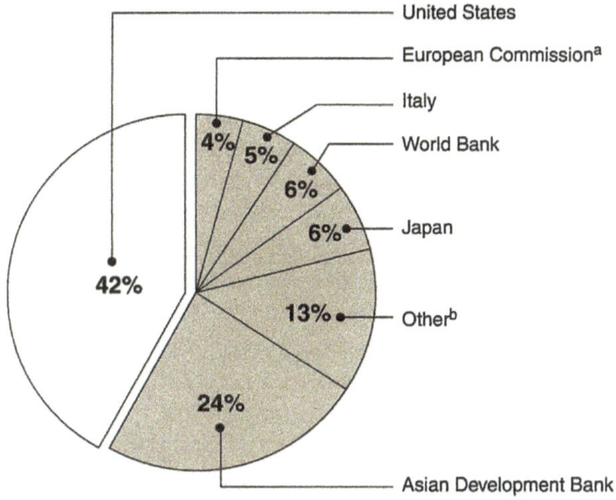

Figure 7.10 Pie-chart breakdown of percentages of donor funds committed for regional and national highways in Afghanistan, 2002–2007. [a]European Commission funding represents contributions by the Swedish International Development Cooperation Agency. [b]Other donors include the Islamic Development Bank, India, Iran, Pakistan, and Saudi Arabia. *GAO analysis of data from the Consultative Group for the Transport Sector and USAID*—http://www.gao.gov/assets/280/278015.pdf.

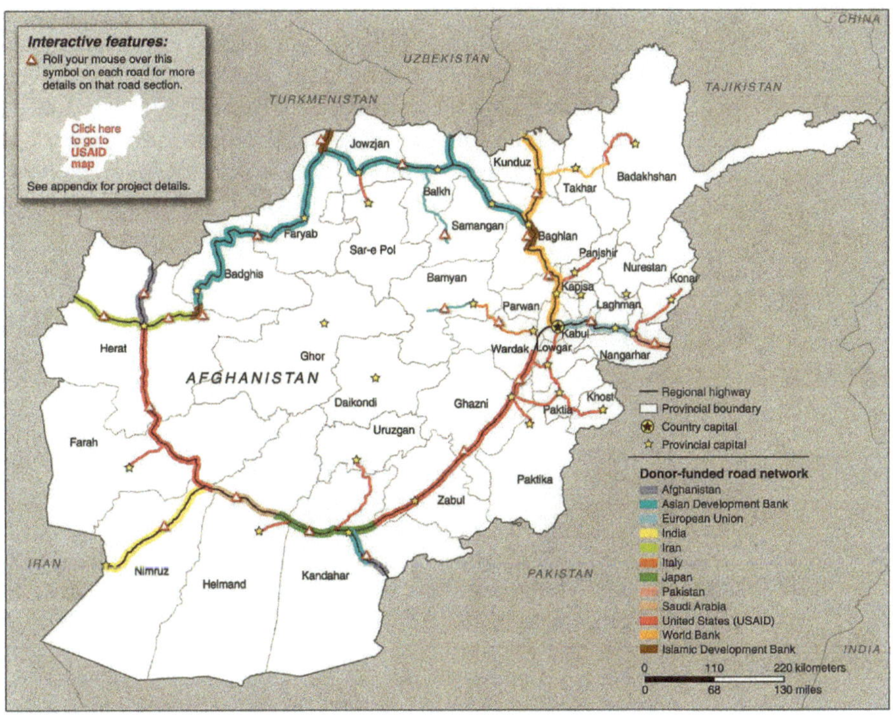

Figure 7.11 Distribution of donor-funded regional and national highways in Afghanistan. Several of the most essential road links, such as the Kabul to Kandahar and the Herat to Dileram were USA and Japan funded, but the all-important Kabul-to-Tajikistan route was taken care of by the World Bank. The only real problem area left is the completion of the ring road in the northwest around Faryab and Badghis, which is the responsibility of the Asian Development Bank. *GAO analysis of data from the Consultative Group for the Transport Sector and USAID.*

Road building in Afghanistan can be enormously expensive, costing over a million and a half dollars per kilometer, but with efficiency measures and good security the cost can be pared down to only $80,000/km. The cheapest road in simple terrain, secure environment, easily available material, and minimal asphalting cost only $47,000/km. In recognition of the necessity for adequate maintenance of the highways after they are built, not something Afghanistan has ever been known for particularly, US Agency for International Development (USAID) has developed training procedures and instruction manuals to help in these procedures. By 2011 USAID and the US military had rehabilitated some 1626 km of regional and national highways using these procedures (Figure 7.12 (A,B)).

The World Bank took responsibility for another essential part of the road network, with refurbishment of the Kabul to Kunduz part, to meet a new bridge crossing the Panj tributary to the Amu Darya north of Kunduz at Sher Khan Bandar on the Afghanistan side to Nizhnii Panj and the E123 road on the Tajikistan side. The Afghanistan–Tajikistan Bridge cost $37 million, paid by the USA and built by an Italian contractor under the supervision of the USACE (Chapter 20; Figure 20.4(A,B)). This bridge was completed in August 2007 and relieved the considerable pressure for import–export of war materiel through Afghanistan's northern border. In spite of the great importance of the roads to Afghanistan's future, the problems with the construction of the roads have been enormous and very expensive. Charges of graft, kickbacks, fake and real threats to increase the costs of security and protection money, roads to nowhere, slipshod methods, and fraudulent billing have all been charges leveled against various people and corporations (Cronin, 2010; Risen, 2010; Rubin and Risen,

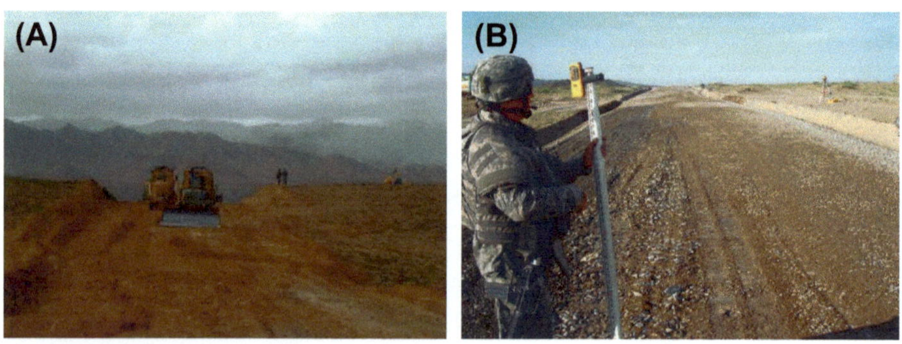

Figure 7.12 (A) Task Force Pacemaker, 864th Engineer Combat Battalion soldiers performing road grading mission in Afghanistan. *(Photograph by Sgt David Roscoe; Special to American Forces Press Service.)* (B) US Army Staff Sgt Troy L. Bohanon, a member of Company A, 864th Engineer Combat Battalion, helping to survey the Khyur to Mest road in Afghanistan. *Photograph by Sgt David Roscoe; Special to American Forces Press Service.*

2011; Wallechinsky and Brinkerhoff, 2012; Nissenbaum, 2012; Shalizi, H., 2012). War profiteering has long been a part of the war games everyone plays/pays through taxes to the governments all over the world and thence to big corporations, who have a history of playing the game very comfortably to their benefit since at least the American Civil War in the mid-nineteenth century, and probably much longer.

To top it all off, in fact building good roads that will last in Afghanistan's most harsh physical environment is even more difficult than most places. The region is subject to an endless suite of ongoing assaults on roadways, courtesy of high seismicity, avalanches of mud, rocks, snow and ice, river valley floods, and endless mud at the wet thaw times of the year (Roscoe, 2007; Anonymous, 2013; Hagen et al., 2010; Shroder et al., 2011).

Railroads

The possibility of railroads through the rough and hostile terrain of Afghanistan has been a major idea and bone of contention in a land of many bones, for a very long time (Shroder, 2007; Grantham, 2013; Henderson, 2011; Norling, 2011; Nelson, 2011; Johnson, 2012a; Anonymous, 2012a). The British had several plans for railroads inside the country in the late nineteenth century, but they all came to naught in the face of an intransigent Amir Abdur Rahman Khan who felt that railroads, if they were to ever come, should only be used in the far interior of the country, away from the borders where they could be used to invade the country. When he learned of the British-built Khojak Tunnel and railroad line through it to New Chaman border post near Quetta, he accused them of plunging a "knife into my vitals". Afghan subjects were forbidden under pain of death to use this railway so that they had to carry their goods over the intervening mountain in British India and deposit them on the other side. Amir Habibullah in the early twentieth century also considered British railroads as a "spear pointed at the heart of Afghanistan". Under the reign of Ammanullah from 1923 to 1929, a small showpiece railroad ran 7 km from central Kabul out to the Darulaman Palace in what is now the far southwest part of the city. Otherwise railroads were kept entirely out of Afghanistan except for short freight yards built only a short way into Afghanistan by the Russians at the Towraghundi border post in the northwest at the Turkmenistan border, and in the north at the Hairatan border post in Afghanistan where the Soviet-built Friendship Bridge from Uzbekistan is located. These potential rail links were all built to the Russian gauge of 1520 mm (4 ft 11 5/6 in).

In 2010 a new line at the same gauge as that in Uzbekistan was built 75 km from Hairatan to near the airport in Mazari Sharif (Figure 7.13). The first freight services began on the line in August 2011, handled by the

Figure 7.13 Guarding the diesel engine and train from Uzbekistan on the new railroad service to Mazar-i-Sharif. *Photograph from Google Earth™.*

Uzbekistanian national railway, Uzbekistan Temir Yullari, until the new Afghanistan national railway can take over in a few years (Henderson, 2011; Johnson, 2012; Norling, 2011). The work was carried out by the China Metallurgical Group Corporation, in connection with their Aynak copper mine south of Kabul. This is the first part of plans to extend the $4 billion, 900-km railway line east to Kunduz and then on south along the Kunduz River and its tributaries to Bamiyan. The Shebar Pass (2987 m) out of Bamiyam could be tunneled and then the railroad would run down the Ghorband River valley into the Shomali Plain north of Kabul and from there on to the Aynak copper mine; eventually the line is to be extended all the way east to Jalalabad and the Torkam gate to the Khyber Pass in Pakistan where a different track gauge of 1676 mm (5 ft 6 in) was in operation. Where such gauge incompatibilities exist, the technique of bogie exchange can be set up where the trucks of the cars are switched out by lifting the car body with a crane and setting it back down on trucks of the different gauge.

In addition to China, the countries of Iran, Pakistan, and India all have government or corporate plans for separate railroad activities in Afghanistan. For example, although it is not at all popular with Washington, in 2006 Iran began extending a standard gauge line (1435 mm or 4 ft 8½ in) from Mashad east 202 km to Herat. Similarly, India is planning a railroad to connect its newly planned Hajigak iron mine near Bamiyan to western Afghanistan, where it would tie in with a new Iranian line to their Chabahar port facing out into the Gulf of Oman and the Indian Ocean. This would allow Indian freight in and out of Afghanistan, without any contact with the archenemy, Pakistan. Part of their proposed rail line would parallel the road that India built recently in Nimroz Province in the southwest from Dilaram to Zaranj on the Iran–Afghanistan border (Nelson, 2011; Norling, 2011) (Figure 7.14(A–D)).

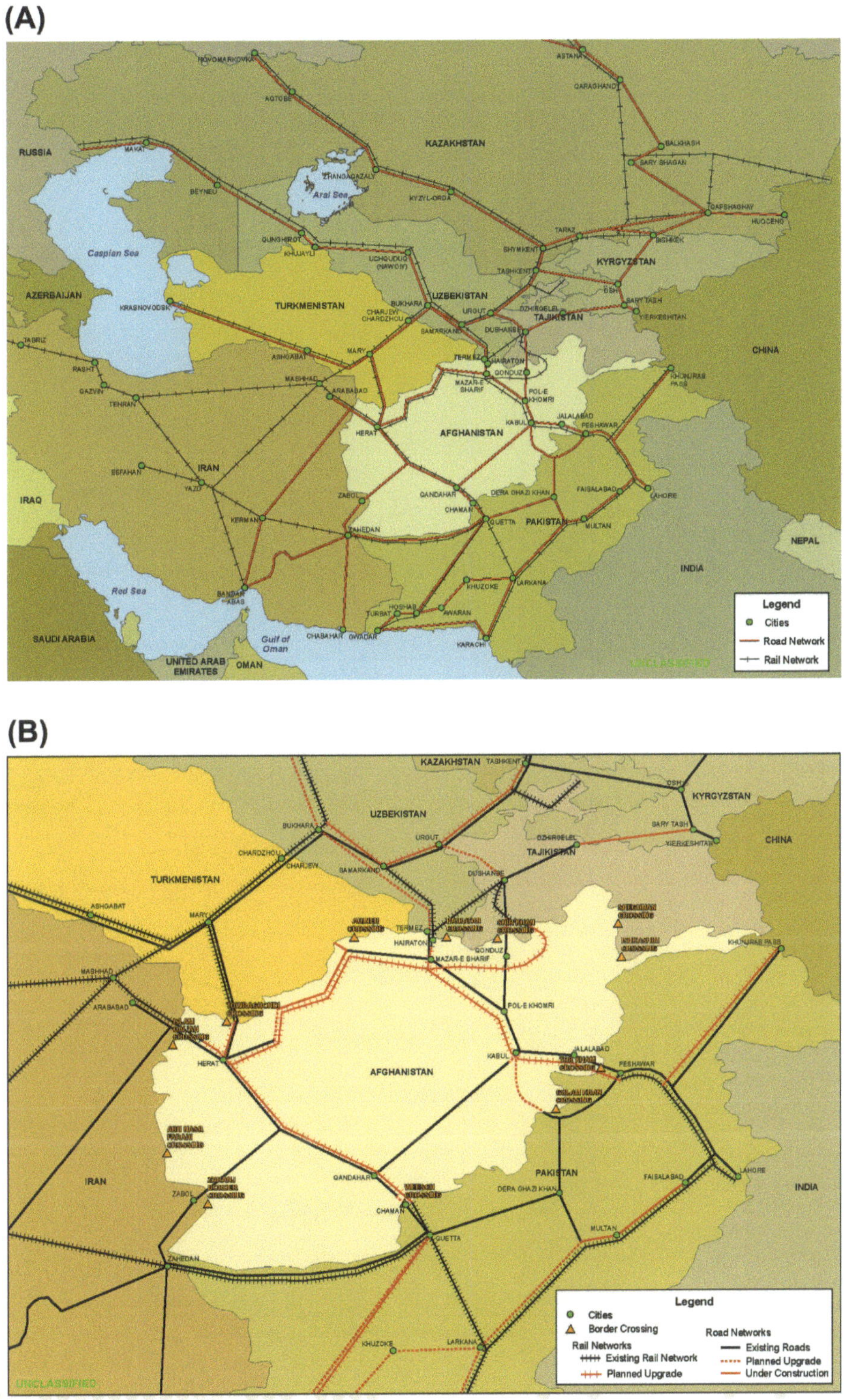

Figure 7.14 Diagrammatic road and railway networks existing and planned for Afghanistan and its neighboring countries.
(A) Small-scale, greater regional road and rail plan *(after* Starr, *Kuchins, 2010)*.
(B) Moderate-scale regional road and rail plan *(map after* Starr, *Kuchins, 2010)*.

(C)

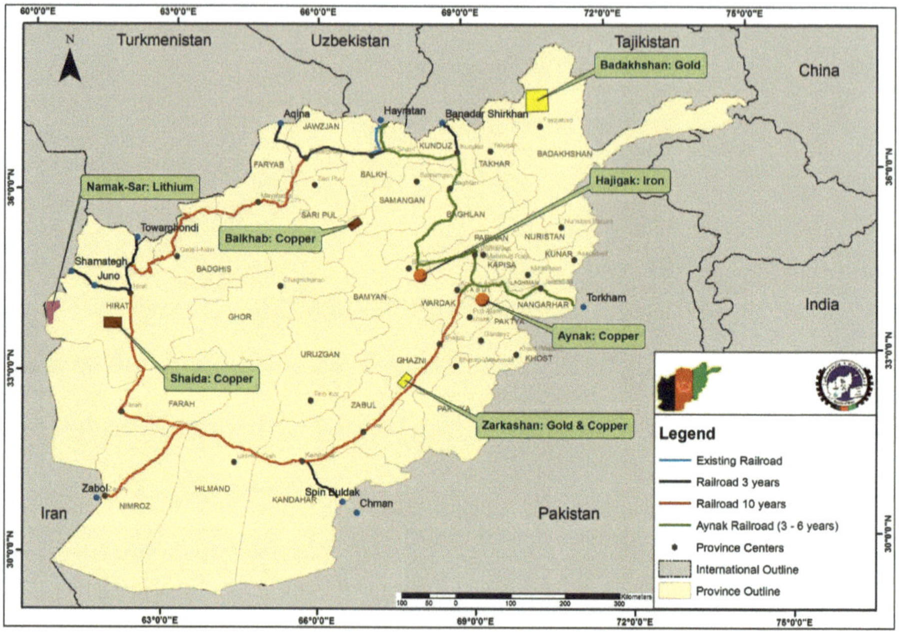

(D)

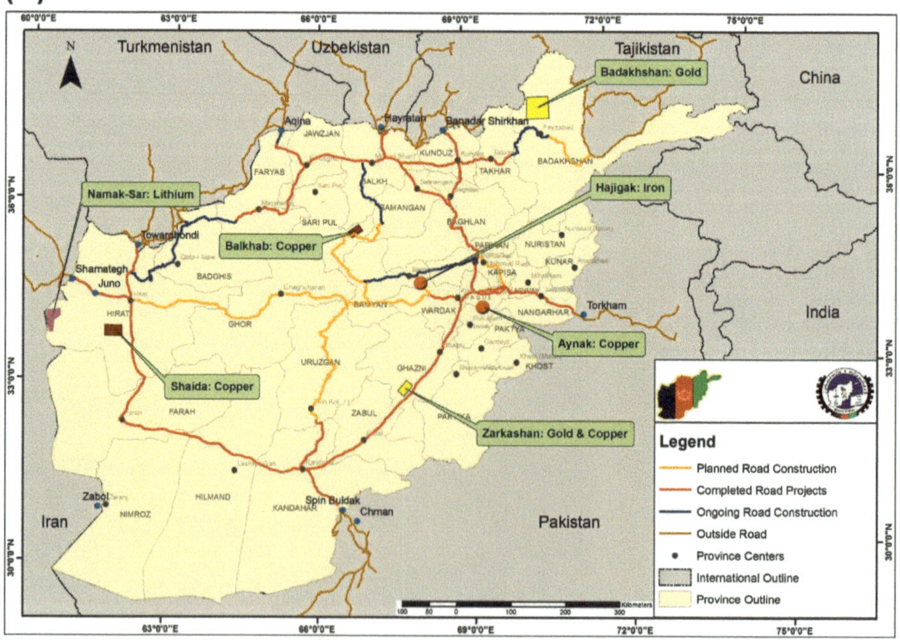

Figure 7.14 (C) Afghanistan Ministry of Mines and Industries plan for railroad infrastructure upgrades. (D) Afghanistan Ministry of Mines and Industries plan for road infrastructure upgrades.

Figure 7.15 Soviet-built, natural gas pumping plant near Sheburghan in northern Afghanistan. *Photograph taken by Hamidullah Amin followed by threats from Soviet guard, c. 1976; part of Atlas of Afghanistan Project under direction of J. Shroder.*

Pipelines

Pipelines generally carry oil and gas, but they can also carry water-borne slurries of metal ores for some distances. The issues of pipelines in Afghanistan has a long history, with energy fluids going both north across the Amu Darya from Shindand into Turkmenistan (Figure 7.15), as well as in the 1980s coming south from Uzbekistan to Bagram to help fight the Soviet–Afghanistan War. These old pipelines have been in disrepair and disuse for decades, since the end of the Soviet–Afghan War.

The idea of a Trans-Afghanistan Pipeline (TAP) or a Turkmenistan–Afghanistan–Pakistan–India pipeline (TAPI) had been broached first by the Unocal oil company in the mid-1990s to move gas across Afghanistan from the Daulatabad field (later also the Iolotan–Osman field) in Turkmenistan alongside the western ring road in Afghanistan from Herat and Kandahar and out into Pakistan through Quetta and Multan, with other connections on to India through Fazilka in Punjab as well (Coll, 2004; Foster, 2008; Ali and Shroder, 2011; Mehdudia, 2010, 2012; Laskar, 2012; Anonymous, 2012b). Various permutations of the idea such as alternate routes through Iran (Figure 7.16), as well as through central Afghanistan to Kabul and Peshawar have been proposed, but the great desires by both Pakistan and India to obtain greater energy supplies have been a strong push to bring the pipeline into reality, although a good deal of discussion over rates of transit fee has ensued recently. The latest estimates of pipeline cost are $7.6 billion, which the four TAPI nations would have to attract commercial investors to finance, construct, and operate.

Electrical Power Grid

Historically, Afghanistan never had electrical power coverage for more than about 22 percent of its population. Only about 70 percent of the original transmission and distribution systems survived the decades of war, with large

Figure 7.16 Map of proposed Turkmenistan–Afghanistan–Pakistan–India Pipeline (TAPI), together with proposed Iran–Pakistan–India Pipeline (IPI). *After Foster, 2008.*

quantities of copper wire and other materials being stolen and sold off as scrap in Pakistan. The existing power supply infrastructure that remained in the country has been extensively damaged and fragmented. The distribution networks in the cities, where still available, are generally rundown and in bad repair because the money and expertise to fix problems are in short supply. Existing power plants frequently can only use a portion of their full potential because of lack of spare parts and maintenance. In practice, many Afghans who can afford it have come to rely upon hydrocarbon-fired generators, which are however, expensive, noisy, and polluting. The continued rise in price of kerosene, gasoline, and diesel makes the use of such generators increasingly expensive. Five to eight years ago only around 7 percent of the population was served with electricity, but by 2011 the number had risen to 10–20 percent (estimates vary), with 60 percent of all connections in Kabul (Figure 7.17). Demand far exceeds the supply with rampant load shedding and programmed outages. Satellite images of Afghanistan at the nights show only a few flyspecks of light in an otherwise stygian darkness everywhere (Fischer, 2011; Rose, 2011; Zorpette, 2011).

Air Transport, Airports and the Afghan Air Force

The history of airplanes in Afghanistan goes back almost a century during the reign of King Amanullah in 1921, when not so long after it was founded, the Soviet Union introduced military aircraft into the country. In 1919 Amanullah had secured from the new Soviet leader, Vladimir Lenin, the promise of an aviation unit being sent to Kabul. The Afghans had experienced being bombed by the British during the Third Anglo–Afghan War and did not care much for the

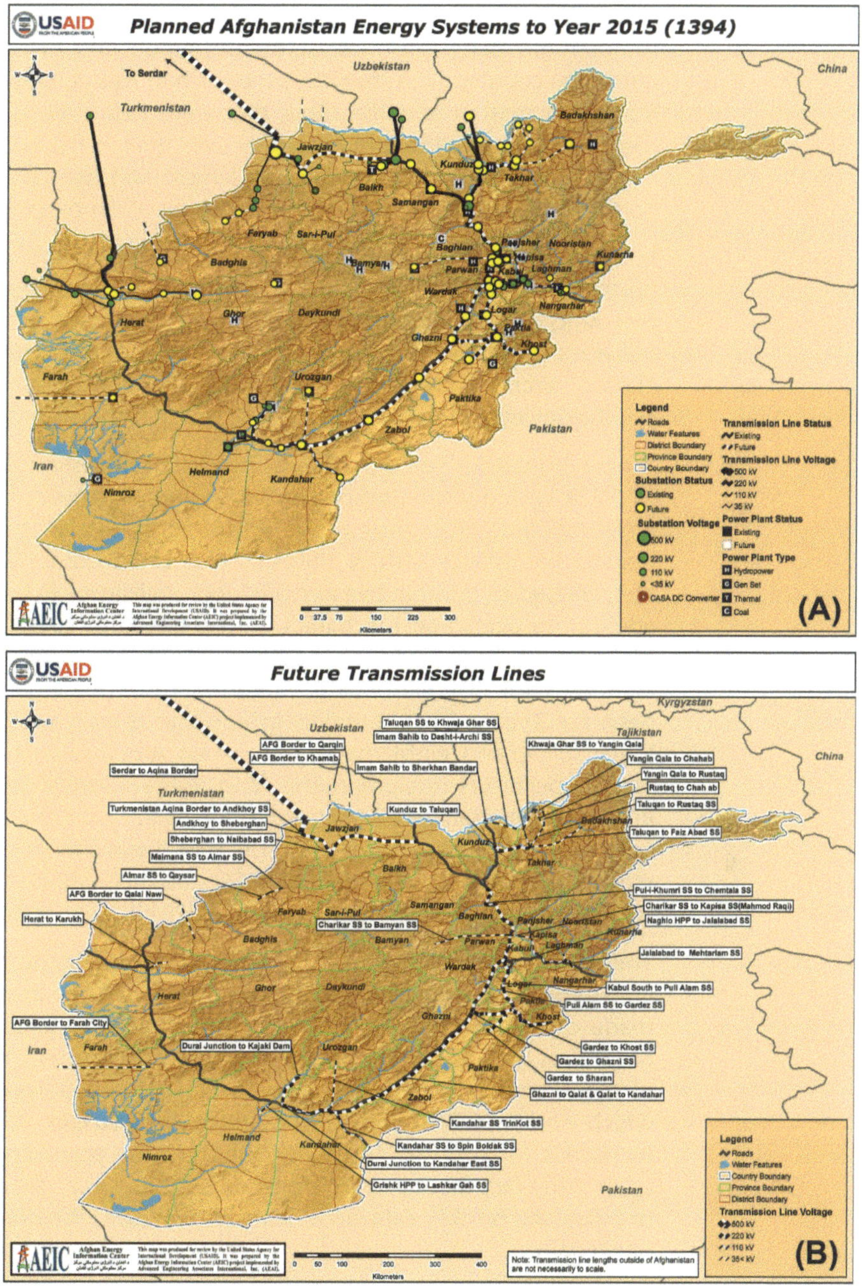

Figure 7.17 **Envisioned transmission lines alignment for Afghanistan.** (A) Planned Afghanistan energy systems to year 2015 (1394). (B) Future transmission lines. *After AIEC, 2011.*

experience, so as a matter of national pride they wanted some similar capabilities. Two years later the Soviets sent three aircraft—a Farman, a Nieuport, and a Sopwith, along with about two dozen Russian personnel—by train to Termez (modern-day Uzbekistan) just across the Amu Darya River on the northern border of the country. Unfortunately, the Farman was destroyed before ever getting inside Afghanistan and the other two were delivered by caravan in damaged pieces, so it is not thought that they ever flew in the country. Meanwhile, the Afghans were busily constructing an airfield and hangar in the Sherpur area of northern Kabul, not far from the present-day US Embassy and Camp Eggers.

In 1922 the Soviets sent several more Nieuports and other aircraft to their northern border with Afghanistan, but halted there when it became apparent that King Ammanullah's government was suspicious of their motives and did not want the Soviet personnel to enter the country. In addition, several Caproni "Scout" aircraft were sent from Italy but they too were damaged en route just like the Soviet aircraft, so it is doubtful too that they ever flew in Afghanistan.

A new Afghan Air Force (AAF) was officially established 3 years later on August 22,1924 with a small number of British aircraft arrived from British India. Two Royal Air Force pilots flew in two Bristol fighter biplanes from Peshawar, which marked the official birth of the AAF. Not to be outdone, the Soviets immediately offered to bring in five aircraft—four Polikarpov R-1s and a Junkers F 13, in addition to pilots and mechanics. In the spring of 1925, the first Afghan pilot, Muhammad Ihsan Khan, an observer, Muhammad Hashim Khan, and an aircraft mechanic, Ghulam Dastgir, returned from training in Italy where they had been for 3 years. That summer, in addition there were five Soviet pilots, one Indian pilot, and eight mechanics, when Afghanistan had its first air parade where spectators, including the King, were allowed to go aloft. Four more Russian R-1 aircraft arrived safely from Termez, with two others having crashed en route, but these were replaced within 3 months. In the fall, the first flying school in Afghanistan opened with 15 students, while another 10 went to the USSR for training (Anonymous, n.d. circa 2010; Cooper and Ali, 2003).

The Soviets ultimately provided 13 aircraft and pilots to build up the new AAF and throughout the 1920s to further develop flight in the country, they also made cash subsidies, and provided other transport and communication technicians, as well as carrying out the laying of some telephone lines. About half of the aircraft provided were biwing Polikarpov R-1 reconnaissance bombers, which were a Russian copy of the British Airco DH.9A. King Amanullah wanted to have a force of 36 airplanes, but that was exceedingly difficult to do up against the fact that all spare parts, oil, and fuel had to be brought in by camel or horse and donkey caravan at least 350 km from the north through mountainous terrain.

The earliest Civil Aviation Service also began in the latter half of that decade when the fledgling AAF conducted liaison and mail service with royal visits to

Kandahar and Jalalabad. Air and passenger service between Kabul and Tashkent was undertaken by a Soviet airline in 1928. Some of these flights were flown by the Junkers F 13 which belongs to the AAF. As the AAF continued to develop in the late 1920s there were 24 pilots available: 12 Russians, 10 Afghans, and two Germans. The Soviet copy of the DeHaviland DH-9A, the R-1s, were the most serviceable aircraft in 1928 and that year 16 more were delivered into Afghanistan. In addition, as a result of his visit to Europe, King Ammanullah received a number of other aircraft as gifts: a Junkers G 24 passenger plane from Germany, a Potez 25 from France, and a Bartel BM-4b trainer from Poland, and he purchased two more Junkers F 13s from Germany as well.

But Ammanullah's many other reforms led to the breakout of the First Afghanistan Civil War in 1928–1929, in which the capital, Kabul, was taken over by the forces of Habibullah, a Tajik from the north, which was fairly catastrophic for the nascent air force. In the midst of the civil war chaos, the British mounted a rescue operation for their own diplomatic mission, and that of a number of other countries as well. Most of the Afghan aircraft were destroyed in the civil war and the AAF was largely out of commission until rebuilding began.

Only the four R-1s that had made it through the war and the one Junkers F 13 were still serviceable; the Soviet pilots had departed for home. The Afghan service staff kept some of these going for over a decade thereafter. Thus it was not until 1937 that a serious effort began to rebuild the air force so that by about a year later some 30 new planes were in service, consisting mostly of British Hawker Hind and Italian IMAM Ro. 37 aircraft. Of the first eight Hawker Hinds built for Afghanistan, four aircraft were fitted with Rolls-Royce Kestrel V engines, plus the other four were fitted with Kestrel UDR engines; 28 planes were sent in total between 1937 and 1939. The Afghans also purchased 16 Italian IMAM Ro.37s, which were airplanes similar to the Hawker Hinds, and were biplanes of mixed construction, with two seats, and a 560 hp Fiat A.30 inline engine that could do 300 km/h (162 knots). The Ro.37 had a 7000-m ceiling, 3000-m climb in 11 min, and over a 1200 km (750 miles) endurance. It had three machine guns (two in the nose and one on the side), twelve 15-kg bombs, and good agility. The Ro.37 was later fitted with the 600 hp Piaggio P.IX radial engine, which had a better reliability than the Fiat, so this was the main version produced. In the spring of 1940, in Kabul existed the AAF headquarters, the flight school with five Breda trainers, and three AAF squadrons (No. 1 – 12 Hawker Hinds; No. 2 – 9-10 Ro. 37; No. 3 – 10 Hawker Hinds). Some of the Hawker Hinds remained in the AAF inventory until 1957, and one from Afghanistan still flies in the famous Shuttleworth Collection of flying antiques in England.

In the post-World War II of 1947, the AAF was redesignated as the Royal Afghan Air Force (RAAF), a title it held through the 1960s when the RAAF consisted of about 100 combat aircraft from the USSR, including MiG-15 fighters, Il-28 light bombers, transport aircraft, and a few helicopters. Also by that time, a number of Afghans were off receiving flight training in the Soviet Union, the USA, India, and several European countries. In 1973 the *coup d'etat* engineered by Daoud Khan permanently ousted King Zahir Shah from power, as well as the "Royal" from the AAF name and brought in even greater Soviet assistance and aircraft, but even so, at first the great expense of a national air force meant that it was neither very advanced nor well maintained.

As the Soviets became more influential in the late 1970s under President Daoud, the AAF began to become a more powerful arm of the Afghan military and by 1978 it actually had a strength of about 180 warplanes, including 80 MiG-17 "Frescoes" and MiG-15UTI "Fagots", about 20 Su-7BMK "Fitters", and 30 Il-28 "Beagle" light bombers. In terms of helicopters, they had about 19 Mi-4 "Hounds" and Mi-8 "Hips", along with some 12 newly delivered Mi-24 "Krokodil" (Hind) gunships, which for Afghanistan at the time were state-of-the-art. In mid-1979 an additional 18 Mi-25D Krokodil helicopters, especially designed for export from the USSR, were added to the mix. The AAF in the late 1970s also had a number of reasonably well-developed airfields, including the large facility at Bagram in the north of Kabul, the Kabul International Airport, Shindand near Herat, the airports near the cities of Mazari Sharif, Herat, Jalalabad, and Kandahar, as well as minor landing facilities at Farah, Ghurian, Zaranj, Ghazni, Kunduz, and Bazai-Bumgaz.

The organization of the AAF in 1988 included:

- 322nd Air Regiment, Bagram Airbase—three fighter squadrons with 40 MiG-21s.
- 321st Air Regiment, Bagram Airbase—three fighter/bomber squadrons with Su-7 and Su-22s.
- 393rd Air Regiment, Dehadi Airbase (near Balkh [Mazari Sharif])—three fighter/bomber squadrons with MiG-17s.
- 355th Air Regiment, Shindand Airbase (south of Herat)—three bomber squadrons with Il-28s and one fighter bomber squadron with Mig-17s.
- 232nd Air Regiment, Kabul International Airport—three helicopter squadrons with Mi-4, Mi-6, and Mi-8s, with another squadron detached to Shindand.
- 377th Air Regiment, Kabul International Airport—four helicopter squadrons with Mi-17s and Mi-25s.
- Unknown air regiment, Kabul International Airport—two transport squadrons with An-2, An-26/30, and one transport squadron for VIPs, an Il-18, and a dozen An-14s.
- Unknown attack Helicopter squadron, Kabul International Airport—15 Mi-24s.
- Unknown attack Helicopter squadron, Jalalabad—15 Mi-24s.
- Air Defense Forces consisting of 92nd and 99th Surface to Air Missile Regiments and an unknown Radar Regiment at Kabul, and the 66th AAA Battalion at Kandahar.

The total number of AAF personnel at that time was about 7000, with about 240 fixed-wing combat aircraft of the types noted above, as well as 150 helicopters and possibly 40 or more Antonov transport aircraft. Most of these were noted by some observers to be either already obsolete or obsolescent; whereas others thought that the AAF was at its strongest in the 1980s and early 1990s and even was viewed by some as a possible threat to neighboring countries. Some of this uncertainty may have been about exactly which were only Soviet activities or aircraft, and which were Afghan, because there was so much exchange between the two groups. About 5000 Czechoslovak and Cuban military advisers also were attached to the AAF, along with large numbers of Soviet personnel. The few available reports from Russian sources who had been there was that Afghans were excellent fliers but terrible and undisciplined pilots and officers. In other words, in terms of training and reliability, the quality of pilots and other staff was low, with the result that they were often denied access to advanced aircraft. Furthermore, the Afghan pilots and officers were seen as not tending to care very much about tactics or weapons parameters either. But some of them did succeed in flying as defectors in several late model Mi-24D helicopter gunships into Pakistan where it was discovered that the planes had been equipped with electronic equipment designed to help with the difficult flying against the mujahideen in the rugged, mountainous terrain.

With the departure of the Soviets in 1988–1989, the AAF began a rapid decline in terms of operational capability and after the collapse of the Najibullah government in 1992 in the civil war the AAS broke into factions, with parts going to the different entities between the Taliban, the Northern Alliance/United Front, and various warlords. By the end of the 1990s, the Taliban actually still held five MiG-21MFs from the 1960s and 10 Sukhoi-22 fighters from the 1970s, as well as six Mi-8 helicopters, five Mi-35s, five L-39Cs, six An-12s, 25 An-26s, a dozen An-24/32s, an IL-18, and a Yakovlev transporter. Their civilian air service contained two Boeing 727A/Bs, a Tu-154, five An-24s, and a DHC-6. The aging aircraft could sometimes take to the air, maintenance permitting, but the pilots were also aging and new ones were not being trained. In 1995 the Taliban in Kandahar did succeed in using one of their MiG-21 aircraft to force down a privately owned Ilyushin Il-76 piloted by Russians to land in Kandahar where the crew was interned for over a year. Finally, by dint of trickery and some minimal force the crew managed to overpower their captors, restart the plane in the knick of time, and fly to freedom.

Following the attack on the USA on September 11, 2001, by the 9th of October the USA was able to declare air superiority over the AAF remnants run by the Taliban, just 2 days after the invasion had begun. In fact the US Coalition operations in the fall of 2001 succeeded in destroying most of the remainder of the Taliban Air Force of Afghanistan, such as it was. Afghanistan was then without an air force until a US-led international effort began in 2004–2005 to rebuild air service in the country that became the basis for the Combined Air

Power Transition Force. In the course of rebuilding the Afghan National Army, plans were made to carve out an Air Corps of some 7000 members. The initial plan approved in 2007 was for a fully self-sustaining and operationally capable Afghan National Army Air Corps (ANAAC) organized into three wings—one each for presidential airlift, rotary-wing, and fixed-wing operations.

By October 2008 as a result of numerous donations and aircraft refurbishments by various national donor states, the Afghanistan aircraft inventory included 21 rotary-wing and 10 fixed-wing that consisted of 18 MiG-17s, 3 MIG-35s, 6 An-32s, 2 An-26s, and 2 L-39s. A variety of technical training courses are now being offered in Afghanistan and new groups of Afghan pilots have been trained in USA. The new headquarters of the ANAAC was opened in the North Kabul International Airport cantonment area, newly cleared of unexploded ordnance and mines.

It is thought by many that in the absence of railroads or other reliable ground transportation, that air power is essential for security and good governance in Afghanistan. Air power appears to be the best means by which the government may touch the lives in a positive way of many Afghans living in the remote and inaccessible villages deep in the Hindu Kush. This has obvious implications to Coalition forces in terms of the ongoing conflict: only time will tell.

Air transport in modern Afghanistan evolved out of the many stutter-step changes that have occurred in the previous nine decades of growth, war, and stagnation or catastrophic steps backward. The rebuilding phases following the Coalition invasion in 2001 ultimately led to much refurbishment. Foreign airlines such as Turkish Airlines, Gulf Air, Indian Airlines, Pakistan International Airlines, all returned to Afghanistan, and Ariana Afghan Airlines, the national airways, resumed service, along with several private companies such as Kam Air, Pamir Airways, and Safi Airways. Some 53 airports have been opened, with about 19 having at least some paved runways (Figure 7.18). The Kabul International Airport (Figure 7.19) is the country's largest airport and the main center for international civilian flights. Kandahar, Herat, and Mazari Sharif are all also international airports, as indeed Ghazni will also become in 2013. At least 10–12 other domestic airports exist, as well as more than a dozen helicopter access airfields. The Kandahar Airfield, which was built originally by the Americans in the 1950s as part of the Helmand Valley Authority, was designed originally as a way station for long-distance propeller-driven aircraft just prior to the onset of large-scale jet traffic world-wide in the 1960s and thereby became a bit irrelevant, except when winter weather in Kabul diverted air traffic south to lower, warmer elevations on occasion (as noted in Chapter 1, I first landed there in 1973 from New Delhi instead of the scheduled Kabul). Nowadays, the Kandahar Airport serves mainly as the US Coalition Airbase as well.

The largest airfield in Afghanistan is Bagram, some 42 km north of Kabul. In 2013 this base was dominated by the North Atlantic Treaty Organization

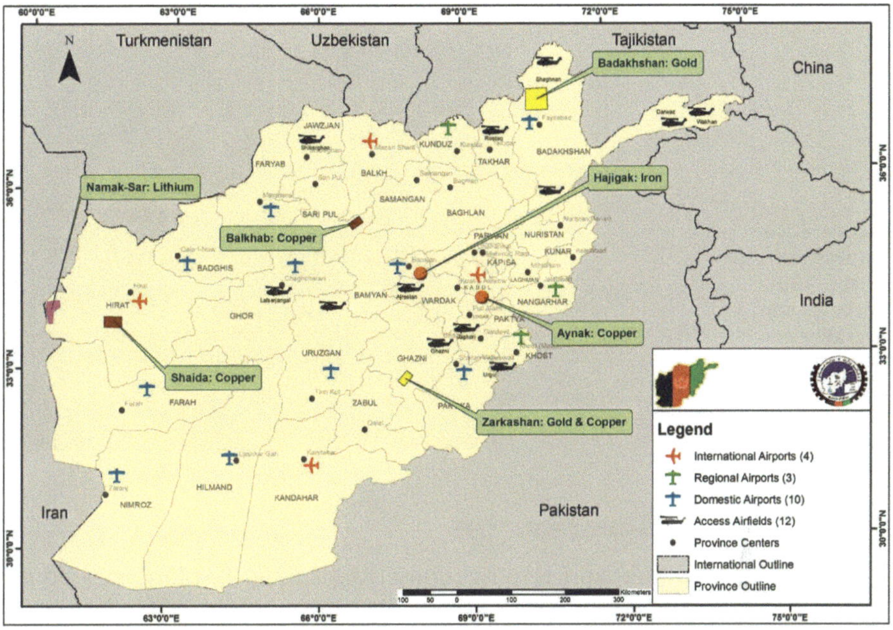

Figure 7.18 Airports in Afghanistan. *After Ministry of Mines, c. 2010.*

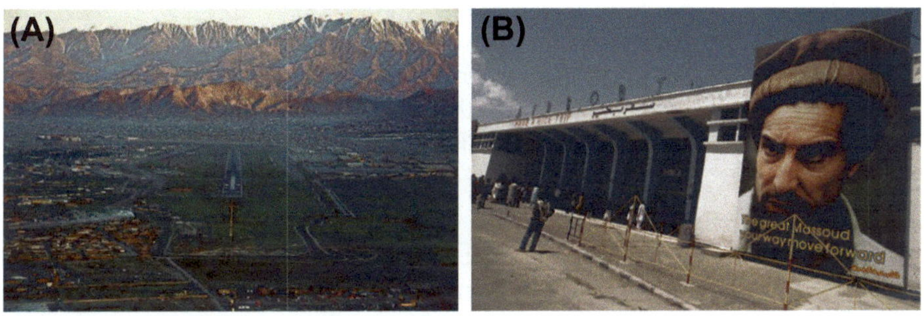

Figure 7.19 (A) The Kabul International Airport from the southeast approach. *(Photograph by Abdul Raqib, Google Earth™)*. (B) The new departure building is on the left, with the exterior façade. *Photograph by David Sepeau, Google Earth™.*

(NATO)-led forces of International Security Assistance Force (ISAF) and had heavy military traffic, especially helicopters and large military transport such as the Boeing 747s, C-5 Galaxy, and C-17 Globemaster III military cargo planes, as well as many other jet aircraft (Figure 7.20). Once the USA and ISAF military forces depart, except for whatever holding forces they leave behind, the base will revert back to the AAF, who will be expected to maintain air control over the eastern regions of the country.

Figure 7.20 **Bagram Airbase with C-130 Hercules cargo aircraft parked below the control tower.** *Photograph by Artur Zielinski, Google Earth™.*

Seaports

As a landlocked country, Afghanistan has long-desired access to the sea, particularly during its periodic political crises with Pakistan when the border has been closed, a country through which much of Afghanistan's imports have long had to travel (Dupree, 1973; Barfield, 2010). The seaport of Karachi, Pakistan, has long served as Afghanistan's only access to the sea, imperfect as that might have been for many decades. In recent years, however, two other port cities have begun to open up new opportunities, even though neither is completely perfect because both are still in other countries; Gwadar in the far southwestern part of Baluchistan Province in Pakistan, and Chah Bahar in extreme southeastern Iran.

Gwadar began life as a small and shallow port owned by the Sultanate of Oman over a century ago but was sold to Pakistan in 1957. Even though it occupied a highly strategic location at the mouth of the Persian Gulf, the port languished until in 2002 the Chinese government signed a deal with Pakistan to pay $198 million of the required $248 million of Phase I of rebuilding Gwadar as a major port city with three multipurpose ship berths by 2006 (Haider, 2005) (Figure 7.21). Phase II at $600 million was to include nine additional berths, a bulk cargo terminal, a grain terminal, and two oil terminals, but these stalled. Several low capacity highways across southern coastal and central Baluchistan connect the port to the rest of Pakistan, and an improved new highway to the extreme northwest corner of Baluchistan at Saindak has been under construction since at least 2005, with a plan to develop access from the north to the states in Central Asia, presumably through Afghanistan. A pipeline from the port was also planned. Management of the port was deeded to the Port of Singapore Authority for 40 years, but the failure of the Government of Pakistan

Figure 7.21 Gwadar Port in southwestern Baluchistan, Pakistan, with new ship unloading facilities. *Photograph by Skybolt; from Google Earth™.*

to reconstruct sufficient heavy transport highways out of Gwador, predatory real estate speculation, and other problems led to transfer of the port authority to the Chinese Overseas Port Holdings company, who are expected to greatly improve the commercially troubled undertaking (Walsh, 2013).

Iran and India watched the development of the Gwador port with considerable unease; India because of fears of further encirclement by China, and a greater strategic depth created for the Pakistani Navy, and Iran because of its competition for access to the states of Central Asia, as well as to Afghanistan. Iran and India's response to Gwadar was to tacitly compete with Pakistan in capturing access routes and energy-related trade with Afghanistan, Kazakhstan, Kyrgyzstan, and Tajikistan. Accordingly, India built the highway connecting Zaranj on the Iranian—Afghanistan border on into western Afghanistan to Dilaram and the ring road, and the two countries signed the Roadmap to Strategic Cooperation by which India is to help build the port of Chah Bahar, and other road link connectors through Iran, Afghanistan, and Tajikistan, ultimately to China. Multiencirclements with India on both sides of Pakistan add to the crisscrossing competitions to establish spheres of influence and strategic depths.

Iran has warm relations with the Central Asia states as well as with President Karzai, who is not happy with Pakistan for supporting the Taliban, so this gives the Iranians a partial upper hand with Pakistan. They are also giving Afghanistan a substantial discount on port fees, with the exception of oil tankers, so that it reflects a healthy relationship for Afghanistan, but terrifies the Pakistanis for fear of the loss of importance of Gwador, once billed as a coming "Rotterdam" of South Asia. The Gwador–Chah Bahar competition is heating up, but however it goes, it appears to favor Afghanistan the most, but only if the jostling for influence continues at fairly relaxed paces.

Coalition Military Supply Routes

The fighting of any war requires a tremendous amount of supply to keep the weapons, materiel, fuel, food, and the myriad of other logistical requirements to maintain the effort. Military logistics, or what is generally known as combat service support, addresses highly uncertain conditions of the war effort wherein protecting one's own supply lines, or attacking those of the enemy, are fundamental. Distances are also critical, which in the case of the war in Afghanistan, is excessively complex and lengthy, with much of it traversing what is essentially the enemy territory of Pakistan (Figure 7.22) (Radin, 2011).

In 2008, about 90 percent of the military supplies to support US and Coalition forces arrived by ship at the port of Karachi, Pakistan, on the Indian Ocean. From there the military supplies were off-loaded and transported by truck across Pakistan to the two main access points into Afghanistan: the Chaman–Spin Baldak crossing point in the south, and the Torkham crossing point in the north. Most of the trucks that are driven across Pakistan are in the hands of the

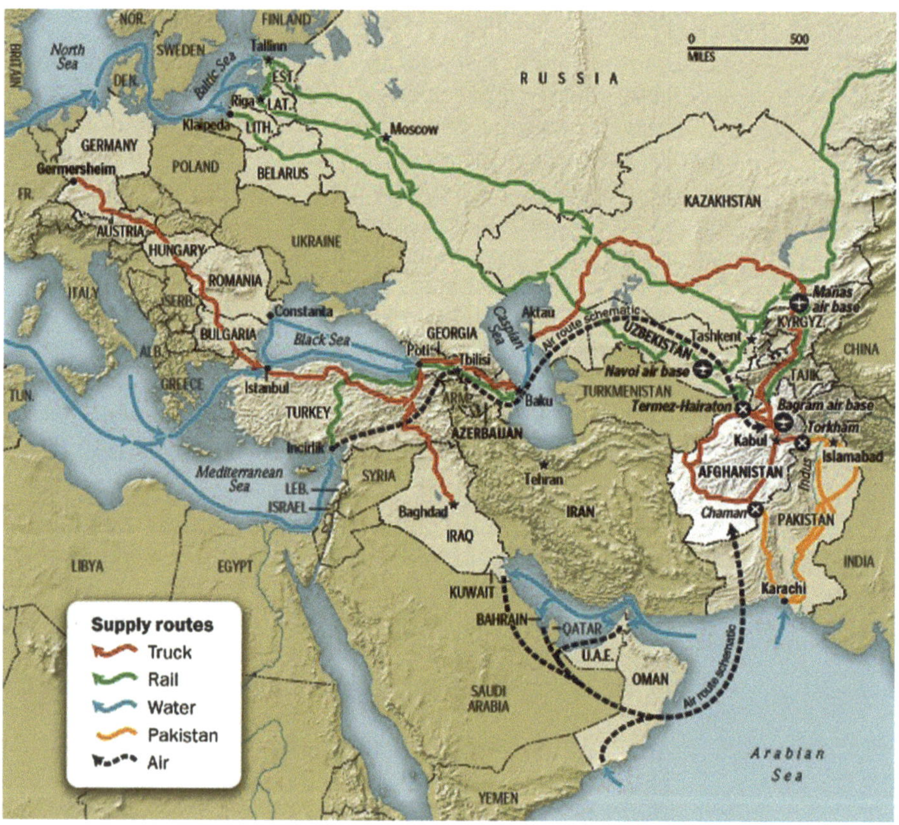

Figure 7.22 **The supply routes of the Coalition forces by overland truck, railway, ship, and air (air routes are diagnostic only) into and out of Afghanistan.** *Map after National Geospatial Intelligence Agency and Gene Thorp; Washington Post July 2, 2011.*

so-called Pashtun mafia (particularly members of the Shinwari and Afridi tribes) that is a kind of guild responsible for much of the transport in both the countries. Much of this trucking mafia is riddled with Taliban supporters, with the result that transshipments can be highly vulnerable (Whitlock, 2011).

The southern route of transport passes through the province of Baluchistan and the city of Quetta, before traversing the Khojak Pass into the Chaman border crossing into Spin Baldak, Afghanistan, and north to Kandahar. The northern route is through the province of Khyber Puktunkhwa (formerly North-West Frontier Province) into Peshawar and the Jamrud Fort gate to the Khyber Pass, at the top of which is the almost completely lawless, drug smuggler's town of Landi Kotal, before descending out of the pass down to the Torkham gate into Afghanistan. From there the transport is through the city of Jalalabad and on up the torturous road through the Kabul River Canyon to Kabul.

Most of the military supplies that arrive by these two routes through Pakistan are noncombat materials such as food, water, construction supplies, and fuel, the latter of which is especially vulnerable to attack from Taliban saboteurs who have blown up tanker trucks repeatedly on both supply routes. In February 2009, a key bridge on the Khyber Pass route was also blown up by Taliban insurgents but the blockage was only temporary as the Pakistani military is highly efficient at installation of Bailey bridges and other temporary structures to keep the roads open. On other occasions, for example, in 2010 when helicopters from the US-NATO ISAF twice strafed Pakistani border posts at Kurram, the Government of Pakistan closed the Torkham gate for a time, which in 1 day forced over 150 trucks off the road. The same thing happened in November 2011, when ISAF helicopters inadvertently killed a number of Pakistani border guards and the Government of Pakistan closed the border crossings once again (Jones, 2010; Anonymous, 2008).

To decrease its dependence on the unreliable routes through Pakistan (Anonymous, 2008), the US military and Coalition forces have greatly expanded their use of supply lines overland through Russia and Central Asia to deliver supplies along several routes known collectively as the Northern Distribution Network (NDN). The goal of the NDN is to bring supplies in from the north through Central Asia, Russia, and the Caucasus region. Since 2008 the US-led Coalition has worked out deals with a number of countries for transit rights and infrastructure improvements in: (1) the Baltic seaports of Estonia, Latvia, and Lithuania; (2) the Russian railroad network from the Baltic states to Central Asia; (3) the Caucasus states of Georgia and Azerbaijan with their ports and railroad networks from the Black and Caspian seas into Central Asia; and (4) the Central Asian states of Kazakhstan, Uzbekistan, Kyrgyzstan, and Tajikistan with railroad networks from Russia and the Caspian Sea into Afghanistan. Some of the new deals also include air transit rights so that some deliveries are air freight. These routes are much more stable than those through Pakistan, but are more circuitous and thus far more expensive. In addition, many are subject to other restrictions, such as other national interests, including shifting Russian policy demands. The far southern route avoids Russia but is subject to several on- and off-loading procedures for different transportation modes along the way, which limits capacity and adds greatly to expense. Since 2009, the NDN has been ratcheting up supply delivery and by the end of 2011 it is expected to be above 75 percent of the total supplies delivered, perhaps even higher if the Pakistan border is closed for any reason, which it often is (Anonymous, 2011).

Essential military supplies come into Afghanistan by other routes, especially by cargo aircraft into the Bagram Airbase north of Kabul, but also into the Navoi Airbase in Uzbekistan directly north of Afghanistan, from whence the trucking into Afghanistan is much safer and shorter. Some air cargo also lands at other bases in Afghanistan as well.

The Kyrgyz government had announced that it would close its US Airbase at Manas but new negotiations and the payment of three times more money has allowed the base to continue to function, mainly as a troop transit center. In addition, reports have also indicated that some of the NATO–ISAF countries have thought of shortening their supply lines by utilizing transport through Iran, but this is not viewed with favor by the USA, mainly because of Iranian support of terrorism elsewhere in the Middle East (Tynan, 2009).

As the US military draws down its forces in the country preparatory to presumed near full departure by the end of 2014, vast quantities of materiel have to be removed by some of the same routes it was brought in on (Hodge, 2013). Because the Pakistani routes are so unreliable and dangerous, new staging grounds for departure are being established north of the Hindu Kush at Camp John Pratt in Mazari Sharif, which is adjacent to the commercial airport, as well as the new rail line through Hairatan and across the road and rail bridge into Uzbekistan. In early 2013, the routes out were being tested to see if the routes, permits, and agreements were all functioning as well as they had on the way in. Securing export permits from the GIRoA was likely to a source of considerable friction because of customs rules of export that either did not exist or had not been followed when the materiel was imported in the initial invasion and later materiel augmentation. The Afghanistan Custom's and Revenue Department is likely to have its hand out for considerable revenue enhancement in this process.

References

(All web references accessed February 2013.)

AEIC, 2011. Afghanistan Energy Infrastructure Map Book. Afghan Energy Information Center; Ministry of Energy and Water; USAID (US Agency for International Development); 19 p. http://www.afghaneic.org/library/maps/Mapbook/high/Afghanistan%20Energy%20Infrastructure%20Map%20Book.pdf.

Afghanistan Ministry of Mines. http://mom.gov.af/Content/files/Afghanistan%20Civil%20Aviation%20Infrastructre.jpg.
Ali, S., J.F. Shroder, Jr., 2011. Afghanistan's mineral fortune: Multinational influence and development in a post-war economy. Institute for Environmental Diplomacy and Security, James Jeffords Center for Policy Research, University of Vermont, Burlington, VT, 24 pages.
Anonymous, September 7, 2008. Pakistan "Blocks" Fuel to US Forces. Aljazeera.
Anonymous, n.d. But probably 2010. Building Air Power for Afghanistan: A Short History of the Combined Air Power Transition Force and the 438th Air Expeditionary Wing.
Anonymous, 2011a. Coalition Supply Routes. http://www.nps.edu/programs/ccs/.
Anonymous, 2011b. Military logistics. http://en.wikipedia.org/wiki/Military_logistics.
Anonymous, June 7, 2012a. Chinese Engineers Arrive in Afghanistan to Plan Railway. ToloNews, 1 p. http://tolonews.com/en/afghanistan/6498-chinese-engineers-arrive-in-afghanistan-to-plan-railway.
Anonymous, May 23, 2012b. Historic Agreements Bring Long-Awaited TAPI Pipeline Closer to Reality. Asian Development Bank, 1 p. http://www.adb.org/news/historic-agreements-bring-long-awaited-tapi-pipeline-closer-reality.
Anonymous, 2013. Road Improvement in Afghanistan. United Nations Operations. 1 p.
Barfield, T., 2010. Afghanistan: A Cultural and Political History. Princeton University Press, Princeton, NJ, 389 p.
Coll, S., 2004. Ghost Wars: The Secret History of the CIA, Afghanistan, and bin Laden, from the Soviet Invasion to September 10, 2001. The Penguin Press, 695.
Cooper, T., Ali, K.S.S., October 29, 2003. Afghanistan, 1979–2001; Part 1; Afghan Air Force in the late 1970s. Indian Sub-continent Database.
Cronin, X.A., 2010. The Asphalt Ribbon Rebuilding the Kabul to Kandahar Highway. The Louis Berger Group, 200 p. http://www.louisberger.com/OurProjects/Asia/~/mediAfghanistan.pdf.
Dupree, L., 1973. Afghanistan. Princeton University Press, Princeton, NJ, 760 p.
Fischer, H., 2011. Expansion of Electricity Supply. German Development Cooperation with Afghanistan, German Embassy, Kabul, 2 p. http://www.ez-afghanistan.de/fileadmin/content/fact-sheets/englisch/2012-01-energie-sektor-e.pdf.
Foster, J., 2008. A pipeline through a troubled land: Afghanistan, Canada, and the new energy game. Foreign Policy Series; Canadian Centre for Policy Alternatives; 3(1); June 19; 17 pages.
GAO, 2008. Afghanistan Reconstruction: Progress Made in Constructing Roads, but Assessments for Determining Impact and a Sustainable Maintenance Program are Needed. Report to Congressional Committees. US Government Accountability Office. 55 p.
Grantham, A., 2013. Railways of Afghanistan. http://www.andrewgrantham.co.uk/afghanistan/.
Gregorian, V., 1969. The Emergence of Modern Afghanistan: Politics of Reform and Modernization, 1880–1946. Stanford University Press, Stanford, CA, 586 p.
Hagen, E., Shroder Jr., J.F., Lu, X.X., Teufert, J.F., 2010. Reverse engineered flood hazard mapping in Afghanistan: a parsimonious flood map model for developing countries. Quaternary International 226 (1–2), 82–91.
Haider, Z., 2005. Baluchis, Beijing, and Pakistan's Gwadar port. Georgetown Journal of International Affairs. 6 (1), 95–103. http://journal.georgetown.edu/wp-content/uploads/6.1-Haider.pdf.
Henderson, B., December 21, 2011. Afghanistan Opens First Ever Train Route. The Telegraph, 2 p. http://www.telegraph.co.uk/news/worldnews/asia/afghanistan/8970647/Afghanistan-opens-first-ever-train-route.html.
Hodge, N., February 8, 2013. U.S. sets stage for exodus from Afghanistan. Wall Street Journal, 3 p. http://online.wsj.com/article/SB10001424127887323452204578292094085822794.html.
Isphani, M.Z., 1989. Roads and Rivals: The Politics of Access in the Borderlands of Asia. I.B. Tauris & Co. Ltd, Publishers, London, 286 p.
Johnson, K., March 11, 2012. Afghanistan Opens First Major Railroad. Southeast Missourian, 2 p. http://www.semissourian.com/story/1824648.html.

Jones, Keith, October 1, 2010. Pakistan Shuts Key Afghan War Supply-Route. World Socialist Website. http://www.wsws.org/articles/2010/oct2010/paki-o01.shtml.

Kuchins, A.C., 2011. A truly regional economic strategy for Afghanistan. The Washington Quarterly. 34 (2), 77–91. http://csis.org/files/publication/twq11springkuchins.pdf.

Laskar, R.H., April 18, 2012. TAPI Project: India, Afghanistan Fail to Agree on Transit Fee. The Navhind Times. http://www.navhindtimes.in/business/tapi-project-india-afghanistan-fail-agree-transit-fee.

Mehdudia, S., December 11, 2010. TAPI Project Will Be the New Silk Route, Says Deora. The Hindu, 2 p. http://www.thehindu.com/business/Economy/article946191.ece.

Mehdudia, S., January 25, 2012. TAPI Project Takes a Step Forward. The Hindu, 2 p. http://www.thehindu.com/news/national/article2832202.ece.

Nelson, D., November 2, 2011. India Plans 'World's Most Dangerous Railroad' from Afghanistan to Iran. The Telegraph, 2 p. http://www.telegraph.co.uk/news/worldnews/asia/india/8862583/India-plans-worlds-most-dangerous-railroad-from-Afghanistan-to-Iran.html.

Nissenbaum, D., February 10, 2012. Roads to nowhere: program to win over Afghans fails. Wall Street Journal (US Edition). http://online.wsj.com/article/SB10001424052970203554104576655280219991322.html.

Norling, N., 2011. Afghanistan's Railroad Frenzy. Central Asia – Caucasus Institute Analyst; Johns Hopkins University, 3 p. http://www.cacianalyst.org/?q=node/5629/print.

Radin, C.J., December 4, 2011. Analysis: the US-Pakistan relationship and the critical factor of supply. The Long War Journal. http://www.longwarjournal.org/archives/2011/12/us_pakistani_relatio.php.

Risen, J., November 5, 2010. War Reconstruction Fraud Draws Big Fines. New York Times, Asia Pacific.

Roscoe, D.E., 2007. Army Engineers Work to Connect Afghanistan One Road at a Time. US Army, 1 p. http://www.army.mil/article/4446/Army_Engineers_Work_to_Connect_Afghanistan_One_Road_at_a_Time/.

Rose, A.J., May 25, 2011. Afghanistan Electricity: After Years of Rebuilding, Most Afghans Lack Power. Huffington Post, 4 p. http://www.huffingtonpost.com/2010/07/19/afghanistan-electricity-a_n_650713.html.

Rubin, A.J., Risen, J., May 1, 2011. Costly Afghanistan Road Project is Marred by Unsavory Alliances. New York Times, Asia Pacific. http://www.nytimes.com/2011/05/01/world/asia/01road.html?pagewanted=all.

Schubert, F.N., 1991. U.S. Army corps of engineers and Afghanistan's highways 1960 — 1967. Journal of Construction Engineering and Management 117 (3), 445–459.

Shalizi, H., May 27, 2012. Afghan Road Lifelines Blocked by Graft, Kickbacks. Reuters. 3 p.

Shroder Jr., J.F., 1983. The U.S.S.R. and Afghanistan mineral resources. In: Agnew, A. (Ed.), International Mineral Resources a National Perspective. American Association for Advancement of Science, pp. 115–153.

Shroder Jr., J.F., 1989. Afghanistan resources and Soviet policy in Central and South Asia. In: Hauner, M., Canfield, R. (Eds.), Afghanistan and the Soviet Union: Collision and Transformation. Westview Press, pp. 101–119.

Shroder, J., 2007. Afghanistan's development and functionality: renewing a collapsed state. Geojournal 70, 91–107.

Shroder Jr., J.F., Weihs, B., Schettler, M., 2011. Mass movement in northeast Afghanistan. Journal of Physics and Chemistry of the Earth 36, 1267–1286. http://dx.doi.org/10.1016/j.pce. 2011.03.003.

Starr, S.F., Kuchins, A.C., 2010. The Key to Success in Afghanistan: A Modern Silk Road Strategy. Central Asia-Caucasus Institute and Silk Road Studies Program, Washington, DC. http://www.silkroadstudies.org/new/docs/silkroadpapers/1005Afghan.pdf

Toynbee, A.J., 1961a. Between Oxus and Jumna. Oxford University Press, Oxford, UK, 211 p.

Toynbee, A.J., April 1961b. Impressions of Afghanistan and Pakistan's North-West Frontier. International Affairs 37, 158–163.
Tynan, Deidre, May 10, 2009. Uzbekistan: Karimov Gives Washington the Air Base It Needs for Afghan Operations. Eurasianet.org. http://www.eurasianet.org/departments/insightb/articles/eav051109a.shtml.
Wallechinsky, D., Brinkerhoff, N., January 3, 2012. USAID Road Building in Afghanistan... A Study in Waste. AllGov (Everything our government really does). news=843825.
Walsh, D., February 1, 2013. Chinese Company Will Run Strategic Pakistani Port. New York Times, p. A6 http://www.nytimes.com/2013/02/01/world/asia/chinese-firm-will-run-strategic-pakistani-port-at-gwadar.html?_r=0.
Whitlock, Craig, July 2, 2011. U.S. Turns to Other Routes to Supply Afghan War as Relations with Pakistan Fray. Washington Post. http://www.washingtonpost.com/world/national-security/us-turns-to-other-routes-to-supply-afghan-war-as-relations-with-pakistan-fray/2011/06/30/AGfflYvH_story.html.
Zorpette, G., July 5, 2011. Struggling for Power in Afghanistan. New York Times, 3 p. http://www.nytimes.com/2011/07/06/opinion/06zorpette.html?pagewanted=all&_r=0.

In this Chapter

8

Hazards and Disasters in Afghanistan

Abstract

The natural hazards that affect people and infrastructure in Afghanistan are considerable and have produced numerous disasters in the past. Earthquakes, floods, droughts, sandstorms and dust storms, rapid wet debris flows, snow avalanches, landslides, and extreme weather events are the chief threats. A number of emergency plans and disaster management operational systems have been implemented to attempt to control the negative results of these hazards. New plans to address the hazardous effects of climate change are also underway.

Keywords: Afghanistan Disaster Knowledge Network; Afghanistan Flood Hazard Map; Droughts; Dust storms; Earthquakes; Extreme weather events; Floods; Landslides; National Adaptation Programme of Action for Climate Change (NAPA); National Disaster Management Plan (NDMP); National Emergency Operations Centre; Rapid wet debris flows; Sandstorms; Snow avalanches.

Some 2.2 million years ago near the beginning of the Pleistocene Ice Ages, a large stratovolcano about 60 km in diameter at its base loomed above the outskirts of what much later became Ghazni, Afghanistan. This huge volcano, with a summit icecap and glaciers coming down its sides may have been over 6000 m high and possibly resembled Mt. Kilimanjaro in East Africa. Great pressure and heat had been building up inside the volcano for a long time, when finally the top of the mountain blew off, it left behind a gigantic caldera hole (~25 × 50 km) in the ground, Dashti Nawar, the flat bottom of which ended up at an elevation of about 3100 m (See Chapter 5 with its section on lakes). Kohi Safed peak (4750 m) on its northwest flank is a snowcapped remnant of the volcano's flanks. Although plentiful volcanic ash would have certainly spread downwind where it is preserved in the Siwalik sedimentary rocks of the Himalayan foothills, the most impressive associated event was the huge mass of ash, boulders, and muddy water from the melted glaciers. This unstable mass began from the volcano's summit to produce a massive lahar or volcanic mudflow that roared downhill across an ancient landscape, all the way to the ancestral Indus River area of the time.

Natural Resources in Afghanistan. http://dx.doi.org/10.1016/B978-0-12-800135-6.00008-8

Interestingly enough, when wet lahars race across a landscape they commonly pick up loose rocks on the surface over which they travel, and incorporate these remnants into their matrices. Early toolmaking human precursors were around in those days, perhaps *Homo habilis* or *Homo erectus,* because the stone tools they left behind were incorporated into the rocky debris of the Bain Boulder Bed in Pakistan, where they have been collected for over a century (Khan et al., 1985) (Figures 8.1–8.3).

The Dashti Nawar lahar from what is now Afghanistan that produced the Bain Boulder Bed far back in antiquity in what is now Pakistan is an example of a

Figure 8.1 Outcrop of the lahar diamictite of the Bain Boulder Bed in Pakistan that occurs as the cliffed bed on the horizon that dips off to the right. Note for scale the figure of Javed Khan, who first studied the unit using paleomagnetics, who is standing below the cliff in the lower center. *Photograph by the author in 1983.*

Figure 8.2 Boulder of low density volcanic pumice in Bain diamictite. *Photograph by the author in 1983.*

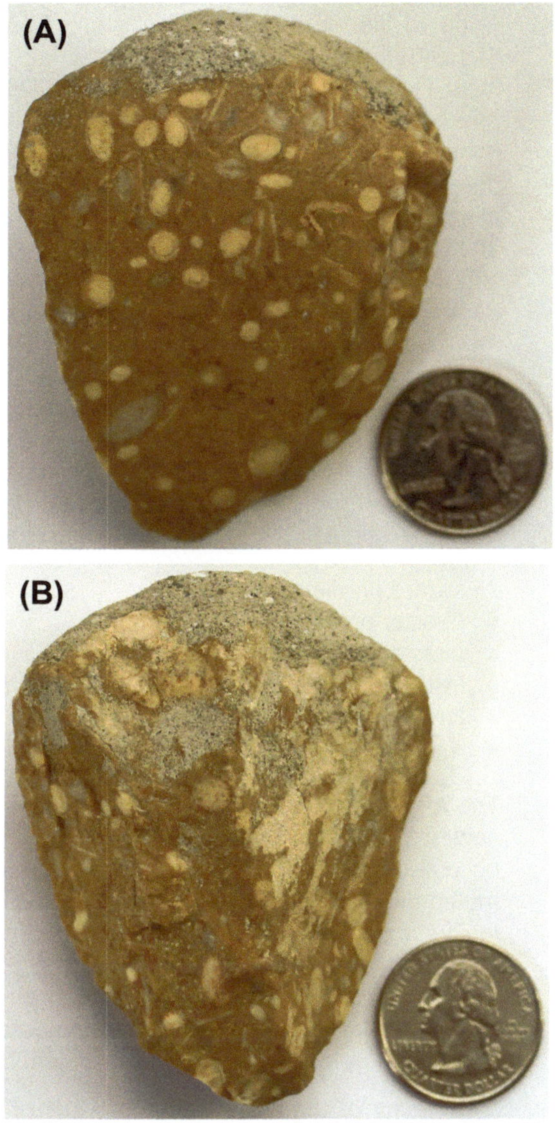

Figure 8.3 Photograph of early Olduvan–Acheulian type of stone tool collected by the author in 1983 from the Bain Boulder Bed at Bain Pass in Pakistan. (A) Rounded side; (B) Obverse chipped side with >3 faces. The stone tool was crafted from a river-rounded pebble of *nummulites*-bearing chert from the Eocene-aged limestone characteristic of the region. The tool had been worked into a small palm-sized piece, dropped to the ground by its last user and picked up from the ground surface by the lahar that rolled over it about 2 million years ago. Such tools had been recognized there by early British geologists, although they interpreted the deposit as glacial in origin (Morris, 1938). *Photographs for the author by John P. Ficenec.*

striking natural hazard that is no longer a threat to anyone in that part of the world. Although marginally active volcanoes (Kohi Sultan) do exist in far northwestern Baluchistan close to the border with Afghanistan, the volcanoes of the region have diminished in size and activity up to the present time so they are no longer a hazard worth considering.

Other natural hazards such as earthquakes, floods, droughts, sand and dust storms, rapid wet debris flows, snow avalanches, and landslides, however, are highly active and worthy of considerable attention because they are so damaging to life, limb, and infrastructure. In fact, not long after the invasion of Afghanistan by the Coalition forces in late 2001, a National Disaster Management Plan (NDMP) was put forward in 2003 by the United Nations Assistance Mission in Afghanistan (UNAMA) to assist the Government of the Transitional Islamic State of Afghanistan to improve the management and coordination of disasters through establishment and facilitation of multisector and multidisciplinary programs and activities for enhancing capacity at all levels (Anonymous, 2003a). From this and other efforts by UNAMA, the Asian Development Bank, and the Independent Administrative Reform and Civil Service Commission of the Islamic Transitional Government of Afghanistan, a new strategy was put forward for relevant institutional strengthening. This also led to the Afghanistan Disaster Knowledge Network (http://www.saarc-sadkn.org/countries/afganistan/about_adkn.aspx) that by 2010 had developed a complete and well-conceptualized plan, which included plentiful input from the South Asian Knowledge Network, with useful input from outside consultants (Gupta, 2010). The plan included a reasonably complete listing of all relevant parts required to function at as high a level as possible in the event of a disaster, such as:

- hazard profiles;
- disaster profiles;
- disaster management framework;
- institutions (relevant to disaster management);
- laws/policies/plans;
- major initiatives.

The Afghanistan National Disaster Management Authority (ANDMA), and the newest NDMP have assessed the hazards in Afghanistan and judged them to be predominantly categorized into four broad categories—geological, hydrometeorological, biological, and technological. The first two are of the most interest here, as well as being the most common, with the result that much further information is needed about earthquakes, floods, flood-induced landslides and the so-called mudflows, the sand and dust storms, extreme heat and cold, avalanches, and droughts. The technological hazards and the biological pest attacks and epidemics do not need to concern us further here.

A summary of the chief natural hazards of the country (Gupta, 2010) include as a very major problem, the fact that the northeastern parts of Afghanistan are subject to medium and large earthquakes because most epicenters in the past have occurred there (Figure 8.4). This includes the city of Kabul, where the dense population and generally shoddy construction add to the vulnerability. Dry landslides can be accelerated by seismicity, and when hydrometeorological factors are added in so that moisture is increased, a variety of wet landslides and rapid wet debris flows can result as well. This establishes a continuum of such processes wherein the mean velocity of the moving mass can be graphed

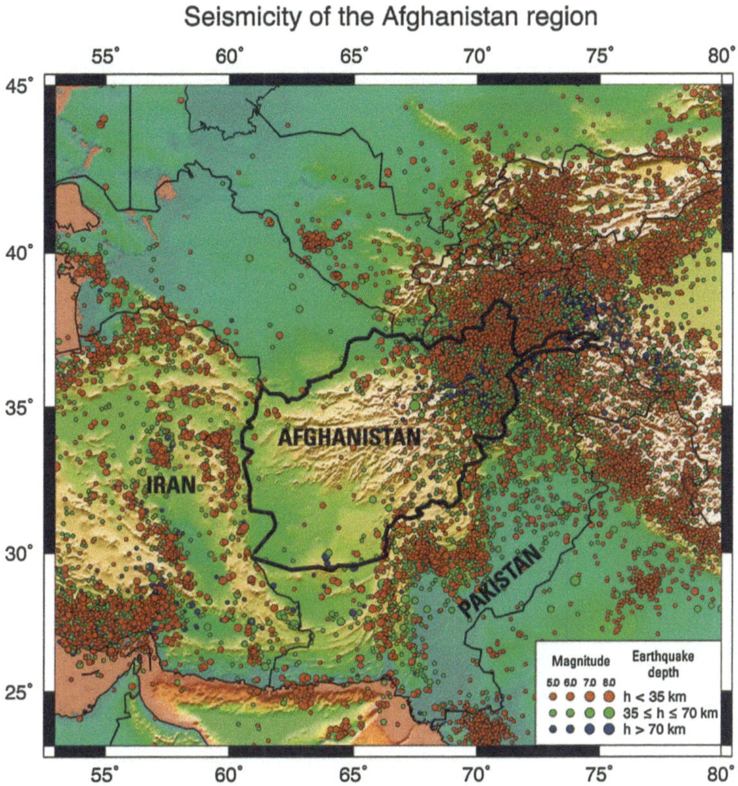

Figure 8.4 Preliminary instrumental seismicity of the region of Afghanistan and neighboring countries arranged by color for the depth of the earthquakes. Red circles are shallow focus, green is intermediate, and blue is deep focus; size of circles is magnitude. *After Benz et al. (2005).*

against the sediment concentration with water amount so that many different recognizable categories of different mass movements and related floods can be established (Pierson and Costa, 1987) (Figure 8.5).

In Afghanistan, for example, winter flooding begins in January and continues into May. The southwest, most arid part of the country, is on the edges of human livability, with the result that it is highly drought-prone, whereas the western and central belts are highly flood affected. Climate extremes and difficult geophysical seismic conditions produce inherent vulnerability, which combined with limited development activities and the common violence stretch the coping mechanisms of communities there close to collapse. In addition, several southwestern and a few northern provinces have the dual effects of both flood and drought. The addition of several hazards as fairly high probabilities in a single local provincial area, leads to the designation of multihazard high risk zones (Figure 8.6).

Vulnerability of populations to hazards is exacerbated by poverty, higher population density, illiteracy, and lack of access to health facilities (Gupta, 2010). Poverty headcount rate in this context is measured as high if >44%, medium

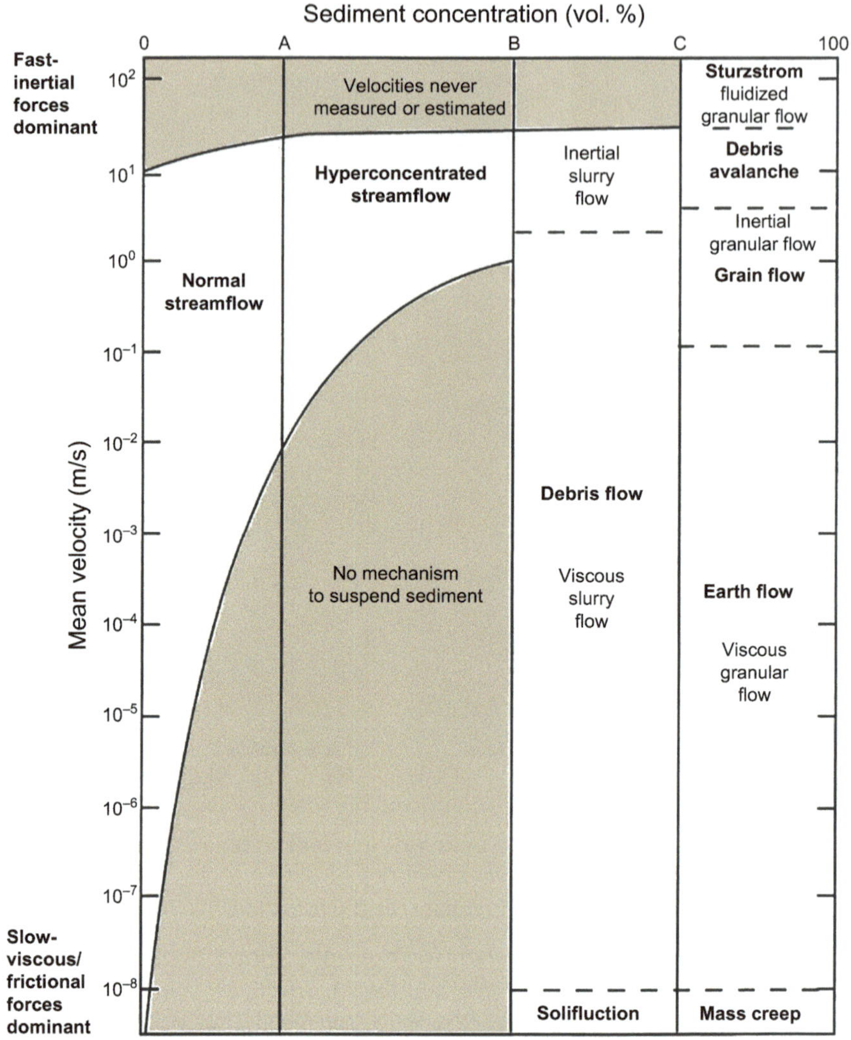

Fluid type	Newtonian	Non-newtonian		
Interstitial fluid	Water	Water + fines		Water + air + fines
Flow category	Streamflow		Slurry flow	Granular flow
Flow behavior	Liquid	Plastic		

Figure 8.5 Diagram from Pierson and Costa (1987) showing relationships between concentration of sediments in water that range from normal flowing muddy water up through various concentrations of sediment–water mixtures in different sorts of wetter debris flows to drier sturzstroms (high speed rock avalanches). Volcanic lahars are a form of rapid, wet debris flow, for example.

as 35–44%, and low as <35%. Population density is measured as high if 100 or greater, medium if 40 to 99, and low if less than 40. Illiteracy rate is measured as high if <15%, medium if 15–30% and low if >30%. Access to health facilities by it taking greater than 1 h to get there is high if the population percent is >75%, medium if the population percent is 25–75%, and low if the population

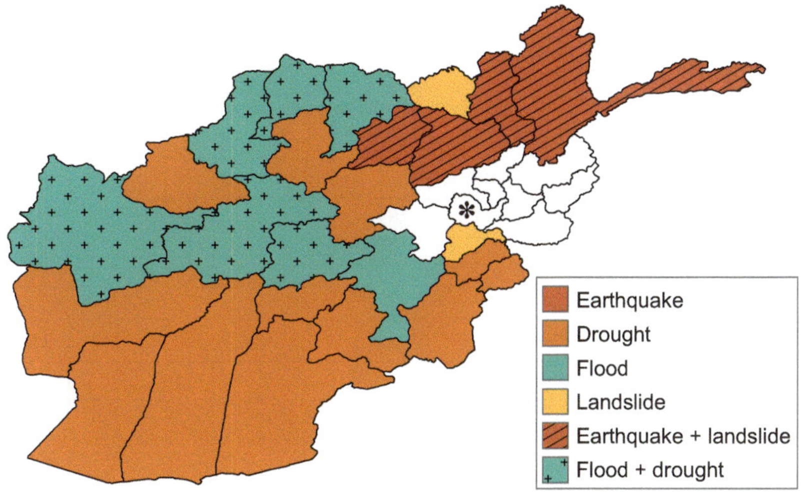

Figure 8.6 Map of the provinces of Afghanistan (as of 2010) with the highest hazard vulnerability. *After Gupta, 2010.*

Vulnerability indicators	Laghman	Nangarhar	Paktika	Khost	Ghor	Daikundi	Urozgan	Zabul	Ghazni	Kunarha	Bamyan	Nooristan	Takhar	Kunduz	Sar-e-pul	Badghis	Kandadahar	Helmand	Paktya	Parwan	Kabul	Kapisa	Baghlan	Faryab	Jozjan	Balkh	Samangan	Badakshan	Logar	Wardak	Farah	Nimroz	Herat	Panjsher
Poverty	H	H	H	H	H	H	L	L	H	H	H	H	M	M	M	M	L	L	H	L	L	L	M	M	M	M	M	M	L	L	M	L	M	L
Population density	H	H	L	H	L	L	L	L	M	M	L	L	M	H	L	L	L	L	M	H	H	H	M	M	M	M	L	L	M	M	L	L	L	L
Illiteracy	H	M	H	M	M	M	H	H	M	M	M	M	H	M	H	H	H	H	L	L	L	L	M	M	M	L	M	M	M	M	M	M	L	M
Access to health facility	M	M	M	L	H	H	H	H	M	M	M	M	L	L	M	M	M	M	L	L	L	L	M	M	L	L	M	M	M	M	M	M	L	L

Note - In the above Table, H (High), M (Medium) and L (Low) legends have been allotted based on the following vulnerability criteria:

- Poverty (Poverty Headcount Rate, > 44% - High, Poverty Headcount Rate, 35 to 44% - Medium, Poverty Headcount Rate, < 35% - Low).
- Population Density (per sq.km), Pop. Density, 100 and > - High, Pop. Density, 40 to 99 - Medium, Pop. Density, < 40 - Low.
- Illiteracy (Illiteracy Rate, < 15%, - High, Illiteracy Rate, 15 to 30% - Medium, Illiteracy Rate, > 30%, - Low).
- Access to health facility (taking > one hour), population % > 75% - High, population % 25 to 75% - Medium, population % > 25% - Low.

Figure 8.7 Plot of vulnerability indicators by province to indicate vulnerability profiles from high, medium, to low in Afghanistan. *After Gupta, 2010.*

percent is >25% (Figure 8.7). If the vulnerability indicators are combined with selected simplified natural hazard phenomena, certain provinces may be designated as priority regions (Figure 8.8).

An institutional framework for disaster management has been built up for the whole country of Afghanistan as a kind of tiered pyramid with the National Disaster Management Commission (NDMC or National Commission) at the top. Below this a series of other administrative structures arranged from the

Vulnerability indicators	Laghman	Nangarhar	Paktika	Khost	Ghor	Daikundi	Urozgan	Zabul	Ghazni	Kunarha	Bamyan	Nooristan	Takhar	Kunduz	Sar-e-pul	Badghis	Kandadahar	Helmand	Paktya	Parwan	Kabul	Kapisa	Baghlan	Faryab	Jowzjan	Balkh	Samangan	Badakshan	Logar	Wardak	Farah	Nimroz	Herat	Panjsher
Population density	H	H		H										H						H	H	H												
Poverty	H	H	H	H	H	H			H	H	H	H							H															
Illiteracy	H		H				H	H					H		H	H	H	H																
Access to health facility					H	H	H	H																										
Hazard vulnerability																																		
Earthquake+ Landslide													H										H				H	H						
Flood + Drought					H		H																	H	H	H							H	
Drought				H				H			H				H	H	H	H	H												H	H		

Figure 8.8 Sixteen high priority provinces that have multiple high vulnerability indicators. *After Gupta, 2010.*

international organizations who deal with disaster response, relief, rehabilitation, and mitigation paired up with the ANDMA, who together work with the national lead ministries of the government and other nongovernmental organizations to look after provincial level disaster management committees and commission, who in turn do the same for the district level disaster management committees, and on down to the local community-based organizations and development councils.

The NDMC is tasked with:

- Determining major directions of disaster preparedness and management.
- Convening meetings when any major disaster strikes.
- Declaring national emergency situations and their end, when appropriate.
- Adopting measures to save lives and eradicate disaster causes.
- Taking steps for prevention, relief, recovery, and rehabilitation using available resources.
- Allocating funding from the National Emergency Fund.

A National Emergency Operations Centre was established in 2005 and tasked with emergency operations, communications and warnings, requests for additional resources from neighboring areas, overseas support, development of policies and plans, and issuing emergency information and providing damage assessments. A sector-wise cluster approach has been established to leverage disaster management functions from the various ministries that deal with shelter, education, health, water and sanitation, nutrition, agriculture and livestock, and disaster risk reduction. A National Disaster Risk Reduction Plan (NDRRP) has been established with six principal components for: (1) institutionalizing and mainstreaming disaster risk reduction; (2) establishing safe schools; (3) safe hospitals; (4) resilient cities; (5) setting up early warning systems; and (6) measuring preparedness for effective response. All of this has been set up with milestone measures and target dates of responses by the years 2013, 2015, and 2020. The key roles and responsibilities of all the concerned ministries under the NDRRP were also spelled out (Gupta, 2010).

In sum, the establishment of the complex bureaucratic structures and physical facilities to deal with hazards and disasters in a war-torn, poorly developed, and rather backward country like Afghanistan is a noble objective that is bound to have some beneficial effects when called upon in the event of rise of emergency situations. The actual implementation of what has been established will be another matter, however, given the traditional recalcitrance to make many decisions when things get out of hand, as they so often do in Afghanistan. The country can be proud that it has established such a comprehensive and well thought-out plan to cope with emergencies that is better than any such plans in many neighboring countries. Still, the proof will be in the outcomes after the next major disaster; probably there will be a fairly steep learning curve and it will take time to sort out many problem areas of disaster management. The beginning seems bright enough, however.

A view of the chief natural hazards in Afghanistan adds further information to understand the great problems the country faces from its natural environment. The natural hazard stressors are among the strongest or most commonly difficult in the world and are deserving of considerable study in their own right to understand their causes and to seek solutions for their mitigations.

Earthquakes

Afghanistan has a long history of major earthquakes; in fact it has had more recorded large magnitude earthquakes (>7–8 magnitude) than anywhere else in the world (Wheeler and Ruckstales, 2007; Shroder et al., 2011b). The reasons for this so commonly devastating seismicity have to do with the fact that Afghanistan lies within the actively tectonically deforming, southern portion of the Eurasian plate (Dewey, 2006). Thus the northward motion of the Indian and Arabian plates at velocities of some 3–4 cm/year (Wheeler and Rukstales, 2007) produces enormous cumulative stresses in the crust as various subunits, or "blocks" jostle about, rupture, and cause considerable seismic unrest, especially in the northeast part of the country where earthquakes happen frequently, but also to the west of the country where the seismicity is considerably less. The "Seistan block", for example, in southwest Afghanistan (Ambraseys and Bilham, 2003) is the bedrock subunit beneath the Seistan basin that has subsided to host the Helmand River delta and its endorehic *hamuns* where the river's waters end up before evaporating away (*See* Chapter 5). These blocks are fragments of continental and oceanic crust that have collided with, and been sutured onto the southerly edges of the Eurasian plate next to the north Afghan platform rocks (Wheeler et al., 2005) (See Chapter 2).

At the present time the Hari Rud fault along the south rim of the north Afghan platform is the active boundary with the southern accreted arc rocks and other crustal fragments to the south (Figure 8.9). As the Indian plate moves northward to the east past the rocks of Afghanistan, the two regions are separated by what is known as transpressional motion, wherein rocks are squeezed up and out of the crust, while the overall sense of motion is left-lateral shear (so named as one stands on one side of a crustal block and observes the motion across the fault moving to either one's left or right). The faults along which this shear is expressed include the Chaman fault that passes within a few kilometers directly to the west of Kabul, as well as the Central Badakhshan fault and other faults that run through the middle of the Badakhshan Pamir ranges.

Early on after the invasion of Afghanistan by the Coalition forces in late 2001, the US Geological Survey was tasked with a number of different kinds of assessment of seismic hazard in Afghanistan in order to gain a more complete understanding of the problem:

- Compilation of regional catalogs of historical and instrumental earthquakes to prepare maps.
- Development of a seismotectonic map.
- Conducting studies of seismic source zones and development of a Quaternary fault map leading to paleoseismic mapping.
- Development of a regional seismic hazards map.

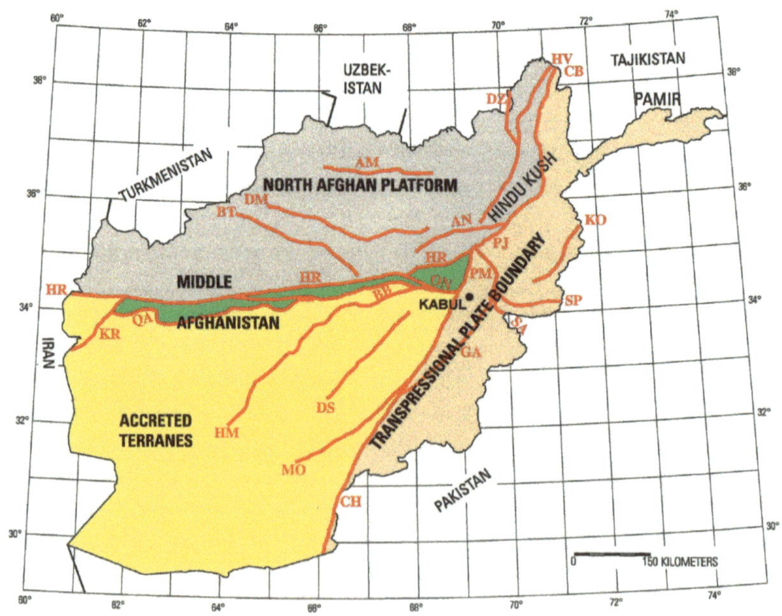

Figure 8.9 Tectonic regions of Afghanistan showing three main tectonic regions. North Afghan Platform in pink; Accreted Terranes in yellow; and Transpressional Plate Boundary in tan. Main fault zones discussed in text in red (HR—Hari Rud; CH—Chaman; PM—Paghman; CN—Central Badakshan; DZ—Darvaz; PJ—Panshir).

With the seismicity catalog, the seismotectonic framework, and the new paleoseismic data that were developed (Figure 8.10), a seismic hazard map could be constructed (Figures 8.11). The narrow, red striped bands (IX) running southwest–northeast are zones along the surface traces of the Chaman–Panshir–Central Badakhshan faults, as well as the Darvaz fault, along which if a strong earthquake occurs, violent shaking will happen with peak accelerations of 65–124 %g, with people thrown forcibly to the ground, monuments and columns falling, and most unreinforced masonry buildings undergoing heavy to very heavy structural damage. Away from these regions a progressive diminution of earthquake intensity occurs that is symbolized with various shades of browns, until the green of seismic intensity V occurs, with a peak acceleration of only 3.9–9.2 %g, and but moderate shaking, with the swinging of suspended objects, the falling over of precarious objects, but only negligible damage to unreinforced masonry buildings (Boyd et al., 2007).

Destructive earthquakes have been known in Afghanistan for more than four millennia, with more than 7000 Afghans having been killed in the past 15 years alone (Boyd et al., 2007). The Nahrin earthquake of May 1998 about 300 km

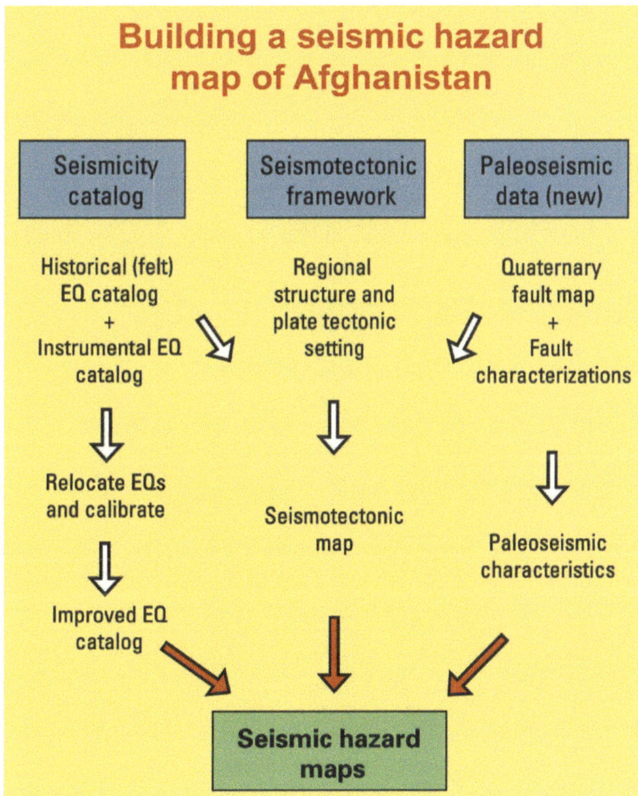

Figure 8.10 Generalized diagram that shows the data sources required and the interim steps needed to develop a seismic hazards map of Afghanistan. *After Benz et al. (2005).*

north of Kabul killed more than 4000 people, which is a kind of warning that more such events will occur close to population centers and lifelines in future, with attendant destruction of property and loss of life.

Tens of thousands of lives are routinely lost in other large earthquakes in the neighborhood around Afghanistan, often because of what are termed "cold pour" concrete construction where the cement is allowed to set before the next batch so that only weak junctions form, or no undeformable triangles are produced in walls with cable or rebar reinforcement, so that building floors "pancake" to kill people. Unreinforced brick and stone buildings vibrate in earthquakes with the same intensity as the ground so that such buildings collapse in Afghanistan, even while landslides knocked loose rush down slopes to overwhelm and kill people. Centuries ago, people seem to have noticed that buildings with layers of wood interspersed between the piled up and mud mortared stones of many walls did not collapse as much in earthquakes, with

(A)

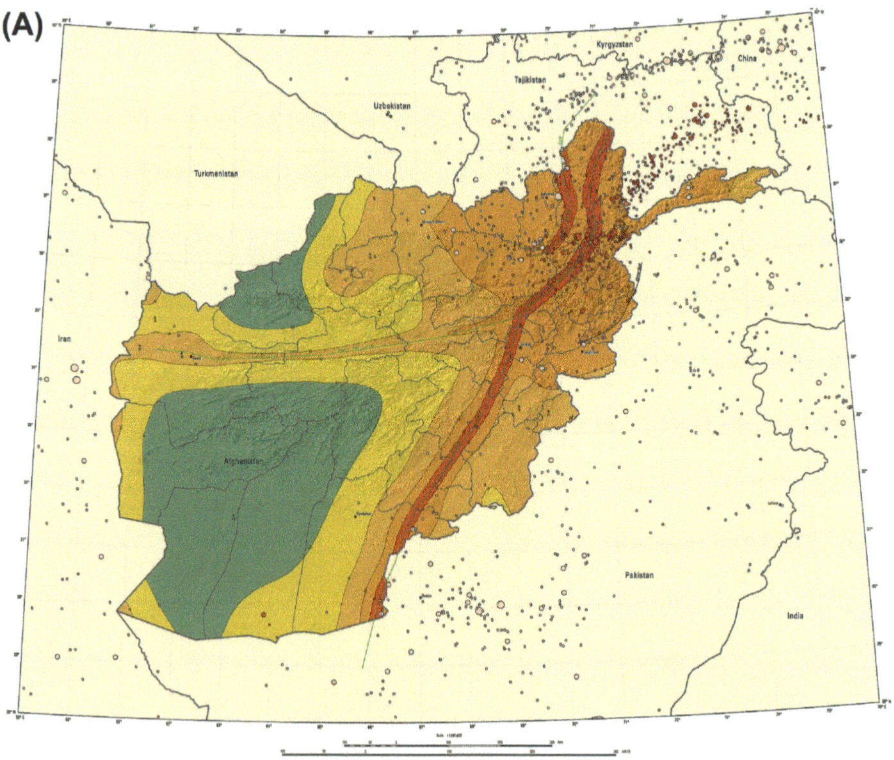

Figure 8.11 (A) Earthquake hazard map for Afghanistan *(After Boyd et al. (2007))*; (B) Expanded legend for the earthquake hazard map for Afghanistan. The peak accelerations in percent of gravitational acceleration (%g) for each zone are provided in color and the seismic intensity in Roman numerals from I to X, with written descriptions of potential damages.

(B)

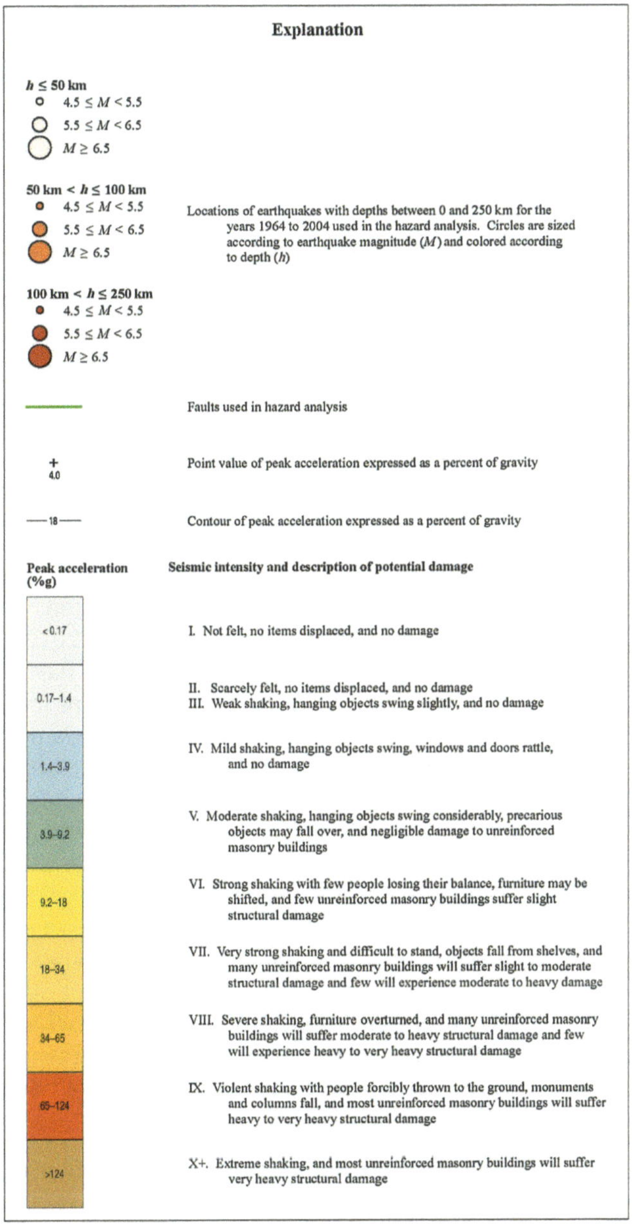

2 percent probability of exceedance in 50 years corresponds to a ground-motion return time of approximately 2500 years.

Figure 8.11 —cont'd

the result that such construction is quite common in many mountain villages. Houses in Afghanistan with the common barrel-vault or dome-vault roofs of mud bricks are death traps where wooden roof beams are rare. Improved construction standards and techniques, as well as scientific assessments of seismic hazard to guide the construction, will be required to improve matters into the future.

Landslides

The high mountains, steep slopes, highly fractured bedrock overlain by weak sediments, torrential rain storms, and high seismicity in so much of Afghanistan mean that slope failures in the country are plentiful and of many diverse types (Shroder, 1989; Shroder and Weihs, 2010; Shroder et al., 2011a,b). High-speed rockfalls and rock slides are profuse below cliffs that occur throughout the high rugged mountains, and failures occur repeatedly of the common sedimentary rocks and sediments, especially where shales and unstable loesses occur in the northern part of the country.

Loss of property and life from landslides and snow avalanches has also been extreme in the country, with many tens to hundreds killed in recent years (Anonymous, 2010, 2012a,b; Rubin, 2012). In addition, a number of places occur in Afghanistan where landslides have blocked rivers and caused inundation upstream. Such landslide dams are inherently unstable if the impounded lake grows large enough because the dam may not hold and abrupt landslide lake outburst floods (LLOF) can result. Lake Shewa in Badakhshan, for example, formed the second largest such landslide dam in the world, with serious implications should it ever burst out into the Panj River tributary to the Amu Darya (See Chapter 5 on lakes in Afghanistan). At the present time large quantities of water are issuing from the middle front face of the dam and several small slumps of dam face material have occurred in recent years. The Lake Shewa dam front has stood for thousands of years without total failure, however (Shroder and Weihs, 2010), but it still needs monitoring. In other cases the common seismicity is related to rock slides and the destruction of travertine dams that produce floods downstream, some of which seem to have produced various kinds of myths that have been passed on down through history (Bourrouilh-le Jan et al., 2007).

Mass movements of nearly all types occur in Afghanistan but in the high relief of the rugged Pamir and Hindu Kush mountains of northeastern Afghanistan, mass movement threats to lives and property necessitated intensive study by the author and his students to elucidate future problems for development. Twenty-two different mass movements in bedrock in the Badakhshan Province of northeastern Afghanistan were studied (Shroder et al., 2011b), including large rockfalls and rock slides, along with massive slope-failure complexes with many types and rates of movement. The Lower Munjan slope-failure complex, for example, is a large (~6 km long; ~11 km wide), multilobate mass of failed rock fragments with pronounced hummocky topography that once dammed the river valley it fills (Figures 8.12 and 8.13). The southern Badakhshan region in which it occurs has had at least 21 magnitude ≥7 historical earthquake epicenters clustered around it since 1842 (Wheeler and Ruckstales, 2007; Shroder et al., 2011b), which suggests a high likelihood that seismic acceleration was a main cause for the failure. A plot of the slope exposure directions or aspects of all these bedrock slope failures in northeast Afghanistan showed the highest number of failures facing in a southeast direction (Figure 8.14), which was a result of steep fault scarps that face in that direction, so that the high seismicity characteristic of Badakhshan would have accelerated the failures down those scarp slopes.

Where higher altitudes prevail in the region, ice-cemented and ice-cored rock glaciers are also common and overlie some of these mass movements, which provide evidence that some of these slope-failure complexes are quite ancient, because it takes considerable time for these rock glacier features to form. Inasmuch as seismic energy sources in the Eastern Hindu Kush are maximal in southern Badakhshan, and relief, slope angles and precipitation all increase from west to east as well, the causes of the pervasive mass movement are plentiful enough, although direct cause and slope-failure effect are generally not known. Some weak sedimentary lithologies downfaulted into, or draped across crystalline rocks, have also failed in that region in the past. Some intermixed tills also occur in these landslides but are not easily differentiated, even with analysis on the ground.

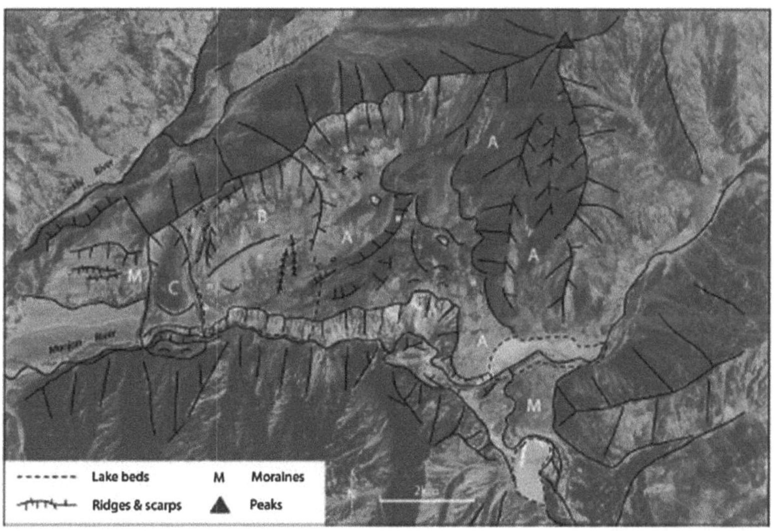

Figure 8.12 Sketch map of Lower Munjan slope-failure complex overlain across satellite imagery from Google Earth™, showing lobes A, B, and C that have failed at different times.

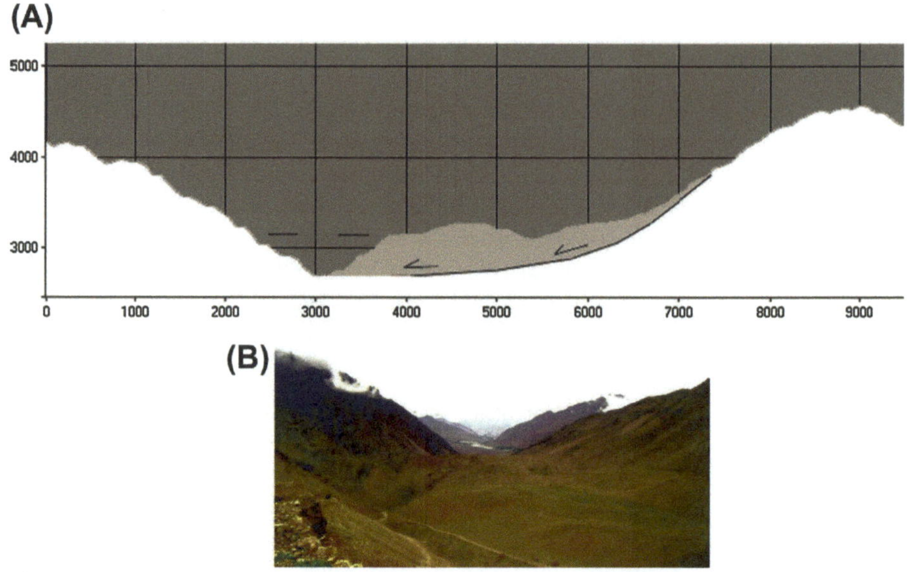

Figure 8.13 (A) Approximate Northeast–Southeast cross section through Lower Munjan slope-failure complex showing likely slip surface into the river valley, with dashed lines indicating former cross-valley landslide dam now removed by subsequent failure, and draining of lake into which sediment had accumulated; (B) Photograph by Romin Fararoon from Google Earth™ looking southeast across the hummocky surface.

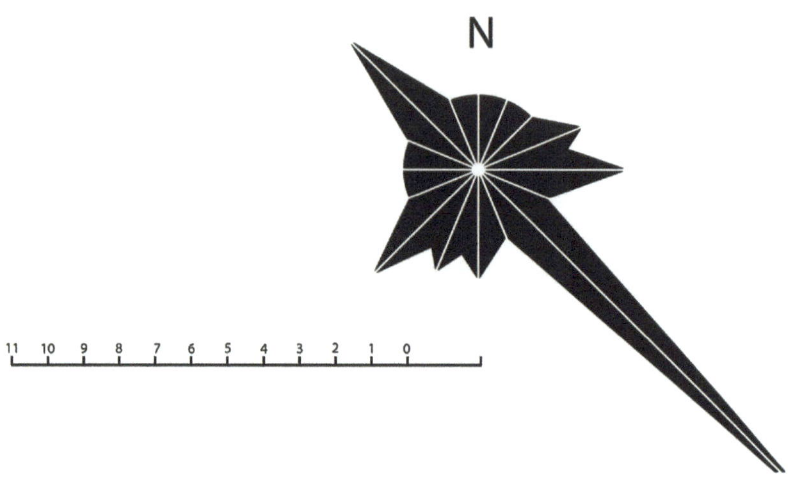

Figure 8.14 Rose diagram of landslides in northeast Afghanistan. The strong orientation to the southeast is a result of major southeast-facing fault scarps, and high and steep valley sidewalls there that have been glaciated in the past as well.

Using high resolution satellite imagery and digital elevation models, geomorphologic parameters were assessed on these landslides to characterize spatial organization structures related to zones of erosion, areas of deposition, and further hazard potential. Analyses indicate that many of the massive slope

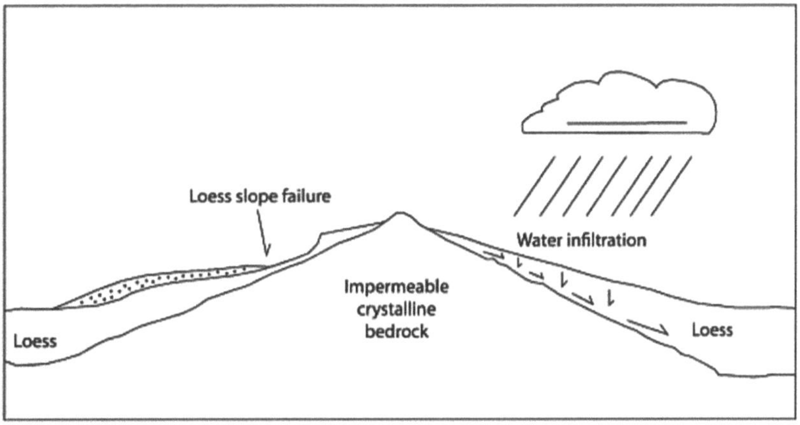

Figure 8.15 Diagram of typical loess draped over crystalline bedrock in northeast Afghanistan. Microscopic to mesoscopic (connected pore spaces to joints and fissures) routes of water infiltration shown as subsurface barbed arrows on right are commonly expressed over time through elutriation of fine clastics to become megascopic pipes and tunnels that can be meters in diameter and highly conducive to subsequent ground failure and mass movement of the loess.

failures in Afghanistan can be characterized and differentiated into various process domains and chronologic-development zones with their different impacts upon the landscape. Mass movements in Afghanistan can exhibit unique topographic signatures that can be used to better assess hazards in other mountain areas, especially where landslide-dam breakout floods threaten. Development of roads, bridges, buildings, and irrigation networks should be done with care in those regions of Afghanistan that have plentiful landslides.

Mass movements in northeastern Afghanistan also include plentiful failures in loose, silty loess that was emplaced originally as wind-blown dust (Shroder et al., 2011a). The loess region in northeastern Afghanistan occurs in the Badakhshan and Takhar provinces and was likely created by dust blown to the east from the Karakumsky Desert and the alluvial plains of northern Afghanistan. This loessic dust was deposited against the largely crystalline rocks of the Hindu Kush mountain range that rises up along the eastern half of Afghanistan as a result of transpressional tectonism (Figure 8.15). The loess overlies less permeable crystalline and sedimentary bedrocks such as Triassic granite, Proterozoic gneiss, and Miocene and Pliocene clastics in the area with the largest concentration of slope failures. Thirty-four loess slides and flows were mapped and analyzed using remote satellite imagery over digital elevation models on Google Earth™. This source enabled location, classification, and measurement of failures. Findings revealed that most failed slopes faced north, west, and northwest (Figure 8.16). This trend can be explained possibly as different moisture contents resulting from the primarily westerly wind direction, which may cause more precipitation to be deposited on west-facing slopes, and the

Figure 8.16 Rose diagram showing the slope-aspect frequency of loess slope failures in northern Afghanistan. The pronounced higher frequency of failures to the north through the west aspects is apparently because of the effects of westerly storms interacting most with west and north-facing slopes.

sun's position during the hottest part of the day. Additionally, the easterly rising Hindu Kush range may cause more slope area to face west in the study region. Other contributing factors could be the very high seismicity of the area, which may cause rapid dry fluidized loess flows, and landscape modification by humans. Several loess slope failures appear to be generated by water concentration through irrigation ditches and possible rutted tire tracks, which can create tunneling between the loess and its less permeable bedrock. Causes and effects of loess failure in Afghanistan need to be investigated in more depth. Further study may lead to the adoption of more sustainable and safe farming practices and more informed housing locations, which may prevent loss of property, crops, and transport routes.

As noted above (Figure 8.5), a wide variety of slow to mostly rapid, wet debris flows also can occur in Afghanistan, especially where extensive deforestation and overgrazing have stripped off vegetation to degrade watersheds and slow safe infiltration of precipitation (Shroder, 2012). Instead in the degraded watersheds the water entrains profuse sediment (muddy debris, boulders) so that it moves into the drainage ways as a series of mobilized sediment and boulder pulses that sweep down and spread out over land lower down (Figure 8.17) (Shroder, 2013). After such events the sediment dries out and hardens or sets up almost like concrete so that it commonly cannot be farmed again for many years.

Figure 8.17 (A) Photograph by the author in summer 1978 of a rapid wet debris flow that swept across the road to Bamiyan in the Ghorband River valley, overwhelming farm fields and houses; (B) Rapid wet debris flow in motion with the front edge about to reach the village in the middle distance. Thirty-five years later the scars on the landscape from the event are still visible upon the landscape.

Floods

The flood-prone rivers of Afghanistan have long been a severe problem in need of a solution. For example, from March to May, 2012, in eight provinces in northern Afghanistan alone, as many as 160 people were killed, 81 injured, and 8533 houses were destroyed, which adversely affected 22,167 families (Table 8.1). At the same time, the destruction included more than 126 wells, 644 canals and dams, 406 small bridges, and 454 km of road. Such destruction is all too common on an annual basis, and speaks volumes towards what is needed to protect people and their infrastructure in Afghanistan (Palau, 2012). In fact, Afghanistan is ranked as the second highest in the world after Bhutan in terms of flood exposure and deaths (Beekma and Fiddes, 2011).

Table 8.1 Flood damages in northern Afghanistan; March to May 2012. ANDMA; National Emergency Management Operation Center (iMMAP & USAID)

Damages	Badghis	Baghlan	Balkh	Faryab	Herat	Jawzjan	Saripul	Takhar	Total
People killed	4	13	19	22	7	2	51	42	160
People injured	0	9	0	0	4	12	6	50	81
Families affected	7032	6225	1800	850	820	508	4682	250	22,167
Houses destroyed	369	475	130	1809	223	208	2525	2794	8533
Wells destroyed	39	23	13	–	–	–	–	51	126
Canals & dams destroyed	23	142	12	200	129	–	–	38	644
Small bridges destroyed	14	53	–	–			107	64	406
Roads destroyed (km)	96	126	–	212	–	20	–	–	454

More than a half century ago in the times right after World War II, when the demands of understanding the problems with the flood regime of the Helmand River in southwest Afghanistan were necessitated by the Helmand Valley Project, the US Geological Survey was tasked with building a flood forecasting mechanism for that river, and later for the whole country. The standard science of flood prediction in the twentieth century was devised to use the statistics of past floods to calculate the probabilities of future such events. A sufficiently long, continuous historical record of at least 50 or more years would enable the annual river discharge records to be rank ordered and plotted on logarithmic graph paper as a regression line so that the rank of each peak discharge could be used to calculate a recurrence interval for floods. Once the ranges of possible flood stages and maximum discharges were known for each river, then more precise map plots could be drawn on flood plains to show the average expected annual flood, or the 10-year, 50-year, 100-year flood limits, and so on, as is done in the developed parts of the world as the standard statistical, or probability-based methods of prediction and planning for floods.

In Afghanistan, the US Geological Survey in the 1950s began to branch out all over the country to build river discharge measurement sites with surveyed cross-sectional profiles so that a proper statistical record of flood stages could be collected and used later in flood prediction. These records thus began in the late 1940s and 1950s, and continued up through the 1970s until the Soviet invasion of 1979. The termination of the river discharge measuring program and flood stage modeling caused by the following three decades of war meant that the limited 20–30 year record was too short to allow the statistical methodology of flood prediction modeling to work properly. Nevertheless the Afghanistan Information Mapping Service, an agency set up by the UN and U.S. Agency for International Development (USAID) to provide essential map information in the country, did its best by drawing small-scale maps with a series of rather crudely constructed parallel lines along river courses to indicate some sort of a potential flood hazard there. Of course this was not terribly constructive, as people already knew that low-lying areas near the essential river courses were prone to some periodic inundation; some other technique was needed to predict floods in a more accurate fashion that would better enable floodplain zonation. Similarly the World Health Organization (2010) produced a flood hazard distribution map (Figure 8.18) that constructed a flood hazard index as very low, low, medium, high, and very high at a very small-scale so that in terms of planning purposes only a gross overall distribution could be obtained.

Accordingly, an Afghanistan Flood Hazard Mapping (AFG-FHM) Project was introduced in 2008 by personnel from the North Atlantic Treaty Organization (NATO) that is headquartered in Brussels, Belgium. The NATO C3 (Consultation,

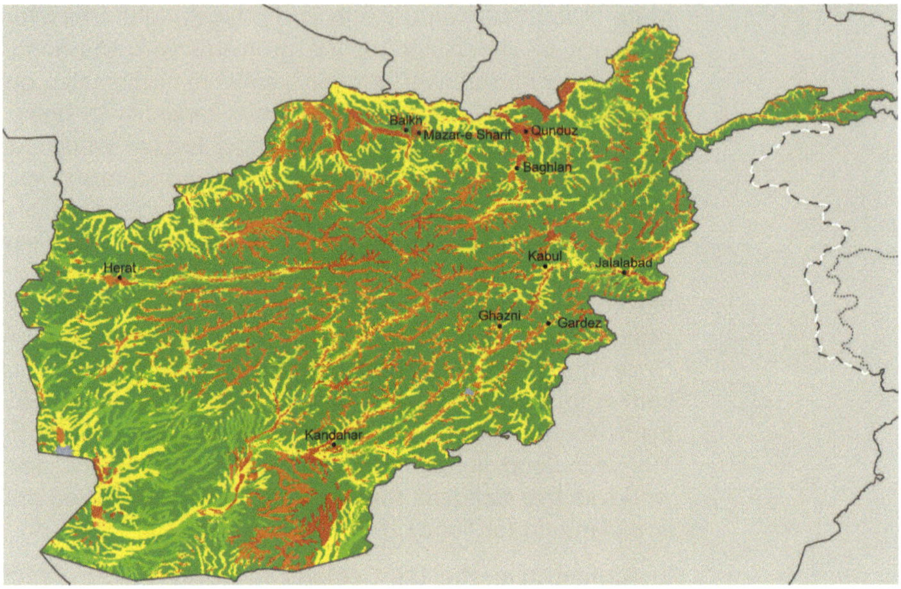

Figure 8.18 Flood hazard distribution map for Afghanistan. Where dark green indicates a very low hazard; light green is low; yellow is medium; orange is high; and red is very high hazard. *After World Health Organization, 2010.*

Command, Control) Agency provides essential communications, intelligence, surveillance, and reconnaissance missions for its military mission in Afghanistan and elsewhere. The NATO C3 Geo team under the direction of Emlyn Hagen was tasked with developing a new methodology for flood forecasting in Afghanistan. The result (Hagen et al., 2010) was a new procedure utilizing a kind of cellular automata that were overlain virtually upon the landscape to which various amounts of water were applied by computer to fill up and flow to the next automaton cell. This new procedure was thought to be able to work also in many other developing countries where prior flood modeling was always inadequate or nonexistent because of the lack of historical flood data.

The AFG-FHM Project achieved success with the classified Digital Terrain Elevation Data (DTED) models DTED1 and DTED2, which are high resolution digital elevation models made by the US government but which are not available to the public because they are detailed enough to enable route plotting, targeting, and other factors likely to be of altogether too much use to those who would seek to bring harm to Afghanistan, rather than the humanitarian assistance they so desperately need. Even though the original digital data sets are not available to the public, the AFG-FHM itself is available and it and various other variations derived from it can be downloaded from the Civil-Military Overview website at www.cimic.web and numerous other websites associated with USAID (Figure 8.19).

AFG-FHM was created using a Gauckler-Manning-Strickler hydraulic model, calibrated using observed past flood events that were available, plus remote sensing observational data from the Dartmouth Flood Observatory. While quite good on

Figure 8.19 Flood hazard map of Afghanistan based on classified large-scale DTED altitude data showing four areas of test-case rivers as irregular-shaped black spots in the center (Chagcharan), north (Balkh Ab), far northeast (Farqar), and far east (Kunar). The test-case rivers in the AFG-FHM were assessed by a Nebraska team for validity of the flood modeling and were found to be reasonably adequate, although the areas of widespread alluvial fans were judged to be somewhat problematic because of the difficulties of predicting flooding in those environments where rapid wet debris flows are so common.

most river courses in Afghanistan because it showed where the floods of different depths and times could occur (Figures 8.20 and 8.21), it was not as good on the many alluvial fans in the country, where the kinds of floods produced by the rapid wet debris flows, unfortunately can jump out of their banks or avulse without a predictable mechanism. The parsimonious flood map model and general methodology employed to create the AFG-FHM, however, was thought to be a robust enough method to generate extreme inundation outlines.

Some of the map derivatives that have been made from the AFG-FHM include the CAP 2012–AFG-FHM (Figure 8.22) (Anonymous, 2012c), and individual estimated risk maps made for different provinces (Ghor, Samangan, etc.) (Anonymous, 2012d,e). These larger-scale maps show flood risk according to high, medium, and low values, as well as plots of each individual settlement and its risk, and estimates of percent of population at different levels of flood risk (Figure 8.23). A detailed accounting of educational facilities throughout the country that were

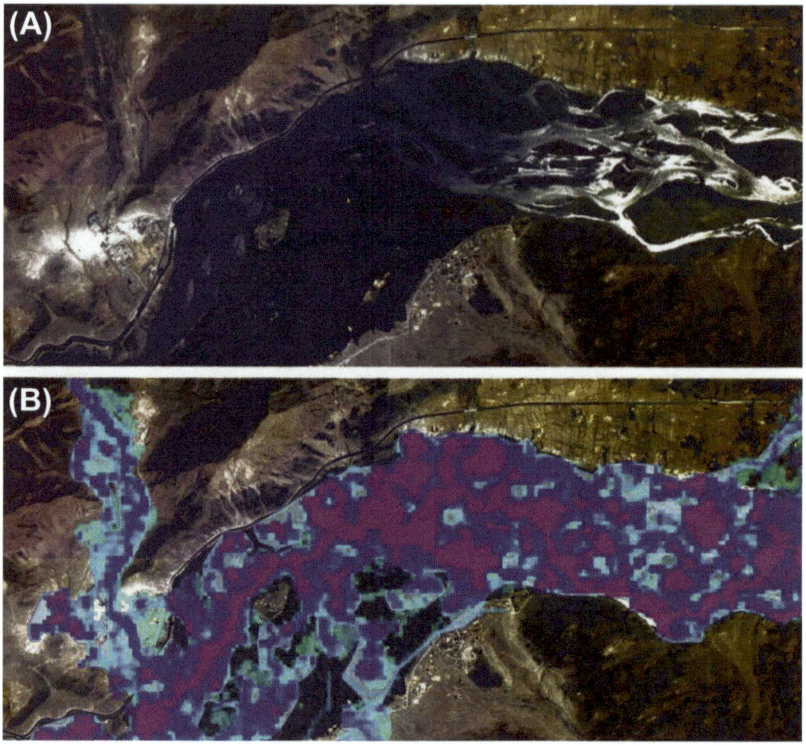

Figure 8.20 (A) An aerial image of Kunar River in the eastern part of Afghanistan; the area shown is approximately 5 by 2 km; (B) The same area overlain with the AFG-FHM Grid image. The maximum inundation depth is 8.7 m in the river center line; with the minimum depth of 0.01 m. The maximum inundation width shown here is 774 m.

exposed to flood risk was developed from the AFG-FHM as well, so overall the original Hagen et al. (2010) effort produced a reasonable return investment in humanitarian assistance for Afghanistan. It will be a blessing indeed, if this new technology brought into application for a beleaguered place, is able to profit Afghanistan from this new science.

Droughts

The Afghan paradox of water, of course, is too much or too little of it, and hardly ever in the right places either. The question of what can be done is a major issue of great significance for the future of the country, especially as the climate changes, with probable negative consequences in both directions—floods and droughts. In fact, Afghanistan shows one of the highest shares of people exposed to drought, as it is fourth in the world behind Jordan, Iraq, and Ecuador (Beekma and Fiddes, 2011). Droughts cause widespread humanitarian crises, often bringing about famines, violent conflicts, and displaced populations (Hall, 2012). For example, when the lower Helmand River dried up entirely in 2001, along with all the *hamun* lakes there (*See* Chapter 5), thousands of people had to leave the region and seek life elsewhere; many of the

Figure 8.21 An aerial image of the Balkh Ab river in north-central Afghanistan showing the AFG-FHM Grid plotted over it with some registration error. The yellow arrows on the floodplain show where the grid should be, and the green arrows show places higher on the valley sides where flooding would not likely occur.

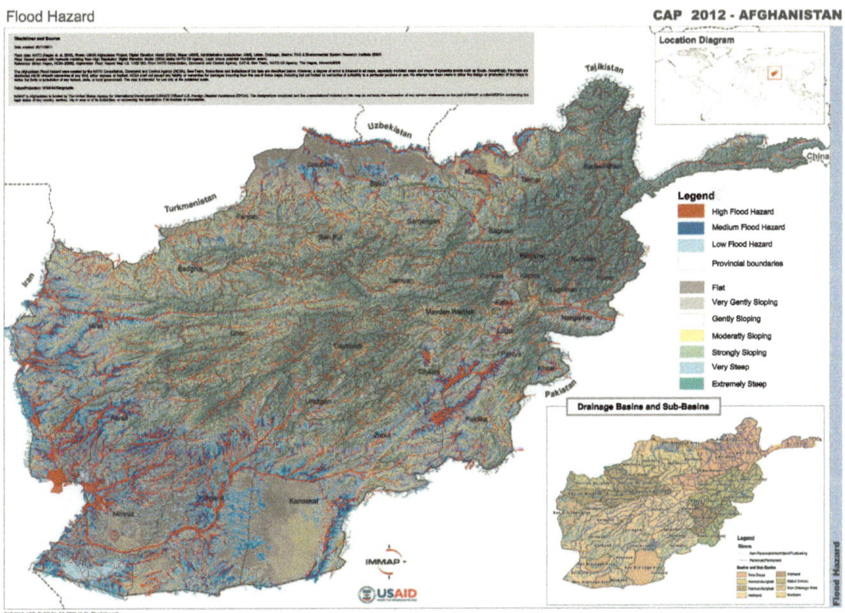

Figure 8.22 CAP 2012 flood hazard map of Afghanistan showing high (red), medium (blue), and low (light blue) flood hazard areas. The map online is completely scale adjustable, with a robust zoom function that enables public examination down to a local level of about 1:100,000 scale. *After iMMAP and USAID; Anonymous (2012c).*

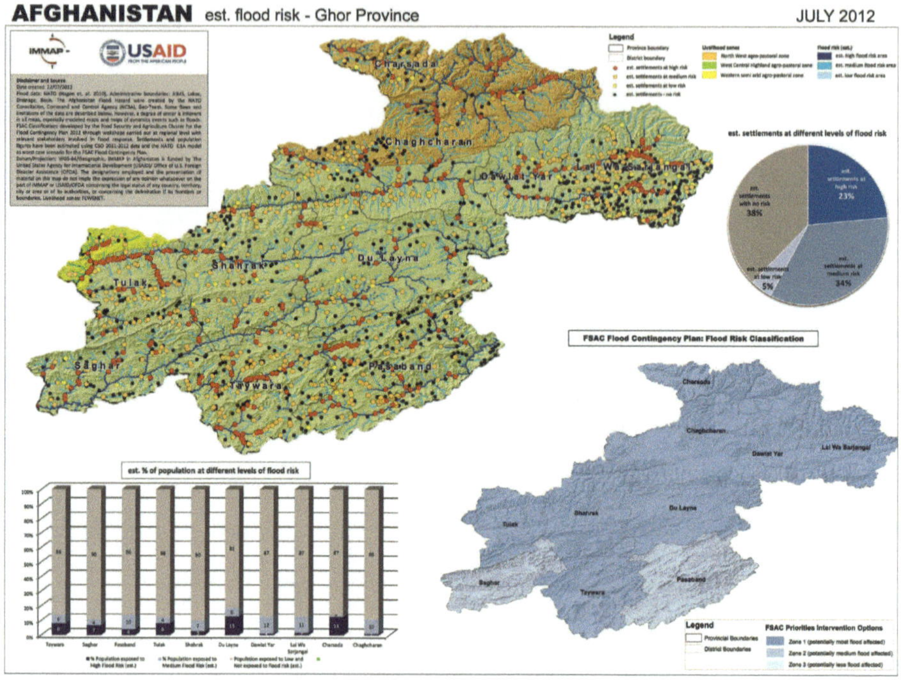

Figure 8.23 Map of estimated flood risk for Ghor Province in west-central Afghanistan, with individual settlements plotted at high risk (red spots), medium risk (orange spots), and low risk (yellow spots), and no risk (black spots). High risk settlements are 23% of the total, medium risk are 34%, low risk are 5%, and no risk are 38%. Flood risks for population percentages were included, as well as flood contingency plans. *After iMMAP and USAID; Anonymous (2012d).*

refugees ended up in Kabul, where they only added to the influx of refugees seeking asylum from the many other perils to life that repeatedly assault them in that region. In fact, unsustainable population growth in any area can be attributed to an estimated 80 percent of water scarcity challenges associated with economic development worldwide (Lavender, 2011).

Because over 80 percent of the annual precipitation of Afghanistan falls as snow in the predominant mountain ranges of the country, the winter snowfall is absolutely essential and underpins the viability of much of the country's irrigated crops. About 45 percent of the annual wheat crop is irrigated, with the remaining 55 percent being rain-fed. The severe drought of 2007–2008, for example, was particularly hard on the northern rain-fed wheat crop where low precipitation and high temperatures dramatically reduced the winter snowpack prematurely from its usual condition (Figure 8.24). As a result the satellite vegetative index data for the region (Figure 8.25) indicated huge reductions in winter grain and pasture growth for the year, which implied widespread losses of crops across northern, northeastern, and western regions (Anonymous, 2008).

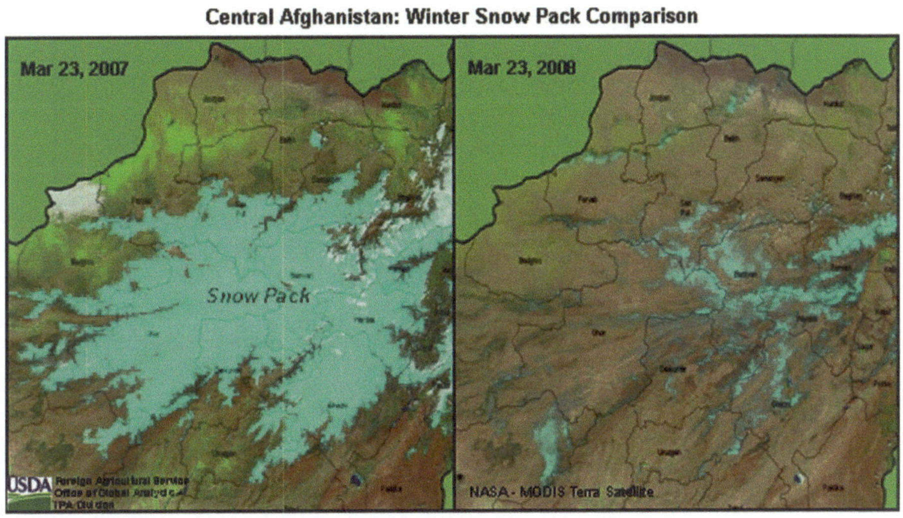

Figure 8.24 The winter snowpack conditions in Afghanistan on 23 March 2007 and 2008 showing the dramatically reduced state of the spring snowfall. *After Anonymous (2008).*

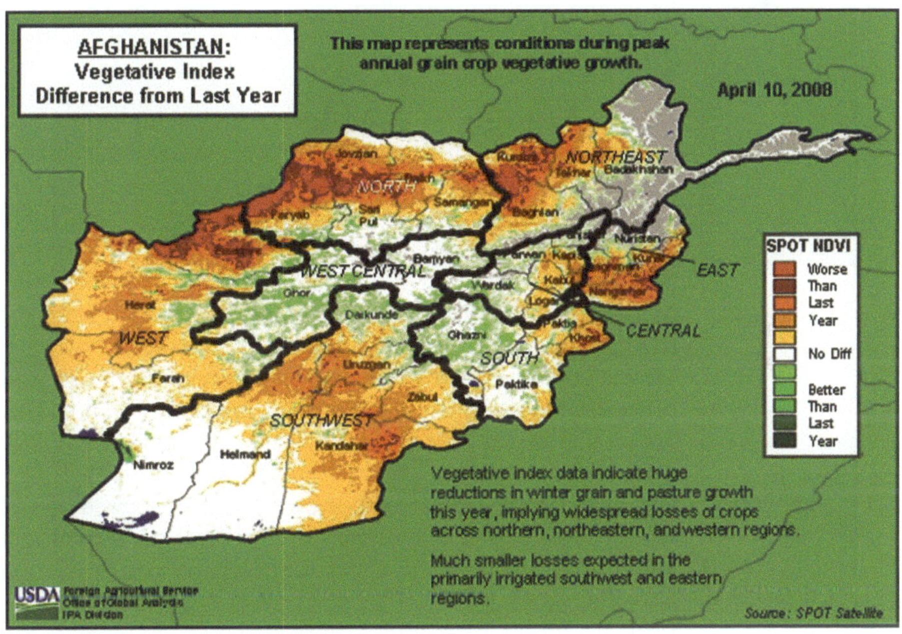

Figure 8.25 The normalized vegetative difference index (NVDI) for Afghanistan in spring 2008 (10 April 2008) from SPOT satellite imagery that shows five levels of difference (orange to tans, browns and red) of vegetation that were worse than the previous year. *After Anonymous (2008).*

In recent decades several seasons of drought have occurred repeatedly. Afghanistan has historically undergone climate cycles of about 15 years, of which two to three have generally been drought. In recent years, however, a marked tendency has developed for the drought cycles to occur more frequently than the historical model predicts. The period of 1998–2005/6 marked the longest and most severe drought in the known climatic history of the country. In fact it has been estimated (Economic Cooperation Organization, 2013) that drought in Afghanistan is likely to be regarded as the norm by 2030, rather than as a temporary or cyclical event. The results of the long-term drought in Afghanistan continued on in 2008/-09, and 2011 (IRIN, 2011; Riechmann, 2011; UNEP, 2012e), although the winter of 2012 resulted in drought-relieving heavy snows that also took their toll in snow avalanches; sometimes the Afghan people just cannot seem to win for loosing so much, no matter what.

Sandstorms

Sandstorms or dust storms can occur when strong winds blow over loose soil or sandy ground so that huge quantities of loose material are mobilized and carried along in the air currents. Because desert regions have so little vegetation to hold the soil in place, such regions have plentiful loose sand and dust on the surface that is easily mobilized when the winds blow. When such storms begin the resulting dirt filled air masses carry the coarser sand particles bouncing along within a few meters of the ground surface, whereas aloft everything is clouds of finer silt and clay dust.

Causes of these storms in Afghanistan are multiple and include localized winds, frontal storms, or the more synoptic and regional effects, as well as special monsoon-related storms. The localized whirlwinds or dust devils are brought about by solar heating of the desert surface so that the warmer air rises. As the speed of the rising air increases during strong heating, the whirlwinds will increase in velocity on the ground like mini tornados and stir up the dust. In some cases this will also pull down stronger winds from the upper atmosphere to mobilize further sand and dust on the surface. The midlatitude cyclonic storms are types of frontal storms that are borne generally eastward by the planetary Westerly winds. These common storms are caused by the dynamic interaction of the warm tropical and cold, polar air masses so that the warm tropical air is lifted cyclonically into the atmosphere where it mixes along warm and cold fronts, always with a great deal of surface wind that mobilizes sand and dust storms. Finally in the summer season in Afghanistan when monsoonal low pressure zones occur along Afghanistan's eastern border, air pressure differences between central Asia's northern plains and Afghanistan's southern deserts force strong winds to blow south between the central mountains of Afghanistan and the mountains of eastern Iran and into the lower pressure zones of the summer monsoon (Figure 8.26). These winds can reach very high speeds up to 180 km/h during the 120-day period between June and September when ground temperatures can be >50 °C so that the wind of 120 days can be an exceptionally hot blast from the north of Afghanistan that mobilizes huge quantities of sand and dust.

Figure 8.26 Approaching dust and sand storm in Sangin, 4 May 2011, near Combat Outpost Shukvani, Musa Qala, Helmand Province, Afghanistan. *Photograph by Cpl. Bryan Nygaard; Defense Video and Imagery Distribution System (DVIDS).* http://www.dvidshub.net/image/523013/sandstorm-sangin.

Sandstorms and dust storms in Afghanistan are notorious for causing damage. Not only do the fine particulates gain access into homes through tiny cracks and crevices, but the grit gets into machinery to cause damaging abrasion or clogging of engines. Agricultural fields first have their plants wilt to death from the moisture stress, and then they are stripped of their vegetation and the soil eroded away. Livestock can be overwhelmed by the heat and dust, and human respiratory and eye infections overwhelm local hospitals. In 2003, for example, during a particularly strong wind of 120days in western Afghanistan, as many as 120 villages were evacuated after the worst sandstorms in living memory engulfed houses, destroyed crops, and contaminated or filled in water supply conduits (Anonymous, 2003b). About 12,000 people in 57 villages were very badly affected, and at least 300 families fled the region. Wells and irrigation canals were filled in and livestock died due to lack of food or water.

Extreme Weather Events

Other weather events that occur as hazards in Afghanistan include extremes of heat and cold. The exceptionally hot summers have already been mentioned for southwestern Afghanistan, but if temperatures commonly over 50°C, were not bad enough, when the wind blows over 180kmph too, the effect is a blowtorch on crops, animals, and people. Temperatures as hot as 48°C have even been recorded in the generally cooler northern lowlands as well (Vaccaro, 2001).

The opposite extreme is the cold winters with blizzards right out of central Asia. The town of Chagcharan at 2183m altitude in central Afghanistan, for example, has winter temperatures as low as −46°C (Vaccaro, 2001). Because Afghanistan has so little fuel for extra fires beyond cooking, staying warm in such winter temperatures is most difficult and many people every winter succumb to the cold in their largely unheated homes, if they even have houses to live in. Heavy snows and cold temperatures in 2008 in northeastern Afghanistan killed 161 people, and put 800 families out of their homes (Anonymous, 2008). In the winter of 2012 as many as 40 people also died from the cold in one part of the country, including 22 children less than five years old. Sixteen other children elsewhere died as well. Many refugees from one disaster, war event, or another, do attempt to live in tents through the winter, with 30,000 in Kabul alone during the winter of 2011–2012 (ToloNews, 2012). It is a most difficult existence (Figure 8.27).

Climate Change

Afghanistan is characterized by such a precarious balance between low levels of precipitation and primary agricultural production that any climate change tips that delicate balance quickly into desperation. Despite the absence of good long-term climate records, the available data and trends from neighboring countries indicate that mean annual temperature has increased by 0.6 °C since 1960. Increases have been most pronounced during the autumnal months, with a significant increase in the number of exceptionally hot days and nights. Changes in precipitation regimes have tended to vary more between regions than the temperature does, with mean rainfall decreasing slightly overall at a moderate rate of 0.5 mm/month. Spring rainfall has been decreasing about 2.7 mm/month (6.6 percent per decade) (Savage et al., 2008; Anonymous, 2009). Temperatures in Afghanistan are predicted to increase 2–6.2 °C by the 2090s, with the temperature increases also negatively affecting water availability because of consequent increases in evaporation. On the other hand, average rainfall is expected to have a small increase in certain areas of the north, perhaps 10–20 mm by 2090, but with significantly drier winters in the south (Savage et al., 2008). In other work, the Climate Research Unit of the University of East Anglia, UK (Islam et al., 2011), recognized that during the last 50 years average temperatures in Afghanistan have increased by about 0.8–1 °C,

Figure 8.27 A cold winter in Kabul where poor people commonly have little winter clothing to protect themselves against low temperatures. *After AFP—Getty Images, with permission.*

whereas average precipitation has decreased by 90–100 mm/year. Temperature variability has also amplified, with extreme temperatures exceeding long-term maximum and minimum temperatures by 30 percent.

The climatic computer models that have been run for Afghanistan suggest that the country will be confronted by a range of new and increased climatic hazards, especially drought-related desertification and land degradation (Savage et al., 2008). Drought is expected to become regarded as the norm in much of Afghanistan by about 2030. Flood impacts are expected to be amplified due to more rapid snowmelt in spring as a result of the predicted higher temperatures in combination with further land degradation, devegetation, and land mismanagement (Figure 8.28).

The socio–economic impacts of climate change in Afghanistan are expected to have a significant negative impact on economic growth in the country (Anonymous, n.d.). The reason, of course, is that the people of Afghanistan have only a very weak capacity to cope with such change, which is not helped much at all by the very poor past record keeping of climate observation that makes it so difficult to model or predict extreme weather events. Even the reasonably predictable contemporary hazardous weather events in the country keep many poor Afghans in a desperate cycle of poverty. Climate change will create additional stresses for these specific vulnerable groups, especially in agriculture and water resources management that are most likely to be severely impacted.

Figure 8.28 Photograph of land degradation in Sar-Rostaq, Takhar Province in northeast Afghanistan where the removal of much of the prior pistachio forests from the loess-mantled uplands and attempted rain-fed cropping has resulted in landslides and soil erosion that pass down into the agricultural lowlands to overwhelm fields and irrigation systems with sediment. *Photo by Wali Modaqiq (2008) for UNEP; by permission.*

In recognition of the problems that they already have with existing weather and climate, the government of Afghanistan has formed a National Adaptation Programme of Action for Climate Change (NAPA) (Anonymous, 2009). The NAPA working group has identified all of the hazards discussed above and compiled a sensitivity matrix of them that indicated that the droughts, floods, and rising temperatures presented the greatest hazards to ecosystems and livelihoods in the country. NAPAs vision for the nation is to increase awareness amongst its citizen shareholders about the effects of climate change on their lives, as well as to develop specific action that will build capacity to respond effectively to current and future threats from climate change. NAPA's objectives are to: (1) identify and prioritize projects and actions that will help communities adapt to climate change; (2) seek overlapping and mutually reinforcing synergies with existing multilateral environmental agreements and development activities, with emphases on both mitigating and adapting to the expected adverse effects of climate change; and (3) bring climate change considerations into the national planning process. A total of 51 potential activity options for better adapting to climate change were identified by the NAPA working group and as of early 2013, were in the process of being worked up into usable forms, with various farming groups and others being brought into country-wide planning processes in so far as possible.

The main climatic hazards identified for example, include: (1) periodic drought; (2) floods due to untimely and heavy rainfall; (3) flooding due to thawing of snow and ice; (4) temperature rise; (5) frost and cold spells; (6) hail, thunder and lightning; and (7) monsoon and the 120-day winds (Table 8.2). With this as a data collection matrix, the impacts of these climatic hazards could be noted as loss of lives and livelihoods, health problems, duration of the impacts, food security and agriculture issues, environmental effects such as biological diversity and forestry, and water availability, quality, and accessibility. These seven main climatic hazards also had their multiple effects assessed on the three main necessities that they control, such as the many and varied ecosystem services, livelihood activities, and the means of livelihood (Table 8.3). The resulting matrix of hazard sensitivity guided information collection, as well as allowing additive intercomparison of the multiple resource groups in the development of a preliminary exposure index compared to the impact index, thus allowing useful quantitative comparison. The seven sectors of agriculture, forestry and range, biodiversity, health, energy, waste, and water resources could then be assessed in terms of temperature rise, change in precipitation patterns, changes in evapotranspiration and salinity, and droughts to see how they will fare during climate change (Anonymous, 2009) (Figure 8.29).

Table 8.2 Narrative description of climatic hazards and some of their impacts (After Anonymous, 2009)

Climatic Hazards	Description	Loss of Life and Livelihoods	Human Health	Duration	Food Security and Agriculture	Environmental Effects (Biological Diversity, Forestry)	Water Availability, Quality, and Accessibility	Trends
Periodic drought	Decrease in productivity of crops; forced migration; changes in livelihood; decrease in amount of exports; and financial losses.	Group of livestock herders including Kuchi, irrigated agriculturalists and dryland farmers are affected; and around 10,000 casualties per year of severe drought.	Malnutrition, spread of diseases such as malaria, sishnaia, cholera, typhoid, tenasagenata, asscaris, and diarrhea.	Up to 8 years	Drought has generated an estimated loss production of: 75% wheat, 85% rice, 85% maize, 50% potatoes, and 60% of overall farm production, between 1998 and 2005.	Pistachio, pine nut, wild almond, and conifer forest production negatively affected. Waterfowl sanctuaries such as Dashti Nawar, Ab-i-Estada, and Kol-i-Hashmat Khan have dried up. Wildlife being displaced.	Decreased availability of under ground water, springs and karezes dry-up, flow of major rivers collapses, degradation of watersheds and drop in level of water reservoirs and dams.	Increasing frequency and intensity.
Floods due to untimely and heavy rainfall	Collapse and sedimentation of irrigation canals; destruction of agricultural lands; loss of crops and livestock; collapse of dwellings; spread of epidemic diseases; destruction of infrastructure such as roads and bridges; and damage to the national economy.	Approximately 750 casualties per year.	Increased incidence of cholera, typhoid, diarrhea, and malaria.	3 months	Loss of 10% of agricultural production and gardens that are located alongside rivers and in high risk areas.	Soil degradation; loss of natural forest (riverine forest being particularly affected); increased levels of water siltation and sedimentation; and displacement of wildlife.	Infrastructure damage around US$ 300 million as a result of a severe flood.	Increasing frequency and intensity.

Continued

Table 8.2 Narrative description of climatic hazards and some of their impacts—cont'd

Climatic Hazards	Description	Loss of Life and Livelihoods	Human Health	Duration	Food Security and Agriculture	Environmental Effects (Biological Diversity, Forestry)	Water Availability, Quality, and Accessibility	Trends
Flooding due to thawing of snow and ice	River levels rise; destruction of riverside agricultural and nonagricultural (forest, range, etc.) lands; landslides; soil erosion; destruction of infrastructure such as bridges and gabions.	Around 100 casualties per year.	Increased incidence of cholera, typhoid, diarrhea, and malaria.	4 months	2% damage of agricultural land alongside rivers.	Soil degradation; loss of natural forest (riverine forest being particularly affected); increased levels of water siltation and sedimentation; and displacement of wildlife.	Destruction of riverbanks, associated agricultural land and infrastructure at a value of around US$ 400 million as a result of a severe flood.	Increasing frequency
Rise in temperature	Increase in levels of incidence of diseases that affect humans, agriculture, and livestock; habitat changes affect wildlife; and changes in vegetation cover and associated grazing patterns.	Around 1000 casualties per year	Increased incidence of malaria, leshmania, typhoid, and diarrhea.	3 months	Decreased agricultural, livestock, and horticultural production.	Less productivity of natural system, displacement and changing of wildlife habitat.	Increasing evapotranspiration rates; reduction in water level.	Increasing frequency and intensity.

Table 8.2 Narrative description of climatic hazards and some of their impacts—cont'd

Climatic Hazards	Description	Loss of Life and Livelihoods	Human Health	Duration	Food Security and Agriculture	Environmental Effects (Biological Diversity, Forestry)	Water Availability, Quality, and Accessibility	Trends
Frost and cold spells	Degradation of fruits, crops, vegetables, and health disease, poor economy and increasing of poverty.	Loss of fruits and potatoes; approximately 300 indirect casualties per year.	Illnesses associated with cold weather.	3 days, two times per year	20% of gardeners in the country are affected, particularly those with horticultural crops.	Affects forest rehabilitation and afforestation programs (particularly nurseries and saplings).	Low impact.	Increasing frequency and intensity.
Hail, thunder and lightening	Destruction of crops (particularly horticultural crops); human and livestock losses; and outflow/gush from floods.	Approximately 150 casualties per year.	Illnesses associated with cold weather.	Around 20 days	Up to 20% loss in horticultural and crop production.	Low impact.	Low impact.	Increasing frequency and intensity.
Monsoon and 120-day winds	Desertification; degradation of agricultural lands and crops; destruction of infrastructure; air pollution; spread and transmission of diseases and respiratory problems; sedimentation of irrigation systems and springs; local and national economy negatively affected.	Around 10 casualties per year.	Eyes, respiratory and skin diseases.	120 days	Decrease in horticultural and crop production, degradation of rangeland and reduced livestock production.	Desertification; decreased plant cover.	Losses (temporary and more permanent) of infrastructure, siltation of water sources, decreased quality of water.	Increasing frequency and intensity.

Table 8.3 Sensitivity matrix for resources and groups to climate change (After Anonymous, 2009)

	Climatic Hazards									
	Periodic Drought	**Floods Due to Untimely and Heavy Rainfall**	**Flooding Due to Thawing of Snow and Ice**	**Rise in Temperature**	**Frost and Cold Spells**	**Hail, Thunder, and Lightening**	**Monsoon and 120-Day Winds**	**Preliminary Exposure Index**	**Weighting-Area Affected/ Importance**	**Final Score**
Ecosystem Services										
Soil water content	5	2	4	4	1	2	3	60	1	60
Water from irrigation	5	3	1	3	1	1	2	46	0.8	37
Fuelwood	4	3	1	2	1	2	3	46	0.8	37
Grazing	5	3	2	4	1	2	3	57	0.7	40
Fodder	4	3	2	3	1	3	3	54		0
Wild fruit production	4	1	1	3	3	2	2	46		0
Wildlife	4	3	2	4	1	2	2	51		0
Medicinal plants	4	3	2	3	1	3	2	51	0.4	21
Pollination	4	3	1	1	4	4	3	57	0.2	11
Livelihood Activities										
Food crops	4	4	4	3	3	4	3	71	1	71

Table 8.3 Sensitivity matrix for resources and groups to climate change—cont'd

	Climatic Hazards									
	Periodic Drought	**Floods Due to Untimely and Heavy Rainfall**	**Flooding Due to Thawing of Snow and Ice**	**Rise in Temperature**	**Frost and Cold Spells**	**Hail, Thunder, and Lightening**	**Monsoon and 120-Day Winds**	**Preliminary Exposure Index**	**Weighting-Area Affected/ Importance**	**Final Score**
Market crops	5	4	3	3	5	4	3	77	0.5	39
Livestock	5	3	2	4	2	1	2	54	0.9	49
Wood	4	3	1	2	1	2	3	46	0.8	37
Daily wage labors	5	2	1	3	1	1	2	43	0.8	34
Useful Insects	5	2	1	4	4	3	4	66	0.2	13
Trade	4	4	3	3	3	3	2	63	0.5	31
Means of Livelihood										
Dryland farmers	5	4	3	4	1	3	4	69	0.7	48
Irrigated agriculture farmers	5	4	4	3	4	4	3	77	1	77
Livestock herders	5	3	3	4	2	2	2	60	0.9	54

Continued

Table 8.3 Sensitivity matrix for resources and groups to climate change—cont'd

	Climatic Hazards									
	Periodic Drought	**Floods Due to Untimely and Heavy Rainfall**	**Flooding Due to Thawing of Snow and Ice**	**Rise in Temperature**	**Frost and Cold Spells**	**Hail, Thunder, and Lightening**	**Monsoon and 120-Day Winds**	**Preliminary Exposure Index**	**Weighting-Area Affected/Importance**	**Final Score**
Pastoralists (Kuchi)	5	4	3	4	1	3	3	66	0.5	33
Poppy growers	2	3	3	1	3	4	2	51	0.8	41
Impact index	89	61	45	62	42	52	53			
Weighting-area affected	1.0	0.8	0.6	0.8	0.7	0.4	0.4			
Final score	89	49	27	50	29	21	21			
High priority	Weighting scale ranging from 1 (low impact) to 5 (high impact)									

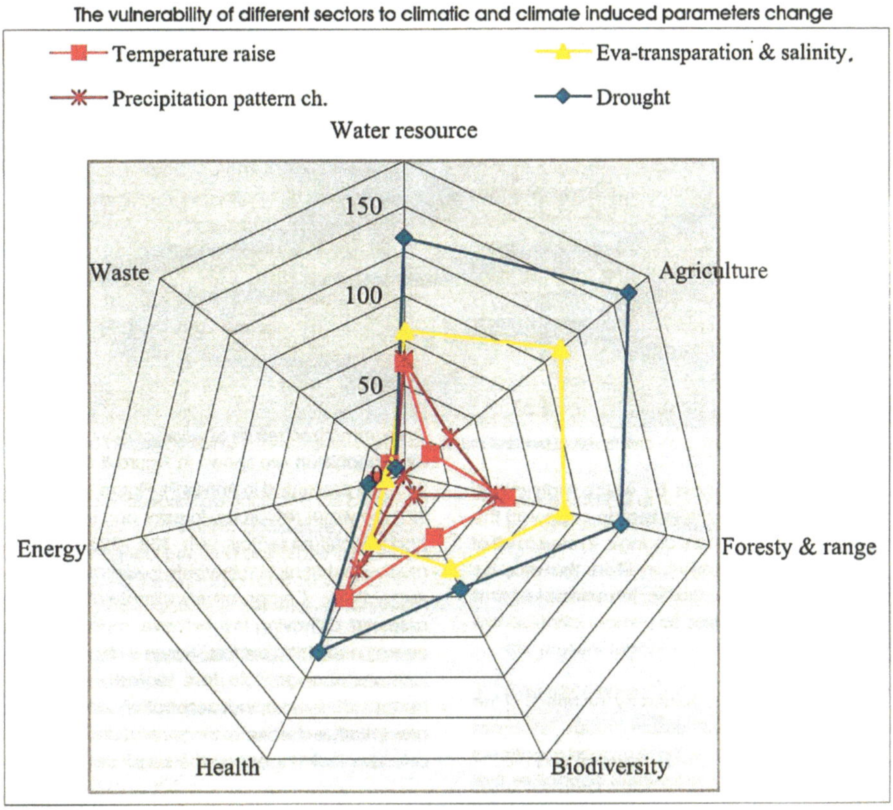

Figure 8.29 Plot of vulnerability of different sectors to change of climate or climate-induced parameters. *From Anonymous (2009).*

References

Ambraseys, N., Bilham, R., 2003. Earthquakes in Afghanistan. Seismological Research Letters 74, 107–123.

Anonymous, n.d. Afghanistan: review of socio-economic impacts of climate change. HTSPE. International Programme Management. 2 p. http://www.htspe.com/climate_change.htm.

Anonymous, 2003a. Afghanistan: National Disaster Management Plan. Department for Disaster Preparedness (DDP); Transitional Islamic State of Afghanistan; United Nations Assistance Mission in Afghanistan (UNAMA), 67 p. http://www.andma.gov.af/Contents/295/Documents/140/Afghanistan_NDMP_Final_Jul_04%5B1%5D.pdf.

Anonymous, 2003b. Killer Sandstorms Ravage Afghanistan. The Statesman, India. http://www.accessmylibrary.com/coms2/summary_0286-23927197_ITM.

Anonymous, August 12, 2008. Afghanistan: Severe Drought Causes Major Decline in 2008/09 Wheat Production. United states Department of Agriculture (USDA) Foreign Agricultural Service; Commidty Intelligence Report, 6 p. http://www.pecad.fas.usda.gov/highlights/2008/08/Afghanistan%20Drought/.

Anonymous, 2009. Afghanistan: National Capacity Needs Self-Assessment for Global Environmental Management (NCSA) and National Adaptation Programme of Action for Climate Change (NAPA). United Nations Environment Programme (UNEP); National Environmental Protection Agency, Afghanistan; Global Environmental Facility. 125 p.

Anonymous, February 10, 2010. Afghanistan Avalanches Kill at least 165 in Salang Pass. BBC News. http://news.bbc.co.uk/2/hi/south_asia/8506033.stm.

Anonymous, 2012a. Afghan Quakes: Landslide Killed 80, Says Governor. BBC News, Asia. http://www.bbc.co.uk/news/world-asia-18406216.

Anonymous, January 20, 2012b. Heavy Snow, Avalanches Strike Afghanistan. Accuweather. http://www.accuweather.com/en/weather-news/heavy-snow-avalanches-strike-a/60462.

Anonymous, 2012c. Flood Hazard. CAP 2012 – Afghanistan. iMMAP; USAID (Scale~1:12M). https://afg.humanitarianresponse.info/sites/afg.humanitarianresponse.info/files/Flood_Hazard.pdf.

Anonymous, 2012d. Afghanistan: Estimated Flood Risk – Ghor Province (July 2012). iMMAP; USAID. http://reliefweb.int/map/afghanistan/afghanistan-est-flood-risk-ghor-province-july-2012.

Anonymous, 2012e. Afghanistan, UNEP Launch USD $6 Million Initiative to Help Communities Adapt to Effects of Climate Change. http://www.unep.org/newscentre/default.aspx?DocumentID=2697&ArticleID=9300.

Beekma, J., Fiddes, J., 2011. Floods and Droughts: The Afghan Water Paradox. Afghanistan Human Development Report 2011; Center for Policy and Human Development (CPHD); Kabul University (UNDP – Afghanistan). Research Paper; 29 p. http://www.cphd.af/Downloads/Background%20Papers/Flood%20and%20drought%20-%20The%20Afghan%20water%20paradox.pdf.

Benz, H., Machette, M., Sipkin, S., Wheeler, R., 2005. Assessing the Seismic Hazards of Afghanistan. USGS Fact Sheet 2005-3038 . http://pubs.usgs.gov/fs/2005/3038/.

Bourrouilh-le Jan, F.G., Akram, B., Schvoerer, M., 2007. Band-e-Amir lakes and Dragon Valley (Bamiyan): myths and seismicity in Afghanistan. In: Picardi, L., Masse, W.B. (Eds.), Myth and Geology. Geological Society, London, pp. 121–132. Special Publication No. 273.

Boyd, O.S., Mueller, C.S., Rukstales, K.S., 2007. Preliminary Earthquake Hazard Map of Afghanistan. USGS Afghanistan Project Product No. 156; US Geological Survey Open-File Report 2007-1137 . http://pubs.usgs.gov/of/2007/1137/.

Dewey, J.W., 2006. Seismicity of Afghanistan and Vicinity. USGS Afghanistan Project Product No. 28; US Geological Survey Open-File Report 2006-1185. http://pubs.usgs.gov/of/2006/1185/downloads/pdf/of06-1185_508.pdf.

Economic Cooperation Organization, 2013. Afghanistan. Institute of Environmental Science and Technology, 8 p. http://eco-iest.org/research-programmes/trans-boundary-protected-areas-tbpa/afghanistan/.

Gupta, M., 2010. National Disaster Mangement Plan, 2010: Afghanistan. Afghanistan National Disaster Mangement Authority; 27 pages. http://www.andma.gov.af/contents/295/Documents/226/Updated_Afghanistan_NDMP.pdf.

Hagen, E., Shroder Jr., J.F., Lu, X.X., Teufert, J.F., 2010. Reverse engineered flood hazard mapping in Afghanistan: a parsimonious flood map model for developing countries. Quaternary International 226 (1–2), 82–91.

Hall, M., 2012. Drought & Food Insecurity in Afghanistan. Civil – Military Fusion Centre, 6 p. https://www.cimicweb.org/Documents/CFC%20AFG%20Humanitarian%20Assistance%20Archive/CFC_Afghanistan_Food-Insecurity_Jan2012.pdf.

Islam, F., Hove, H., Perry, J.-E., International Institute for Sustainable Development, 2011. Afghanistan: Review of current and planned adaptation action. Review of Current and Planned Adaptation Action: South Asia. Afghanistan. pp. 42–49. http://www.preventionweb.net/files/25653_afghanistan.pdfhttp://www.cakex.org/sites/default/files/asia%20-%20south%20asia%20regional%20and%20country%20profiles.pdf.

IRIN, 2011. Analysis: Afghanistan drought conditions could spell disaster. UN Office for the Coordination of Humanitarian Affairs. http://www.irinnews.org/report/93781/analysis-afghan-drought-conditions-could-spell-disaster.

Javed Khan, M., Opdyke, N.D., Shroder Jr, J.F., 1985. Bain diamictite; lithology, age, and origin. Geological Bulletin, University of Peshawar 18, 53–64.

Lavender, L., 2011. When Demand Surpasses Supply: Water Scarcity in Afghanistan. Civil – Military Fusion Centre, 6 p. https://www.cimicweb.org/Documents/CFC%20AFG%20Social%20Well-being%20Archive/CFC_Afg_Water_Scarcity_Sept11.pdf.

Modaqiq, W., 2008. Cover photograph on UNEP Final Joint Report, February, 2009. Afghanistan: National Cacity Needs self-assessment for Global Environmental Management (NCSA) and National Adaptation Programme of Action for Climate Change (NAPA). United Nations Environmental Programme, Nairobi, Kenya; 128 pages. http://www.unep.org/dgef/Portals/43/publications/Afghan_NCSA_and_NAPA_2009.pdf.

Morris, T.O., 1938. The Bain Boulder-bed: a glacial episode in the Siwalik Series of the Marwal Kundi Range and Shekh Budin, North-West Frontier Province, India. Quarterly Journal of the Geological Society 94, 385–422.

Palau, R.G., 2012. Potential Flood Scenarios & Afghanistan's Disaster Management Framework. Afghanistan Resource Desk; Civil – Military Fusion Centre, 7 p. https://www.cimicweb.org/cmo/afg/Documents/Afghanistan-RDPs/CFC_Afghanistan_Flood-Management_Mar2012.pdf.

Pierson, Costa, J., 1987. The rheological classification of subaerial sediment-water flows. Reviews in Engineering Geology 7, 1–12.

Riechmann, D., 2011. Afghanistan faces drought, hunger. Huffington Post. http://www.huffingtonpost.com/2011/12/03/afghanistan-drought_n_1127029.html.

Rubin, A.J., March 6, 2012. Avalanche and Attacks Add to Woes of Afghans. New York Times. http://www.nytimes.com/2012/03/07/world/asia/thaw-brings-avalanche-and-attacks-in-afghanistan.html?pagewanted=all&_r=0.

Savage, M., Dougherty, B., Hamza, M., Butterfield, R., Bharwani, S., 2008. Socio-Economic Impacts of Climate Change in Afghanistan: A Report to the Department of International Development. Stockholm Environment Office, Oxford, UK, DFID CNTR 08 8507.

Shroder Jr, J.F., 1989. Slope failure: extent and economic significance in Afghanistan and Pakistan. In: Brabb, E.E., Harrod, B.L. (Eds.), Landslides: Extent and Economic Significance in the World. A.A. Balkema, Rotterdam, Netherlands, pp. 325–341.

Shroder Jr., J.F., Weihs, B.J., 2010. Geomorphology of the Lake Shewa landslide dam, Badakshan, Afghanistan, using remote sensing data. Geografiska Annaler 92A (4), 471–486.

Shroder, J.F., Schettler, M.J., Weihs, B.J., 2011a. Loess failure in northeast Afghanistan. Physics and Chemistry of the Earth 36, 1287–1293. http://dx.doi.org/10.1016/j.pce.2011.03.001.

Shroder, J.F., Weihs, B.J., Schettler, M.J., 2011b. Mass movement in northeast Afghanistan. Physics and Chemistry of the Earth 36, 1267–1286.

Shroder Jr, J.F., 2012. Afghanistan: rich resource base and existing environmental despoliation. Environmental Earth Sciences. 67 (7), 1971–1986. http://link.springer.com/article/10.1007/s12665-012-1638-7.

Shroder, J.F., 2013. Ground, aerial, and satellite photography for assessing geomorphology and geomorphic change. In: Shroder, J., Bishop, M.P. (Eds.), Remote Sensing and Geographic Information Science in Geomorphology. Treatise on Geomorphology, vol. 3. Academic Press, San Diego, CA.

Vaccaro, C., October 4, 2001. Rugged Landscape Creates Extreme Afghan Weather. USA Today. http://usatoday30.usatoday.com/weather/world/2001/afghanistan-climate.htm.

ToloNews, 2012. Severe Cold Kills 40 Afghans, Health Ministry Says. Tolonews.com. http://www.tolonews.com/en/afghanistan/5403-severe-winter-cold-kills-40-afghans-health-ministry-says.

UNEP, 2012e. Afghanistan, UNEP launch USD$6 million initiative to help communities adapt to effects of climate change. http://www.unep.org/Documents.Multilingual/Default.asp?DocumentID=2697&ArticleID=9300&l=en.

Wheeler, R.L., Rukstales, K.S., 2007. Seismotectonic Map of Afghanistan and Adjacent Areas. USGS Afghanistan Project Product No. 154; US Geological Survey Open-File Report 2007-1104 . http://pubs.usgs.gov/of/2007/1104/.

Wheeler, R.L., Bufe, C.G., Johnson, M.L., Dart, R.L., 2005. Seismotectonic Map of Afghanistan, with Annotated Bibliography. US Geological Survey Open-File report 2005-1264. 31 p. http://pubs.usgs.gov/of/2005/1264/pdf/OFR-1264.pdf.

World Health Organization, 2010. Afghanistan: Flood Hazard Distribution Map. http://www.who-eatlas.org/eastern-mediterranean/images/map/afghanistan/afg-flood.pdf.

Part II

Introduction to Resources: Bones of Contention or Solutions to Interminable War?

With the geology and geography of Afghanistan as a framework or context of this volume upon which to hang some of the history of the incessant warfare there, we can next consider the rich natural resources of the country. These are either the bones of contention for additional war as other outside powers attempt to take over the resources, or conversely, the resources are the most likely solution to develop the country and help it out of its grinding poverty. In this Part II of the book the delineation and nominal fixing of the borders of Afghanistan with its neighbors; China, British India/Pakistan, Persia/Iran, and Russia/USSR/Turkmenistan/Uzbekistan/Tajikistan—these regions are first discussed because the border fixing between them is so germane to an extended discussion about resources and where those resources lie geographically.

Strikingly, early on in the history of Afghanistan and the USA, several American geologists were offered excellent terms to find and develop resources in the country but the distance from markets and the intervention of World War II precluded that eventuality. Instead the arrival of the Cold War confrontation with the Soviets and their excellent and numerous geologists allowed mapping and natural resource assessment throughout Afghanistan in the 1960s–1970s. This resulted in the discovery of a number of world-class deposits of copper, iron, chromite, hydrocarbons, rare earths, and other valuable materials that if brought on stream to market would potentially affect world prices of these commodities, as well as produce important new revenue sources to the government of Afghanistan.

The history of the discovery of rich resources, however, has rarely been good for the country in which the commodities were found, with the result that concepts have emerged in recent years concerning resource wars that are fought to gain control of resources somewhere, or the resource curse wherein countries are ruined by the violence and corruption that springs up around such resource discoveries. Afghanistan would seem to be squarely in the middle of such a muddle, with both resource curses and resource wars swirling it ever deeper into violence. In part to forestall some of this, while at the same time bringing the very best of modern remote-sensing technology into the mix, the US Geological Survey brought in hyperspectral remote sensing to cover most of the country with high-resolution, radiometric, magnetic, and gravimetric imagery and measurement that enabled detailed resource assessments. The results of this analysis, together with the original good work of the Russian geologists, and some limited field work by the USGS, the British Geological Survey, and others, all added to the resource data bank until finally the Pentagon and NATO officials recognized that the many rich resources of Afghanistan might actually be the answer to the incessant violence. If the people of Afghanistan could finally profit from the rich resource base that they inherited but did not know how to exploit without outside help and investment, then the future of the country might actually be robust. As neighboring China and India bid successfully on the copper, iron, and oil deposits of Afghanistan, several hundred other deposits were readied for bidding as the country began to undergo its first resource rushes from the world community of resource-extraction industries. Within the past decade new concepts of resource corridors were recognized also as a characteristic of many developed countries in the West, the ideas of which had been successfully transferred to a number of other undeveloped countries as well. Newly applied to Afghanistan, the idea of resource corridors included a mix of a new transport network of roads and railroads, coupled with ancillary support industries, around core mineral extraction locations that collectively promised to spring Afghanistan loose from its centuries of abject poverty and backwardness. Resource development in Afghanistan thus became the seeming answer to the insurgency against the Karzai government; if the resources could be gotten out of the ground and the country raised out of its poverty, perhaps the incessant repetitive wars could come to an end.

The exposition of all the issues of resources in Afghanistan provides the essential background to understand very pivotal factors that will drive much of the unfolding happenstance in Afghanistan. Once the resource genie is out of the bottle in the country, a whole new history of development is most likely to occur, but with unknown regional consequences. Collateral events are likely to reverberate into neighboring countries with a variety of good and bad effects, not the least of which will be many environmental issues that must be faced. But the many good results from resource extraction will likely transcend the bad, in the sense that the great poverty of the region may be partly assuaged, at least for a time.

In this Chapter

9

Afghanistan Border Fixing

Abstract

The international borders of Afghanistan were defined in late nineteenth century when the British Empire, in collaboration with the Russian and Persian empires, forced the present-day borders onto the greater Afghanistan region. Prior to border demarcation through foreign border commissions (British, Russian, Persian, Chinese), almost no Afghans were involved. Parts of the existing borders have remained highly contentious to this day, with the Durand-Line border between Afghanistan and Pakistan that was drawn to deliberately split the Pashtun tribes, being the most heated argument of the present time. Basically Afghanistan has opposed this border, whereas official Pakistan, wants to keep the border as is. This is precisely because the border does effectively divide the fractious Pashtuns and better enable the Punjabi-dominated government in Pakistan to keep control of its portion of the tribes. Partition of Afghanistan has been discussed repeatedly, particularly when conditions and politics have become extreme and violent.

Keywords: Afghanistan–Amu Darya border; Afghanistan–Chinese border; Afghanistan–Pakistan border; Afghanistan–Persian border; Durand Line; National borders; Partition of Afghanistan.

National borders matter, even while they fixate and highly annoy people who must deal with their effects constantly. Delineating and demarcating them, smuggling across them, escaping over them to live and fight another day, fighting for the resources encompassed by them; all are the human condition writ large upon the landscape, whether visible from outer space or not. In some cases national borders actually can be seen from astronauts in spacecraft, especially where they are lit up at night with continuous spotlights to keep out smugglers (India and Pakistan), deliberate forest cutting (US and Canada), or ungrazed next to overgrazed (northwestern Afghanistan, Negev Israel–Egypt).

Borders can both facilitate or constrain the human interactions across them in trade and in conflict, in peace and in war. Borders come in many different landscape types: natural oceans and lakes where such occur, equally natural rivers and mountain ranges in other situations. Even forests and deserts can serve, although generally artificially surveyed lines are superimposed across such features. Four types of international boundaries occur: (1) antecedent, where

Natural Resources in Afghanistan. http://dx.doi.org/10.1016/B978-0-12-800135-6.00009-X

preexisting natural features such as rivers, bays, lakes and mountains exist; (2) subsequent, which were set corresponding to their different ecumenes after the settlement areas of different groups expanded and met; (3) superimposed, where the boundary is imposed across a region by an outside force through treaty so that existing cultural landscapes may be cut through; and (4) relic, where the defunct boundary may still be visible on the landscape but its actual absence is the result of political changes (Rodrigue, 1989). The perception of some of these boundaries may be in the eye of the beholder in certain cases, as where a superimposed boundary is thought of as a relic by people on which it is superimposed, but regarded as an iron rule by others with their own vested interests. European powers often delineated boundaries on maps without consideration of the economies or cultures of the people whom they were colonizing. These delineations are in many cases reverberating down through history to the detriment of all concerned. As the following will show, Afghanistan is a case in point.

Four types of boundary dispute can arise: (1) positional disputes; (2) territorial disputes; (3) cultural disputes; and (4) resource disputes (Anonymous, n.d.a). Positional disputes can arise when a feature such as a natural river boundary changes its location by erosion and deposition so that one side gains land and the other loses it. Some states may armor or fix their river banks against this, such as Takistan and Uzbekistan did against the Panj and Amu Darya, whereas Afghanistan has largely been too poor to do this and has lost much land as a result. Territorial disputes arise where one culture claims part of the territory of another based on a history of past migration or emigration. Cultural disputes can arise over differences in language, religion, politics, and ethnicity. Resource disputes increase as populations expand and demand more and more of certain resources such as surface water that flows from one territory or another, or hidden ground water that migrates underground, or oil and gas.

The present-day borders of Afghanistan were delineated in the late nineteenth century, mainly as a result of the "Great Game" rivalry between Russia and Britain and the desire to create a buffer zone where the borders of the two expanding empires were not contiguous so that the two great powers of the time could avoid direct conflict with each other. Afghanistan now shares borders with Turkmenistan (744 km), Uzbekistan (137 km), Tajikistan (1206 km), China (76 km), Pakistan (2430 km), and Iran (936 km) (Figure 9.1), pressed on it by outsiders, which led directly to unresolved border disputes with Pakistan and Iran that continue to be great sources of unfair friction to be discussed below. Afghanistan's rugged and mountainous frontiers, together with its water-starved desert lowlands significantly affected the border conflicts in past times and will certainly produce problems in future as the climate changes. No major conflicts, however, exist at this point in time along Afghanistan's northern desert, river, and mountain peak boundaries with the Central Asian states, or with China. In 1946, the midchannel of the Amu Darya and Panj rivers was fixed as the boundary between the USSR and Afghanistan, even though these rivers have notoriously mobile channels (Figure 9.2). That original agreement with

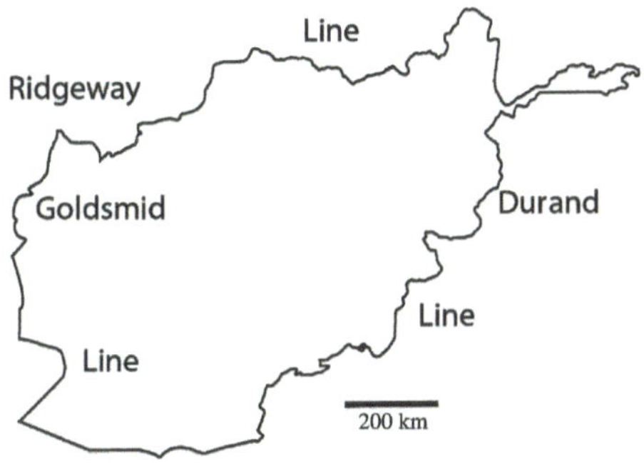

Figure 9.1 Borders of Afghanistan showing Durand Line on the east and south, Goldsmid Line on the west, and Ridgeway Line in the north.

Figure 9.2 Photograph of abandoned Soviet armored personnel carriers put to alternate good armor use at Hairatan on the Amu Darya to protect the south side of the river in Afghanistan against the depredations of the eroding river. The middle thalweg (deepest part of the channel) is supposed to serve as the border but the river is notorious for lateral migration by alternate erosion and deposition. With a much higher and more flexible budget on the north side, the USSR originally, and now Uzbekistan and Tajikistan further upstream on the Panj tributary, could afford to armor their river banks with rock riprap and other materials, whereas the partly nonfunctional and poverty-stricken GIRoA could not afford to pay for armoring materials, except for this cast-off war wreckage for which the Afghans had little other use. *Photograph by David Adamex from Google Earth™.*

the USSR has remained in effect with the successor states that derived from the breakup of the Soviet Union.

Afghanistan-Central Asian Borders

As noted, the northern border of Afghanistan with the republics of central Asia that formerly belonged to the USSR was a direct product of the so-called "Great Game" of nineteenth-century intrigue and diplomacy between Great

Britain and Russia. The joint Afghan Boundary Commission was constituted as an Anglo-Russian venture whose task was to accurately delineate the frontier between northern Afghanistan and Russia's central Asian territories to replace a line drawn crudely in 1873 on vague and inaccurate maps (Balland, 1990; U.S. Department of State, 1983; Yate, 1888).

The two governments of Great Britain and Russia opened negotiations in March, 1869, following the Russian military drive into central Asia as they captured Tashkent, Bokara, and Khiva. Beginning with the far eastern boundary of Afghanistan, begrudging general agreement occurred with the British proposal that the northern frontier of Afghanistan could be fixed relatively easily along the Amu Darya, although the Russians did note that rivers make bad boundaries and mountain ridges are better because they are a more natural barrier. Thus some two-thirds of the initial border determination along the river was reasonably easy and the other one-third in the northwestern part of the country was across a great deal of monotonous desert and steppe terrain of the Karakumsky Desert and was initially not thought to be worth a great deal of effort in further delineation. In the Badghis Province portion of this area, for example, the vague agreement noted only that the Afghan frontier was "well known and need not here be defined". Such imprecision only invited later misinterpretation and border disputes, with the result that not long after this was agreed, the issue needed to be revisited, this time with a full joint boundary commission in the field. Accordingly, from 1882 to 1888 the British and Russian field parties worked at delineating the border where there were no natural landmarks or landform features along which to construct any sort of obviously marked political boundary.

In the meantime, Russian stalling to first precisely define a boundary before it could actually be demarcated on the ground enabled their army to make haste and reinforce its local positions, as well as to take new ground as long as they still could. Such blatant manipulation considerably surprised the British commissioners under the direction of Lieutenant Colonel J.W. Ridgeway who observed such activity in the field. In any case, the joint boundary commission set to work from a point near the Persian border and they moved northeastward toward the Amu Darya to install boundary posts where they could and to negotiate their way around complex irrigation networks in other places. Land swaps were made where disputes arose until finally the group converged toward completion of the task. They had installed a total of 79 boundary posts along the 394-mi (630-km) length of the northwest border, which later came to be known as the "Ridgeway Line", to compare it to the Durand Line that set off the boundary with British India (later Pakistan).

The border along the middle of the Amu Darya river channel was a reasonable decision as long as the river remained in place, but of course the high discharge variations on the Nile-sized river caused considerable lateral migration of the banks with attendant problems of border location. The northern side of the river banks were later artificially hardened and stabilized for decades by the USSR and later by the governments of the Central Asian republics, but the poorer and less knowledgeable Afghans have been unable to keep up with the many channel changes, and so have lost considerable land as a result as the river has moved dominantly south (Figure 9.2). Ephemeral islands in the course of the Amu Darya have changed hands on numerous occasions and were the source of border discussions throughout the twentieth century. Bridges built in the late twentieth and early twenty-first centuries across the Amu Darya and Panj now allow relative ease of cross-border access (Figures 9.3(A,B) and 9.4(A,B)).

Figure 9.3 (A) View of the Friendship Bridge border crossing at Hairatan across the Amu Darya to Termz, Uzbekistan SSR, shortly after its construction and inauguration in 1982 by the Soviet and Afghanistan governments. *Photograph by Roman Ozersky and Sergei Sevruk, Afghanistan Today, Planeta Publishers, Moscow, USSR, 1983.* (B) The Friendship Bridge in the twenty-first century. *Photograph by Marco Giudicci; from Google Earth™.*

Figure 9.4 (A) The new bridge at the Shir Khan Bandar Bridge and border crossing over the Panj River tributary to the Amu Darya and into Tajikistan. The $38 million bridge was built in 2010 by an Italian company under the supervision of the US Army Corps of Engineers. *Photograph by Thiarmen; from Google Earth™.* (B) View looking south across the road deck of the Shir Khan Bandar Bridge between Afghanistan and Tajikistan. *Photograph by Dr Khalid Arian; from Google Earth™.*

Finally in the late nineteenth century, the far northeastern part of the boundary needed the attention of the joint border commission because many of the small principalities straddled both sides of the border and needed to be dealt with in some fashion. Amir Abdur Rahman Khan was convinced with an annual subsidy of 50,000 rupees to give up some of his rights to lands north of the Amu Darya and to take on at least nominal administration of the Wakhan Corridor panhandle that was constructed between Russian holdings in Tajikistan to the north from British holdings to the south in the Chitral parts of British India. The remote boundary east of Lake Zorikol was still in need of demarcation as late as 1893 and finally that too had border posts installed up to the Chinese Pamir mountains. The work was spread over a stretch of 172 km (108 miles) by a joint British and Russian field party that did the surveying work in the summer of 1894 so that the agreement could be signed on March 11, 1895.

For many years people would cross the northern frontier clandestinely in both directions to escape Soviet control or to make war on the Soviets but the progressive tightening of border-crossing controls was established though guard posts and frequent patrols on the Russian side. During the Soviet–Afghan War of the 1980s, some indications in 1981 suggested that the Soviet Union had annexed the Wakhan but that was later shown to be untrue. In response to Chinese territorial claims in the region, the existing boundaries were reconfirmed in the 1980s.

The Afghanistan-Chinese Border

Although the Afghanistan boundary with British India to the south and Russian Central Asia to the north was forced on the Afghans by those external great powers (The Geographer, 1969; Anonymous, n.d.b), the resulting short part of it with China was not regarded with great seriousness by anyone in the nineteenth century. In the mid-twentieth century, however, the Chinese government in Beijing decided to better determine the boundaries it inherited. Accordingly on November 22, 1963 a new boundary agreement between the two countries was signed in Beijing. That following summer a joint commission representing the two countries went into the field to demarcate the boundary in the high mountains of the Wakhan Corridor. After the summer field observations, a final protocol was signed between the two governments in Kabul in March, 1965.

The 76 km length of this border is all in the high mountain watershed boundary of the Little Pamir that divide the waters of the Karachukur to the east from the Aksu to the west, which is tributary to the Amu Darya forming the northern border of Afghanistan with the Central Asian Republics (formerly of the USSR). The boundary traverses a high ridge in the Little Pamir which is all above 5000 m (16,400 ft) in altitude, except where it is pierced by a few passes such as the Wagjir Pass at 4923 (16,147 ft) altitude that is traversable in summer.

The Wagjir Pass was last used semiofficially in 1949 by caravans from Kashgar, as well as by Chinese pilgrims to Mecca but has been effectively closed to border traffic since then except for a few smugglers from time to time. Weapon smuggling to the Mujahideen was reputed to have come through here from China immediately following the communist coup d'etat of 1978 (US embassy oral report to author, 1978). A new road has been bulldozed in from the Chinese side, and at the present time it is closed to foreigners; recently it has had a reputation as a low-intensity drug-smuggling corridor to bring opium grown in Afghanistan into China in the summer months (Figures 9.5 and 9.6). Because the Chinese built the all-weather, paved Karakoram Highway just 90 km as the crow flies east-southeast of here through the Kunjerab Pass (4693 m–15,397 ft), they would have little difficulty in bringing a new transport highway and rail route through the Wagjir in future.

Figure 9.5 View in Google Earth™ to the northeast from Afghanistan into China of Wakhjir Pass into the wide pamir valleys beyond. The newly bulldozed road to the top of the pass at 4963 m (16,147 ft) shows as a light snow-covered line just the left and above the pushpin marker of the pass. A similar wide valley of a tributary to the Chakmaktin Kol valley in the Wakhan Corridor exists just out of site below this view.

Figure 9.6 Bactrian camels at a watering site in the grassy pamir valley near the habitation of Baza'l Gonbad some 43 km and 1000 m lower in elevation from Wakjir Pass to their east. *Photograph by Arthis; from Google Earth™.*

Afghanistan-Pakistan Border—The Durand Line

The highly contentious border between Afghanistan and Pakistan was established originally in the nineteenth century by the British to promulgate their policy of divide and rule over the altogether unruly Pashtun tribes that occurred on both sides of the border (Rubin and Siddique, 2006; Qaseem, 2008; Bajoria, 2009; Behn, 2012; Siddique, 2012; Rosenberg, 2012; Hodge, 2012). Throughout the nineteenth century the two major powers at play in the "Great Game" in Asia were the British who were moving north and the Russians who were moving south. Ever since Tsar Peter the Great of Russia in the early eighteenth century had decried the need for a warm-water port, the British had feared that the place desired by the Russians was to be in the Indian Ocean, perhaps

Karachi in the heart of their dominions. The formation of a buffer state between their empires was conceived by the British as a classic maneuver to keep their borders apart, with the Durand Line as a product of that effort.

Indian Civil Service agent, Henry Mortimer Durand (1850–1924) was assigned the task of delineating the southern border of Afghanistan in the 1880s as the British sought to survey the region and establish the locations of the tribal lands in enough detail to facilitate accurate delineation of spheres of influence between the British and Russian empires. The western and northern borders of Afghanistan were determined by a dialogue between the countries of Persia (Iran) and Russia with the participation of the Afghans, but in the delineation of the southern demarcation that became the Durand Line, the Amir of Afghanistan was taken into confidence by the British.

In somewhat contentious negotiations with the British, Amir Abdur Rahman Khan wanted to settle his boundaries and asked for a British mission to Kabul in October 1888 but the request was not granted until 1893 when the government of India sent a mission headed by then Sir Mortimer Durand, Foreign Secretary. The map of the proposed new boundary presented to the Amir made him quite apprehensive and he complained that cutting him away from some of the frontier tribes would injure his prestige and weaken him in the eyes of his former subjects. He predicted that if some of the Pashtun tribes were removed from his dominions that they would become engaged in constant fighting and plundering against the British, and in the event of incursions by any foreign enemy the tribes would be the Britishers' worst enemies. Indeed his prediction did prove true in later years. But in actual fact, some of what the British ended up delineating was an approximation of what had been conquered and controlled by the Sikhs prior to the British takeover anyway.

Nonetheless, in spite of his initial objections, the Durand Line agreement was reached and signed on November 12, 1893 by Amir Abdur Rahman Khan and Sir Henry Mortimer Durand. The locations of the sphere-of-influence demarcation were placed upon maps and then later the actual demarcation of the boundary was set in place by pillars erected where the border was not contested locally. Many tribesmen, seeing their loss of autonomy, however, opposed the demarcation from the very beginning so no pillars were erected in some places. In later years some of these pillars were pulled down in disgust, or moved to the imagined advantage of one side or another.

The Durand Line has remained contentious for most of its length for over a century, through many changes of government on both sides of the border. Various elements of the Durand Line Agreement had to be repeatedly reaffirmed with the British Raj under various new treaties in 1894, 1895, 1905, 1919, 1921, and 1930. After the departure of the British from the subcontinent in 1947, the elimination of the Durand Line and the creation of a new "Pashtunistan" independent of the newly created Pakistan was proposed by government

Figure 9.7 Looking into Afghanistan over the Torkam Gate at the bottom of the famous Khyber Pass. Note the national flags of the two countries on the left. *Photograph by Agence France–Presse/Getty Images (Published in Wall Street Journal, 2012).*

discussions in Kabul that all prior border agreements should be null and void. The Afghan government took a strong stand against the creation of Pakistan at the partition of India, and especially at the re-establishment of the Durand Line; the sphere-of-influence demarcation as a new international border. The government of Afghanistan voted against recognition of Pakistan as a new nation, and under the strong influence of Pashtun majority leadership, has since the 1947 partition of British India, consistently not recognized the Durand Line as its international border, which view continues up to the present time. This issue reverberated back and forth across the border between Afghanistan and Pakistan thereafter. The US government has stood by this border as a fixed entity, however, which has annoyed the GIRoA under President Karzai. The famous Torkam Gate from the Khyber Pass of Pakistan into Afghanistan that allows passage back and forth between the two countries is periodically closed in wrangles over the border and other issues (Figure 9.7).

Afghanistan-Persian Border

The national border between Afghanistan and Iran has long been rather more peaceful than the highly contentious Durand-Line border with Pakistan, but nonetheless, a certain amount of problematic confrontation has occurred there for centuries, with the result that some sort of delineation was deemed advisable in the late nineteenth century (Yate, 1888; Balland, 1990; U.S. Department of State, 1983; Hanaz, 2012; Mogelson, 2012; Anonymous, n.d.c). The boundary line runs in a general north-to-south direction some 936 km (582 miles) through the basin-and-range country of what is known as the "Iranian Plateau". The political subdivision of this land goes back to the eighteenth century when two rival powers emerged from the remnants of the Safavid empire, the Durani empire on the east, and the Qajar empire on the west. Once Dost Mohammad Khan, the Amir of Kabul, had taken Herat in May 1863, it became necessary to establish delineated borders between the empires. Delineation was undertaken in three stages that corresponded to three different sectors of the frontier, and was done entirely by ostensibly neutral arbitration commissions, British at first, and then again in the twentieth century by Turkish authorities.

The area of southern Afghanistan at the delta mouth of the Helmand where it emptied into the *hamun* basins of the Seistan was the region most in need of attention because of the vital water supply that irrigates a dense population of farm villages. The area is of considerable strategic importance and in the nineteenth century was a much disputed border area where both Kabul and Tehran struggled for the allegiance of local chieftains. Following the death of Dost Mohammad not long after he conquered Herat, the Persians advanced over 440 km (250 miles) up along the Helmand River deep into what is now Afghanistan, occupying as far as Nadi Ali by 1865. When Sher Ali Khan took the Kabul throne in 1868 the threat of war between Persia and Afghanistan over this Persian incursion caused entreaties to the British to help solve the problem of the borders. Accordingly the Seistan Arbitration Commission, comprised of four British officers under the command of General F.J. Goldsmid, undertook the task and visited the region for six weeks starting in February 1872.

A glance at any map of Afghanistan can show the result where a substantial jog eastward to the detriment of Afghanistan was created from a straighter north-south line that might be expected. Instead the "Seistan proper" on the west was awarded to Iran, and the "outer" Seistan on the east, with many fewer people was given to Afghanistan. The clear favoring of Persia over Afghanistan in this decision was well recognized by Sher Ali, who refused to ratify the arbitration so that the British had to give him an indemnity of 1.5 million rupees in compensation, which finally bought him off enough for him to give consent in October 1873. Even so the selection of any distinct topographic markers of the border in the Seistan region had not been possible because of the ephemeral nature of endlessly varying water levels, shifting shorelines and migrating channels in the otherwise vast flat area, with the result that the border was actually highly indistinct and likely to be subject to further disputes, especially insofar as a fair division of waters in times of drought would be an additional point of stress in future, which has happened repeatedly.

Flooding and delta distributary channel change in 1896, coupled with great drought in 1902, led to armed skirmishes across the border and again new requests to the British to arbitrate once again. This time the Persians pressed for even more Afghan land to be awarded to them but the revived Seistan Boundary Commission under the command of British Colonel A.H. MacMahon demurred and set up 89 boundary posts to definitively demarcate the border. In addition they granted to the Persians a right to one-third of the Helmand

River water flow at its entry point into the Seistan, in addition to holding both countries equally responsible for the proper adjudication supply of irrigation waters to the existing canal network, regardless of whatever natural shifts might occur in the river's delta distributary branches.

Upstream along the middle Helmand, however, the Afghans after 1946 began an extensive irrigation development project designed to bring some of their desert lands of the Dashti Margo (Desert of Death) under cultivation, a fact that greatly bothered the Iranians (the name "Persia" had been changed to "Iran" in 1935). A new tripartite group of neutrals was named to arbitrate in 1951, the Helmand River Delta Commission (*Komisyuni Deltyi Helmand*), with Canada, Chile, and the United States as the external participants. This commission redivided the river waters but Iran actually wanted more so the recommendation languished until 1971 when yet another drought occurred. Finally in 1977 a new water division treaty was signed, seemingly putting to rest the century-long border dispute. The lower Helmand, however, dried up entirely in the pervasive drought of 2001 causing more distress and thousands of refugees (Figure 9.8(A–D)). Since the last water division treaty, in fact border incursions by the Iranians over greater access to water have occurred right up through 2008. It is thus likely that the predicted increasing droughts and progressive desiccation of the region over coming decades will be likely to increase tensions in the region in future. The area is already highly politically unstable as the result of drug smuggling from Afghanistan into Iran, as well as people smuggling wherein Afghan refugees attempt to begin the long route of escape to Europe or the Persian Gulf to find work (Hanaz, 2012; Mogelson, 2012; Anonymous, n.d.c).

The northwest part of the Afghanistan–Persian border was marked originally along the course of the Hari Rud river, where after reaching the border area as it flows west from Herat, it swings abruptly north. Border delineation here was accomplished in 1889 by Major General C.S. MacLean, the British consul general for eastern Persia at the time. He set 39 boundary posts in the region to delineate a border that has been remarkably free of conflict since then.

The central part of the border between Afghanistan and Persia was left unsettled up into the 1920s, when finally in June 1928 Persia and Afghanistan decided to set that part of their border. Accordingly they had a Turkish commission under the command of General Fahreddin Altay demarcate a 400 km (250 miles) section of the border to connect the Seistan section with the Hari Rud section. Although only minor disputes have ever occurred along this largely desert section of the border, the future may bring other difficulties if the Afghans ever decide to begin wholesale extraction of lithium or other valuable salts from the Namaksar dry lake basin through which the border is established.

(A)

➤ Satellite imagery of the Sistan basin, 1976

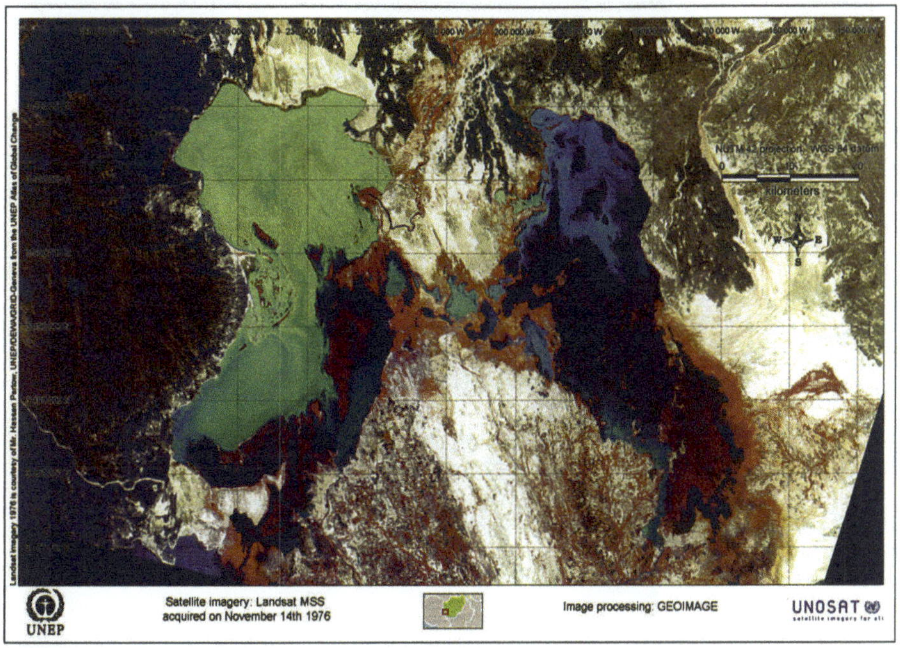

(B)

➤ Satellite imagery of the Sistan basin, 1987

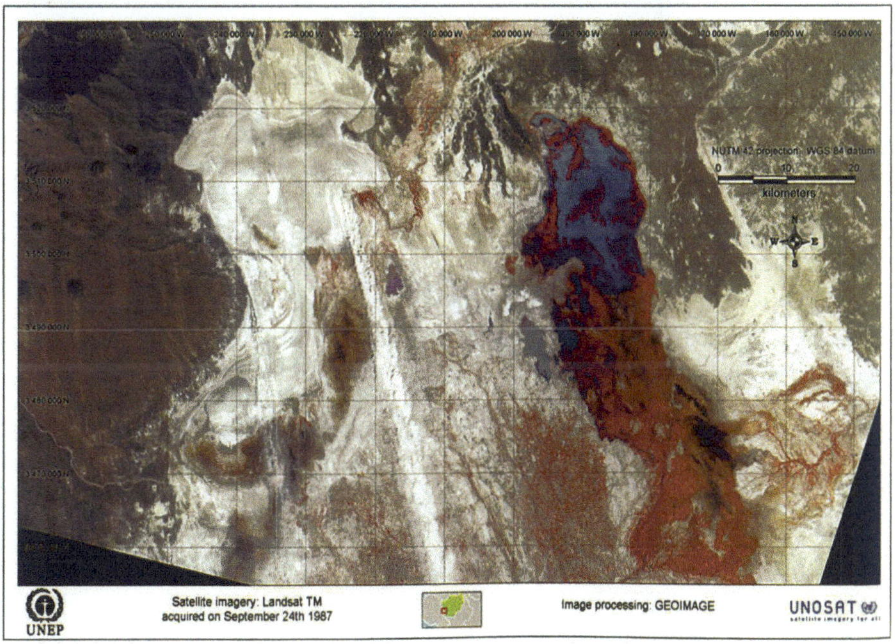

Figure 9.8 Satellite images of Afghanistan–Iran border area showing configuration of some of the Helmand *hamuns* at the outer edges of the inland delta of the Helmand River that wax and wane depending upon the annual flow down from the Central Hindu Kush mountains. The flow is first into the eastern (right) Hamooni Puzak inside Afghanistan, and then to the western (left) Hamooni Saberi that is bisected east to west by the border between the two countries (Afghanistan is to the top), before flowing south into the Darya Cheyye Seistan in Iran, and finally counterclockwise back into Afghanistan into the salt flats of Gaudi Zirreh in the extreme southwest, lowest part (455 m) of the country. A—1976; B—1987; C—1998; D—2001. *Imagery is from the United Nations Environmental Program (2003).* Afghanistan: Post-Conflict Environmental Assessment.

(C)

➤ Satellite imagery of the Sistan basin, 1998

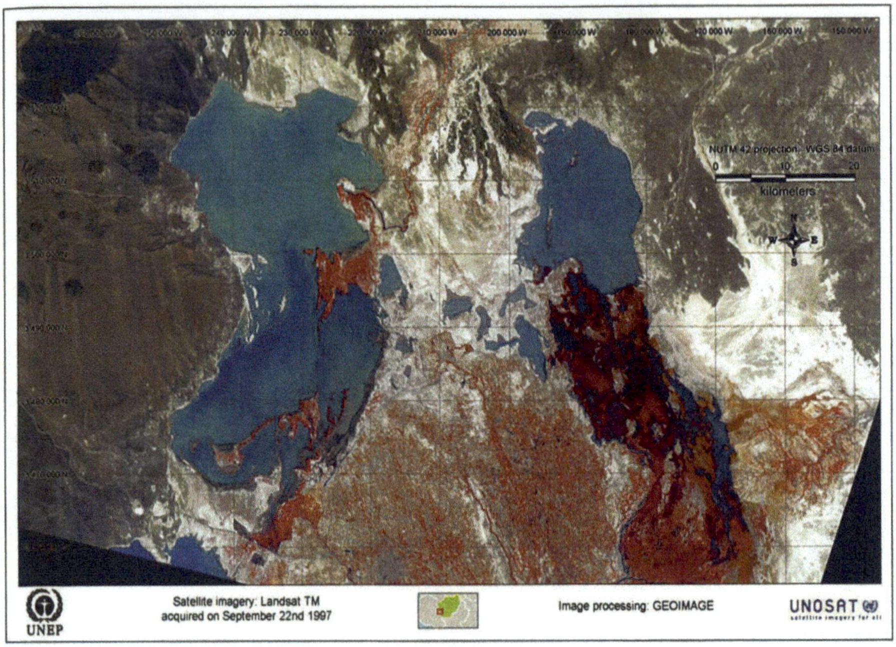

(D)

➤ Satellite imagery of the Sistan basin, 2001

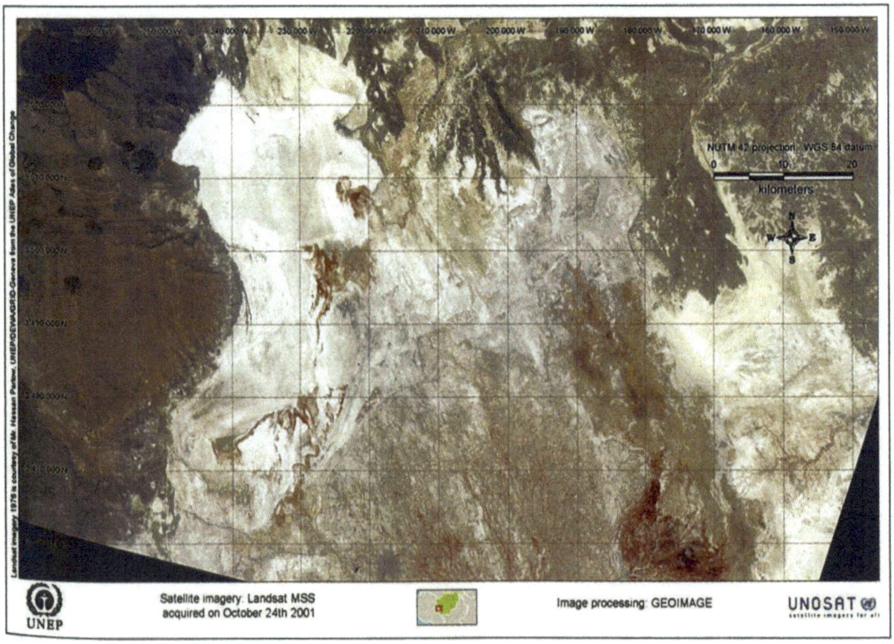

Figure 9.8—cont'd

Partition of Afghanistan

As the above pages show, Afghanistan was stitched together as a geopolitical state by Great Britain and Russia in their late nineteenth century Great Game in order to build a better buffer between their colliding empires; but that which was stitched together can just as easily be unraveled. For more than a century since the initial compressive suturing of the diverse ethnicities and geographies together, equally centrifugal schemes or reasons for disentanglement have also been presented.

When the country was first being created inside its present boundaries in the last third of the nineteenth century, the Russians expressed several times to the British that rivers made bad boundaries and that they really wanted the Hindu Kush mountains as the real wall to divide their territorial possessions in Central Asia, rather than the rivers of the Panj and Amu Darya (Oxus) (Curzon, 1889). This did not happen, of course, as the British wanted a rather more extensive buffer zone, and even pressed the unwanted Wakhan Corridor of mountains and wide grassy "pamir" valleys on an initially unwilling Amir Abdur Rahman Khan in order to create a wider buffer zone. Then when the Soviets invaded Afghanistan in 1979, a number of Western scholars thought it likely that annexation of northern Afghanistan would occur, with a new national border likely to be drawn along the Hindu Kush axis, exactly as the Russians had indicated a century before (Krakowski, 1987; Shroder, 1989; Meer, 1986). The author saw maps drafted in Pakistan for newspaper consumption during the Soviet occupation that had a dismembered Afghanistan reconstructed in exactly that fashion (Figure 9.9). In 1986 the idea of partition was presented to the people of Pakistan in a newspaper as the solution

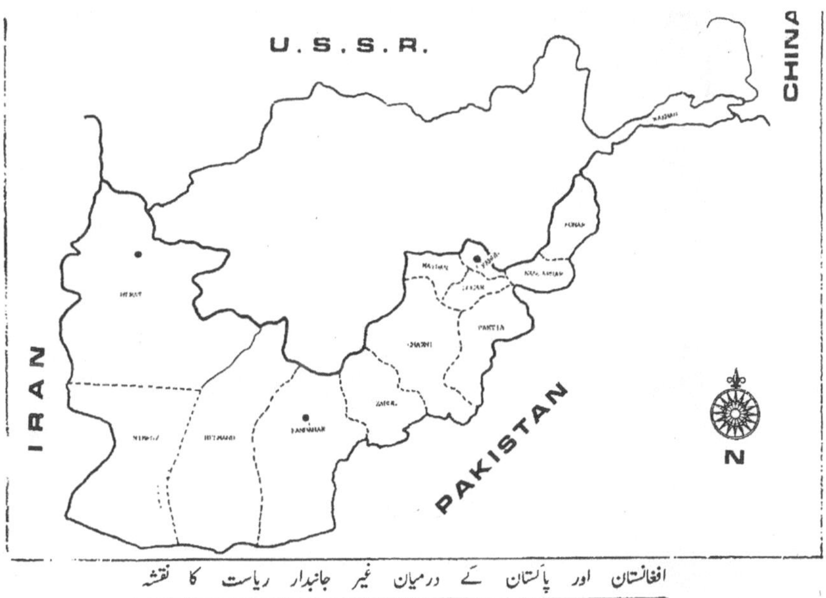

Figure 9.9 Map of Afghanistan made in Pakistan in 1986 illustrating an imagined partition wherein northern parts of Afghanistan are carved out to presumably one day join with Soviet Turkmenestan, Uzbekistan, and Tajikistan. *Jang; January 8, 1986.*

to many of their ethnic and other problems, with the idea that the big Pashtun chunk of southern Afghanistan would be likely join with the Northwest Frontier Province (now Khyber-Puktunkhwa Province) and reunite the Pashtuns split apart by the British. Needless to say, those of the Punjabi ethnicity who dominate the political scene in Pakistan were not too happy to see that development because it meant that they might not be able to control things so well in future. A communist north Afghanistan and a Pashtun tribal south would have been the result of such a partition at that time. When the Soviets departed in 1988–89 without accomplishing their political goals, the idea of partition went into abeyance for a time.

The idea of partitioning Afghanistan because of the intense ethnic discords and the ongoing great pressures exerted from the different Pashtun-dominated governments in Kabul is an idea, however, that just does not go away, especially given all the long historical Pashtun distaste for the Durand Line, their desire for a reunited Pashtunistan homeland, as well as the Pashtun-dominated Taliban who use their safe havens across the Durand Line in Pakistan to continue their attacks on Afghanistan. Accordingly, a new map was published in 2006 that reflects boundaries, not as they were drawn by colonial powers, but as the local populations would like them to become as the people struggle to emerge from the humiliations and defeats of the nineteenth century (Figure 9.10) (Peters, 2006; Nazemroaya, 2006). In this rather fanciful scheme, the Durand Line is gone, and Pakistan is the biggest loser as Afghanistan has incorporated all the Pashtun heartland inside its borders, leaving only the Punjabis and the Sindhis

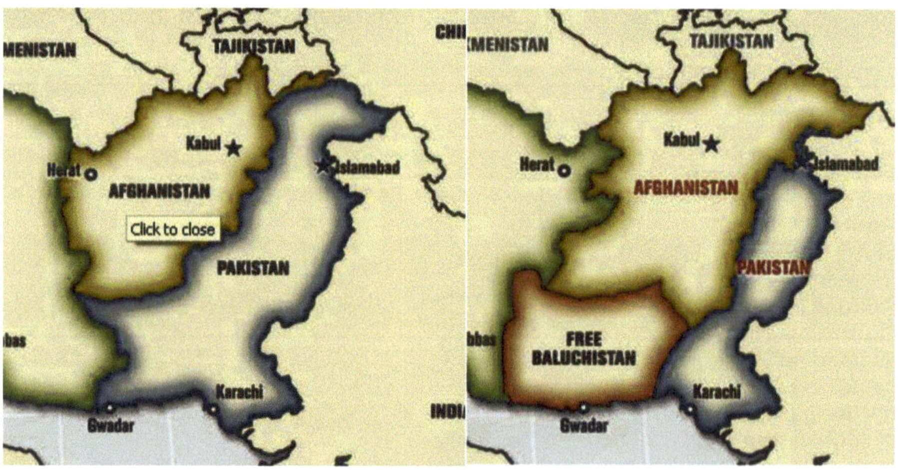

Figure 9.10 **A fanciful map of a redrawn Afghanistan and surrounding countries of the Middle East and South Asia, which incorporates the Pashtun-dominated, Khyber–Puktunkwa Province (former North-West frontier Province or NWFP) of northwestern Pakistan into Afghanistan, whereas at the same time Afghanistan is losing some of its Farsiphile (*Farsiwan*–Persian speakers) territory in the west into a newly drawn Iran.** *Map by Ralph Peters.*

in a dramatically truncated Pakistan. Afghanistan has lost parts of Farsiphile western Afghanistan to Iran, although still somewhat nonsensically retaining the Shiite Hazara homeland in the central Hindu Kush (logically the Hazaras have far more in common with the Shiite Iranians and so should have been included with the redrawn Iran in this fanciful redrawing of the borders). Nevertheless, the idea of a differently partitioned Afghanistan reflects some of the possibilities afoot that have been considered. This model stirred up a good deal of controversy in Pakistan when it was first discovered by them as they read about "Blood Borders" in the US Armed Forces Journal (Peters, 2006).

The idea of the phoenix-rising partition of Afghanistan resurfaced once again in 2010 and 2011 as a way for the USA to get out of Afghanistan in a sort of face-saving gesture of withdrawal from a losing war (Blackwill, 2010, 2011; Lucas, 2010; McElroy, 2010; Tellis, 2012; Smith, 2012). In the proposed plan by a former US Ambassador to India and National Security Adviser in 2003–2004, Robert Blackwill suggested that the Taliban should be allowed to control the south and east, the Pashtun heartland of Afghanistan, whereas the more northern Tajik, Uzbek, and Hazara minorities would be left as the only parts of Afghanistan to be defended any further. This latest resuscitated idea for the partition of Afghanistan, to once again split the most troublesome Pashtuns away from the other Afghanistan ethnicities, was widely derided on all sides, by the Kabul and Islamabad governments and Taliban alike, as well as being denied as a realistic plan by the US Ambassador in Kabul.

Nevertheless, this latest round of partition talks led to ideas of three cases for partition and three against, as follows (Fisher, 2010):

Cases for Partition

1. *Salvage northern Afghanistan*—Taliban will inevitably control most of the Pashtun south but US air power and special forces, along with Afghan army and like-minded nations, could ensure that north and west Afghanistan do not fall to jihadi extremists. Partition was thought the best outcome consistent with vital national interests and US domestic politics.
2. *Route to withdrawal*—The Taliban would seek to reconquer the entire country, but US, allied military, and growing Afghan army capabilities could stop this. Renewed civil war is likely without long-term US military presence, but unlikely with such long-term commitment. Partition specters might even produce a change of heart in Pakistani military support for Afghan Taliban.
3. *Decentralization*—Decentralizing the Afghan state so that local regions have greater autonomy might overcome the failing experiment in centralized democracy. This might produce Taliban control in some areas and unstable, ill-regulated strongman or warlord governance in others. Centralized governance matches neither real internal distributions of power nor local notions of legitimacy. No effective military solution is possible if intended political goals are so badly misaligned with underlying social and political frameworks.

Cases against Partition

4. *Afghan nationality*—Some think the people of Afghanistan have a strong sense of nationhood despite ethnic heterogeneity, foreign invasions, the rise of Islamic fundamentalism, and weakness of the Afghan state. Partition might be attractive for geopolitical strategists, but it is unlikely the Afghan people will countenance the idea.
5. *Worsens situation in southern Afghanistan*—Partition would leave many non-Pashtuns at the mercy of the Pashtuns in the southern part and risks complete denial of human rights to women in Pashtunistan. The plan smacks of neocolonialism, moving from the classic notions of divide and rule to divide and depart, both of which the British practiced in the subcontinent with disastrous consequences.
6. *Stoking such nationalism is way too unpredictable*—The partition argument has an unrepentant sense of hubris with no credible basis existing for saying what the character of successor states in Afghanistan would be like. No reason exists to think that stoking more nationalism might not create unintended instabilities, including the demonstration effect it could have on Kashmir, or that this solution could possibly reduce the characteristic regional violence.

The borders of the present country of Afghanistan thus have a long history of imprecision and uncertainty, now exacerbated by fogs of war and likely failures of prediction. What was created in the nineteenth century could certainly be reconfigured in the twenty-first, but the uncertainties are many and the likelihoods of success seem too few and unpredictable. One thing is certain—the region's classic volatility is most likely going to move in unfortunate directions unless a game changer such as a huge and rich resource base can possibly improve conditions for the average people of the region (Hoodbhoy, 2013).

Once the specters of border reconfiguration are raised, however, just as Wright (2013) has done recently for the whole of the Middle East as well, then future border configurations in southwest Asia grow ever more murky. Certainly major forces for border reconfigurations exist all across the Middle East and South Asia, just as do the conservative forces for maintaining the status quo of fixed if porous borders. The questions throughout these regions are whether nationalism is stronger than older forces of identity during times of conflict or tough transitions. As Wright (2013) has pointed out, however, other factors can keep a region from fraying, such as good governance, decent services and security, a common enemy, or equitably shared resources. Afghanistan does have a chance to remain a country of shared alliances between disparate ethnicities and religious variations, to share effectively in the resource wealth. The operant question at hand remains whether the idea of an integral nation state of Afghanistan can possibly trump the forces of dispersion.

References

Anonymous, n.d.a. Boundaries, Borders and Disputes: The origin of boundaries. Nature Gallery (Geography), 7 p. http://www.pacificislandtravel.com/nature_gallery/boundries.htm.

Anonymous, n.d.b. Afghanistan border crossings. Caravanistan, 5 p. http://caravanistan.com/border-crossings/afghanistan/.

Anonymous, n.d.c. But circa 2007. Iran and Afghanistan. Institute for the Study of War, Washington, DC, 9 p. http://www.understandingwar.org/iran-and-afghanistan.

Bajoria, J., 2009. The Troubled Afghan—Pakistani Border. Coucil on Foreign Relations, 6 p http://www.cfr.org/pakistan/troubled-afghan-pakistani-border/p14905.

Balland, D., circa 1990. Boundaries of Afghanistan. Encyclopaedia Iranica. IV (Fasc. 4), 406–415. http://www.iranicaonline.org/articles/boundaries-iii.

Behn, S., October 2, 2012. Afghanistan—Pakistan Border Tensions Flare Amid Bombardment Claims. Voice of America; News, Asia, 2 p.

Blackwill, R.D., July 21, 2010. America Must Give the South to the Taliban. Financial Times, 1 p. http://www.ft.com/intl/cms/s/0/7caa5128-94f3-11df-af3b 00144feab49a.html#axzz2JANmp21D.

Blackwill, R.D., January/February 2011. Plan B in Afghanistan: why de facto partition is the least bad option. Foreign Affairs. 42–50. http://www.foreignaffairs.com/articles/67026/robert-d-blackwill/plan-b-in-afghanistan.

Curzon, G.N., 1889. The fluctuating frontier of Russia in Asia. The Nineteenth Century 25 (44), 267–283.

Fisher, M., August 10, 2010. Should Afghanistan be Broken into Smaller Countries? The Atlantic Wire (Atlantic Monthly Group), 2 p. http://www.theatlanticwire.com/global/2010/08/should-afghanistan-be-broken-into-smaller-countries/23404/.

Hanaz, Paula, 2012. The Politics of Water Security between Afghanistan and Iran. Independent Strategic Analysis of Australia's Global Interests; Future Directions International; Strategic Analysis Paper. http://reliefweb.int/sites/reliefweb.int/files/resources/The%20Politics%20of%20Water%20Security%20between%20Afghanistan%20and%20Iran%20-%20March%201%202012.pdf.

Hodge, N., December 10, 2012. Afghans probe corruption at borders. Wall Street Journal. 3 p. http://online.wsj.com/article/SB10001424127887324024004578171410335390372.html.

Hoodbhoy, N., January 19, 2013. Redrawing the Regional Boundaries of Afghanistan? Balouch Resolution Sponsor, US Congressman Rohrbacher Meets Afghan Opposition, Thomas Ruttig. Aboard the Democracy Train (book advertisement), 7 p. http://www.aboardthedemocracytrain.com/redrawing-the-regional-boundaries-of-afghanistan.

Krakowski, E., 1987. Afghanistan: the geopolitical implications of Soviet control. In: Klass, R. (Ed.), Afghanistan and the Soviet Union: The Great Game Revisited. Freedom House Pub, New York, pp. 161–185.

Lucas, G., July 30, 2010. Leaks Trigger New Debate in US about Partition or Other Options for Afghanistan. DW (Deutsche Welle). http://www.dw.de/leaks-trigger-new-debate-in-us-about-partition-or-other-options-for-afghanistan/a-5851472.

McElroy, D., September 13, 2010. NATO Urged to Allow Partition of Afghanistan. The Telegraph, 2 p. http://www.telegraph.co.uk/news/worldnews/asia/afghanistan/7998057/Nato-urged-to-allow-partition-of-Afghanistan.html.

Meer, Waris, January 8, 1986. Afghanistan and Pakistan (in Urdu). Jang.

Mogelson, L., October 21, 2012. The Scariest Little Corner of the World. New York Times Magazine, 13 p. http://www.nytimes.com/2012/10/21/magazine/the-corner-where-afghanistan-iran-and-pakistan-meet.html?pagewanted=all&_r=0.

Nazemroaya, M.D., 2006. Plan for Redrawing the Middle East: The project for a "New Middle East". Global Research (Centre for Research on Globalization), Montreal, Canada, 8 p. http://www.globalresearch.ca/about.

Peters, R., June 2006. Blood borders: how a better Middle East would look. Armed Forces Journal; http://www.armedforcesjournal.com/2006/06/1833899.

Qaseem, A.S., 2008. Pak-Afghan Relations: The Durand Line Issue. Institute of Policy Studies, Islamabad; 4 pp. http://www.ips.org.pk/the-muslim-world/986-pak-afghan-relations-the-durand-line-issue.html.

Rodrigue, J.-P., 1989. The Geography of Transport Systems. Types of International Boundaries; 1 p. http://people.hofstra.edu/geotrans/eng/ch5en/conc5en/types_boundaries.html.

Rosenberg, M., October 28, 2012. When Afghans Look to Border with Pakistan They Don't See a Fixed Line. New York Times, Asia Pacific, 3 p. http://www.nytimes.com/2012/10/29/world/asia/in-afghanistan-comment-on-border-brings-tension.html?_r=0.

Rubin, B., Siddique, A., 2006. Resolving the Pakistan—Afghanistan Stalemate. Special Report 176 United States Institute of Peace, Washington, DC. 20 p. http://www.usip.org/publications/resolving-pakistan-afghanistan-stalemate. http://www.voanews.com/content/afghanistan_pakistan_border_tensions_flare_amid_bombarment_claims/1518887.html.

Shroder Jr, J.F., 1989. Afghanistan resources and Soviet policy in Central and South Asia. In: Hauner, M., Canfield, R. (Eds.), Afghanistan and the Soviet Union: Collision and Transformation. Westview Press, pp. 101–119.

Siddique, A., October 24, 2012. Border Talk Crosses the Line in Afghanistan. Radio Free Europe; Radio Free Liberty, 3 p. http://www.rferl.org/content/afghanistan-durand-line/24749671.html.

Smith, G., January 24 and September 6, 2012. U.S. Envoy in Kabul Denies Partition Rumors. The Globe and Mail, 2 p. http://www.theglobeandmail.com/news/world/us-envoy-in-kabul-denies-partition-rumours/article1359670/.

Tellis, A.J., June 22, 2012. Pakistan's Impending Defeat in Afghanistan. Carnegie Endowment for International Peace, 3 p. http://carnegieendowment.org/2012/06/22/pakistan-s-impending-defeat-in-afghanistan/c6sn.

The Geographer, May 1, 1969. International Boundary Study; No. 89. Afghanistan—China Boundary; (Country Codes: AF-CH) U.S. Department of State; Office of the Geographer; Bureau of Intelligence and Research. http://www.law.fsu.edu/library/collection/limitsinseas/ibs089.pdf.

U.S. Department of State, September 15, 1983. Afghanistan—U.S.S.R. Boundary; No. 26 (Revised)(Country Codes: AF-UR) The Geographer; Office of the Geographer; Bureau of Intelligence and Research. http://www.law.fsu.edu/library/collection/limitsinseas/IBS026.pdf.

Wright, R., September 29, 2013. Imagining a Remapped Middle East. The New York Times.

Yate, Major C.E., 1888. Northern Afghanistan. William Blackwood and Sons, Edinburgh.

In this Chapter

10

Lost Resource Opportunities

Abstract

Since the very first explorations of the Afghanistan territory by the British in the 1840s, knowledge of plentiful resources in the country kept accumulating and was reasonably well understood by the external science community at least. In the late nineteenth century, British geologist Griesbach was responsible for mapping and describing the geology and resources in the country, as also were a number of American petroleum geologists in the early twentieth century. Nothing came of these efforts, however, in part because of great distances to market, lack of interest in the home countries, and fears of a looming war. After World War II the American-led Helmand Valley Authority produced the huge and expensive Helmand Valley Project as a major Cold War counter effort to Soviet inroads in Afghanistan but the earlier good resource opportunities that had been revealed, were lost to later incompetent follow-up assessments.

Keywords: American geologists Fox and Clapp; Aloys Arthur Michel; British geologist Griesbach; Erroneous assumptions of resource paucity; Helmand Valley Project; Russian geologist Obruchev.

The history of geological and resource exploration in Afghanistan is fraught with all sorts of turns and twists, intrigues, and double deals, perhaps as one could expect in such a place where the threat of outsider manipulation has been ever-present for millennia. Perhaps in some warped way this would seem to befit the tortured terrain where continental fragments have been rafted across the paleo-Indian Ocean (Tethys Seaway) over the millions of years and sutured onto the underbelly of the Asian Plate. The great plate-tectonic migrations discussed in so many introductory geology texts, produced in Afghanistan a plethora of rock types and resources that came from somewhere else. But these rocks and resources are neither easily understood, nor do they give up their secrets without a struggle in the complex terrain. Still, the eye of the average geoscientist is drawn to these enigmas because to do otherwise would be to give up without finding answers, an anathema to those who would seek to understand what could be significant. Such relinquishment might feel right in

Natural Resources in Afghanistan. http://dx.doi.org/10.1016/B978-0-12-800135-6.00010-6

the face of such hostile, vertiginous, and remote terrain, but it would be wrong; indeed many have come from afar over the centuries and looked with interest and expertise. Others with inadequate training have come too, but their vision may have led to warped views that forever condemned the country to less than its geological due or rightful impacts on world resource markets.

As far as we know, the British Raj initiated the first geological and resource analysis in Afghanistan nearly two centuries ago in their series of exploration and military excursions into the country (Section II and Drummond, 1841; Hutton, 1846; Griesbach, 1881, 1885, 1886a,b, 1887a,b, 1892; Hayden, 1913). In general the later British rational for the geological exploration was to locate resources that could be used as trade for British industrial products. Their terrain exploration and mapping was very militaristic, however, with topographic specialists on constant lookout for water sources and fodder for their animals, while easier routes of travel in the event of hostilities were sought as well. Area handbooks or gazetteers of local information were produced to enable easier return to difficult places in the event this was ever needed. Early on the Afghans became suspicious of all this nosing around, especially the topographic mapping. One Afghan in the nineteenth century viewed the mapping with much alarm as British topographers mounted every hill; "With the British mapping everything in sight, oh what will become of us?" he declared with dismay.

Of course the topographical mapping by the British military was largely to counter the Russian conquests in Central Asia; the Great Game was playing out in earnest as the two great powers vied for control of real estate at the conflicting edges of their expending dominions (Hopkirk, 1990). In the course of these activities, a Viennese-trained, Austrian-British geologist, C.L. Griesbach, joined the Geological Survey of India in 1878. He also attained the rank of Captain in the 6th Battalion Royal Fusiliers and in March 1880 was deputed on special service to southern Afghanistan where he was present at the battles of Girishk (July 14, 1880), Maiwand (July 27, 1880), and at the siege and battle for Kandahar (September 1, 1880). From 1884 to 1886 he was appointed as the official geologist to the Afghan Boundary Commission, and while in Afghanistan traveling around, looking at resources, he discovered considerable coal and some iron ore. Although he did report on the basic geology he encountered, some of his reports on resources were suppressed by the British Raj because the authorities in India were alarmed at the richness of the deposits "bordering the Russian frontier", and decided that it would not be desirable for him to make these facts known publicly because such resources would be highly valued by the Russian government of the time. His rich resource report was never published (Lee, 1996).

From January 1888 to July 1889, Griesbach's pricey services were made available to Amir 'Abd al-Rahman Khan in Kabul to discover mineral resources. The Amir could see, however, that Griesbach was more interested in Afghan politics and mapping Afghanistan for the benefit of the British than in the task for which he was ostensibly employed (Kakar, 1979). The two often played chess late into the night, with the captain always loosing, perhaps wisely diplomatic, while discussing international politics. Aware of Griesbach's military and probable intelligence background, the Amir hoped to induce Griesbach to make favorable recommendations to his government in India, but Greisbach seemed only interested in protecting India and not particularly in saving Afghanistan from Russian designs. The captain did suggest that British troops be brought in to fight alongside the Afghans against the Russians, but the Amir did not think that this would work too well, given the past warfare by the British against the Afghans.

After Griesbach left Afghanistan, shortly to direct the Indian Geological Survey, Frank A. Martin was employed by the Amir to look after a number of related issues, especially to report on rock samples and do assays of promising ores. Martin did submit to the Amir a plan for exploitation of the mineral resources but nothing was to come of it. Martin developed the distinct impression that the Amir was very interested in all possible resources in the country and wished to know where everything was located. He did not appear, however, to want to have any possible deposits worked or mined because he feared the cupidity of his powerful neighbors in that if he actually opened and started working any mines, this would give practical evidence of the richness of the resources and he would thereby loose his country as it would be taken away from him by force for those resources (Martin, 1907).

This deep geological interest by the British in Afghanistan lasted right up into World War II (WWII), with a number of reports produced by personnel seconded from British India into Afghanistan to do special reports (Fox, 1943a; Gee and Seth, 1940). The coal fields and outcrops of salt diapirs in the north of Afghanistan were of special interest at this time. Later during the Cold War with the Soviet Union, these particular resources became of special interest in the west, especially the salt. This was primarily because the labile rock salt of the region is highly mobile under pressure from the overlying petroliferous rocks so that penetrative salt migration upwards can produce mushroom-shaped, salt-dome hydrocarbon traps of considerable worth, just as has happened in the rich oil and gas fields of the Gulf Coast, USA.

This strong British interest in economic geology passed right on into the late twentieth and early twenty-first centuries as well, with experts from the UK being brought in to investigate oil, gas, and metalliferous ores right up through the recent post-9/11 decade. For example, in 1977 economic geologist, Richard H. Sillitoe, from London was employed by the US Embassy to advise them about resources when he was out from the UK to have a look at the metal prospects (Sillitoe, 1977), as well as Soviet oil and gas drilling in the north. He met with me for a time to impart whatever knowledge he could. He expressed an opinion that the Himalayan orogeny, which also produced the Afghan Hindu Kush, did not seem to be particularly prolific in generating ore bodies, at least near the surface where they could be mined with present-day technologies (Sillitoe, 1978, 1979). The older rocks in Afghanistan, which were there long before the rise of the Hindu Kush - Himalaya (HKH), were another story, however. Similarly, after the post-9/11 invasion of Afghanistan by Coalition forces, the British Geological Survey arrived to look after the exceptionally rich Aynak copper deposit in the very old rocks just south of Kabul in Logar province.

Other countries also evinced an interest in these resources early on in the twentieth century, with French, German, and Russian interest being the strongest (Furon, 1924; Akram, 1947; Kaever, 1968; Obruchev, 1927), and all of whom maintained interest and expertise on through the twentieth and into the twenty-first centuries. The most special importance, however, was the interest by both Soviet and American geologists prior to WWII. Once the Afghans regained from the British the right to conduct their own foreign affairs after the conclusion of the third Anglo-Afghan War in 1919, a Soviet delegation arrived in Kabul. The Soviet interest was the most striking, in that Vladimir Obruchev (1927) was working on the mineral "riches" of Afghanistan only a few years after the Communist takeover in Moscow in 1917–1918, so it was clear from the earliest days exactly on what the Soviet eyes were trained for many years.

Vladimir Afanasevich Obruchev was born in 1863 and trained at the St. Petersburg Mining Institute in 1886, so he became one of Russia's greatest geologists in the nineteenth and early twentieth century as he explored all over central and eastern Asia. In his long life (1863–1956) he published over a thousand papers, which is an extraordinary rate of production. Obruchev received many of Russia's and the Soviet Union's highest honors, even being lauded with a stamp in his name and likeness in 1963 at the height of the Cold War (Figure 10.1). While a professor at the Moscow Mining Institute in the 1920s he undertook a first Russian look at the geology of Afghanistan, concluding that the country had much to offer. His paper (Obruchev, 1927) even had drawings of underground mining and the old wooden oil derricks on its introductory page (Figure 10.2), but, of course, no such features were actually in existence in Afghanistan until much later.

Having fought three brutal wars with the British Raj on their southeastern doorstep in the past century, and well aware of Russian militaristic expansion, and even direct aggression in the Panjdeh Incident on their northwestern border in 1885 where over 600 were killed, the Afghans were not about to let either great colonial power of the time to have exclusive rights to its suspected mineral wealth. Instead they turned to a rising power that was far, far away, and therefore apparently safe from the temptations of colonial greed, the Americans.

In spite of numerous prior diplomatic rebuffs from the USA to set up formal diplomatic exchanges between the two countries that began as far back as 1921, the government of Afghanistan kept trying to get Washington to pay attention to them (Poullada, 1981). Finally the springboard seems to have been the incessant peregrinations of the geology community in their never-ending

Figure 10.1 Four kopek postage stamp from the USSR in 1963 bearing the likeness of Academic V.A. Obruchev (1863–1956), who was one of the early "greats" of Russian exploratory geology. He traveled through and studied vast stretches of Siberia, Central Asia, and even Afghanistan, where he wrote about "fossil (mineral) riches" of the country in 1927. Obruchev was much decorated during his lifetime and his name was applied to a mountain range, a mineral, a crater on the moon, and an oil and gas research vessel. His huge granite tombstone in Moscow's most famous Novodevichy cemetery has a full likeness of his head carved into it above a crossed geological hammer and quill pen at the base.

search for resources. Thus the Government of Afghanistan under King Zaher Shah in 1936 granted an exclusive 75-year oil concession, as well as mineral exploration rights that could lead to exclusivity, to an American company, the Inland Exploration Oil Company of New York, which was a consortium of several wealthy individuals, as well as other companies including Texaco and Seaboard (Anonymous, 1937; US Department of State, 1937; Clapp, 1939).

This resource and diplomatic activity pivoted around Frederick G. Clapp (1879–1944), who as the Petroleum Adviser to the Imperial Government of Persia, first went to Afghanistan in 1934 upon invitation from the Royal Government of Afghanistan (RGA). While in Afghanistan he began to develop ideas about resource exploitation when he was shown oil seeps. At one point Clapp observed about 200 Afghans crudely attempting to dig an oil well next to a river in Tir Pol west of Herat where oil seeps occurred (Figures 10.3 and 10.4). In 1936 he returned to visit coal outcrops in the Ghorband River valley (Figure 10.5).

He and the Honorable C. Hart, former American Minister at Tehran, became vice presidents of Inland Exploration and in 1936 traveled to Berlin to meet with Afghan Foreign Minister, who was in Germany temporarily to sign the agreement with the Americans. The concession contract specified that the company should work in five provinces of its choice, begin drilling within one year, and build an oil pipeline for export through Iran rather than British India.

Figure 10.2 Facsimile page on the right in Russian of V.A. Obruchev's (1927) article in Novyi Vostok (New East) on the "fossil riches" of Afghanistan showing a cartoon at the top right of miners at work inside a mountain and out on the surface a forest of the old fashioned wooden oil derricks, even though no real underground mining or oil-well drilling was underway in Afghanistan at the time. The journal name "New East" was published on it also in French as the Review of the Russian Association for Oriental Studies, later known as the Scientific Association of Soviet Orientologists.

Figure 10.3 The Governor of Herat and about 200 men digging for oil some 3 km east of Tirpul (Tir Pol), which itself is about 90 km east of Herat in western Afghanistan close to the Iranian border. Frederick G. Clapp, an MIT-trained geologist (1901 graduate) who traveled the world in search of oil and gas, was working in Iran in 1934 and came over to nearby Afghanistan to have a look at the geology, where he took this picture in July of that year. Hand digging for an oil well into the water table is not an especially recommended procedure for success!

Figure 10.4 **Picture of an oil seep into the ground water that was exposed by digging near Tir Pol, Afghanistan, in 1934.** *Photograph by Frederick G. Clapp.*

Figure 10.5 Photograph of coal seams at Chahr Deh-ye in the Ghorband River Valley taken by Frederick Clapp during one of his visits to Afghanistan in 1936.

The payments from Inland to the RGA were rather less lucrative than the ones given to the Iranians, presumably because of landlocked Afghanistan's great distance from the sea. Clapp reported to the US Department of State that the German Foreign Office and the Russian Embassy in Berlin did everything they possibly could to hamper the final negotiations for the concession. The Russian emissaries, finally realizing that it was impossible to block the concession, made a final appeal to the RGA to grant them a small strip of land on the Afghan-Russian Frontier on which they could also look for oil but this was disregarded. On the other hand the Germans went on to quickly form The Afghanistan Mines Ltd, an Afghan–German company to survey and exploit the mineral resources in the country, as well as a number of other projects to extend their influence (Gregorian, 1969). By 1939 the Nazi regime had successfully inserted hundreds of Germans into Afghanistan where they worked and advised throughout the government.

In 1937 Clapp then led about nine personnel from Inland Exploration to Kabul (Figure 10.6), and included his own wife as well as his son and daughter, which is a reflection of the ease and relative safety the family must have felt as all traveled around the country in search of resources and tourist adventures (Figure 10.7). Clapp noted in the diplomatic correspondence that their intention was to pipe Afghan oil to a point in Iran in the vicinity of Birjand, a few miles from the Baluchistan frontier of British India, where it would meet with oil being piped out of Iran, before both would be sent on to the Persian Gulf. The Inland

Figure 10.6 Group portrait of four (on the left) of Inland Exploration Company representatives in Kabul in 1937. Frederick Clapp, the leader of the group is fourth from the left. Note his high-topped, lace-up boots, which were a characteristic field dress from the nineteenth century.

Figure 10.7 Clapp family portrait standing in the Duab Valley as they are dressed in Afghan finery (embroidered quilted *chapan* on left, *posteen* sheepskin coats on other three—note the rich embroidery on Frederick G. Clapp's *posteen*). From left to right: son Edward Gardner Clapp; wife and mother, Helen Drew Ripley Clapp; Frederick Gardner Clapp; and daughter Clara Francis Clapp. Afghanistan in 1937 was a remote and difficult in which to travel, then as now, but it was safe enough for such family travel even so.

Exploration Company was stipulated to build its own port on the Persian Gulf at some place a few miles from the city of Gwadar, close to the Baluchistan border. A long-term plan was also to build a railroad to run from the Persian Gulf to the Afghan oilfields. Needless to say, once some of these plans leaked to the British government, considerable adverse comment about American intentions came from British officials and commercial circles.

Ernest F. Fox, who was also part of Clapp's group, set out separately in 1937–1938 to travel into Badakhshan and mountainous central and southwestern Afghanistan (Ghor and Hazarjat) in search of more mineral resources. Unfortunately he said little of geological or mineral-resource substance in his book describing his travels in Afghanistan (Fox, 1943b), probably because of company proprietary interests in not letting any outsiders know what he really found or thought. In any case, by 1938 considerable opposition to Inland's efforts had mounted over the delicate diplomatic issues, distance from the sea, overall difficulty of working in Afghanistan, need for large numbers of technically trained Americans, and the absence of any official American Embassy in Kabul to look after them. With the rumblings of an impending WWII upon them, the company decided abruptly to scuttle all its plans in Afghanistan. On June 19, 1938, the RGA was given written notice of the hasty and quite unexpected cancellation of Inland Exploration's concession, much to the dismay and shock of the Afghan officials, who attributed Inland's precipitous withdrawal to devious political motives.

The one good result that resulted from Inland Exploration's flurry of activity in Afghanistan was a growing awareness in Washington that greater diplomatic recognition needed to be established between the two countries. The result was that from May 1935 to 1936, William Harrison Hornibrook served as a nonresident US Envoy or Minister Plenipotentiary from Tehran to Afghanistan. Then Louis Goethe Dreyfus served from 1940 to 1942, until the Kabul Legation was established. The American Legation at Kabul was opened on June 6, 1942, with Charles W. Thayer as Chargé d'Affaires *ad interim*. Subsequently Cornelius Van Hemert Engeret, who had previously been assigned in Tehran, took over in Kabul from 1942 to 1945 during the height of WWII. The American Legation was not elevated to full Embassy status until June 5, 1948, when Ely E. Palmer presented his credentials to the RGA (US Department of State, 1937; http://history.state.gov/countries/afghanistan).

During WWII Afghanistan was fairly well caught between the major powers and unable to do much with its resources, even its river water. The Afghans had to be very careful because their neighbor Iran had been invaded and divided by Britain and the Soviets for being too pro-German, and Afghanistan had hundreds of Germans (teachers, military officers, engineers, doctors) working there on technical projects (Gregorian, 1969). In 1941 the RGA received an ultimatum from the British and Soviets to expel all Axis subjects (Germans, Italians,

Japanese), to which Kabul acceded. The war devastated the Afghan economy in other ways as well. Modernization stopped dead in its tracks, a victim of fleeing capital that went to India, as well as because of the departing Germans, who had been leaders in expertise, and has been planning on bringing in plentiful machinery and money as well (Barfield, 2010). Instead the RGA's reform and modernization programs were interrupted, development plans were abandoned or postponed, the economy was fractured, and inflation as much as 20 percent was rampant. The answer to these problems in the Afghan eye was to bring in American expertise and money, even though the immediate prewar effort with Inland had been so devastatingly aborted. Thus the arrival of the US Legation in 1942 was regarded as a major development that would get Afghanistan back on track.

With the oil and mineral resource development seemingly derailed for a time, the issue of water for the always arid land took on a renewed importance after the war, and one of the first things focused upon was the Helmand River. That name in old Persian actually means "abundant water", as the river drains some 43 percent of the country, but it actually only has about 11 percent of the total flow of all the rivers in Afghanistan (Favre and Kamal, 2004). The river rises in the mountains just west of Kabul and flows southwest to the Iranian border, while being the essential water resource for southwestern Afghanistan from a time far back into prehistory. In fact at one point, thousands of years ago, it fed an enormous Lake Seistan ringed with ancient cities where now only *hamun* marshes, dry lakes, and salt flats exist as pitiful remnants.

Irrigation ditches had been dug from the Helmand for centuries. As far back as 1910–1914, irrigations works had been constructed on it by the RGA. In the late 1930s Japanese engineers were working there, as well as officials from the German government who provided assistance. Their departure in 1942 necessitated the Afghans going it alone during WWII, but progress was quite limited because their plans required use of modern machinery and engineering equipment far beyond what the Afghans themselves could supply (Michel, 1959). Foreign currency reserves accumulated by the RGA from agricultural exports during the war then enabled the government to seek foreign assistance once the war was finished. Only the victorious USA after the war appeared still quite well equipped, sufficiently remote, and disinterested enough to supply the necessary expertise and engineering skills to meet the need. Accordingly, the Morrison-Knudsen Company (MK) of Idaho, who had built the famous Hoover (Boulder) Dam across the Colorado River in the 1930s, seemed the perfect match. MK was hired by the RGA in 1946 as a sort of adjunct to the Ministry of Public Works, inasmuch as all sorts of road- and dam-building projects around the country were proposed to them. But bid estimates by MK were judged too high by the Afghans and ultimately after a few years, MK became exclusively concerned only with the Helmand Valley Project (HVP) (Michel, 1959; Clapp-Wincek and Baldwin, 1983).

The objectives of the HVP (Elkins, 2009) were to:

- Broaden irrigation in the Helmand Valley and thereby expand agricultural activity.
- Provide flood protection.
- Introduce new crops and agricultural products, including a dairy-farming industry.
- Resettle some 550 landless and nomadic (Kuchi Pashtun) tribal families into the region, which was designed to lessen friction with Pakistan while enhancing control over a difficult and fractious, floating population.
- Elevate standards of living.
- Reduce foreign exchange expenditures by reducing imports of food, fiber, and animal feed.
- Provide new electricity to farms and processing facilities.
- Establish new domestic industries by using locally produced raw materials to produce textiles.
- Improve transportation and communications networks.

To accomplish these lofty and overambitious goals on the cheap, as it were, MK was drawn into a rather circumscribed situation wherein the prior work done had to be incorporated into the new, which did not always make sense. Thus the original canal alignments were technically not the best, so half way into the rehabilitation work MK decided that a big new dam and reservoir system were required to meet the RGA's huge (and likely unrealistic) irrigation objectives. Earlier people had noted some problems existed in the area with certain gypsum-rich soils characteristic of many deserts, and MK also wanted to know whether or not there really would be enough water to irrigate the projected area of development. Both of these issues could be solved with the detailed ground surveys MK proposed, but the RGA would have none of it, so both parties suggested that a 20 percent ±margin of error or "slop rate" either way would be sufficient to solve any problems in the absence of expensive surveys (not a particularly good or scientific way to do business!) (Clapp-Wincek and Baldwin, 1983; Elkins, 2009). Whether or not that 20 percent margin of error was really ever achieved in retrospect is not an issue that seems to have ever been discussed much.

Thus by 1948 the HVA Project had begun to mushroom out of all realistic physical, competent administrative, or justifiable financial proportion so that it became necessary to submit a development loan request to the United States Export-Import Bank. This then changed the ante from a simple construction contract with a private company, into a direct government-to-government relationship, which is quite another issue. Furthermore in typical negotiations for finalizing any loan or most kinds of grant proposals, the requested amount was subsequently cut down far below what the RGA was comfortable with. This then left the Afghans committed to a project without adequate resources and unable to do a proper job in the comprehensive and coordinated manner that they at least aspired to; even with the rather sparse administrative capacity they exhibited at the time.

In the early 1950s completion by MK of the Dahla Dam (1952) across the Arghandab River tributary to the Helmand, and the Kajakai Dam (1953) across the mainstem of the Helmand, raised the stakes to new levels. The US sponsorship of the loan through the Export-Import Bank put its national prestige on the line, while the RGA was uncertain if it could even meet the terms of the loan. Amazingly enough also, about 20 percent of total RGA expenditure in the 1950s went into funding the HVA. Similarly MK had stuck itself with badly placed canals in unsurveyed lands with problematic soils, impermeable in places with thick caliche ($CaCO_3$-cemented) layers prone to water logging and salinization, all of which added even more potential for failure over time. From the 1950s on then, assistance from the USA in the Helmand Valley was making the best out of a bad situation that kept getting worse.

The situation was not seen as all bad, however, given the abundance of blame that could be laid on all sides. Land under cultivation almost doubled, farm income went up by a factor of 10, and traditional cropping began to shift towards more commercial crops. Nonetheless, the detractors in later years (Elkins, 2009) noted a bottom line that:

- Politically the HVP failed to live up to its billing as a modernization force.
- Diplomatically it failed to build a strong relation with the USA, and is reputed to have even pushed Afghan relations leftwards toward the Soviets.
- Economically it failed to produce the benefits envisioned.
- Ecologically soil degradation and salinization increased.
- Sociologically and culturally the HVP failed to produce the broad social changes envisioned.

Into this mixture of the good, the bad, and the ugly in the late-1950s stepped a fresh young doctoral student from Columbia University, Mr Aloys Arthur Michel, to evaluate the situation for his dissertation. He was an economic geographer and did a superb job of assessing all pertinent elements of the Afghan economy, as far as he could, given the data at hand at the time. His evaluation of the direction of Afghan economic development included assessments of the mineral resource sector, agriculture, industrialization such as it was, and transportation. A section in his dissertation on "Disadvantages of Mineral Development at the Present Time", was quite negative about that sectoral element. This negativity is hard to reconcile given the reality in the field, even if fine details about resources were not known with any degree of certainty. To assess this negative viewpoint as it reverberated in American government circles up through ensuing decades into the twenty-first century, requires further background evaluation.

Absent any definitive documents about their views of Afghan resources inside the US Department of State in Washington, or its Kabul embassy during the middle to late twentieth century, it is difficult to know their real opinions, official or otherwise. Perhaps that is just it, no position, formal or otherwise, was being taken, other than maybe a general sense that there was not much in Afghanistan even worth considering. After the abortive flurry of interest in the 1930s by Inland Exploration, perhaps the embassy could be forgiven for its lack of interest, but nevertheless it is worth noting that so little effort seems to have been made to find out more.

To be sure, periodically in the 1960s and 1970s a visiting businessman would occasionally come by and collect rock samples from various places, but this interest seemed to go nowhere. In fact at one occasion at the US Embassy in the late 1970s, the author was shown a collection of rock samples that had been obtained from the outcrop by an American corporate executive or geologist who had passed through. Finding that he was prohibited by the RGA from exporting the specimens by normal post or in his luggage, he sought to have them shipped by diplomatic pouch to his lab, something the embassy refused to do as he was a private citizen, but they kept the materials in their back map room. In the collection that the author observed were specimens of iron ore from the well-known, rich deposits at Hajigak near Bamiyan.

Indeed the restrictive behavior of the RGA in terms of such rather ordinary rocks made little sense, other than their general inability to determine what might be valuable or not. The result seemed to be a general opinion prevalent that foreigners were somehow trying to get away with something of value that should be prohibited. In another case at the Ministry of Mines and Industries, the author asked to see examples of whatever ores or gem minerals they might have in their possession. A large metal vault in the basement was presented, with signed paper seals pasted across it, but no other locks. The seals were forthwith cut in the presence of the signatories and inside the vault were stacks of galvanized metal trunks laden with lapis lazuli from the famous Sari Sang mines in Badakhshan, other semiprecious and precious stones, as well as miscellaneous specimens. Whether the trove was a legitimate collection made by government officials, or materials confiscated at the airport or border posts was not made clear, but the semisecret handling of the materials seemed unusual. In another instance the author was taken through the back alleys of old Kabul and up a narrow, circular staircase to a roof in the sun where a plate of spectacular finger-sized, uncut emerald specimens from the Panshir Valley was presented for inspection. The issue at hand presented to the author was to smuggle them out of the country because little legal export was allowed without copious bribes. The proposal seemed simple enough, yet it had several recipes for disaster implicit in it, so it was declined.

Thus the overall impression of resources available in Afghanistan in the late decades of the twentieth century was that the US Embassy did not think that there was much of value in the country, whereas many of the Afghans themselves seemed to think quite otherwise. Many informed the author that they thought they were sitting on a treasure trove that was much desired by foreigners, but who were telling them that they had little of value. The US Embassy, logically to many Afghans as the best place to realize their resource-acquisition ambitions, seemed relatively disinterested, or overly pessimistic about distance to market, lack of necessary infrastructure or technical training, and administrative naivety. To be sure, the embassy did test the resource waters from time to time, bringing in a "resource attaché" from New Delhi or further afield, but such short junkets apparently did little to convince the embassy of any potential value of the resources in the country.

Apparently then, it was into this persistent dichotomy of mostly negative opinions about resources in Afghanistan that Aloys Michel stepped when he arrived in the country in 1957 for his 6-month tour. But not being a geologist or resource specialist himself, Michel had to rely upon the opinions of others for his information. In the course of some 107 interviews he conducted, six were with Afghans about resources (a professor of geology at Kabul University, two geologist engineers in the field, three bureaucrats in the Ministry of Mines and Industries), and four resource-cognizant bureaucrats at the US International Cooperation Administration (USICA) (International Cooperation Administration, a precursor to US Agency for International Development (USAID)). From this he constructed his rather negative assessment of natural resources in Afghanistan and returned home to complete his doctoral studies. In his end and summary chapter he noted that it was clear that certain lines of development were "pretty well foreclosed, at least in the present state of knowledge of Afghanistan's resources" (emphasis his). He went on to note that only coal and perhaps petroleum offered any hope, unless some ores of sufficient quality and quantity to overcome transport handicaps could be discovered. His careful equivocations offer a typical scholarly hedge in case something new was discovered, but he entirely ignored over a century of knowledge about such world-class, rich deposits as the Hajigak iron, or the Ainak copper that had actually been mined in antiquity, among many other deposits. He emphasized only agriculture as a hope for the future, probably echoing the pervasive party line at the ICA and in the US embassy. In any case, his massive 441-page dissertation work (Michel, 1959) was never formerly published but a copy was preserved in the library of the prestigious National Academy of Sciences directly adjacent to the Department of State in Washington, D.C. We cannot know for certain, but given the evidence of opinions observed by the author with considerable skepticism in the US embassy in Kabul in the 1970s (Shroder, 1983, 1989; Shroder and Assifi, 1987; Shroder and Watrel, 1992; Ali and Shroder, 2011), we may presume that his dissertation was read over sufficiently by officials to be incorporated as gospel into some official thinking. Apparently, once such thinking becomes embedded in the fabric of bureaucratic officialdom, it is most difficult to dislodge.

Together with the ever-accumulating problems with the HVP with all of its negativity concerning agriculture, the problems for the future of Afghanistan were beginning to add up to what eventually became some fairly insurmountable difficulties by the end of the decade of the 1970s. American responses over the prior three decades to multiple requests and overtures from what had long been a stable and strongly pro-western Afghan government were indifferent and even niggardly (Poullada, 1981). The reasons for the American failure or response were complex, but were thought by former American diplomats to have included: (1) considerable indifference to the strategic importance of Afghanistan; (2) woeful ignorance about the cultures, languages, and lands of Afghanistan; and (3) a strange and quite unjustifiable appeasement of the Soviets in hopes that they would restrain themselves voluntarily. In this fashion the opportunities for the USA to have participated in extensive resource extraction in Afghanistan seemed to have gone unrecognized and withered away, rather than being recognized by the US government. Other countries and the United Nations were not being quite so shortsighted, however (Neilson and Gannon, 1976; Shareq et al., 1977; Abdullah et al., 1980; Vejlupek, 1980).

In the absence of an intelligent response from the Americans, in the last remaining few years of the Afghan monarchy, the RGA turned its resource and geology questions toward Europe (Germany, France), but especially toward the Soviet bloc (Shroder, 1983, 1989; Shroder and Assifi, 1987; Shroder and Watrel, 1992; Ali and Shroder, 2011). With its attention on resources in Afghanistan accordingly diverted elsewhere for decades to come, the US quite missed the resource ball when it failed to note the arrival of ~250 Soviet geologists in the country in the 1960s and 1970s. The resulting Soviet collaborations all through the government of Afghanistan, but especially in the Afghanistan Geological Survey, detailed a wide variety of mineral resources in the country (Neilson and Gannon, 1976; Shareq et al., 1977; Abdullah et al., 1980; Vejlupek, 1980). The details of these resources are related in the following chapter 11.

References

Abdullah, S.H., Chmyriov, V.M., Dronov, V.I., 1980. Geology and Mineral Resources of Afghanistan, vols. 2. Geological Survey of Afghanistan, Kabul.

Akram, M., 1947. Bibliographie analytique de l'Afghanistan. Centre de Documentation Universitaire, Tournier & Constans, Paris, 112 p.

Ali, S., Shroder, J.F., 2011. Afghanistan's Mineral Fortune: Multinational Influence and Development in a Post-War Economy. Institute for Environmental Diplomacy and Security, James Jeffords Center for Policy Research, University of Vermont, Burlington, VT, 24 p.

Anonymous, July 17, 1937. Inland exploration company acquires all Afghanistan for an oil search. The Petroleum Times, 78–79, 121–122.

Barfield, T., 2010. Afghanistan: A Cultural and Political History. Princeton University Press, Princeton, NJ, 389 p.

Clapp, F.G., February 27, 1939. Explorations in Iran and Afghanistan. The Oil Weekly, 71–72.

Clapp-Wincek, C., Baldwin, E., 1983. The Helmand Valley Project in Afghanistan. Unpublished A.I.D. Evaluation Special Study No. 18; U.S. Agency for International Development. http://www.dtic.mil/cgi-bin/GetTRDoc?AD=ADA518306.

Drummond, J., 1841. On the mines and mineral resources of Afghanistan. Journal Asiatic Society of Bengal 10, 74–93.
Elkins, J., 2009. Impediments Precluding Success of the Helmand Valley Project. Human Terrain System – Research Reachback Center (US Army), 7 p. https://ronna-afghan.harmonieweb.org/Lists/Submitted%20Content/Attachments/14/RRC-SIR-09-0012%20-%20Helmand%20Valley%20Project.pdf.
Favre, R., Kamal, G.M., 2004. Watershed Atlas of Afghanistan. FAO and AIMS, 183 p. http://aizon.org/watershed_atlas.htm.
Fox, C.S., 1943a. The Coal Fields of Afghanistan. Geological Survey of India Special Report to Afghan Government. British Institute Library, Kabul, 54 p.
Fox, E.F., 1943b. Travels in Afghanistan 1937-1938. MacMillan, New York, 285 p.
Furon, R., 1924. Les ressources miners de l'Afghanistan (Est). Revue Scientifique (Paris) 62, 313.
Gee, E.R., Seth, T.C., 1940. The Salt Resources of Afghanistan. Geological Survey of India Special Report to Afghan Government. British Institute Library, Kabul, 185 p.
Gregorian, V., 1969. The Emergence of Modern Afghanistan: Politics of Reform and Modernization 1880-1946. Stanford University press, Stanford, CA, 586 p.
Griesbach, C.L., 1881. Report on the geology of the section between the Bolan pass in Biluchistan and Girishk in southern Afghanistan. Memoirs Geological Survey of India 18 (pt 1), 60.
Griesbach, C.L., 1885. Afghan and Persian field notes (1). Records Geological Survey of India 18, 57–64.
Griesbach, C.L., 1886a. Afghan and Persian field notes (2). Records Geological Survey of India 19, 48–65.
Griesbach, C.L., 1886b. Field notes from Afghanistan; (No. 3) Turkestan. Records Geological Survey of India 19 (pt 4), 235–267.
Griesbach, C.L., 1887a. Field notes from Afghanistan; (No. 4) Turkestan to India. Records Geological Survey of India 20 (pt 1), 17–26.
Griesbach, C.L., 1887b. Field notes; (No. 5)—to accompany a geological sketch map of Afghanistan and north-eastern Khorassan. Records Geological Survey of India 20 (pt 2), 93–103.
Griesbach, C.L., 1892. The geology of the Safed Koh. Records Geological Survey of India 25 (pt 2), 59–109.
Hayden, H.H., 1913. The geology of northern Afghanistan. Memoirs Geological Survey India 39, 1–97.
Hopkirk, P., 1990. The Great Game: On Secret Service in High Asia. John Murray, London, 562 p.
Hutton, J., 1846. Notes on the geology and mineralogy of Afghanistan. Journal of Natural History. 6, 562. http://history.state.gov/countries/afghanistan.
Kaever, M., 1968. Bibliographie der Afghanistan—Literature, 1945–1967, Geologie und Palaontologie. In: Arbeitsgemeinschaft Afghanistan und Deutsches Orient-Institut (Deutsche Orient Stiftung), vol. 1, pp. 131–154.
Kakar, H.K., 1979. Government and Society in Afghanistan: The Reign of Amir 'Abd al-Rahman Khan. Modern Middle East Series, vol. 5, University Texas Press, Austin & London, 328 p.
Lee, J.L., 1996. The Ancient Supremacy: Bukhara, Afghanistan and the Battle for Balkh, 1731–1901. E.J. Brill, Leiden, Netherlands, 695 p.
Martin, F.A., 1907. Under the Absolute Amir. Harpar & Brothers, London, 388 p.
Michel, A.A., 1959. The Kabul, Kunduz, and Helmand Valleys and the National Economy of Afghanistan. Foreign Field Research Program, Sponsored by Office of Naval Research, Report No. 5. National Academy of Sciences—National Research Council, Washington, D.C., 205 p.
Neilson, J.M., Gannon, P.J., 1976. Mineral Evaluation Project Afghanistan. United Nations Development Program. Unpublished Report AFG/74/00Z vols. 2, United Nations Library, Kabul.
Obruchev, V.A., 1927. Fossil riches of Afghanistan (in Russian). Novyi Vostok 231 (16), 226–231.
Poullada, L.B., 1981. Afghanistan and the United states: the crucial years. The Middle East Journal 35 (2), 178–190.

Shareq, A., Chmyriov, W.M., Stazhilo-Aleekseev, K.F., Dronov, V.I., Gannon, P.J., Lubemov, B.K., Kafarskiy, A.Kh., Malyrov, E.P., Rossovskiy, L.N., 1977. Mineral Resources of Afghanistan, second ed. United Nations Development Program Project. AFG/74/012, 419 p.

Shroder, J.F., 1983. The U.S.S.R. and Afghanistan mineral resources. In: Agnew, A. (Ed.), International Mineral Resources: A National Perspective. AAAS Selected Symposium, vol. 90. American Association for Advancement of Science, pp. 115–153.

Shroder, J.F., 1989. Afghanistan resources and Soviet policy in Central and South Asia. In: Hauner, M., Canfield, R. (Eds.), Afghanistan and the Soviet Union: Collision and Transformation. Westview Press, pp. 101–119.

Shroder, J.F., Assifi, A.T., 1987. Afghan mineral resources and Soviet exploitation. In: Klass, R. (Ed.), Afghanistan and the Soviet Union: The Great Game Revisited. Freedom House Pub, New York, pp. 97–134.

Shroder, J.F., Watrel, R.H., 1992. Mineral resources in Afghanistan (summary of report for USAID). AACAR Bulletin (Association for the Advancement of Central Asian Research) 5 (2), 9.

Sillitoe, R.H., 1977. An evaluation of the exploration potential and procedure for porphyry copper deposits in Afghanistan. Unpublished Report for United Nations Revolving Fund for Natural Resources Exploration. October, Kabul, 67 p.

Sillitoe, R.H., 1978. Metallogenic evolution of a collisional mountain belt in Pakistan: a preliminary analysis. Geological Survey London 135, 377–387.

Sillitoe, R.H., 1979. Speculation on Himalayan metallogeny based on evidence from Pakistan. In: Farah, A., deJong, K.A. (Eds.), Geodynamics of Pakistan. Geological Survey of Pakistan, Quetta, pp. 167–179.

US Department of State, 1937. Foreign Relations of the United States Diplomatic Papers, 1937. The British Commonwealth, Europe, Near East and Africa, Afghanistan, pp. 597–614. http://images.library.wisc.edu/FRUS/EFacs/1937v02/reference/frus.frus1937v02.i0024.pdf.

Vejlupek, M., 1980. Geologie a Nerostne Suroviny Afghanistanu (Geology and Unexploited Resources of Afghanistan), vol. 22. Geologicky Pruzkum, Prague, Czechoslovakia, 16–18.

In this Chapter

11

Discovery of Rich Resources

Abstract

In the mid-twentieth century, constant diplomatic pressure during the Cold War by the Soviets against neighboring Afghanistan finally led to governmental acceptance of a major effort by some 250 Russian geologists to map the overall geology of the country, and to find valuable mineral resources that could be exploited, perhaps to the benefit of Afghanistan, but certainly to reward the finder Soviets. The two decades of Soviet mapping and analysis that took place in the 1960s and 1970s led directly to the discovery of resources that were rich by any standard. Valuable deposits of natural gas, oil, coal, cement, copper, and iron ores were all discovered and plans made for their future major exploitation.

Keywords: Cement; Coal; Copper; Gas; Iron ore; Oil; Russian-Soviet geological exploration; Topographic mapping.

Once the government of Afghanistan started having trouble with the Americans on several levels in the 1950s—the failures of the USA to support them militarily, the problematic Helmand Valley Project, the difficulties with resource extraction, and the pressures from the north by the Soviets; finally all were too much to ignore by the Royal Government of Afghanistan (RGA). Furthermore, the expansion of the Cold War after World War II (WWII) quickly led to the means to provide the RGA with the development aid and expertise it had sought from the prewar Germans (Barfield, 2010). Thus in the mid-1950s the RGA entered into a new military agreement with the USSR that provided it with the necessary military equipment and training to protect itself in uncertain postwar situations against any external or internal pressures—or so it naively thought. This caused the USA to quickly increase its own assistance programs in a strong Cold War rivalry that the Afghans played with a fine degree of skill to their own advantage as the competition between the two sides increased. Roads and airports were built, bridges and dams constructed, and farm assistance provided. Various advisers were inserted by both sides into many levels of the government, although the Soviets were far more successful at it than the Americans or their NATO allies.

Natural Resources in Afghanistan. http://dx.doi.org/10.1016/B978-0-12-800135-6.00011-8

At some point in the 1950s, the US pointed out to the RGA that in order to develop the country, build the roads and dams, and improve the overall infrastructure and the most vital agriculture, the country would need a program to construct high-quality, large-scale, topographic contour maps. In those days such maps were produced from reasonably low altitude, overlapping stereoscopic aerial photographs, which in the laboratory enabled a technician with a mapping instrument to see the land in three dimensions to draw contour lines of equal elevation. The Afghans lacked such capabilities entirely but the Americans offered to do the job with their aircraft and advanced mapping laboratories in California. Fairchild Aerial Surveys, a Howard Hughes corporation that often did clandestine work for the US government from time to time, took on the task. This project came to the attention of the Soviet government, which took strong issue with the Americans flying aircraft anywhere near their borders. Accordingly the Soviets were allowed to take over by flying their own mapping aircraft over the northern one-third of Afghanistan, whereas the Americans took on the southern two-thirds. American aircraft flying over or along the borders with either Iran or Pakistan at the time were viewed without local suspicion; after all, U-2 spy aircraft were being routinely flown over the Soviet Union from Peshawar, Pakistan, and American aircraft were common over all the nearby border regions in South Asia.

Thus it was that in the late 1950s and early 1960s, topographic maps at scales of 1:50,000, 1:100,000, and 1:250,000 were produced by Fairchild and delivered to Kabul for conversion by the Afghan Cartographic Institute (ACI) into their own map series. The Fairchild map plates and air photo negatives were delivered and promised as supposedly the only copies in existence, but of course another full set went clandestinely to the US Department of Defense where their map catalogue suddenly increased by orders of magnitude in quality and quantity. Simultaneously on the ground, flotillas of experts in geographical nomenclature scoured the countryside to produce a gazetteer of geographical naming conventions amongst the plethora of languages and multiple names for the same features. Thus Jalakot and Jalalabad were used for the same city, or Spin Ghar and Safed Koh for the same mountain range. By the early 1960s the entire country had been mapped in detail for the first time (Glicken, 1960), even though the Soviet and American maps had compatibility problems along the boundaries where the two versions were supposed to overlap and join.

Ever since British days in the early nineteenth century and thereafter, the Afghans always had a problem with foreigners mapping their country. It was not a complicated issue because they saw it for what it was, mapping was a means by which outsiders (spies!) could find out what was there, and thereby perhaps lust after whatever it was in some fashion that was most likely to cause harm to the people of Afghanistan. Nonetheless, throughout the 1960s these topographic maps were available on a nearly one-per-day basis in Kabul so that the various embassies assigned one of their Afghan worker "fixers" to go to the map office day after day to accumulate a whole set. After Soviet advisors moved progressively into the

ACI in the 1970s this practice was stopped, but not until complete sets had gone to various European embassies, the West Germans in particular, but also the British and French to an unknown extent. US Agency for International Development (USAID), on the other hand, obtained a full set of the US Department of Defense (USDOD) map catalogue to use in its development projects, and although the Americans were not supposed to show them to the Afghan office people who worked there too, it was a rather loose operation with little real security of the kind that is most common with intelligence or military operations.

All the while the various assistance projects of the Cold War were unfolding, Americans did not seem to care that the Soviets had been pouring geologists into the country. To be sure, the US embassy did support the Kabul University (KU) Seismic Station, ostensibly to study the terrible earthquake situation that Afghanistan has been subject to for its entire existence. Not mentioned, however, was the fact that the station was really also in the business of watching nearby nuclear testing at Semipalatinsk in the Kazakstan Soviet Socialist Republics (SSR), as well as the Lop Nor dry lake atomic test site in the Taklamakan Desert of China. As the only American geologist (with a background in physical geography as well), I was asked by the US embassy to look after the KU seismic station, pay the three Afghan workers, and make weekly deliveries of the collection of seismograms to the embassy for delivery by diplomatic pouch to Washington. The compensation to the author from the embassy and the US Geological Survey was an old but free International Scout truck with 4-wheel drive, and a modest travel budget.

But I was mainly in Afghanistan as the Director of the National Atlas of Afghanistan, a USAID-sponsored project, being run through the US National Science Foundation for quality control, to collect as much mapped data as feasible, as well as to make topical maps where possible. Collecting information and maps in the course of this work led me directly into a confrontation with the Soviet advisors and their Afghan communist sympathizers. In the course of the Atlas-making process the discovery that so many Russian geologists were working in Afghanistan in the 1970s was a revelation to me—there must be a good resource base of some kind! The first good clue was in the United Nations library in Kabul where some unusual reports surfaced, including one by an author, P. J. Gannon (Neilson and Gannon, 1976), who participated in some fashion as an editor or critic with the Russians in the search for, and mapping of, the mineral resources in Afghanistan. This initial information led me to the discovery of works in progress that eventually became available when they were published (Shareq et al., 1977). Evidence that the resources in Afghanistan were really rather good, and that they had been downplayed by more than just the Americans for many years (Abdullah et al., 1980), was supported by other evidence all around that just kept accumulating.

In fact what had happened was that after their unfortunate experience with the Inland Exploration Company, and after they had waited out WWII, the RGA first asked the new United Nations in 1949 for assistance with their oil exploration. Nothing came of this entreaty to the UN, so in 1950 the Soviet Union offered to step into the breach, but once again the RGA refused. Instead the government next invited the French to explore for oil in Afghanistan. Unfortunately because France was then a NATO ally, the Soviets were able to successfully pressurize Kabul to cancel the French contract. In 1954, in spite of renewed requests from the Soviets to let them initiate oil exploration, the Afghan government then tried again and brought in neutral, non-NATO Sweden to look for oil. By 1956, possibly because hard-rock Sweden was not known to have a great deal of expertise in such endeavors, little progress had been made. The following year the RGA was finally pressured into accepting the Soviet-bloc country, Czechoslovakia, operating with a Soviet loan to begin the first exploration. Oil strikes were reported directly thereafter in 1958.

At that same time, the USA tried to get back into the oil and gas game in Afghanistan, at least to a limited extent. The US Geological Survey was asked by RGA to help provide assistance in writing a new Petroleum Law that would provide background for legislation to help private businessmen who wished to work in the budding industry. This so-called "wildcatting law", modeled in part on the way the USA had so successfully developed its own oil and gas industry in the early twentieth century, was presented to the RGA (Cerkel and Miller, 1958), but was never implemented in any fashion. This was probably because such private ownership of drilling results and personal profit making without outright government control and total ownership was not exactly something that either the communist Soviets or the RGA were particularly enamored of.

In any case the Czech drilling team was quickly replaced by personnel from Soviet Techno-Export, an engineering/development organization with Komitet gosudarstvennoy bezopasnosti (KGB) involvement and intelligence-gathering capabilities. By 1960 the Soviets were reporting important and rich oil fields in the vicinity of Sheberghan in northwest Afghanistan (Figure 11.1(A,B)). The following year they provided some $200 million in additional credit for further oil and gas exploration and development, which led in 1963 to an additional agreement for the surveying and exploration for other mineral resources throughout the country. This was the agreement that led to the next two decades of intensive resource research and geological mapping by Russian geologists across Afghanistan that directly produced the important geological maps at multiple scales in the late 1970s (Shareq et al., 1977).

The US embassy sought to keep up in the late 1960s with what was happening with the Soviet-backed, mineral resource exploration but was only marginally successful with a bit of its effort (Sweetwood, 1968). Soviet control over the Ministry of Mines and Industries in Kabul was growing stronger and little information was getting out. A propaganda booklet on the Soviet work on resources in Afghanistan was published (Anonymous, 1973), but the work

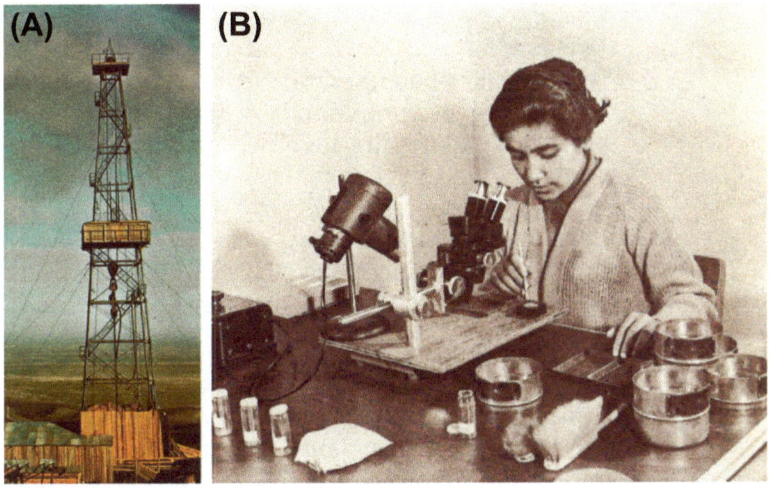

Figure 11.1 (A) Oil rig in northern Afghanistan being used for exploration for oil and gas *(after Anonymous (1961))*. (B) Young Afghan female technician, probably about 1960, examining geological samples, possibly from oil drilling. The picture is particularly striking because of the lack of veiling and her short hair, which are reflections of the relative freedom of women in Afghanistan at the time, particularly in the bigger cities or universities, where this picture was probably taken. *After Anonymous (1961).*

was so superficial and laden with language mistakes that the idea of a scanty resource base was perpetuated. In 1974–1976 two Canadian geologists were brought to Afghanistan by the United Nations to Afghanistan to assess the resource exploration progress (Neilson and Gannon, 1976) but reported critically about concealed documents and attempts to impede their research. Their report, however, was quite favorable about the potential resources but was rejected by the government and ultimately suppressed to only a file copy in the UN library in Kabul (Shroder and Assifi, 1987), which at least I was able to read without interference, but could only transcribe by pencil.

In 1977 the 419-page survey of Afghan resources was published under UN authorization, but because it was confidential it was not made available until several years later. The document and maps were identified as the work of Soviet and Afghan geologists, condensed and translated into English by a Russian and edited into reasonable English by the Canadian P.J. Gannon (Neilson and Gannon, 1976). Strikingly after the communist coup in Kabul in 1978 and numerous defections from the government bureaucracy, it was also learned that the Soviet advisors had been systematically deluding the Afghan government itself about the resources in the country. Mr H. Afzali, Director of the Afghanistan Geological Survey (Afzali, 1984), was not even told about some of the resource discoveries. In some cases two sets of reports were prepared, one more pessimistic for Afghan consumption, and another more realistically positive for the Soviet government.

Finally after the communist coup of 1978, however, the 1977 resource inventory was released in some fashion and became available for assessment. A World Bank report (Kavalsky et al., 1978), however, although it was quite favorable about the prospects of future Afghan resource development, had to be held back after the coup because although the resource situation had not changed, the political and economic climate was dramatically altered and resource development was now seen as much less probable, at least in their capitalist eyes. Indeed events were to play out over coming decades in exactly that fashion, so much so in fact, that many people later forgot about the initial resource optimism in Kabul that was at hand in the late 1970s and early 1980s.

The 1978 coup and Soviet invasion the following year, for example, did set off an escalation of resource rhetoric that was unmatched theretofore. The abortive Taraki-Amin regime only lasted a few years but quickly designated the northern petroliferous region around Sheberghan as a principle base for industrial expansion (Anonymous, 1979). The previous Daoud government that they had killed in office had planned a number of Soviet-supported projects, which were then reaffirmed:

- $30 million in petroleum equipment.
- $22 million for detailed mapping of the north.
- $600 million for the Aynak copper ore smelter and a second natural gas fertilizer plant.
- $50 million for a first rail and vehicle bridge across the Amu Darya from Termez on the north to Hairatan in the south (see Figure 9.3(A,B); Figure 11 2).
- $5 million for renovation of the Sher Khan river port on the Amu Darya.
- A new oil refinery.

The 1977 report (Shareq et al., 1977; Abdullah et al., 1980) was 419-pages long and divided into three main parts. Part I at 48 pages discussed the main geologic features, including stratigraphy, tectonics, and igneous rocks, and was

Figure 11.2 A first Soviet diesel engine with Babrak Karmal crossing into Afghanistan on the Friendship Bridge in spring 1982. *After Mas'ud et al. (1983).*

followed by 22 pages of the metallogenic features. Part II summarized 1432 specific mineral that were precisely located, described, and categorized as *deposits, occurrences*, or *showings*. Another 306 mechanical mineralogical haloes were suggested as potential occurrences of specific minerals. The exhaustive catalogue of the mineral locations comprised some 70 percent of the report. Part III attempted to stratigraphically define the different ground-water aquifers in the country, and showed 198 specific water localities, including 112 areas of mineralized waters and 86 fresh-water springs and wells in the desert and steppe regions. Overall the report included 11 excellent maps:

- geologic map (1:2,500,000 scale);
- tectonic map (1:2,500,000 scale);
- magmatic complexes (1:2,000,000);
- metallogenic zones (1:5,000,000);
- mineral deposits, occurrences, showings (1:2,000,000); and
- maps showing:
  - fuel minerals and ferrous metals;
  - nonferrous metals;
  - tungsten and tin;
  - rare and precious metals;
  - nonmetallic minerals;
  - hydrogeologic map; and
  - fresh and mineral waters

These maps and reports were prepared by the Russian and Afghan geologists according to international standards, which thereby lent credibility to their propositions. In aggregate the five mineral-deposit and mineral occurrence maps listed above (1–5) show an impressive amount and variety of valuable resources that were available (Figures 11.3–11.5). In fact, several of the deposits (copper, iron, rare earths, etc.) were judged by the Canadian team and the World Bank (Neilson and Gannon, 1976; Kavalsky et al., 1978) to be world-class deposits that would be immediately commercial if brought on stream and would also be likely to affect the world price at the time. A more detailed look at a number of the different resources is instructive in understanding what the issues were at the time when they were first discovered, and what has developed since then. In some cases with a particular resource the issues are now fairly moot; in other cases things have developed so that the issues are likely to be major players in the years to come—only time will tell.

Natural Gas

The issue of this energy source in Afghanistan is replete with discovery, transport, and usage problems, as well as historical changes that are complex but important to the future development of the country because they impact

Mineral Resources in Afghanistan
As Compiled by Afghan and Soviet Geologists
As of 1977 (Deposits, Occurrences and Showings)

58 solid combustible materials: 45 coal, 3 lignite, 9 peat, 1 shale.

898 metallic minerals:
85 ferrous metals, 69 iron, 2 manganese, 14 chromium, 241 copper, 92 lead and zinc, 7 aluminum, 4 molybdenum, 136 tungsten, 174 tin, 12 bismuth, 141 mercury, 1 cadmium, 27 beryllium, 44 lithium, 11 cesium and ribidium, 32 tantalum and niobium.

4 radioactive elements: uranium, thorium and rare earths.

105 precious metals: 95 lode gold, 5 placer gold, 5 silver.

Total of 118 non-metallic minerals, including 36 chemical raw materials:
8 sulphur, 7 fluorite, 17 barite, 2 celestite, 2 borosilicate.

2 mineral fertilizers.

80 other non-metallic minerals:
22 muscovite, 26 asbestos, 7 talc, 2 magnesite, 6 graphite, 17 gypsum.

14 salts.

20 precious and semi-precious stones:
1 ruby, 7 emerald, 2 kunzite, 1 garnet, 1 lazarite,
4 serpentine, 5 tourmaline.

23 electronic and optical minerals:
18 quartz, 5 calcite.

69 industrial materials:
5 limestone, dolomite and marble quarries;
20 facing and ornamental quarries;
15 sand and gravel quarries;
29 cement raw materials:
8 limestone and marl quarries; 4 limestone and dolomite fluxes;
3 refractory clays; 7 clays for brick, roofing tile, etc.;
4 porcelain and pottery clays; 3 glass raw material and siliceous sandstone.

Prepared by M.S. Noorzoy.
Source: Abdullah Shareq et al., eds. *Mineral Resources of Afghanistan* (Kabul: Ministry of Mines, 1977): 73-79.2d. ed. Translated from Russian by G.M. Bezulov.

Figure 11.3 List of mineral resources in Afghanistan. *Compiled by Noorzoy (1987) from data in Shareq et al. (1977); Abdullah (1980).*

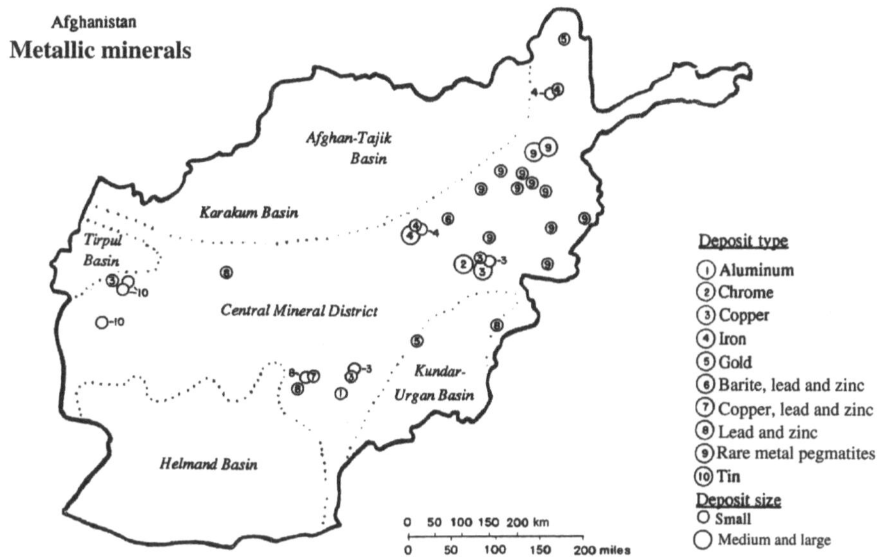

Figure 11.4 Map of the known metallic minerals in Afghanistan in 1977. Note their concentration in a Central Mineral District of dominantly crystalline (igneous and metamorphic) bedrock, surrounded by sedimentary basins on all sides. *After Shareq et al. (1977); Abdullah et al. (1980).*

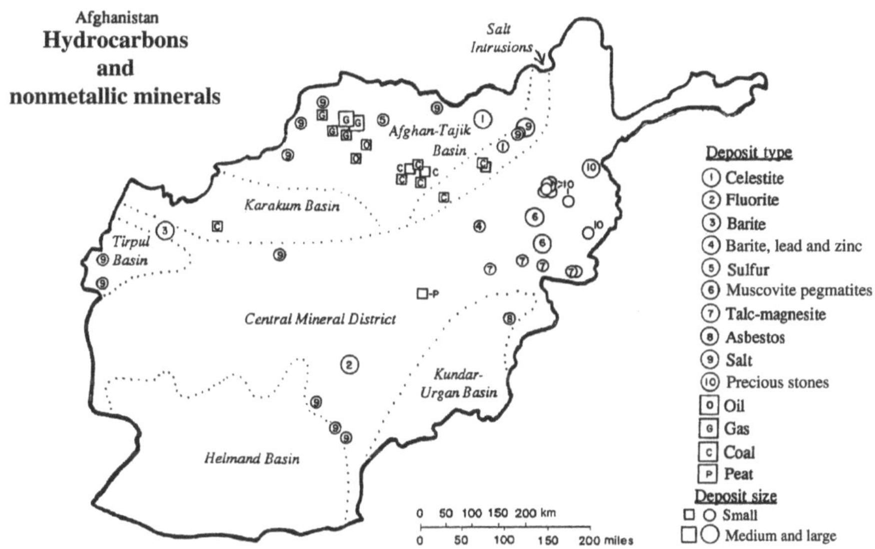

Figure 11.5 Map of the known hydrocarbons and nonmetallic minerals in Afghanistan in 1977. Many of the sedimentary basins had not been explored at all at this time, and this remains true up to the present time because of lack of security, great expense, and remoteness. *After Shareq et al. (1977); Abdullah et al. (1980)*

how the resource can be used or marketed. The history of the resource in Afghanistan is that the discovery of more than 15 oil and gas fields in northern Afghanistan by the Soviets from the 1960s to the mid-1980s was a significant development for the county's energy future, yet only a small proportion of the gas was ever used in Afghanistan for domestic purposes (Shroder and Assifi, 1987). At peak production in the late 1970s-early 1980s, Afghanistan

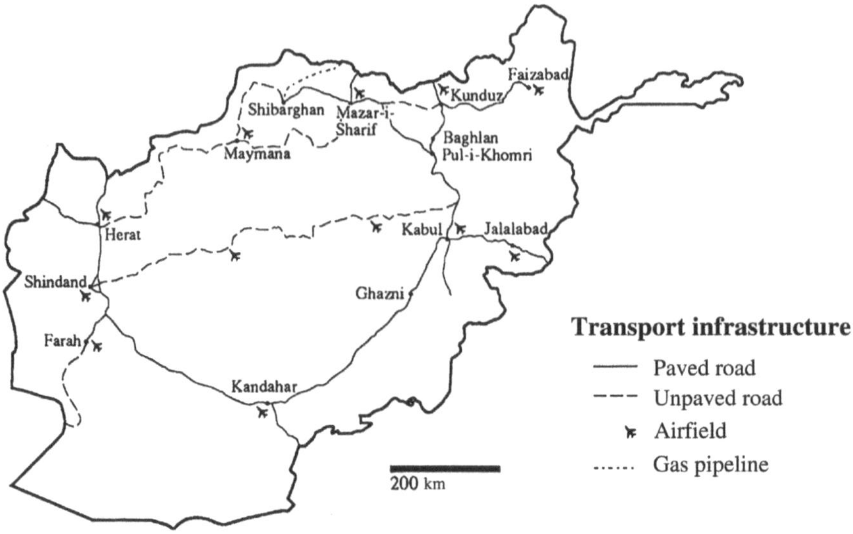

Figure 11.6 Transport infrastructure of Afghanistan as of the middle 1980s. Note gas pipeline in north of Afghanistan that shipped gas from Afghanistan north into the Soviet Union in the 1970–1980s. *After Hauner and Canfield (1989).*

sent 70–90 percent of its natural gas output into the Soviet Union through a subsidiary pipeline under the Amy Darya and into the Uzbekistan SSR and the national pipeline grid of the USSR (Figure 11.6). In an interesting accounting and pricing manipulation, all the dials and gauges to measure the gas flow were on the north side of the river where they could be read only by the Soviet buyers. Such a lopsided marketing system where the buyer tells you how much you are selling and then sets the price for however much you do not know you delivered, smacks of really rapacious capitalism. To add to the irony, the price actually paid was well below world market rates, or even local agreements by the Soviets with Iran, as an example. Two gas fields developed by the Soviets included Khwaja Gogerdak and Yatimtaq near Sheberghan 120 km west of Mazari Sharif reached a production in the mid-1970s of 275 million cubic feet per day (Mmcf/day). By 1978 when their newer Djarquduk field came onstream the output went up to 385 Mmcf/day. The Soviet Union was able to construct several pipelines in Afghanistan to deliver fuel. The first was beneath the Amu Darya (river) border between the two nations for the gas, and the second was a small diameter pipeline from north to south to deliver diesel fuel to their troops at the Bagram airbase north of Kabul. In 1978 the Soviets had estimated that the Afghanistan proven and probable natural gas reserves were perhaps up to 5 trillion ft^3 (Anonymous, 2013). In any case, this cozy deal for the Soviets to extract Afghan's natural gas at rock-bottom, dirt-cheap prices was over when the last Russian gas adviser ignominiously departed in 1989 and Afghanistan's second Civil War erupted before the Taliban took over. Production then shut down in the 1990s and these pipelines then went into disuse, once the Soviets withdrew following the conclusion of the Soviet-Afghan War.

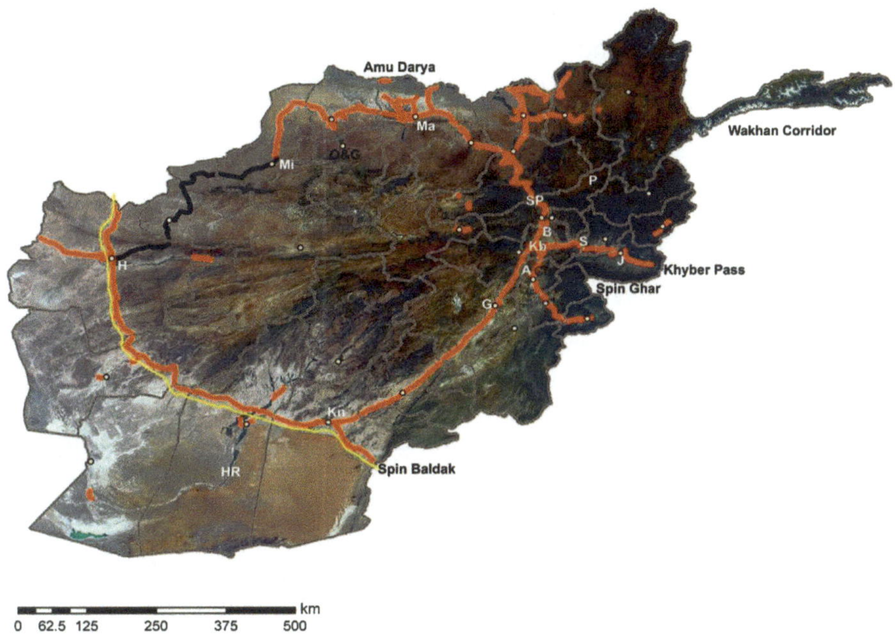

Figure 11.7 **Proposed gas pipeline in the west of Afghanistan (yellow line) alongside the outer western part of the so-called Ring Road (red and black lines).** H—Herat; HR—Helmand region; Kn—Kandahar. Base map is a satellite image mosaic.

During the 1990s when the Second Afghanistan Civil War was raging and Taliban government emerged in 1996 into a moribund economy, some excitement reigned in the American Unocal Company because they thought a long-awaited gas pipeline project from Turkmenistan and across Afghanistan to Pakistan, and possibly on to India, would then be easier to implement (Ali and Shroder, 2011). The Taliban were greatly interested because millions of dollars in transit fees to ship gas would have been paid to them. Unocal hired the Afghanistan Studies Center at the University of Nebraska at Omaha (UNO) to educate largely Pushtun Afghan workers in the trades and technical skills necessary to construct the pipeline alongside the western highway from Turkmenistan to Herat, Kandahar, and on south to Spin Buldak on the Pakistan border (Figure 11.7). UNO set up an Unocal training facility on a huge 56-acre compound at the junction between the A1 Kabul to Kandahar highway and the A75 from Kandahar to Spin Buldak, coincidently close to one of the compounds of Osama bin Laden (Coll, 2004). Some fairly illiterate Taliban and others even came to UNO at the behest of the US Department of State to listen to a lecture by the author on using remote-sensing and geographic information systems and remote-sensing technology to accomplish their objectives. The project ultimately foundered because of a combination of human rights abuses by Taliban, coupled with the attacks on the US embassies in East Africa by al Qaeda members trained in camps in Afghanistan (Shroder, 2007).

The trans-Afghanistan-pipeline idea to transmit vital energy to the huge markets in Pakistan and India, and further abroad has not gone away, however, and a number of alternate routes have been suggested, some to include Afghanistan and others to avoid the unstable area (Ali and Shroder, 2011). One proposal even has Afghanistan receiving a dollar for every million BTUs of energy transmitted across its territory. The chronic shortage of fuel in South Asia, particularly to India's booming economy, when Iran and Turkmenistan have huge gas surpluses to sell, means that the likelihood of pipelines by various routes in the region is quite high in future.

Oil

In Afghanistan five major geologic basins occur with sedimentary rocks bearing greater or lesser hydrocarbons (gas, oil, coal) (Shroder and Assifi, 1987) (Figure 11.8). Between 1957 when the Soviets first took over oil exploration in Afghanistan, and 1984, by which time it was becoming far too dangerous for them to do any more field work (Moritz, 1981), they had discovered seven oil fields. Most of the wells were shut in, however, to await further developments and during the civil war of the 1990s little production happened. Following 9/11 and the invasion of Afghanistan in late 2011, the US Geological Survey was tasked by the White House with investigating mineral resources in Afghanistan. By 2006 their estimates were for about 1.6US billion barrels of crude oil in the north of Afghanistan (Klett et al., 2006).

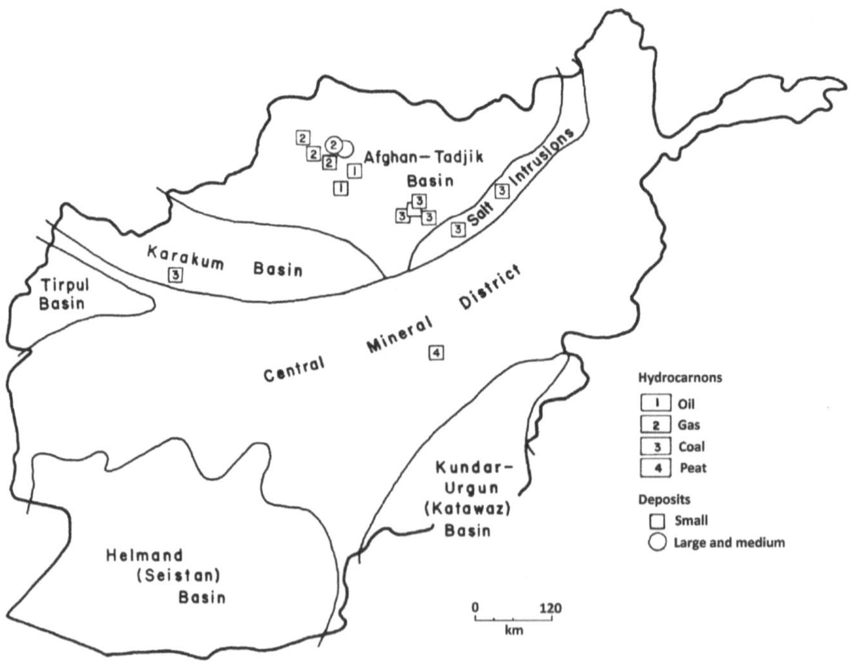

Figure 11.8 Afghanistan hydrocarbon (oil, gas, coal, peat) distribution in the country at minimal amounts. Much more coal occurs than is shown on this map. The basins are all sedimentary and young Cenozoic—dominantly Neogene in age. *After Shroder (1983).*

By August 2011 production of a modest 800 barrels a day of crude came from the Angot field near Sari Pul. Only this oil field is producing at the present, although discovery of another new 1.8 US billion barrel field in the north in 2010 gave Afghanistan much heart that other discoveries would be in the offing. The result has been the opening of the Afghanistan oil and gas sector to exploration by foreign companies, with China's National Petroleum Corporation being the first to ink a deal in December 2011. The Exxon Corporation immediately began negotiations to become involved in oil exploration in Afghanistan as well (Carlisle, 2011; Najafizada, 2010; Peter, 2011; Ahmad, 2012; Ferris-Rottman and Venugopal, 2012), which because it is such a giant multinational firm, leant considerable credibility to projections of future production capabilities of Afghanistan.

Coal

Deposits of coal in Afghanistan were originally considered to be "vast" by the World Bank but when assessed by the US Geological Survey, they were judged, perhaps rather more precisely, to be moderate to potentially abundant (Kavalsky et al., 1978; SanFilipo, 2005). The reasons for the variable opinions were that coal seams have not been much explored, mapped, or drilled out extensively to determine more accurate tonnages that could be expected. Much of the coal is deep or inaccessible, and so is little known about the resource at those depths. In any case the coal of the country has barely begun to be seriously exploited, although production has been underway since before WWII (Figure 11.9).

Figure 11.9 Karkar coal mine in Baghlan Province, Afghanistan, owned by the government. The mine has electricity but the electric railcars were broken down and had to be winched out. The support timbers on the right had to be imported at considerable expense because they are no longer available locally. Lack of proper timbering in some big mines has led to fatal collapses in recent years (Abkhorak mine, Ruyi Du Ab district, September 2013). *after San Filipo (2005).*

Figure 11.10 Dangerous hand-driven artisanal mines or 'dog holes' for coal without roof supports in the Andara area of Baghlan Province. The spoil piles are wasted coal fragments that weather out on the surface and degrade the environment. *After SanFilipo (2005).*

Coal production about eight years ago was about 200,000 Mt/year, which is fairly close to the prewar peak reported in 1978, nearly all of which came from RGA-owned mines. Today only about 20 percent is government production, with the rest being private "artisanal" mines, which tend to be primitive, dangerous, and potentially damaging to the resource and the environment (Figure 11.10).

The distribution of known coal in Afghanistan is predominantly in the north from Herat to Badakhshan and has been mapped in two basic categories: (1) areas of potential shallow coal (less than 1000 m deep) that is Mesozoic in age (older than 65 million years); and (2) areas of potential deep coal (1000–1400 m deep) that is Mesozoic in age. In addition is a large area in southeastern Afghanistan of younger Paleogene sedimentary rocks in the Katawaz Basin that should have natural gas supplies, but it has never been drilled in Afghanistan (Figure 11.11). The same rocks, however, occur across the border in Pakistan, where they host the Sui Gasfield, Pakistan's largest and most essential energy source. In addition to these coal deposits to meet some of the energy needs of Afghanistan in future, most coal beds can also be drilled to produce coal-bed methane gases, which also burn quite well.

Cement

Afghanistan is endowed with huge quantities of the proper kinds of limestone and coal to at least theoretically produce plentiful cement. Modern development after WWII led to the first cement production in 1958, and by 1966 production had climbed to 158,000 short tons (143,336 metric tons). Production up through the late 1970s was up and down from half that figure to approximately equal to it. The erratic production produced scarcities at times, which resulted in a flourishing black market and a 50 percent premium being paid at times over

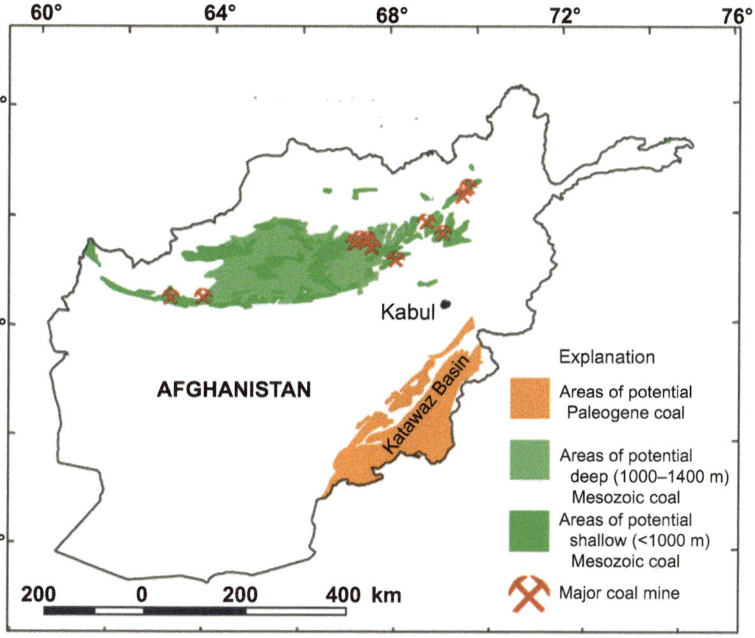

Figure 11.11 Coal deposits in Afghanistan. The Katawaz Basin in the southeast also has the prospects of gas production because similar rocks that occur over the border in the Sui field in Pakistan contain gas. The deeper coal beds (light green) are prospectively valuable for coal-bed methane as well. *After SanFilipo (2005).*

Figure 11.12 The 122-m-long rotary cement kilns of Ghori II and satellite coolers at Poli Khumri in 2010. *After V.G. Mossetti (2011) for USGS.*

the otherwise government regulated price (Shroder and Assifi, 1987). In practice in Afghanistan in the prewar days of the 1970s, these shortages sometimes produced sloppy cement making where not enough cement was mixed with the aggregate sand and gravel, and too much water was added so that subpar or weak concrete was produced that broke down easily later in construction.

In 1977 eight limestone quarries and two cement plants existed, one at Poli Khumri north of the central Hindu Kush (Figure 11.12), and the other at Jebali

Figure 11.13 The kiln of the cement plant at Jabal os Saraj northwest of Bagram in the Kabul Basin in the late 1950s. *After Anonymous (1961).*

Seraj directly south of the mountains (Figure 11.13). Two other plants were planned to come into production in the early 1980s, one in Herat and the other at Kandahar, but neither got started because of the onset of the Soviet-Afghan War. The two that were running, however, in combination with the lofty plans for the others, did result in a most unusual trade protocol signed with the USSR in 1981 whereby an equal exchange of cement ton-for-ton was to be undertaken so that the superior Afghan cement could be sold on the world market by the Soviets in exchange for vital foreign exchange for themselves (Kavalsky et al., 1978; Mossotti, 2011).

Neither of the half-constructed plants at Herat or Kandahar were able to produce cement because of the war, and in retrospect neither has been found to have adequate supplies of raw materials close enough to be able to compete in price with cheaper modern sources of cement across the borders in Turkmenistan, Iran, or Pakistan, so refurbishment of either plant is not feasible. The Jabali Seraj plant, originally built by the government of Czechoslovakia, was the first rotary kiln in Afghanistan in 1957 and produced vast quantities of cement in its time. It was stopped by the Taliban in 1996, being bombed 10 times, with a final direct hit on the ring bearing of the main kiln, which put it permanently out of commission. Multiple visits and testing by cognizent personnel from the US Geological Survey and representatives from the Czech Republic showed that it could actually be put back into working order with adequate renovations (Mossotti, 2011). The cement plants at Poli Khumry are in a far better shape at the present time and one of the two sets of kilns is still operational (Ghori I), although it was dismantled for repair when being visited. The Ghori II plant was started under construction some 20 years ago and was almost complete when visited; a third kiln (Ghori III) is on the drawing board for the same site.

The reconstruction and redevelopment of war-torn Afghanistan is greatly hampered by the lack of high-quality cement even though the country is endowed with plenty of the requisite raw materials of limestone, clay, coal to fire the kilns, and gypsum added to slow down drying time. Afghanistan presently has among the lowest cement production in the world at about 2 kg/person (50,000 Mt/year, where 1 Mt = 1000 kg or 2204.6 pounds), whereas in neighboring Pakistan it is 92 kg/capita, and in a developed country like the United Kingdom is 200 kg/capita. In 2005 the production of Afghan cement was only 16,000 Mt/year, whereas the consumption was 2.5 million Mt, with the balance being imported from Iran (1.8 million Mt), Pakistan (4 million Mt), and Uzbekistan and Turkmenistan (300,000 Mt) (Anonymous, 2011). Demand in Afghanistan has been forecasted to reach as much as 7.2 million Mt per year by 2020, providing the country does not slip back into Taliban-fueled anarchy once the International Security Assistance Forces (ISAF) Coalition departs in 2014.

Keen competition in state-of-the-art cement production and low costs on imports from neighboring Iran, Pakistan, China, Tajikistan, Uzbekistan, and Turkmenistan indicate that only the shorter distances involved in some domestic production are in favor of local cement production by Afghanistan to meet the increasing demand. Probably refurbishment and upgrading of both the Poli Khumri and Jabal os Saraj plants makes sense at this time but production costs will have to be monitored carefully to stay competitive. Otherwise the ever-present Pashtun-dominated trucking import industry will be able to undercut local prices by bringing in cut-rate cement from surrounding more stable countries.

Copper

More peculiar intrigues, corrupt deals, outright combat, and potential environmental and archaeological disruption have occurred around the deposits of this metal than any other in Afghanistan. The copper deposits at Aynak in eastern Logar Province some 20–30 km southeast of Kabul appear to have been mined in antiquity, perhaps as much as 4000 years ago, or even longer (>5600 years ago?) if the speculations of some archaeologists were proven to be true. The issue in this is that plentiful nearby cassiterite tin in placer form could have been used to usher in the Bronze Age when some enterprising early Afghan mixed copper and tin for the first time and his spear or sword ended up being sharper than his enemy's. This speculation is not supported by dateable evidence found yet, but it is an idea definitely worth considering. The early Bronze Age mining site passes upward through the overlying layers to several thousand years later when about 2600 years ago the region became a Buddhist stronghold, monastery, temple site, and mining location, where thousands of stone and stucco statues of the Buddha and other related gods were produced (Huffman, 2012; Kelly, 2012). Over the course of the following several thousand years when the region was overrun by Islamic zealots carrying the new religion, the old ways were lost and the statues disappeared under the incoming wind-borne dust and slope-wash colluvium that covered and preserved them. Now they are known to sit on top of a huge copper deposit, one of the world's great metal hordes that is lusted after by many, and for numerous different reasons. What the future will bring for those who are entangled in the lure of this metal here, no doubt will make interesting times for all concerned. As the Chinese say in a blessing rather than a curse, "may you not have to live in interesting times".

In the 1970s, when planning for mining of the Aynak deposit were first being made, the Buddhist remains were not much known. Instead, the Soviet planners envisioned a mine of the fairly high-grade ore of 0.7–1.2 percent grade of Cu wherein only the mining and concentration of the ore would be done in Afghanistan, whereas the main smelting and refining would be in the Soviet Union (Shareq et al., 1977; Abdullah et al., 1980; Shroder and Assifi, 1987; Kavalsky et al., 1978). Initial measurements of total copper tonnages were about 4.7 million tons, but later exploration by Soviet geologists produced 150 drill holes, 70 trenches, and 9 adits into the deposit that nearly tripled this figure to 11.5 million tons, which is among the world's largest deposits of this metal. Capture of extensive documents about the richness of the deposit by a Mujahideen commando raid on a Soviet convoy from the area in the early 1980s and sent anonymously to me in Nebraska, suggested greater richness of the deposit than we had believed (Shroder and Assifi, 1987). This suspicion was later confirmed in the late 1980s by conversations that I held with Dr. Yuri Gankovsky, the chief Soviet advisor about Afghanistan to the Soviet government; "the world's largest copper deposit", he said at the time. The World Bank estimated that Aynak could capture 2 percent of the world market share of copper when brought on stream (Kavalsky et al., 1978).

All development at the site stopped during the civil war of the 1990s but an al Qaeda terrorist training camp was set up there; US B-52s bombed it out in 2001with the large bunker-buster bombs in retaliation for the 9/11 attacks in the USA (Dalton, 2007). The close proximity of the Aynak site to Kabul lends itself to ease of access, but the dominance of more warlike Pushtun villages in the immediate vicinity, and the propensity for pollution of ground water and air quality so close to the major capitol city, are severe detriments. The current situation of the copper at Aynak will be explored further in Chapter 12 where the especially noted, major resources of Afghanistan are discussed in more detail.

Iron

The large masses of high-grade iron ore (Fe) high in the Hindu Kush near Hajigak Pass on the way from Kabul to Bamiyan have been known to the outside world since the British explorations of the nineteenth century. In the big resource push by the Soviets in the 1970s, they had trenches excavated across the mountain tops so that they could calculate the tonnages and tenor or grade of the ore more efficiently (Figure 11.14). The ore grade averaged a very high 67 percent (many world iron mines operate successfully at far lower percentages) and at the

Figure 11.14 Mountains of iron at Hajigak in central Afghanistan in 1978. The two light-colored scars on the right of the photograph are trenches dug by the Soviet geologists to investigate the grade of the ore. *Photo by the author.*

time some 111 million tons were measured, with an additional 2 US billion tons judged as speculative. When one compares these figures with the total American production over many decades in the Mesabi Range of the north-central USA at 2.4 US billion tons, the value of this deposit is obvious. French geologists looked at the deposit as well, and judged it at the time to be the third largest such iron deposit in the world (Gentelle, 1980; Shroder, 1983).

The high-altitude location and remoteness of the Hajigak site was a detriment to its potential for production, but the fact that it would all be downhill for the ore to an ore reduction or smelter site in some nearby valley would lend gravitational assistance to move the ore by conveyor belt, slurry pipeline, or even truck. The presence of the ore body some distance from population centers is a positive in the sense of greater protection from air or water pollution due to mining, and the local population of less-warlike Hazara people likely adds to the attractiveness of this ore body to exploitation. The current situation of the iron at Hajigak will be explored further in Chapter 12 where the especially noted, major resources of Afghanistan are to be discussed in more detail.

References

Abdullah, S.H., Chmyriov, V.M., Dronov, V.I., 1980. Geology and Mineral Resources of Afghanistan, vols. 2. Geological Survey of Afghanistan, Kabul.

Afzali, H., 1984. Les resources d'hydcarbures de metaux et de substances utiles de l'Afghanistan: aperç general. Chronique de la Recherché Miniere, Paris 460, 29–51.

Ahmad, S., 2012. Afghanistan to Pump Oil in Five Months: Official. Agence France-Presse (AFP). http://www.google.com/hostednews/afp/article/ALeqM5iPVPiHTy-OS6mUZQIKiWgSSfz7Wg?docId=CNG.4b617bb54804cd949714937750b605c4.2b1.

Ali, S., Shroder, J.F., 2011. Afghanistan's Mineral Fortune: Multinational Influence and Development in a Post-War Economy. Institute for Environmental Diplomacy and Security, James Jeffords Center for Policy Research, University of Vermont, Burlington, VT, 24 p.

Anonymous, circa 1961. Afghanistan: Ancient Land with Modern Ways. Ministry of Planning, Kabul, 199 p.

Anonymous, 1973. Geology and Mineral Resources of Afghanistan. Department of Geological Survey, Ministry of Mines and Industries of the Republic of Afghanistan, Kabul, 103 p.

Anonymous, 1979. Afghanistan steps up drive for oil production. The Oil and Gas Journal 77 (14), 64.

Anonymous, 2011. Cement Quality Limestone in Jabul Saraj Parwan Province, Afghanistan. Afghanistan Geological Survey (AGS) Investor Data Package No. 4; Minerals in Afghanistan; Excellent Exploration and Development Potential; 7 p. http://mom.gov.af/Content/files/Jabul_Seraj_Limestone.pdf.

Anonymous, 2013. Oil and Gas in Afghanistan – Overview. Mbendi, Information Services, Capetown, South Africa. http://www.mbendi.com/indy/oilg/as/af/p0005.htm.

Barfield, T., 2010. Afghanistan: A Cultural and Political History. Princeton University Press, Princeton, NJ, 389 p.

Carlisle, T., 2011. Afghan Oil and Gas Sector Opens. The National. http://www.thenational.ae/business/energy/afghan-oil-and-gas-sector-opens.

Cerkel, D., Miller, R.L., May 27, 1958. Petroleum Law for Afghanistan. Prepared under the Auspices of the Ministry of Mines and Indusries of Afghanistan, and the International Cooperation Administration of the USA (US Geological Survey), Kabul, Afghanistan, 49 p.

Coll, S., 2004. Ghost Wars. Penguin Press, NY, 695 p.

Dalton, R., 2007. Mine games. Nature 449, 968–971.

Ferris-Rottman, A., Venugopal, R., September 7, 2012. Exxon Explores 'Very Promising' Oil and Gas Fields in Afghanistan. Business Insider. http://www.businessinsider.com/this-energy-company-hopes-to-find-oil-in-afghanistan-2012-9?utm_source=feedburner&utm_medium=feed&utm_campaign=Feed%3A+clusterstock+%28ClusterStock%29.

Gentelle, P., July–August 1980. Du non-developpement et sous-developpement. Afghanistan. Les Tempes Modernes(408–409): 281–307.

Glicken, M., 1960. Making a map of Afghanistan. Photogrammetric Engineering 26, 743–745.

Hauner, M., Canfield, R.L., 1989. Afghanistan and the Soviet Union: Collision and Transformation. Westview Press, Boulder, CO, 219 p.

Huffman, B., September 24, 2012. Ancient Site Needs Saving Not Destroying. CNN Opinion, 4 pp. http://www.cnn.com/2012/09/22/opinion/afghanistan-buddha-site-mine/index.html.

Kavalsky, B.G., Borthwick, J., Hadad, W., Imam, H., Kundu, A., Meerman, J., Remy, C., Taylor, S., 1978. Afghanistan: The Journey to Economic Development. World Bank Document report no. 1777a-AF; I – The main report: 358 p.; II – Source material and statistics on the economy of Afghanistan: 225.

Kelly, M., September 26, 2012. This 2,600-Year-Old Buddhist Site Will Soon be Turned into a Chinese Copper Mine. Business Insider. http://www.businessinsider.com/mes-aynak-turned-into-copper-mine-2012-9?op=1#ixzz2HYT90giH.

Klett, T.R., Amirzada, A., Selah, A., Mury, S.A., Nakshband, H.G., Hosine, A., Wahab, Aminulah, A., Ulmishek, G.F., Wandrey, C.J., Agena, W.F., Taylor, D.T., Hill, R., Pribil, M., King, J.D., Pawlewicz, M.J., Barker, C.E., Ahlbrandt, T.S., Charpentier, R.R., Pollastro, R.M., Shenk, C.J., March 2006. Assessment of Undiscovered Petroleum Resources of Northern Afghanistan, 2006. US Geological Survey Fact Sheet 2006-3031; 2 p.

Mas'ud, D.K., Sakharov, A., Bar'yalai, M., Polyakov, G., Shevyrina, N., 1983. Afghanistan Today. Planeta Publishers, Moscow, 271 p.

Moritz, F.A., November 24, 1981. Soviets Face Afghan POW Dilemma. The Christian Science Monitor. http://www.csmonitor.com/1981/1124/112446.html.

Mossotti, V.G., 2011. Cement in Afghanistan. In: Peters, S.G., King, T.V.V., Mack, T.J., Chornack, M.P. (Eds.), Summaries of Important Areas for Mineral Investment and Production Opportunities of Nonfuel Minerals in Afghanistan, pp. 1243–1270 (Chapter 16B) http://pubs.usgs.gov/of/2011/1204/.

Najafizada, E., 2010. Afghanistan Says It Locates 1.8 Billion-Barrel Oilfield in Nation's North. Bloomberg News. http://www.bloomberg.com/news/2010-08-15/afghanistan-discovers-1-8-billion-barrel-oilfield-in-north-ministry-says.html.

Neilson, J.M., Gannon, P.J., 1976. Mineral Evaluation Project Afghanistan. United Nations Development Program unpublished report AFG/74/00Z 2 vols. United Nations Library, Kabul.

Noorzoy, S., 1987. Soviet economic interests and policies in Afghanistan. In: Klass, R. (Ed.), Afghanistan and the Soviet Union: The Great Game revisited. Freedom House Pub, New York, pp. 71–95.

Peter, T.A., December 28, 2011. China Wins $700 Million Afghan Oil and Gas Deal. Why Didn't the US Bid? The Christian Science Monitor. http://www.csmonitor.com/World/Asia-South-Central/2011/1228/China-wins-700-million-Afghan-oil-and-gas-deal.-Why-didn-t-the-US-bid.

SanFilipo, J., 2005. Assessing the Coal Resources of Afghanistan. US Geological Survey Fact Sheet 2005-3073; 3 p. http://escweb.wr.usgs.gov/share/mooney/USGS%20coal%20assessment.pdf.

Shareq, A., Chmyriov, W.M., Stazhilo-Aleekseev, K.F., Dronov, V.I., Gannon, P.J., Lubemov, B.K., Kafarskiy, A.Kh., Malyrov, E.P., Rossovskiy, L.N., 1977. Mineral Resources of Afghanistan, second ed. United Nations Development Program Project AFG/74/012; 419 p.

Shroder Jr, J.F., 1983. The U.S.S.R. and Afghanistan mineral resources. In: Agnew, A. (Ed.), International Mineral Resources a National Perspective. American Association for Advancement of Science, pp. 115–153.

Shroder, J., 2007. Afghanistan's development and functionality: renewing a collapsed state. Geojournal 70, 91–107.

Shroder Jr., J.F., Assifi, A.T., 1987. Afghan mineral resources and Soviet exploitation. In: Klass, R. (Ed.), Afghanistan and the Soviet Union: The Great Game Revisited. Freedom House Pub, New York, pp. 97–134.

Sweetwood, C.W., 1968. Afghanistan: Important Mineral Occurrences. Office of Mineral and Petroleum Attaché, American Embassy, Kabul, Map.

In this Chapter

12

Rich Resource Exploitations, Resource Curses, and Resource Wars

Abstract

Recognition that rich resources actually existed in Afghanistan had to await understandings developed in the US Department of State and Pentagon as they looked for an excuse to withdraw from their apparently unwinnable war. Ideas were also developed about resource blessings becoming instead a resource curse leading to future resource war. A new legal framework would also be needed to control resource extraction to protect the country. The chief resources of concern were the world-class Aynak copper deposit, the Hajigak iron, gold deposits, rare-earth deposits, chromite, and gemstones, all of which offer the potential to pull the country up economically by its own resource bootstraps, if only a reasonable peace could be maintained enough to accomplish the task in the midst of many poor government practices.

Keywords: Aynak copper; Chromite; Gemstones; Gold deposits; Hajigak iron; Rare-earth deposits; Rentier state; Resource curse; Resource war; Rich mineral resources.

The US Pentagon is not known particularly for its good scholarship, or for paying much attention to the arcana of academic disputes over resources in remote regions, but it is known rather for its highly effective, brute force methods to overwhelm opposition in such places. For over a quarter century since the demise of the Cold War, therefore, although a few agencies in the US government knew about the Soviet-backed discoveries of rich resources in Afghanistan (Ali and Shroder, 2011; Blackton, 2010; Shroder, 1981, 1983, 1989, 2007, 2012; Shroder and Assifi, 1987; Shroder and Watrel, 1992), the US military seems to have paid little attention to that fact. After invading Afghanistan in late 2001 with little substantive opposition, and then going off to Iraq on a misguided diversion from the White House from the real problem at hand as the Afghan opposition regrouped for counterattacks, the Pentagon finely woke up to the fact that after almost a decade into their longest war, the resources of

Natural Resources in Afghanistan. http://dx.doi.org/10.1016/B978-0-12-800135-6.00012-X

By Jimmy Margulies, The Record, Hackensack, N.J., North America Syndicate

Figure 12.1
Newly discovered mineral wealth in Afghanistan that had just been recognized by Pentagon in 2010 engendered this perceptive political cartoon by Jimmy Margulies in The Record, Hackensack, NJ.

Afghanistan might actually be rich enough to pull the country out of its malaise of endless war (Figures 12.1 and 12.2). The idea, of course, was that resource riches could provide the money to develop the country so well that war would no longer be necessary because the country potentially could take on the airs of Saudi Arabia and build such a robust economy from its resources that the many young gunmen would have new, more lucrative livelihoods otherwise. Too little, too late perhaps—it would have been nice if the Pentagon had paid a bit more attention earlier on to this prospect of rebuilding the Afghanistan economy by its own resource bootstraps; after all, the resource information had been out there decades earlier, if any of the warrior forces or their advisors in Washington had thought to pay attention. The alternative, of course, could also be that instead now that the US and International Security Assistance Force (ISAF) Coalition troops are going home in some discouragement, the Taliban opposition from Pakistan will seek the resource riches for their own perverted pocketbooks, and give even more impetus to the civil war that has long been waged.

The recent media hyperbole about resources in Afghanistan has been notable to many people since the story broke (Risen, 2010a, b; McNeil, 2010; Peter and Arnoldy, 2010; Tran, 2012; Kenny, 2010), partly because it is such old news that has repeatedly popped up since Drummond's story all the way back in 1841 when the British first invaded Afghanistan (Drummond, 1841), and then again and again,

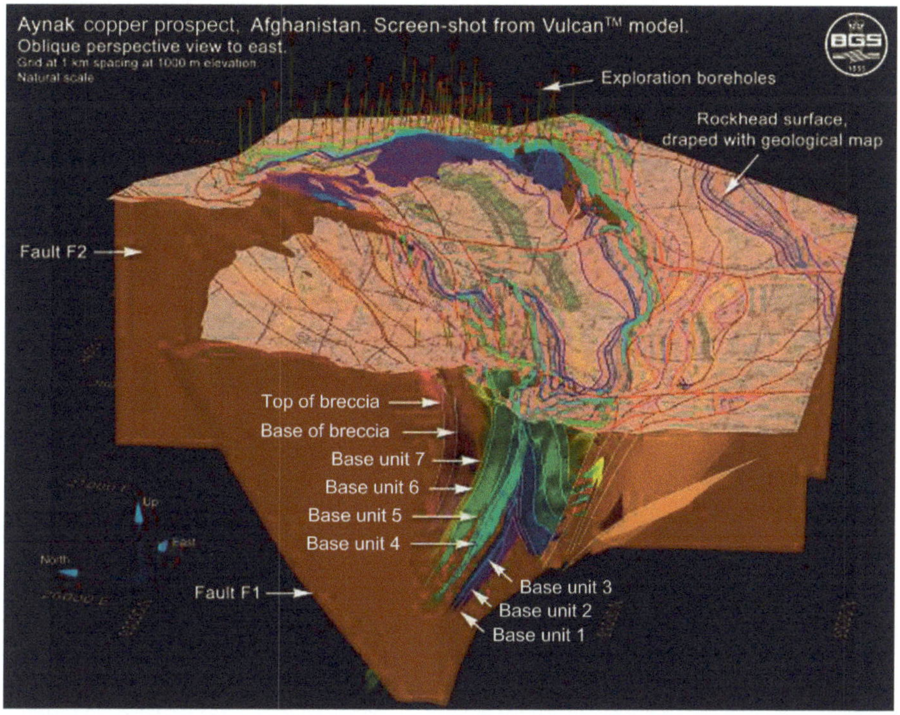

Figure 12.2 Screen shot of Aynak Copper Prospect from Vulcan™ computer model by British Geological Survey, based on 1-km deep boreholes of the ore body obtained in the 1980s. The construction of such a detailed subsurface configuration of the virtual ore body enabled determination of orebody size and tenor, which led to successful bids on a tender offer for the ore. *After British Geological Survey (2007).*

seemingly whenever a new conqueror discovered the facts and wanted to trumpet how great things were going to get for the Afghans once they finally came up out of their underdeveloped and overexploited mire and got to work, instead of playing at war. Yesterday it was throwing out the British or the Russian invaders, today it is the thrall of the radical Islamic vermin to kill the neocrusaders, and tomorrow it might be the rapacious multinational mining companies. Whatever the outcome, the resource-curse blood sport in Afghanistan is now loose, and likely to play out in many unpleasant ways unless solutions are sought and obtained.

Resource Curse

Otherwise known as the "paradox of plenty" or "curse of plenty", this seeming absurdity is a reflection of the fact that countries or regions with an abundance of natural resources, such as hydrocarbons or minerals, have a tendency for lower economic growth or underdevelopment than many others with no particular resources, other than the strong ambitions of their people. The resource-rich countries can be far more violent as well. Afghanistan might be a prime example of this phenomenon (McNeil, 2010; Tran, 2012; Kenny, 2010), although the reviews are mixed on whether or not the country can pull out of its dirt-poor malaise and use its resources to rise above the fray. Mixed cause and effect issues lead to questions about whether or not lack of development is due to historical forces and cultural idiosyncrasies, or perhaps it is because of domestic or foreign machinations to control resource acquisition.

Causes of the resource curse against growth and development are many faceted but can include the following, most of which can apply to Afghanistan:

1. International companies syphon off profits from resource riches.
2. Corrupt governments line their own pockets and do not help the people.
3. Brutal warlords control all mining and take the profits for themselves.
4. Illiterate populations are unaware of game-changing mining issues.
5. Long lead times for resources to come to markets lead to ancillary problems.
6. Excessive government borrowing is common in such situations.
7. Reduction in security and huge costs of policing wild spaces is common.
8. Increased violations of human rights occur.
9. Huge environmental costs develop.
10. Rampant price fluctuations and revenue volatility of mineral materials occur.
11. Great costs of production are a problem.
12. Shipping from remote locations incurs large costs.
13. Societal pressures are considerable for obtaining an immediate share in the profits.
14. It is a great necessity to get local people to buy into the mining.
15. Outside workers are commonly brought in to do the labor rather than the too unskilled local people.
16. Education is neglected because no need is seen for it.
17. Lack of diversification in the economy occurs, along with reduced long-term investment in the infrastructure.
18. Government depends upon resource revenues rather than taxes from citizens, which increases distance between rulers and ruled.
19. Oligarchies or rule by the few are enhanced for those with their hands on the resource wealth.
20. The "Dutch Disease" occurs whereby revenues from resource exports drive up wages and monetary exchange rates so that other tradable goods (agriculture, manufacturing) become more expensive and less competitive to foreign markets.

Closely related to the ideas about resource curses and resource wars, are the economic notions of being what is called a *rentier state*, in which a country derives a substantial part, or all of its national revenues from the sale of indigenous resources to external or foreign entities (Omeje, 2008). Four characteristics generally define whether or not a nation can be identified as a rentier state:

1. Rental situations of resources predominate.
2. The nation's economy relies on substantial rent payments from external sources, thereby not requiring strongly productive, domestic manufacturing, or agriculture sectors.

3. Only a small local labor force is actually used in the generation of the rental monies.
4. The national government is the main, or only recipient of the external rent.

Other characteristics of the rentier state include bloated and inefficient bureaucracies so that a rentier class of bureaucrats develops. Other nearby states may also try to obtain a share of the largesse as a secondary set of rents accrue through the establishment of transportation networks for the resource extraction, or foreign military bases to protect the resources and their transport to market. Rentier states may also attempt to manipulate global political and economic environments by establishing trading restrictions, monopolies, or foreign aid and subsidies in exchange for political influence.

Establishment of a country as a rentier state is generally not seen as a good thing. The country of Afghanistan already has many or most of the proclivities to fall right into becoming a reniter nation; indeed because of the huge transfers of Western capital into the country since the invasion by the US and ISAF Coalition in 2001, in some senses Afghanistan already has become a military type of rentier state. The current resource bonanza, at least until the departure of most of the Western troops by 2014, is likely to contribute more to the rentier mentality and actuality in the near future.

Escaping the Resource Curse

Many ideas pro and con concerning the so-called "resource curse" have been presented, but whether or not the bane exists, escaping its clutches would seem to be a good step in the right direction anyway (Butler, 2004; Humphreys et al., 2007). Four mechanisms to help in the evasion process at an international level, and eight in the domestic arenas can perhaps be applied to some effect in Afghanistan: *International Level*—(1) develop a mechanism for providing access to services of skilled negotiators; (2) develop third-party, natural-resource fund management service; (3) enforce stricter standards on multinational companies; and (4) create and maintain a global or international public information office on revenue management of oil, gas, and other big resources. *Domestic (Afghanistan) Level*—(1) condition the "deal" on likely future prices and other economic circumstances; (2) require transparency of negotiated outcomes; (3) use financial bonding to prevent environmental damage; (4) calculate national wealth correctly; (5) stabilize expenditures; (6) use earning for investment rather than consumption to avoid "Dutch Disease"; (7) take steps to avoid inequities; and (8) strengthen state–society linkages.

In a new approach to dealing with the resource curse (Diamond and Mosbacher, 2013), an idea about a resources-to-cash model has been developed that could bear fruit in Afghanistan if implemented carefully to control corruption, inflation, and other problems. The model works when a government receives large revenue from resource exports so that a certain predetermined proportion of it (ideally at least 50 percent) would be directly disbursed to all the citizens. At that point the government would treat those distributed revenues as income and tax, a certain proportion of it back to the government. This process is actually a clever way to create an active constituency of citizens who are directly affected by the government's management of their resources, instead of the commonly passive populations in corrupt, resource-cursed states characteristic of so many in Africa. In this single distinguished step, the resources-to-cash model would build a broad domestic tax base, which is certainly a fundamental part of any modern, well-governed state. Such methods have already been successful in some 60 developing countries, among which Botswana, India, Mexico, and Panama stand out, so it likely would be possible in Afghanistan to attempt it in well measured and much publicized steps. The likelihood of anything like this actually happening in a poorly managed place like Afghanistan, however, is unfortunately woefully small.

Resource War

In yet another angle on the problems with resource riches, the idea of resource wars has been around for a very long time (Klare, 2002; Renner, 2002). Wars for access to irrigated lands figure prominently in the Bible; more recently we have seen many number of disputes over oil, timber, coltan (columbite–tantalite), blood diamonds, and so forth. In many regions of the world competition over diminishing natural resources has formed the basis for tension and violence. The existing theory about resource war is that in the early decades of this new millennium, wars are, or will be, fought not so much over ideology as during the Cold War, but over access to dwindling supplies of precious natural commodities. The political divisions of the Cold War seem to have given way to a global scramble for oil, natural gas, certain high technology minerals, and especially the most vital resource of all—water. Furthermore as national armies throughout the world define resource security as a primary objective, widespread instability and guerilla resource actions are most likely to follow, especially in those areas where competition for essential materials overlaps with long-standing territorial, religious, and ethnic disputes. In fact in this view, some of the recent explosive conflict between the United States and Islamic extremism could stand revealed as a predictable consequence of a consumer nation seeking to protect vital resources upon which it and its allies depend. The idea of resource war as a driving force is a compelling, if not daunting, message for diplomats, political leaders, and military planners, especially when applied to the blood-sport situation of Afghanistan. In addition, the burgeoning efforts to exploit essential resources help explain some international relations of the present-day. Some have suggested that the world should prepare itself for another century of bloody conflict, especially as water supplies continue to dwindle.

The end of the Cold War diminished the importance of expansive global alliances and massive arsenals. Since then American policymakers have increasingly focused on global competitiveness and the importance of economic strength as vital components of national security. Similarly, other countries have shifted military assets and developed weapon programs designed specifically to protect access to resources considered essential to their national survival. The relentless expansions in resource demand worldwide, the emergence of significant resource shortages, and the proliferation of contests for ownership; all are causing mounting global division, friction, and eventual conflict. The concepts inherent in ideas of resource wars paint a picture in grand strokes of a world facing looming shortages, in some places perhaps sweeping boldly from region to region, pressing leaders to come to grips with the magnitude of the many problems. Increasingly, many people agree that responsible national leaders and military planners must consider resource issues as a critical component in global strategic planning. In this light, the resource and environmental situations in Afghanistan are perhaps exemplary of a certain type of resource war at hand (Brigaldino, 2010).

Alternatively, other scholars note that resource scarcity is in actuality more of a threat multiplier that does not plunge a nation into war on its own, but rather could exacerbate an already deteriorating situation such as prevails over so much of Afghanistan. Such threat multiplication can undermine the weak state, cause economic shock in the brittle economy, and when combined with climate change such as drought, cause mass migration away from the degraded environment (Evans, 2010).

Ideas of resource war have also attracted a certain amount of left- and right-wing rhetoric, conspiracy theorists, religious and end-of-days fanatics, and the lunatic fringe to perhaps plant their verbal minefields on the Internet with their lust for religious world war (Lendman, 2007; Tabb, 2013; Winnail, 2005). Some consider such discussions as a part of the threat industry to sell movies or books, that climate change is not really happening, and that no such thing as resource wars really happens or will happen. The "fracking revolution" in the West that releases hydrocarbons from tight shales seems to some to make the oil of the Middle East irrelevant, thus no resource war or "conflict of civilizations" is any longer likely. Others have seen vast political conspiracies afoot to control the resources of the world and make them available by force to the developed nations that "own" them. But at the end of the day, perhaps, environmentalists see the final resource war as that for the planet's very survival.

Resource Law

The resurgence in identifying and exploiting mineral riches in Afghanistan began in some quarters not long after the post-9/11 invasion and attempted pacification of Afghanistan when it was recognized that resource extraction was the only possible tool to lead to longer-term economic stability for the country (Shahrani, 2013; Anonymous, 2011a; Anonymous, 2013). This vision was to develop an economically vibrant mineral sector that would create jobs, develop infrastructure, generate domestic revenue, and ensure inclusive economic growth for the obvious benefit of all Afghans, because only this process was seen as having the real potential to reduce poverty and sustain long-term growth of the country. Avoidance of many aspects of the resource curse, however, depend almost entirely upon implementation of new legal frameworks for effective and binding social, fiscal, and environmental regimes, as well as the provision of training and capacity building in the relevant agencies and ministries responsible, and particularly the overall improvement of the security situation in the country to make the new laws effective (Anonymous, 2010a; Anonymous, 2011b).

To accomplish the anticipated immense and rapid growth in mineral exploitation, the Government of the Islamic Republic of Afghanistan (GIRoA) realized that the RGA- and Soviet-era procedures of ownership and exploitation of resources by the state were not economically efficient or even viable; they had never worked well previously. Consequently, the Ministry of Mines was restructured from owner/operator to a new and more appropriate role as a regulatory authority who would now promote private investment through the bid tender process whereby outside companies would be awarded exploration and mining contracts. This work, along the entire value chain for extractive industries includes: (1) the establishment of good governance policies, laws, regulations, and frameworks for the resources; (2) training and capacity building for key personnel; (3) restructuring relevant government institutions; and (4) development of resource corridors that will enhance the development of a much-needed and even critical infrastructure.

The new mining laws developed in Afghanistan in 2006 hark back to the early attempts in the late 1950s by the US Geological Survey to establish a means for private individuals to become involved in oil and gas exploration and financing in Afghanistan (Cerkel and Miller, 1958). This effort in report or book form was to help springboard Afghanistan into more modern method of operating with its oil and gas, but ultimately it was very USA-oriented and not terribly sensitive at that time to pro-Soviet or socialist/communist notions of how extractive economies should work in that particular milieu. The result in that world at that particular time, of course, was that the ideas in the book were completely rejected and the book disappeared, except for the used-book stalls in Kabul in the 1960s and 1970s.

As of 2006, however, new regulations were being developed to provide a legal framework for formal geological exploration and mining of minerals through application of mineral rights by private companies. Once obtained, the holder of the mineral rights in Afghanistan would be granted investment security so that the GIRoA could not expropriate the mineral rights without adequate compensation, in accordance with international norms. The new law also grants mineral royalty rates from 5 percent of gross revenue to as much as 10 percent for gemstones. Finally after many decades of clandestine artisanal mining of gemstones and other materials, the law undertook legalization of the ongoing gemstone trade through encouragement of private investment in the mining and government control of the gemstone industry overall.

The ongoing revision of the mining laws has continued for several years because it does not sit well in the old conservative quarters where people are highly nervous that the new rules would give foreign investors far too much certainty and control over development of the resources. The draft mining legislation in 2012, for example, was objected to by the Afghan cabinet as kowtowing to foreign mining interests eager to hijack Afghanistan's natural resources (Rosenberg, 2012). Ashraf Ghani, a senior advisor to President Karzai and new presidential hopeful himself, noted that, "A balance has to be struck so that we can make sure that our patrimony does not become a pot of porridge for others." Ghani, a former finance minister and now World Bank official and professor at Johns Hopkins University, is responsible for overseeing the transition from the largely Western-led reconstruction of Afghanistan to similar programs that are almost entirely under management by the GIRoA. He acknowledged that outside investments in resources and mining by many foreign companies were critical to the effort because their money and expertise would not be available otherwise. But under the original draft legislation, national interests in protecting the unsophisticated civilian population and the natural environment did not seem to be protected enough. Among the new issues with the legislation is to make sure that one-hundred percent of the unskilled staff must be local Afghans, along with a proportion of semiskilled Afghan workers. The infrastructure must be augmented and protected as well.

The reality of mining resources in Afghanistan for foreign companies is that it is seen as one of the riskiest ventures possible. Companies are worried about the lack of the rule of law and the lack of clarity about mining, even more than the security issues or corruption (Wang, 2012). But great care must be taken to ensure that in this uncertain environment, free-wheeling companies in search of a quick profit do not lead to a new source of conflict as has happened in so many other countries, particularly in Africa, for example. At the present time (late 2013 as of this writing) the situation with mining law in Afghanistan is still uncertain.

Overall in Afghanistan, short term profit-making already underway by companies is in oil and gas, as well as the new gold mines. Handled properly, these extraction endeavors will be a clear source of revenue gain for the government as the foreign aid winds down. Long-term investments include the world-class copper and iron, deposits that require the sinking of a huge capital outlay in infrastructure to get the endeavors up and running.

Aynak Copper

The events of the twentieth century concerning Aynak copper were discussed in Chapter 11; here we note the occurrences of the twenty-first century, especially following the invasion of Afghanistan by the Western Coalition following the 9/11 tradgedy. Because the US Geological Survey was immediately involved in the search for Osama bin Laden in the fall of 2001 (Shroder, 2005), their interest in the geology of Afghanistan was heightened, and shortly thereafter they were tasked with an effort to access all of the old geology reports in Moscow compiled by the Soviets. From these and other sources (Orris and Bliss, 2002) the cost of the size of the effort necessary to do all the work required was budgeted at $75 million, which the Bush-Chaney White House cut down to a paltry $5 million. Consequently, the British Geological Survey was brought in to assist on investigation of the Aynak copper prospect (Ali and Shroder, 2011; Anonymous, 2005a,b,c). With access to the high quality, Soviet-era data set of 150 boreholes which are ~1 km deep, 70 trenches, and 9 horizontal adit holes, the British geologists were able to construct first-rate computer models of the underground geometry of the ore body preparatory to planned eventual open-pit mining (Figure 12.2).

From these data a number of somewhat variable estimates of ore reserves and contained metal have been produced, but all agree that the deposit is a world-class ore body that may have an indicated reserve of 240 million metric tons (Mmt) at a grade of 2.3 percent of copper, which would total some 5.52 million metric tons (Mmt) of contained copper metal (Anonymous, 2012a). This is now recognized as Asia's largest deposit, which when its tender offer was put up for bidding in late 2006, attracted a wide variety of international attention. Offers came in from Australia, Canada, China, India, Kazakhstan, Russia, and the USA.

In what was described as a, "perfect storm of tender events", a former official consultant to the Ministry of Mines and Industry on the Aynak Tender, together with a distinguished list of six other contributors from the Ministry of Mines, the US Embassy, the US Geological Survey, and others, wrote a 75 page document outlining the severe problems with the tender award that was made to the Chinese consortium, China Metallurgical Group Corporation (MCC) (Yeager et al., 2009). With an initial bid submission of $1.8 billion, and a front end payment of $800 million to be paid over 5 years, this was the highest remittance ever made to any government in Afghanistan and was three times larger than any of the other bids, but was marred by charges of severe corruption leveled against Minister Ibrahim Adel, who was forced from office as a result (Partlow, 2009). Nevertheless, one of the attractions of the tender offer by MCC, which none of the others provided, was a parallel offer of a wide variety of ancillaries (Synovitz, 2007), such as:

- onsite copper smelter;
- coal mine;
- 400-MW, coal-fired power plant, also to augment Kabul electric supply;
- protective ground water system;
- roads, homes, schools, hospitals;
- new freight railroad from western China, through Tajikistan to Afghanistan and Pakistan;

- help for Kabul University;
- thousands of new jobs;
- $350–400 million in taxes each year to GIRoA.

At the present time (2012–2013) the GIRoA has provided 1700 police officers inside the wire to provide security to Aynak, where otherwise the US Army 10th Mountain Division had been operating outside around the region against the insurgents. Future security directions are not clear, but the awarding of the bid to China may show Afghanistan's desire to diversify its international friendships, and particularly to engage with a partner who has considerable influence with the problem neighbor, Pakistan. Their relationship with China is highly valued by Pakistan, which means that the transportation, industrial, and resource extraction infrastructure brought by the Chinese may not be attacked as much by the Taliban (Begum, 2011).

On the face of it, this Chinese bid sounds highly beneficial to Afghanistan, but a certain amount of distaste for the deal has also emerged. For example, some 5000–10,000 workers are expected to be used but although Afghanistan expects them to be hired locally, the Chinese government has specified that Chinese projects must use Chinese labor, in the same fashion that the US Congress has controlled use of American products in its operations and American airplanes for the workers on its projects (Metz, 2007). This will be a difficult problem for the Chinese to rectify, as well as who exactly provides the security for their mining base buildings and living quarters, once the mine opens. Chinese troops have not been mentioned as of January 18, 2013, nor has the labor problem. It has been noted that from the mercantilest point of view, someone must lose in these transactions, but the ISAF Coalition may not want to provide strong security or cooperate with mining-company security particularly, and if the mine proves as profitable as some think that it might, then the GIRoA may consider itself overexploited (Metz, 2007). Aynak looks, however, as a boon to the local and regional economy in a very barren area that is one claim to fame might be the archaeological ruins of Buddhist relics, temples, and monasteries on it. Nonetheless, with salvage archaeology, perhaps special protection or moving the entire ruin structures, as was done at Abu Simbel in Egypt, might be possible, because the Aynak copper prospect looks like a very good thing to help the battered nation. China must be lauded for linking its great power to the success of the government of Afghanistan, as well as for taking on a project of this magnitude in an environment where success is not at all that certain.

For a project of this magnitude, however, essential problems remain, which have been detailed in a report by a nongovernmental watchdog agency in London, Global Witness, that investigates the links between natural resources, conflict and corruption, and associated environmental and human-rights abuses. The Afghanistan Ministry of Mines replied to this report on its website and then Global Witness responded to the reply (Anonymous, 2012b,c,d). In all, the online dialogue and mild confrontation concerning transparency, renegotiation and recommendations, community engagement, contract comparisons, and ongoing dialogue with civil society is an exceptionally constructive and open discussion that lends far more credibility to the actions of the GIRoA. For example, Minister Wahidullah Shahrani has helped his ministry to establish strong community consultation and grievance redressal procedures, updating the sector regulatory framework for project monitoring, implementing an aggressive capacity development program for contract management, attending to the national cultural heritage issues with the Buddhist archaeological remains, as well as other reforms not detailed. Government legislation has been introduced to set new standards of transparency, public access to information, anti-corruption efforts, and new rules of conduct. Significant development such as this Aynak project in Afghanistan is almost always a good thing and the news on Aynak keeps getting better. All in all, the perfect storm of a few years ago seems to have been calmed by rational actions that provide considerable hope for a country that certainly could use a few rays of sunshine.

In August 2012, however, a rocket attack on Aynak was unleashed by the Taliban. This immediately slowed production at the mine site, but because the Chinese company does not have to pay royalties to the GIRoA until real mining production actually starts, the attack was not viewed as much of a problem. Thus the slowing of development of the mine does make sense until the post-2014 withdrawal of Western Coalition troops will make the future possibilities more clear. In the meantime, China can afford to wait patiently because the ultimate payoff is likely to be rather exceptional.

Hajigak Iron

The Hajigak iron deposit has been known since at least the nineteenth century when local Afghans and British geologists first reported its existence to Amir Abdur Rhaman Khan. The deposit was subsequently assessed by a variety of French, German, Russian, and other geologists, including this author in 1978 (Figure 12.3), which deposit eventually was discovered to be spread out over some 32 km in 16 different ore bodies on the ridge tops between Bamiyan, Parwan, and Wardak Provinces (Anonymous, 2012e; McDonnell, 2006; Nissenbaum, 2011). In spite of the deposit's high grade (~60 percent iron) and huge observable tonnages (1800 Mt) of ore, its remote location, lack of any industrial transport system, and high altitude (c. 3800–4200 m) in the Hindu Kush mountains, the deposit at that time many years ago seemed to have little hope of ever being mining anytime soon. But the decades do pass, and although the remoteness, poor dirt roads, and altitude have not changed much in the interim, the political climate has, and in 2011 the GIRoA awarded a group of companies from India (SAIL or Steel Authority of India, Ltd, National Mineral Development Co., Ltd) and Canada (Kilo Goldmines, Ltd)

Figure 12.3 Photograph from Google Earth™ of northern-most outcrops of Hajigak iron deposit by M. AliStani, taken from about 700 m north of original Soviet-era exploration trenches that are still visible on the ground some half century later after they were constructed (See Figure 11.14). Note modern prospect road construction in the photograph.

with a $6 billion contract to mine the now recognized world-class, Hajigak iron deposit. The deposit is recently estimated to be worth some $350 billion and the deal promised to bring in a steel plant using coking coal from nearby Shabashak in the Daryae Suf district, and 200 km of road, railway, and power lines estimated to cost an additional $10.8 billion. The Hajigak deal is estimated to take at least five years to get up and running (Anonymous, 2012e; McDonnell, 2006; Nissenbaum, 2011). In addition, the fact that another big power in Asia is now forging stronger relations with Afghanistan, as China has been doing at Aynak and elsewhere in the country, can only add to the gravity of damping down the violence and getting the country back on a development track.

Recognition of the huge problem in getting the heavy ore or the refined iron and steel out of Afghanistan to markets in India and elsewhere has led the Indian companies to explore transportation options. Other Indian companies have already been building a 216-km-long road and a dam in western Afghanistan, as Iran is bringing in a railroad to Herat, so one option is to ship ore or refined product out through new linages to the Iranian port of Chahbahar on the Arabian Sea. A better and shorter option is to truck the iron through Pakistan or pump it by slurry pipeline through Pakistan, in spite of the animosity between the two countries. But Pakistani business leaders have urged this course, arguing to the country's military and political leaders that this is one of the only way that their economy can climb out of its doldrums by doing business with the giant Indian economy to the east (Gulati, 2012; Peter, 2012;

Topf, 2012). The transit fees would be a huge boon if the Pakistani government would bow to the inevitability of the competitive necessities.

As a result of the initially unfortunate handling of the Aynak tender offer by the former head of the Ministry of Mines, Ibrahim Adel, discussed in Chapter 11, several other attempts have been made to review and monitor the progress and circumstances of the Hajigak tender. An independent, International Advisory Panel (IAP) on resources development in Afghanistan was created by GIRoA to monitor and render opinions in respect to the procedures and processes used to negotiate the sale and development of natural resources, particularly oil, gas, and metallic minerals, as well as to judge the "fairness" of the terms and conditions of all such agreements and contracts. The transparency of the process was to be paramount. In another effort, a completely independent agency, Integrity Watch Afghanistan (IWA), was set up to publish high-quality research and policy papers, so it designed a program whereby peer-review policy and guidelines could be used to study issues such as resource tender issues and provide reports. This led to an entirely independent assessment of the Hajigak tender process, entirely outside of GIRoA purview (Brice et al., 2012; Stephenson and Safi, 2012; Noorani, 2011; Anonymous, 2010b).

The IAP report, done by Westerners, took issue mainly with an attempt by a small American company Acatco run by an Afghan expatriate to unfairly influence the Bid Evaluation Committee, which the IAP members felt should be referred to the Attorney General. Admittedly Acatco was offering to pay 3–5 more than other bidders for the percent of the per-ton price of the iron and steel, but the IAP held the view that such companies created solely for the purpose of bidding a particular project generally lack operational skills in exploration, mine development, mine operations, steel making, and operation of the infrastructure. They felt that iron ore mines and steel production are enormous industrial undertakings that cannot be designed, built, financed, and put into operation without many years of experience; it is not a business for amateurs and such people should not have been allowed to bid in the first place. The committee made a number of other somewhat technical recommendations, but among the most beneficial, perhaps, was a recommendation for any surplus water developed by the licensees to be shared with the local communities. The most important thing about this monitoring activity is its relative openness and presumable accountability.

The IWA report, which was done by an Afghan who was minimally advised and edited by the author, pointed out that the Afghan government is still not prepared enough to regulate activities associated with large-scale mining. Because the national land ownership and cadastral survey is still in a state of flux after the decadal excursions each into communism and the chaos of the Taliban years (Batson, 2009), the inspectorate department is still not competent to carry out inspections, and the Ministry of Mines (MoM) does not have a provincial presence yet to oversee mining. Although the MoM is doing its best to train people and revamp itself, the considerable number of high-salaried foreign experts who have come and gone are a source of major confusion and puzzlement to the people in the ministries and outside. The MoM has tried to improve the processes, especially at the top, and in English, but it also needs a clear strategy of communication in local languages and outreach to make its transparency initiatives available to the outlying communities and villagers exposed to its influences and impacts. Heightened levels of political contest are likely to erupt as the warlords and other hangers-on jockey for position and attempt to control some of the wealth. Hajigak has brought big resource players at the national level to lock horns and attempt to control the situation, while the corruptive decision making ultimately rests with somewhat unsophisticated and not very wealthy bureaucrats who may be financially, ethnically, legally, or otherwise vulnerable, or subject to intimidation. The only solution is transparency to make sure that the institutions representing the Afghan people have high integrity and resist fiscal temptation. Although Afghan national or ethnic pride is very high on issues like this, Afghans commonly do not understand why the external world holds a rather doubtful opinion that the GIRoA will be able to exhibit very much oversight concerning the slipshod or simply rather wretched civil and environmental control that is so common in almost all developing or rentier countries.

The IWA papers have the potential to be a game changer in the matter of extraction of Afghanistan resources because it brings essential issues out into the open for all to see. Several foreign advisory firms of internationally known, high integrity have been retained to help manage the bidding processes and the subsequent mining, so that the process will be more effective. Although the people of Afghanistan are understandably generally rather opposed to giving up any of their much valued national oversight control to outsider companies, this is probably the only processes that will produce external credibility to potential investors because much of GIRoA is still too incompetent or corrupt to be trusted very far. As a result, the internationally known SRK Mining Consulting that supervises mining sites all over the world (Anonymous, n.d.a), and the leading global law firm of Mayer Brown, and Canadian law firm of Heenan Blaikie, are generally all reputed to have high moral rectitude, honest dealings, and great skill in the mining world so that they can be trusted at the outset to make mining and environmental restoration in Afghanistan. Illumination of the murky mining world anywhere is always a good thing, and these foreign firms are paid very well to look after details like this and so should be held accountable to ensure success in this very difficult political, cultural, and war-torn environment.

Gold

The rare metal gold (Au) has been a favorite of humanity for millennia and needs little introduction, other than the fact that apparently due to global economic uncertainties, its price has risen to unprecedented levels of almost

Figure 12.4 Gold mine in Nor Aaba, Takhar Province, northern Afghanistan. *After Croucher and Sobhani (2011).*

$2000 per troy ounce (~31 g) in recent years. A metric ton (Mt) of gold, which is a common measure of a viable mine's holdings, is thus worth well in excess of $56 million dollars. These high prices make mining speculations in uncertain places such as Afghanistan, possibly quite risky, but potentially extremely lucrative. The idea is to get in to the location, start mining as soon as possible, and be prepared to get out if the times become too unstable, too fast. Gold can even be transported out by aircraft, which adds to its value as concentrated wealth in small volume, albeit great weight. Accordingly, a number of gold mining sites have been designated in Afghanistan, tenders released, and contracts signed. These are the Samti placer and Nor Aaba mine (Figure 12.4), both in Takhar Province in northern Afghanistan abutting the country of Tajikistan, the nearby Badakhshan gold mine in the province of the same name, the Qara Zaghan mine in Baghlan Province, directly south of Takhar Province, and the Zarkashan copper and gold prospect in Ghazni Province southwest of Kabul (Bandler, 2011; Kavanagh, 2012; Jhanmal, 2012; Riechmann, 2010; Williams, 2012). The total amount of the known gold deposits in Afghanistan is thought to be well over 57 metric tons (Mt), which equates to over US $3 billion, certainly a substantial amount which could lift part of Afghanistan by its bootstraps if it were allowed to transpire on this resource alone.

Samti Placer

The most important gold placer in Afghanistan is that of the Samti deposit in Takhar Province west of Badakhshan Province. The Samti deposit (37°34′36″N - 49°49′–69°54′E) is an alluvial placer about 8 km long and 900–1700 m wide. The placer comprises two beds: a lower bed of 25–45 m thickness consisting of pebble-boulder material; the upper bed is 5–20 m thick and is made of sandy loam material. The gold concentrations are thought to be derived from a combination of primary sources and remobilization of the gold in paleo-conglomerates. The characteristic feature of this placer deposit is its spotty pay streak, both in plan and section, with the highest gold content being in the placer's middle segment. Lithologically, the gold-bearing formations are identical with the overlying barren deposits; their upper boundary can only be defined by sampling. Within the placer, there are three zones of possible commercial interest. The Samti Placer carries yellow, dark, and light yellow gold particles with predominantly well-rounded grains. The Samti deposit may contain between 20 and 25 Mt of gold of a high fineness but the ~20 m thickness of the overburden may seriously limit its economic potential. The placer gold mineralization was discovered as two gold-bearing horizons in an alluvial deposit confined to the floodplain and the first terrace of the Panj River.

Nor Aaba Gold Mine

The Nor Aaba gold deposit was eroded out of volcanic rocks millions of years ago and deposited into sediments in Takhar Province near the border with Tajikistan. Reserves are estimated at 24 Mt of gold. The mine was started by Westland General Trading Ltd of the United Arab Emirates, which is owned by an ethnic Uzbek named Haji Abdul Kabir. He has invested about $3.5 million to get the mine started and it was expected to provide $4 million to $5 million per year in royalties to the Government. Unfortunately, several years ago the mining was blockaded by the local people who had been promised work, roads, schools, and clinics, but too many outsiders had been brought to work the deposit instead.

Badakhshan Gold Mine

The Vekadur gold-silver deposit in Badakhshan Province was indicated by the US Geological Survey as having gold-vein mineralization with a podiform orebody, but not much other information was available.

Qara Zaghan Mine

Afghan Krystal Natural Resources Co., which is a local Afghan consortium with an investment of $50 million, is backed by foreign investors from Indonesia, Turkey, the United Kingdom, and the United States, who planned to develop a gold mining project at Qara Zaghan in Baghlan Province, even though the gold reserves were not yet known. Exploration activities were expected to intensify and production to begin in 2013, with the Afghan government earning 26 percent in royalties and taxes from the mine. The mining and mining agreement was facilitated by the well-known banking company, JP Morgan Chase of the United States.

Zarkashan Copper and Gold Prospect

The Zarkashan copper and gold prospect (Figure 12.5(A–C)) is located about 225 km southwest of Kabul just north of Highway A1 in Ghazni Province, with Ghazni city being about 93 km northeast of the project area. In the mid-1960s the German Geological survey conducted a reconnaissance survey there, and exploration in the late 1960s and early 1970s was done by the Soviets and Afghanistan Geological Survey (AGS). More recently the US Geological Survey compiled the historic German and Soviet mapping, geophysics data, drilling and sampling data, and incorporated all into a geographic information system database. Many ancient open-pit and underground workings there show that this area was mined in antiquity. Ore grades obtained by the Soviet and AGS exploration teams show that the skarn mineralization occurred in pockets or as a sheet-like deposit and has a core of relatively high gold grades with a halo of lower grades around the core. The Zarkashan deposit has been contracted for exploration and mining rights by a 50/50 partnership between the Sterling Mining Co. of Kabul and the Belhasa Co. of Dubai. The budget forecast for the site has been calculated at about $64 million for proving out the estimated approximately 7.7 Mt of gold thought to be occurring in the skarn system.

Rare Metals and Rare Earths

Rare metals in Afghanistan include many that are used in a variety of highly desired technological applications in the developed West; far less so in the largely rural and technologically starved countryside of Afghanistan. The rare metal deposits of the country occur in four main deposit types: pegmatites, extinct carbonatite volcanos, mineralized springs, and playas or seasonally inundated, dry lake basins.

The pegmatites are light-colored, quite coarse grained, intrusive igneous rocks that formed originally near the margins of a molten magma chamber in the last phases of cooling and solidification. This so-called "mother liquor" in the old-fashioned parlance to explain to nongeologists, is injected outward into surrounding rocks where it commonly forms dikes or irregular shapes that "squirt" into cracks forced open by the magma pressures. The pegmatites commonly contain quite rare minerals, as well as beautifully colored but clear gemstones (kunzite, tourmaline, topaz, aquamarine) that do not occur elsewhere in the igneous rocks. The most important pegmatite fields occur in the provinces of Laghman, Nangarhar, Kunar, Badakhshan, and Uruzgan, with beryllium, lithium, tantalam and niobium being the most plentiful rare metals that occur. Some cesium, rubidium, and tin also are there.

Carbonatites are an unusual eruptive habit for a volcano, as they commonly can contain suites of rare-earth minerals. Contrary to the name, the elements in

Figure 12.5 (A), (B), and (C) Zarkhashan copper—gold prospect near Ghazni.

these volcanic minerals are neither rare nor earths, but rather these are metallic elements that all but one are more common than gold, silver, mercury, or tungsten. They are also essential components in many high technology products, including electronics, metallurgy, cell phones, wind turbines, batteries, and all sorts of other essential industries. The Khanneshin carbonatite complex in the Registan desert area of lower Helmand Province (Figure 12.6) is now known to host as much as 1 Mt of the light rare earth elements (LREE), which include cerium, lanthanum, neodymium, and praseodymium. The deposit also contains significant quantities of barium, phosphorus, strontium, and uranium (Figure 12.7(A,B)).

Figure 12.6 Oblique aerial view to south-southeast (SSE) over eroded Kohi Khanneshin volcanic area. An area of carbonatite lava flow-rocks occurs the left (east), which overlie a lighter-colored explosive massive fragmental carbonatite tuff. *After M. Antonio; Google Earth™.*

Originally identified by Soviet geologists in the 1970s, reports on the unusual Khanneshin rare earth complex were difficult to obtain or suppressed, but because of the uranium were monitored with satellites for the past four decades through worries that Soviet, Pakistanis, or Iranians would all try to exploit the deposit at some point. Now the rare earths themselves, rather than any of the radioactive elements, provide the greatest real value as highly strategic metals, particularly because the deposit is comparable in grade to the all-important and world-class LREE deposits of Mountain Pass in California or Bayan Obo in China. Inasmuch as China controls 97 percent of world production of rare earths, their importance is obvious. Because of the strategic importance of the rare earths in Afghanistan then, geologists of the US Geological Survey mounted helicopter field trips to the deposit in 2011 surrounded by US Marine guards (Figure 12.8). Because of the high expense with all of the military transport and troops, this little excursion was jokingly reputed to be the second-most expensive such geological field trip ever undertaken, short of going to the Moon! The reason

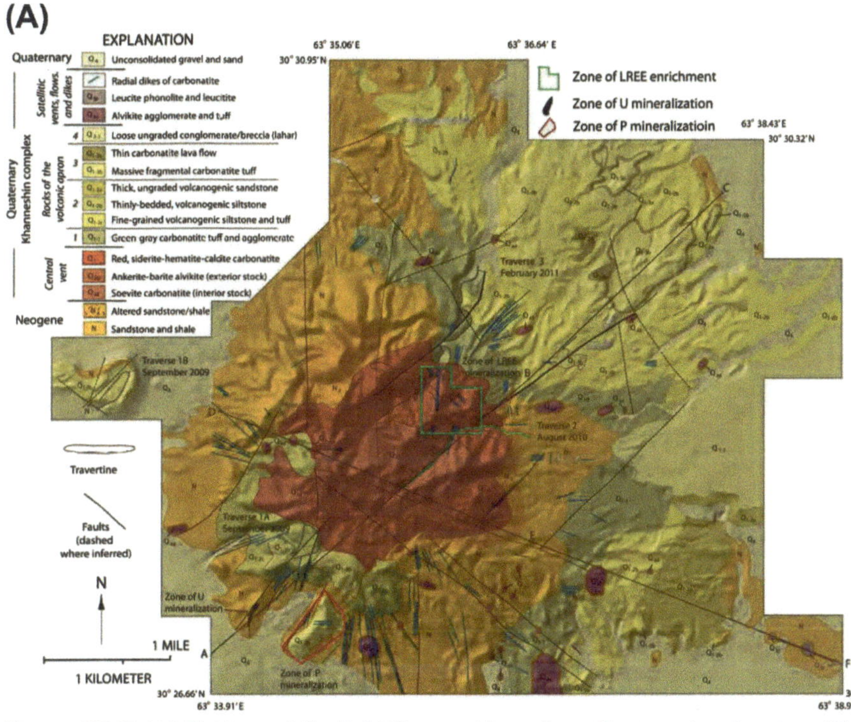

Figure 12.7 (A&B) Maps of the Kohi Khanneshin carbonatite complex showing: (A) the principal divisions of bedrock, as well as the regions rare earth elements (REE), uranium (U), and phosphorus (P) enrichment; and (B) aeromagnetic map of south Afghanistan (Sweeny et al., 2006) showing the permissive tract of U-P-REE mineralization (cb01) in the Helmand. The numbers one to eight refer to other possible carbonatite centers within the tract that show as similar aeromagnetic values. *After Tucker et al. (2011).*

Continued

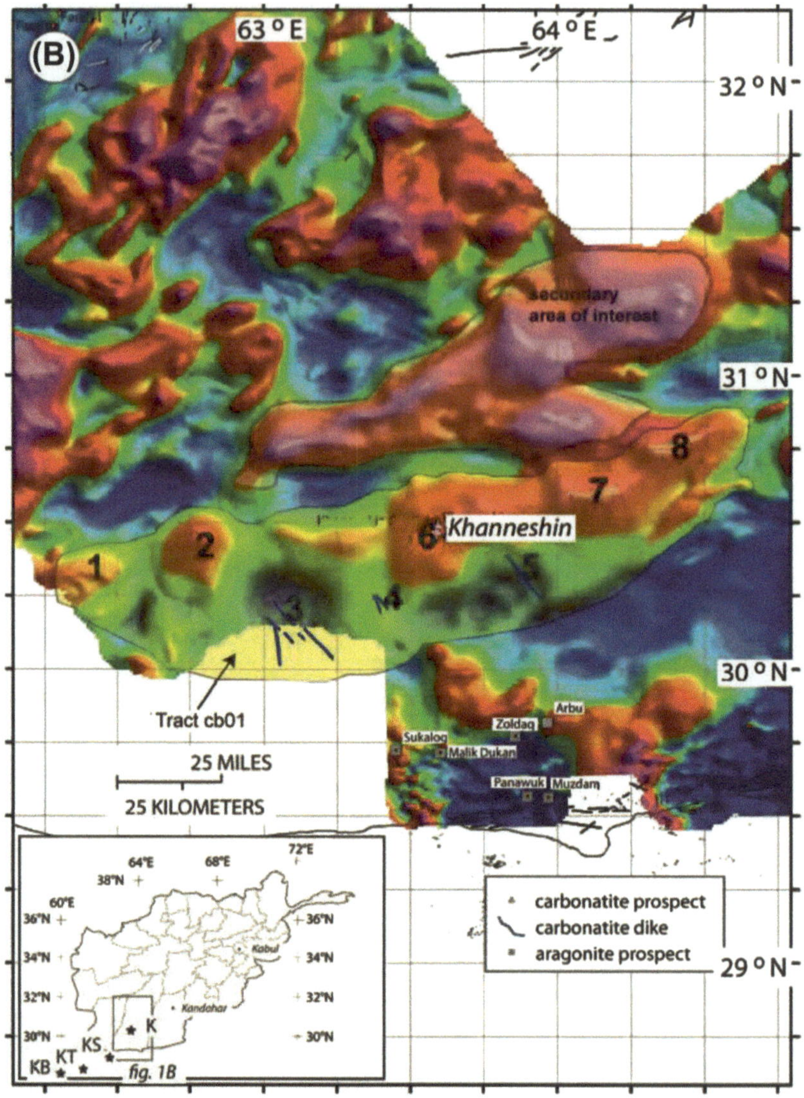

Figure 12.7—cont'd

for such an expensive effort is that the deposit is now able to be judged at being worth about $89 billion dollars, more than enough to capture the world's attention when it comes to market.

The carbonated mineralized springs elsewhere in Afghanistan are of less importance, although their output of lithium, rubidium, cesium, boron, germanium are potentially of commercial value, and they also are enriched in beryllium,

Figure 12.8 Photograph of steep and rugged eroded volcanic terrain of the Kohi Khanneshin carbonatite complex that required extensive protection by US military personnel to keep the Taliban insurgency at bay. *Photograph by R.D. Tucker, USGS and presented in Simpson (2011).*

arsenic, and iron. Several springs near the mouth of the Namakab (salt-water river) River in the Qala-Gorband-Turkman area northwest of Kabul may be attractive targets for mining later.

The rare metal playa-lake deposits of Afghanistan are places where long-term erosion of surrounding rocks has brought metal salts of lithium and boron down to the basins dissolved in the waters, where though later evaporation, the salts are concentrated into potentially commercial deposits. Numerous such salt-laden dry lakes occur around the country in the provinces of Herat, Takhar, Ghazni, Nimroz, and others. In fact the US military at one point was touting Afghanistan as the "Saudi Arabia of lithium" (Anonymous, n.d.b,c; Medlin and Demas, 2011; Risen, 2010b; Tucker et al., 2011).

Chromite

Chromium from ores of chromite is used to make essential stainless steel, tool steel, armor-piercing projectiles, and is used in other high-temperature applications. It is a strategic metal to the USA, because since 1961, the U.S has been 100 percent dependent upon other nations for chromium. The presence of numerous podiform or smaller lenticular bodies of chromite ore occur in Afghanistan south of Kabul in the Dadukhel area of Logar Province, the Deh Yak district of Ghazni, Jugati in Parwan Province, Sperkay and Shandal in Paktia Province, in the Tanai, Jaji, and Mangal districts of Khost Province, and apparently also new reports are out of its occurrence in the Kunar area.

Discovered originally by geologists from Germany and the US Bureau of mines in the 1950s, exploration of some of the chromite deposits continued by various people up through the 1970s when the Soviet geologists took over

(Benham et al., 2009; Anonymous, 2000; Anonymous, n.d.d; Bowley, 2012; Dupee, 2012). Modern assessments have been performed by British, Slovakian, and Afghan geologists. Emplacement of the unusual ores is now understood to have been as original sea-floor, mafic-rich, ophiolite suites that were obducted (thrust) up onto the older Kabul tectonic block and other areas during the Himalayan orogeny.

Most modern mining of chromite for steel hardening in the developed world is from massive ore bodies that are amenable to large open-pit or concentrated underground methods, all of which require massive mining and milling methodologies. In South Asia, however, where such large bodies of chromite do not occur, such as those in southern Africa and elsewhere, for example, the small podiform chromite bodies of South Asia have not attracted much attention from multinational mining companies because of the relatively miniscule sizes in some places size of the individual bodies. Instead, small artisanal, "pick-and-shovel" mines have been developed in many places in Afghanistan. In fact, not long after American soldiers had been deployed to Afghanistan following the 9/11 events and the invasion of the country by the Coalition forces, a former student of the author's, then employed as an improvised explosive device (IED) specialist, emailed the author about such mining. The author suggested possible armor-piercing IED devices for use against Coalition troops, but subsequent intelligence revealed instead that the chromite was being smuggled across the border into Pakistan where it was sold to Chinese investors and transported by truck up the strategic Karakoram Highway and over the Kunjerab Pass (4693 m) to China for use in their steel making instead.

In fact, subsequent elucidation of the *kachakbarari* (smuggling) networks for this and other commodities has revealed that long-established, criminal mining syndicates control the chromite and other cross-border trafficking (Figure 12.9). Despite the presence of 300 armed security guards tasked with securing the

Figure 12.9 Pit left by artisanal chromite mining near Kabul.

mines in the Khost area alone, estimates are that over 6000–7000 Mt of chromite ore production goes out amounting to over $400,000, with lost revenues to the government each day of more than $20,000. Linkages between the criminal mining syndicates and the Haqqani terrorist networks have also developed.

This artisanal mining of the small, or podiform chromite deposits of Afghanistan has recently come under the scrutiny of a watchdog group in Afghanistan to ensure that both the letter of the mining law is being followed, as well as its spirit (Integrity Watch Afghanistan, 2013). Their work to uncover the current facts is commendable, particularly because their actions are so fraught with danger, inasmuch as many of the existent mining participants do not appreciate any criticism at all and may take offense violently. These problems are compounded by mistakes made by the US DOD's Task Force for Business Stability Operation (TFBSO), who also commonly do not much appreciate, or even understand the many cultural/ethnic variations and subtleties in Afghanistan, so that many problems can emerge.

The TFBSO has been assigned the task for the past 3–4 years or so of jumpstarting the Afghanistan mining industry, and in the process in some cases they seem to have fallen in with some people who want to take advantage of weak to no enforcement of mining laws. The result in some cases, as has apparently happened in the Kunar region, has been artisanal mining with little to no governmental supervision or revenue generation. This has been underway in the region for decades with gemstones, and the mining of podiform chromite is now the latest in this sort of activity. For example, the many pods of pegmatites in the profuse granitoid intrusive rocks of Nuristan have long been the main host to diverse gemstones that have been mined without government control for over half a century, so the mineral extraction activity is nothing new; just the fact that it now appears to be chromite that is receiving the attention is new.

The issue at hand with the chromite is that in the Khas Kunar area close to the border with Pakistan the heterogeneous populations of diverse ethnicities have apparently made enough of an accommodation with each other to collaborate in mineral extraction. The Mohmand clan of Pashtuns, however, have laid unofficial claim to most of the chromite deposits in the Khas Kunar region and they have hired local people at 300 Afghanis per day (~$5.27) to crush the ore into gravel-sized pieces, which are then put into sacks for transport. It was reported that some 60–70 tons of the ore per day were smuggled out of the area into Pakistan. Some of the ore was reportedly sold to Chinese traders; other reports were of a Swedish buyer.

Several of the problems associated with this practice of chromite extraction are that the Afghanistan Local Police (ALP) were illegally involved in the process, that the TFBSO had provided training for mineral resource processing and about $13,000 to the police for an ore crusher. A problem in the region, however, is that the ALP are considered by many to be little more than thugs

Table 12.1 Recommendations by Integrity Watch Afghanistan (2013) to reduce or eliminate problems caused by extraction of chromite ore in the Kunar area of eastern Afghanistan

1. Mineral resources are a complex natural gift that should not be turned into a military tool to defeat insurgents.
2. All illegal mining must cease across the country.
3. The GIRoA should investigate all cases of illegal mining, set stringent rules to discourage such, and administer justice in all such cases.
4. The GIRoA must quickly establish a multistructured commission to monitor illegal mining activities quickly and act upon the commission's recommendations.
5. A temporary cessation of illegal chromite mining should be implemented so that the strategic resource can be gradually developed in future into a highly important source of GIRoA revenue.
6. The Wolesi Jirga (Lower House of the Afghanistan Parliament) should create an oversight commission to monitor the mining sector on a quarterly basis for revenues, oversee progress on mining projects, establish oversight over state institutions involved with the mining sector, and publish their reports.
7. Official state mineral law must ensure that mining contracts are not awarded to agents of the government, or to those based upon political connections. Concessions must be awarded only on the basis of, and with a direct reference point to the law.
8. The mineral law and state policy must promote open tenders, and transparent and competitive bidding for small, medium, and large mines.
9. The Afghanistan Local Police should not engage in resource extraction activities but instead should focus on their security mandate.
10. The International Military Coalition should refrain from interfering in sovereign decisions of the Afghanistan State, yet at the same time continue to build governmental capacity, and provide technical assistance to it in the mining sector.

guilty of all sorts of illegal and extra-judicial coercion and nefarious human-rights abuses (Integrity Watch Afghanistan, 2013). The result has been that an important arm of the US DOD and the ISAF forces, the TFBSO, seems to have been co-opted to assist in what they admitted were illegal activities, probably because they thought that the ends were more important than the means. Recognition of these problems in this case has led Integrity Watch Afghanistan (2013) to make a series of recommendations about this chromite mining (Table 12.1), which could be applied far more widely to many other parts of the problematic mining economy of Afghanistan.

Gemstones

Most people have a certain fondness for gemstones, if only because of the passions of royalty to possess them, but especially because of their reputed beauty, rareness, and possibly great value. In South Asia, gemstones also carry considerable worth as a source of quite practical portable wealth in times of political upheaval, something Afghanistan has had more than its share of in many past and present centuries. Fortunately for those so inclined, plentiful gemstone mining locations occur in the country, particularly in the northeast. In fact, archaeological and other evidence from Egypt, throughout many of the countries of the Middle East, southern Europe, and South Asia show that Afghanistan and gemstones were inextricably linked for more than 6500 years of transport and trade. Semiprecious lapis lazuli (*lazward*) from the mines along the right bank of the Kokcha River in southern Badakhshan has even been found in the tombs of pharonic Egypt (Figure 12.10(A–C)).

The characteristic precious stones most desired from Afghanistan include emerald, tourmaline, aquamarine, kunzite, topaz, garnets, and even fluorite. Rubies (*yakut*), which are a red corundum or aluminum oxide are a big favorite (Figure 12.11), but the blue variety are sapphires (*yakut kabut*). Such sapphires have been touted recently from Wardak Province, and even a few examples are known from the famous Jagdalak ruby source in Nangarhar Province, but most are probably brought in from other South Asian mining sources and foisted off as being of local provenance. Spinel or balas rubies (*lal*) are softer than the true corundum-based kind because they are a different composition, but in the old days they were very popular and became famous gems as in such places as the British crown jewels (Figure 12.12). Emeralds (*zamrud*) are a dark-green beryl (Figure 12.13), whereas aquamarines (*neelum*) are a light bluish green beryl (Figure 12.14). Kunzite (*kunzite*) is precious spodumene, which can occur in quite large crystals; those in Afghanistan can come out of the ground as a clear dark-green color, and spontaneously change color to pink or purple with exposure to sunlight (Figure 12.15). Because the original stones can be so large (10–20 cm), spectacularly huge cut stones can be made from the raw product. Tourmaline (*berouche*) can also occur in large crystals that are clear, light blue, pink, or the "watermelon" variety that has a green rind and a pink interior (Bowersox and Chamberlin, 1995; Anonymous, n.d.e; Dupee, 2012; Kim, 2000; Sherin, 2009; Coffman, 2011).

Gemstone mining in Afghanistan has never been very sophisticated, with arduous use of sledge hammers, hand drills, and pick-and-shovel mining being most characteristic. Once portable hand-held diesel drills became available in the past few decades, coupled with dynamite and recycled ordinance explosives, these easier techniques were adopted by many, even though the strong explosive concussions damaged many gems and resulted in much waste. These artisanal mining techniques betrayed the informal and illegal nature of almost all of the gemstone mining efforts, with the result that the GIRoA never was allowed oversight, quality control, or tax revenues. Most of the gemstones have long left Afghanistan illicitly, with 90–95 percent going to Peshawar, Pakistan, for sorting and sending on for cutting in other countries. The total value of stones moved in this black market and clandestine fashion were estimated to be somewhere between $3–160 million, which compares to a total annual world market of about $320 million for uncut colored gemstones.

In recent years the government has taken a more benign approach and provided a private but government-backed, first-ever gemstone trade center in Kabul where miners and venders can sell and exchange their wares, with a 15 percent tax on the goods. Mine operators were relieved to finally have a place to sell their stones legally; in the first few months of operation some six or so years ago, 900 Mt of lapis lazuli alone was delivered to the site. But the bureaucratic

Figure 12.10 (A) Lazurite mineral crystals in the marble matrix and lapis lazuli beads, all from the ancient Sare Sang mine on the right (east) bank of the Kokcha River in southern Badakhshan Province, Afghanistan. (B) Lapis lazuli miners in the Kokcha River valley, southern Badakhshan. Gary Bowersox is one of the chief promoters and vendors of Afghan gemstones; he is the taller gentleman standing in the center with the white shoes and beard. (C) One of the entrances to the lapis lazuli mines of Sare Sang showing dark stains from centuries to millennia of smoke from fires to break the rock, cooking, and light. *Photographs courtesy of G. Bowersox, Geovision, Inc.*

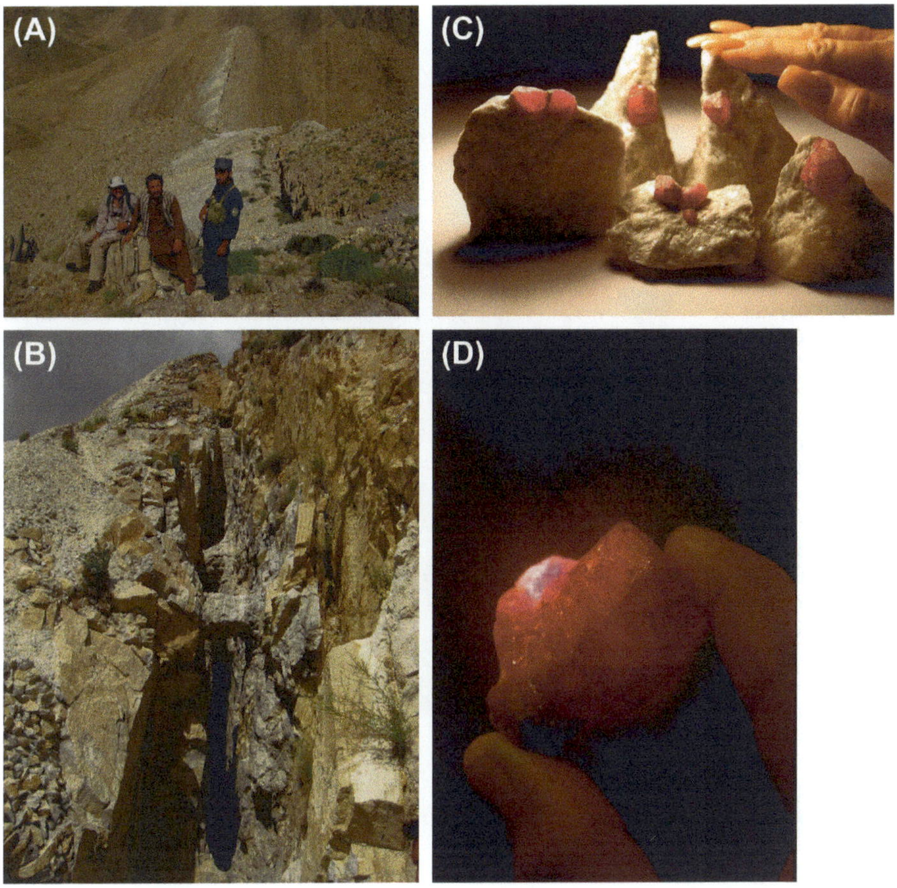

Figure 12.11 (A) Ruby mines near Jagdelak, Afghanistan. These mines are dug in vertically dipping, metamorphic marble layers. (B) Deep ruby mines at Jagdelak that measure a few meters in width where the thin ruby-bearing layer has been carried to depth leaving only a few lateral spokes to prevent collapse, whereas the ruby-bearing blocks between are taken out and hammered into smaller pieces to remove the gemstones. (C) Rubies from Jagdelak in matrix. (D) Ruby of 174 carats from Jagdelak. *Photographs courtesy of G. Bowersox, Geovision, Inc.*

Figure 12.12 Corundum and spinel ruby cabochon jewelry from Afghanistan. *Photographs courtesy of G. Bowersox, Geovision, Inc.*

Figure 12.13 (A) Emerald miners at the Khenj Mine in the Panjshir Valley. The mines are in sheared metamorphic rock, the foliation planes of which dip into the slope at this location. (B) Emeralds in matrix from the Panjshir Valley. (C) Over one million dollars in emeralds being offered in a package sale. *Photographs courtesy of G. Bowersox, Geovision, Inc.*

Figure 12.14 Aquamarine from Afghanistan. *Photographs courtesy of G. Bowersox, Geovision, Inc.*

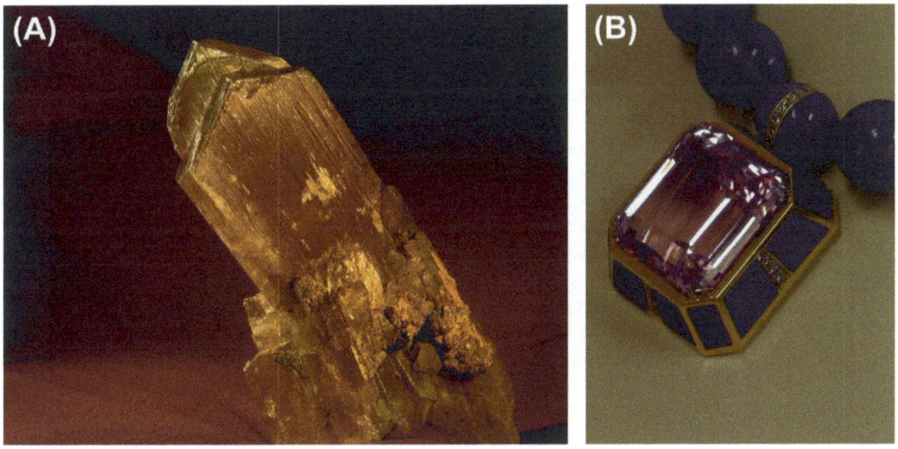

Figure 12.15 (A) Crystal of precious spodumene or kunzite from Nuristan, Afghanistan. Such crystals can grow to many centimeters in length. (B) Faceted kunzite set with lapis lazuli beads. *Photographs courtesy of G. Bowersox, Geovision, Inc.*

paperwork to make these sales is itself arduous, with official sales receipts totaling about a cm thickness, and approval from 17 offices required for sale abroad. Bribes to police and bandits to get the goods from mine to market may add another 25 percent to the price, so the process is not without its impediments even today. Even so in the first years of operation some $2 million in taxes and fees was realized.

Other efforts to improve performance that have been highlighted include capacity building, and education about gemstones, cutting, polishing, gemology, and the creation of more robust quality standards and targeted marketing campaigns. Color and other flaws in natural gemstones have long been treated by special heating, boiling, or irradiation techniques, including sunlight, chemical dying, as well as pressure injections of resin, or even production of complete synthetics. Clearly, much caution and awareness is required for the naive and amateur buyers who come to Afghanistan seeking bargains. In fact actually, gemstone vendors are set up around the gates of all the big military bases of Bagram, Kandahar, and elsewhere to take advantage of the typically short deployments of personnel in search of a going-home present. Nonetheless after decades of war and associated chaos, the popular gemstone trade in Afghanistan seems to be back on track and perhaps subject at last to be at least partially guided and controlled by the government, while also providing the essential revenues.

References

Ali, S., Shroder Jr, J.F., 2011. Afghanistan's Mineral Fortune: Multinational Influence and Development in a Post-War Economy. Institute for Environmental Diplomacy and Security, James Jeffords Center for Policy Research, University of Vermont, Burlington, VT, 24 p.

Anonymous, 2000. Afghanistan Chromite Production by Year. US Geological Survey. http://www.indexmundi.com/minerals/?country=af&product=chromite&graph=production.

Anonymous, circa 2005a. Minerals in Afghanistan: The Aynak Copper Deposit. Afghanistan Ministry of Mines and British Geological Survey, 6 p. http://www.bgs.ac.uk/afghanminerals/docs/aynak_a4.pdf.

Anonymous, 2005b. Aynak Information Package: Part I Introduction. Afghanistan Geological Survey and British Geological Survey, 22 p. http://www.bgs.ac.uk/afghanminerals/docs/tenders/Aynak/PartI_Aynak_Information_Package.pdf.

Anonymous, 2005c. Aynak Information Package: Part II Geological Setting of Aynak and Summary of Exploration – Based on the Final Exploration Report of Western Aynak and Literature on the Kabul Block. Afghanistan Geological Survey and British Geological Survey, 35 pp. http://www.bgs.ac.uk/afghanminerals/docs/tenders/Aynak/PartII_Western_Aynak.pdf.

Anonymous, 2010a. Mining for Sustainable Growth in Afghanistan. World Bank, 2 p. http://web.worldbank.org/WBSITE/EXTERNAL/TOPICS/EXTSDNET/0,menuPK:64885113~pagePK:7278667~piPK:64911824~theSitePK:5929282~contentMDK:22718722,00.html.

Anonymous, June 7, 2010b. Peer Review Policy and Guidelines. Integrity Watch Afghanistan. http://www.iwaweb.org/Docs/Research/Peer%20Review%20Policy%20and%20Guidelines%2019%20June%202010.pdf.

Anonymous, 2011a. A Programmatic Approach to Leverage Extractive Industries for Economic Development (E4D): Background Note. Resource Corridors, 12 p. http://www.eisourcebook.org/cms/files/resource_corridors_background_note.pdf.

Anonymous, 2011b. Exploration in Afghanistan. Ministry of Mines, 1 p. http://mom.afghanistan.af/en/Page/1284.

Anonymous, 2012a. 2011 Minerals Yearbook: Afghanistan (Advanced Release). US Geological Survey, 5 p. http://minerals.usgs.gov/minerals/pubs/commodity/asbestos/myb1-2011-asbes.pdf.

Anonymous, 2012b. Copper Bottomed? Bolstering the Aynak Contract: Afghanistan's First Major Mining Deal. Global Witness, Ltd., London, 83 p. http://www.globalwitness.org/copper_bottomed/index.html.

Anonymous, 2012c. Response to Global Witness' Recent Report on Aynak Copper Project. Ministry of Mines, Islamic Republic of Afghanistan, 2 p. http://mom.gov.af/en/page/8150.

Anonymous, 2012d. Response to Ministry of Mines' Statement of December 2012 on "Copper Bottomed?" Report. http://www.globalwitness.org/library/response-ministry-mines%E2%80%99-statement-december-2012-%E2%80%9Ccopper-bottomed%E2%80%9D-report.

Anonymous, 2012e. 2011 Minerals Yearbook: Afghanistan (Advance Release). US Geological Survey, 5 p. http://minerals.usgs.gov/minerals/pubs/country/2011/myb3-2011-af.pdf.

Anonymous, 2013. The Afghanistan Resource Corridor: Growth Needs to be Inclusive, Sustainable and Leveraged for Impact. (Video). http://www.worldbank.org/en/news/2012/07/02/afghanistan-resource-corridor-video.

Anonymous, n.d.a. Mining Services: An Overview of SRK's Services to the Global Mining Industry. 40 pp. http://www.srk.com/files/File/Brochure/mining-2012-english-a4-lr.pdf.

Anonymous, n.d.b. Rare Metals. Afghanistan Geological Survey. Ministry of Mines; 1 p. http://www.bgs.ac.uk/afghanminerals/raremetal.htm.

Anonymous, n.d.c. Rare-Metal Deposits. Minerals in Afghanistan. Minsitry of Mines and British Geological Survey; 6 p. http://www.bgs.ac.uk/afghanminerals/docs/RareMetals_A4.pdf.

Anonymous, n.d.d. The Chromite Potential of Afghanistan. Minerals in Afghanistan; Ministry of Mines; 4 p. http://mom.gov.af/Content/files/The%20Chromite%20Potential%20.pdf.

Anonymous, n.d.e, but circa 2009. Gemstones of Afghanistan. Minerals of Afghanistan; Ministry of Mines, and British Geological Survey; 6 p. http://www.bgs.ac.uk/afghanminerals/docs/gemstones_a4.pdf.

Bandler, J., May 11, 2011. J.P. Morgan's Hunt for Afghan Gold: A Team of Bankers Starts to Tap the Country's Vast Mineral Riches, with Help from the Pentagon. CNN Money, 8 p. http://management.fortune.cnn.com/2011/05/11/jp-morgan-hunt-afghan-gold/.

Batson, D.E., circa 2009. Registering the Human Terrain: A Valuation of Cadastre. National Defense Intelligence College, Washington, DC, 178 p. http://ni-u.edu/ni_press/pdf/Registering_the_Human_Terrains.pdf.

Begum, I., 2011. Pakistan Perspectives. Morris Anderson, 9 p. http://www.morrisanderson.com/resource-center/entry/Chinese-Growing-Diplomatic-Strategic-and-Economic-Interest-in-Afghanistan/.

Benham, A.J., Kováč, P., Petterson, M.G., Rojkovic, I., Styles, M.T., Gunn, A.G., McKervey, J.A., Wasy, A., 2009. Chromite and PGE in the Logar ophiolite complex, Afghanistan. Transactions of the Institute of Mining & Metallurgy B: Applied Earth Science. 118 (2), 45–58. http://mapdss2.er.usgs.gov/openview/reports/Aynak_Copper_and_Chromium_AOI_Reports/Logar-chromite-reports/Benham%20etal-%202009-Chromite%20and%20PGE%20in%20Logar%20ophiolite%20Afghanistan.pdf.

Blackton, J.S., June 14, 2010. Minerals in Afghanistan? Mais Oui! FP National Security; The Best Defense. http://ricks.foreignpolicy.com/posts/2010/06/14/minerals_in_afghanistan_mais_oui.

Bowersox, G.W., Chamberlin, B.E., 1995. Gemstones of Afghanistan. Geoscience Press, Tucson, AZ, 220 p.

Bowley, G., September 8, 2012. Potential for a Mining Boom Splits Factions in Afghanistan. New York Times, 5 p. http://www.nytimes.com/2012/09/09/world/asia/afghans-wary-as-efforts-pick-up-to-tap-mineral-riches.html?pagewanted=all&_r=0.

Brice, D.P., Coats, S., Felske, B., 2012. Fourth Interim Report on the Hajigak Iron Ore Project Tender. International Advisory Panel on Resources Development in Afghanistan (IAP); Islamic Republic of Afghanistan Ministry of Mines, 5 p. http://mom.gov.af/Content/files/Final%20EXECUTIVE%20SUMMARY%20Interim%20Report.pdf.

Brigaldino, G., 2010. Spinning a Resource War in Afghanistan. 4 p. http://www.towardfreedom.com/home/africa/2021-surprise-surprise-treasures-in-the-dirt.

British Geological Survey, circa 2007. Minerals in Afghanistan: The Aynak Copper Deposit. http://www.bgs.ac.uk/afghanminerals/docs/aynak_a4.pdf.

Butler, N.J., February 26, 2004. Escaping the Resource Curse: Managing Natural-Resource Revenues in Low-Income Countries. BP Global Press. http://www.bp.com/genericarticle.do?categoryId=98&contentId=2017217.

Cerkel, D., Miller, R.L., May 27, 1958. Petroleum Law for Afghanistan. Prepared under the Auspices of the Ministry of Mines and Industries of Afghanistan and the International Cooperation Administration of the U.S.A, (US Geological Survey), Kabul, Afghanistan, 49 p.

Coffman, T., 2011. The gems of Afghanistan. America's North Shore Journal. 3. http://northshorejournal.org/the-gems-of-afghanistan.

Croucher, S., Sobhani, O., December 7, 2011. Gold Worth $1 Trillion Buried in Afghanistan Mountains. International Business Times. http://www.ibtimes.co.uk/articles/263023/20111207/gold-worth-1-trillion-buried-afghanistan-mountains.htm.

Diamond, L., Mosbacher, J., 2013. Petroleum to the people: Africa's coming resource curse – and how to avoid it. Foreign Affairs 92 (5), 86–98.

Drummond, H., 1841. On the mines and mineral resources of Afghanistan. Journal Asiatic Society of Bengal 10, 74–93.

Dupee, M., 2012. Afghanistan's Conflict Minerals: The Crime-State-Insurgent Nexus. Combating Terrorism Center at West Point, 4 p. http://www.ctc.usma.edu/posts/afghanistans-conflict-minerals-the-crime-state-insurgent-nexus.

Evans, A., 2010. Resource Scarcity, Climate Change and the Risk of Violent Conflict. World Development Report 2011: Background Paper World Bank. 23 p. http://siteresources.worldbank.org/EXTWDR2011/Resources/6406082-1283882418764/WDR_Background_Paper_Evans.pdf.

Gulati, M., September 17, 2012. India Seeks to Enhance Its Mining Stakes in Afghanistan – Analysis. South Asia Monitor, 2 p. http://southasiamonitor.org/detail.php?type=eco&nid=3896.

Humphreys, M., Sacks, J.D., Stiglitz, J.E., 2007. Escaping The Resource Curse. Columbia University Press. 408 p.

Integrity Watch Afghanistan, 2013. Chromite Extraction in Kunar, Factor of [In]stability: Case Study. http://iwaweb.org/Docs/EIM/chromite_extraction_in_kunar_factor_of_instability.pdf.

Jhanmal, Z., December 10, 2012. Afghan-Foreign JVs Awarded Copper, Gold-Mining Contracts. Tolo News, 1 p. http://www.tolonews.com/en/afghanistan/8670-afghan-foreign-jvs-awarded-copper-gold-mining-contracts-.

Kavanagh, M., December 6, 2012. Consortium Wins Afghan Mining Permits. Financial Times, 1 p. http://www.ft.com/intl/cms/s/0/64a9b282-3fcb-11e2-b0ce-00144feabdc0.html#axzz2ITZgPjpl.

Kenny, C., December 6, 2010. What resource curse? Foreign Policy. http://www.foreignpolicy.com/articles/2010/12/06/what_resource_curse.

Kim, L., July 25, 2000. Afghanistan's Emerald Heights. Christian Science Monitor. http://www.csmonitor.com/2000/0725/p6s1.html.

Klare, M.T., 2002. Resource Wars: The New Landscape of Global Conflict. Metropolitan/Holt – MacMillan, 304 p. http://us.macmillan.com/resourcewars/MichaelKlare.

Lendman, S., 2007. Resource Wars – Can We Survive Them? Global Research' Centre for Research on Globalization, 7 p. http://www.globalresearch.ca/resource-wars-can-we-survive-them/5892.

McDonnell, P., 2006. The Hajigak Iron Deposit. Minerals in Afghanistan. British Geological Survey and Ministry of Mines, 6 p. http://www.bgs.ac.uk/afghanminerals/docs/Hajigak_A4.pdf.

McNeil Jr, D.G., June 20, 2010. The Curse of Plenty. New York Times. http://www.nytimes.com/2010/06/20/weekinreview/20mcneil.html.

Medlin, J., Demas, A., 2011. USGS Releases Resource Estimate for Afghanistan Rare Earth Prospect. USGS. http://www.usgs.gov/newsroom/article.asp?ID=2936.

Metz, A.E., 2007. The mine at Aynak. The Culture and Conflict Review. 1 (2), 3. http://www.nps.edu/Programs/CCS/Journal/Dec07/Metz_Anyak.pdf.

Nissenbaum, D., November 30, 2011. India wins bid for 'jewel' of Afghan ore deposits. Wall Street Journal. 2. http://online.wsj.com/article/SB10001424052970203802204577067303633783544.html.

Noorani, J., 2011. Hajigak: The Jewel of Afghan Mines. Integrity Watch Afghanistan, Kabul, 32 p. http://www.iwaweb.org/Reports/Hajigak-The_Jewel_of_Afghanistan-2011.html.

Omeje, K. (Ed.), 2008. Extractive Economies and Conflicts in the Global South: Multi-Regional Perspectives on Rentier Politics. p. 257. Aldershot, Hampshire, England.

Orris, G.J., Bliss, J.D., 2002. Mines and Mineral Occurrences in Afghanistan. USGS Open-File Report 02-110; 95 p. http://geopubs.wr.usgs.gov/open-file/of02-110/of02-110.pdf.

Partlow, J., November 17, 2009. Afghan Minister Accused of Taking Bribe. The Washington Post. http://www.washingtonpost.com/wp-dyn/content/article/2009/11/17/AR2009111704198_2.html?sid=ST2009111800831.

Peter, T.A., June 7, 2012. India Eyes Increased Role in Afghanistan. Christian Science Monitor, 2 p. http://www.csmonitor.com/World/Asia-South-Central/2012/0607/India-eyes-increased-role-in-Afghanistan.

Peter, T.A., Arnoldy, B., June 28, 2010. A Wealth of Resources, a Trove of Suspicion. Christian Science Monitor. http://www.sabawoon.com/articles/index.php?page=Afghan_minerals.

Renner, M., 2002. The Anatomy of Resource Wars. Worldwatch paper 162, Danvers, MA, 91 p. http://www.worldwatch.org/system/files/WP162.pdf.

Riechmann, D., December 14, 2010. Afghan Government Approves Gold Mine Deal. Bloomberg Business Week, 1 p. http://www.businessweek.com/ap/financialnews/D9K3I2A80.htm.

Risen, J., June 17, 2010a. World's Mining Companies Covet Afghan Riches. New York Times. http://www.nytimes.com/2010/06/18/world/asia/18mines.html?_r=0.

Risen, J., June 13, 2010b. U.S. Identifies Vast Mineral Riches in Afghanistan. New York Times, 4 p. http://www.nytimes.com/2010/06/14/world/asia/14minerals.html?pagewanted=all&_r=0.
Rosenberg, M., July 23, 2012. Afghan Cabinet Raises Concerns about Mining Legislations, to West's Unease. New York Times, Asia Pacific, 3 p. http://www.nytimes.com/2012/07/24/world/asia/afghan-cabinet-blocks-new-mining-laws.html?pagewanted=all&_r=0.
Shahrani, W., 2013. National and Regional Resource Corridors Program (NRRCP). Ministry of Mines, Islamic Republic of Afghanistan, Kabul, 3 p. http://mom.gov.af/en/page/6355.
Sherin, L.A., December 23, 2009. Afghanistan's Illegal Gem Trade Flourishing. One World, South Asia, 4 p. http://southasia.oneworld.net/news/afghanistans-illegal-gem-trade-flourishing.
Shroder Jr, J.F., 1981. Physical resources and the development of Afghanistan. Studies in Comparative International Development XVI (3–4), 36–53.
Shroder Jr., J.F., Watrel, R.H., 1992. Mineral resources in Afghanistan (summary of report for USAID). AACAR Bulletin (Association for the Advancement of Central Asian Research) 5 (2), 9.
Shroder Jr, J.F., 1983. The U.S.S.R. and Afghanistan mineral resources. In: Agnew, A. (Ed.), International Mineral Resources a National Perspective. American Association for Advancement of Science, pp. 115–153.
Shroder Jr, J.F., 1989. Afghanistan resources and Soviet Policy in Central and South Asia. In: Hauner, M., Canfield, R. (Eds.), Afghanistan and the Soviet Union: Collision and Transformation. Westview Press, pp. 101–119.
Shroder, J., 2005. Remote sensing and GIS as counterterrorism tools in the Afghanistan war: reality plus the results of media hyperbole. Professional Geographer 57, 592–597.
Shroder, J., 2007. Afghanistan's development and functionality: renewing a collapsed state. Geojournal 70, 91–107.
Shroder Jr, J.F., 2012. Afghanistan: rich resource base and existing environmental despoliation. Environmental Earth Sciences. http://dx.doi.org/10.1007/s12665-012-1638-7.
Shroder Jr., J.F., Assifi, A. Tawab, 1987. Afghan mineral resources and Soviet exploitation. In: Klass, R. (Ed.), Afghanistan and the Soviet Union: The Great Game Revisited. Freedom House Pub, New York, pp. 97–134.
Simpson, S., September 30, 2011. Afghanistan: rare earth elements could beat the Taliban (slide show). Scientific American. http://www.scientificamerican.com/article.cfm?id=afghanistan-rare-earth-elements-could-beat-taliban.
Stephenson, J., Safi, A., 2012. Problems Alledged with Major Afghan Mining Deal. McClatchy Washington Bureau, 2 p. http://www.mcclatchydc.com/2012/02/17/139247/firm-alleges-problems-with-major.html.
Sweeney, R.E., Kucks, R.P., Hill, P.L., Finn, C.A., 2006. Aeromagnetic and Gravity Surveys in Afghanistan—A Web Site for Distribution of Data. U.S. Geological Survey Open-File Report 2006–1204 http://pubs.usgs.gov/of/2006/1204/ (accessed 08.08.11.).
Synovitz, R., November 24, 2007. Afghanistan: China's Winning Bid for Copper Rights Includes Power Plant, Railway. Radio Free Europe/Radio Liberty. http://www.rferl.org/content/article/1079190.html.
Tabb, W.K., 2013. Resource wars. Monthly Review. An Independent Socialist Magazine. 7. http://monthlyreview.org/2007/01/01/resource-wars.
Topf, A., March 23, 2012. India Wants to Truck Afghan Ore Through Pakistan. Mining.com, 2 p. http://www.mining.com/india-wants-to-truck-afghan-ore-through-pakistan/.
Tran, M., October 25, 2012. Are Natural Resources a Blessing or a Curse for Developing Countries? The Guardian: Global Development. http://www.guardian.co.uk/global-development/2012/oct/25/natural-resources-blessing-curse-developing-countries.
Tucker, R.D., Belkin, H.E., Schultz, K.J., Peters, S.G., Buttleman, K.P., 2011. Rare Earth Element Mineralogy, Geochemistry, and Preliminary Resource Assessment of the Khanneshin Carbonatite Complex, Helmand Province, Afghanistan. US Geological Survey Open-File report 2011 – 1207; 50 p. http://pubs.usgs.gov/of/2011/1207/pdf/ofr2011-1207.pdf.

Wang, Y., 2012. Experts: Afghanistan Mining Not an Easy Way Out. Medill News Service, Northwestern University. http://medilldc.net/2012/07/experts-afghanistans-mining-industry-is-not-an-easy-way-out/.

Williams, L., October 16, 2012. Afghan Gold Being Chased by Ex-special Forces Entrepreneurs. Mineweb, 2 p. http://www.mineweb.com/mineweb/content/en/mineweb-gold-news?oid=160245&sn=Detail.

Winnail, D.S., 2005. Coming: Resource Wars! Tomorrow's World, 4 p. http://www.tomorrowsworld.org/node/769.

Yeager, J.R., Sediq, M.M., Smith, E., Garner, D., Alief, M.H., Snee, L., Shafer, R., Ritter, D., 2009. The Aynak Copper Tender: Implications for Afghanistan and the West. http://www.scribd.com/doc/22004530/The-Aynak-Copper-Tender-Implications-for-Afghanistan-and-the-West.

In this Chapter

13

Air and Space Technology in Resource Delineation: Peace and War

Abstract

The use of air and space technologies in resource delineation in Afghanistan was a product of the development of new remote-sensing technologies, coupled with favorable circumstances produced by limited vegetation because of the great aridity and human-caused removal of much of the rest. The result was a series of new satellite-image mosaics and other analytical means that enabled the use of hyperspectral remote sensing, in which the spectral signatures of particular minerals of interest can be recognized from high altitude or in space. These data sources, along with such other information as gravity and magnetic maps and other ancillary information, enable the establishment of detailed mineral assessments that can used to acquire appropriate preliminary assessments, and obtain the permits and financing to facilitate the processes necessary to mine the resources.

Keywords: GLIMS project; Gravity anomaly; Hyperspectral remote sensing; Magnetic anomaly; Operation Rampant Lion; Satellite image mosaic; Task Force for Business and Stability Operations (TFBSO).

Some 40 years ago—back in antiquity as students say—when the science of remote sensing was just getting going, the idea was a big surprise that a satellite orbiting hundreds of kilometers above the Earth could literally snap pictures or scan the ground using advanced technologies that could detect all sorts of different electromagnetic radiations. Prior to that time, most geoscientists used aerial photography to figure out exactly what was where on the ground. The science of aerial photogrammetry was developed to a robust degree in WWII, and the science of remote sensing was a similar product of the Cold War to detect what the enemy was doing. Afghanistan was a country that had been flown over repeatedly in the late 1950s and early 1960s to make aerial photographs by both American and Soviet aircraft, which countries were individually responsible for producing topographic maps (Glicken, 1960; Shroder, 1983), but the photographs were secret and not easily available (Figure 13.1(A and B)).

Natural Resources in Afghanistan. http://dx.doi.org/10.1016/B978-0-12-800135-6.00013-1

(A)

(B)

Figure 13.1 (A) Rare aerial photograph of Kabul University as it existed in about 1961. *Photograph by Fairchild Aerial Surveys Company.* (B) Google Earth™ view of 2011 (north is to the top). Among other observations of change over time, note in particular the squatter's houses on the rocky ground to the north that have sprung up in the interim. No sewage or water pipes exist in those locations.

Remote sensing satellite imagery of Afghanistan became a viable alternative, albeit at low resolutions so that the radiations from many small features on the ground were averaged together into a picture element (pixel) some hundreds of meters on a side. The result was the ability to map the entire country of Afghanistan using this new remote sensing imagery that civilians could purchase if they could afford it. With the start-up of the new Afghanistan Studies Center at the University of Nebraska at Omaha, purchase of imagery for the whole country was a necessary expense. (Shroder, 1978).

One of the first jobs undertaken was to build a mosaic of the 70–80 odd images that had to be "stitched" together in some fashion to cover the whole country. In today's technology, this can be done entirely electronically, but in those days each image was printed on photographic paper and the edges of each rectangular image were torn off very carefully to leave a thin "feathered" edge that could be glued down over its matching but neighboring images on all sides. In this fashion, a mosaic of photographic images was built up with carefully but manually registered sections (Figure 13.2). The resulting mosaic measured 1 × 2 m on a side and was mounted on a piece of Masonite construction board and then rephotographed to make multiple copies of the whole. The international borders were carefully transferred by eye and white ink onto the mosaic

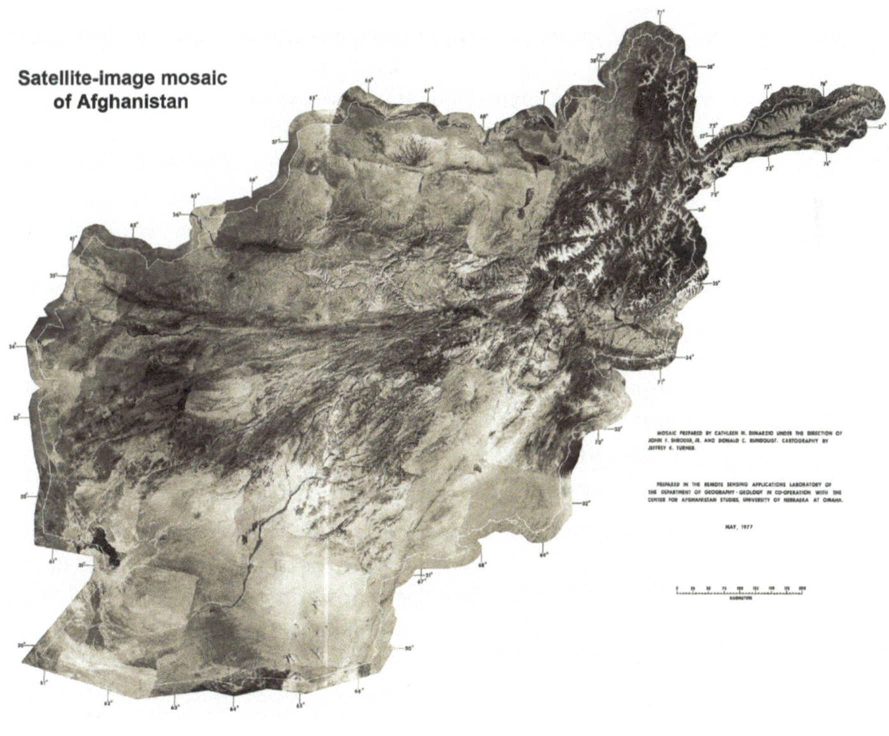

Figure 13.2 Satellite-image mosaic of Afghanistan produced at the University of Nebraska at Omaha in 1977 from mosaicked contact prints of early Landsat imagery. *DiMarzio et al., 1977*

from a set of large-scale US Department of Defense (DOD), Joint Operations Graphic (JOG) series of maps at a scale of 1:250,000 that had been released to the author by the US Department of State under Henry Kissenger's purview for the Atlas project.

A full-sized version of the satellite-image mosaic of Afghanistan was taken by Thomas Gouttierre, Director of UNO's new Afghanistan Studies Center, to Kabul in the mid-1970s to present to the government of Afghanistan as an example of the help being given to them by Americans. In a ceremony presided over by President Daoud and his ministers, Mr Gouttierre unrolled the big mosaic and talked about how Americans were up in space looking down with "cameras" that could make pictures like this. At this time, of course, during the height of the Cold War, it was illegal for anyone to make aerial photographs anywhere in the country. After a pregnant pause while everyone was assimilating what Mr Gouttierre was saying, the voice of a distant government minister piped up, "Can you see our women?" Mr Gouttierre assured the assembled ministers and President Daoud that the sanctity of their women was safe with the low-resolution imagery of the time. This was the beginning of some of the very first uses of satellite imagery for Afghanistan.

In 1977–1978 when I was teaching at Kabul University, running the Kabul University Seismic Station, and acting as the Director of the project to produce a first National Atlas of Afghanistan, a major objective was to gain as much information about the terrain of the country as possible. One method to do this was to travel around the country on the reasonable to rudimentary roads as much possible, as well as trekking off road on foot wherever the danger from bellicose tribal warriors or bandits was minimal. An even better, but quite different method was to take to the skies in US Ambassador Theodore Eliot's airplane. Ambassador Eliot was required by the nature of his job at the time to travel to numerous other cities in the region to serve as the ceremonial American representatives at various dedication, inaugural, and opening ceremonies, among other things. He had a multiseat passenger aircraft flown by two US Air Force officers out-of-uniform, whose job was to ferry him around the country. I was invited on several occasions to be a back-seat passenger with cameras to shoot photographs of whatever interesting features were being flown over, even though this was not quite legal according to the law of that land. The result was a series of low-altitude, oblique aerial photographs that in many subsequent years served well as records to enable analysis of landforms and human landscapes (Figure 13.3(A and B)).

After the communist coup in late April of 1978, the days of the Atlas project in Afghanistan could be seen to be numbered, so the best course of action was to safeguard all the Atlas maps and photographs. The materials were carried by my cook, the late Abdul Waheed, on his bicycle a few at a time across Kabul to the US Embassy and given to the gunny sergeant of the US Marine guard at the gate. From there the vital Atlas materials were shipped back to UNO via

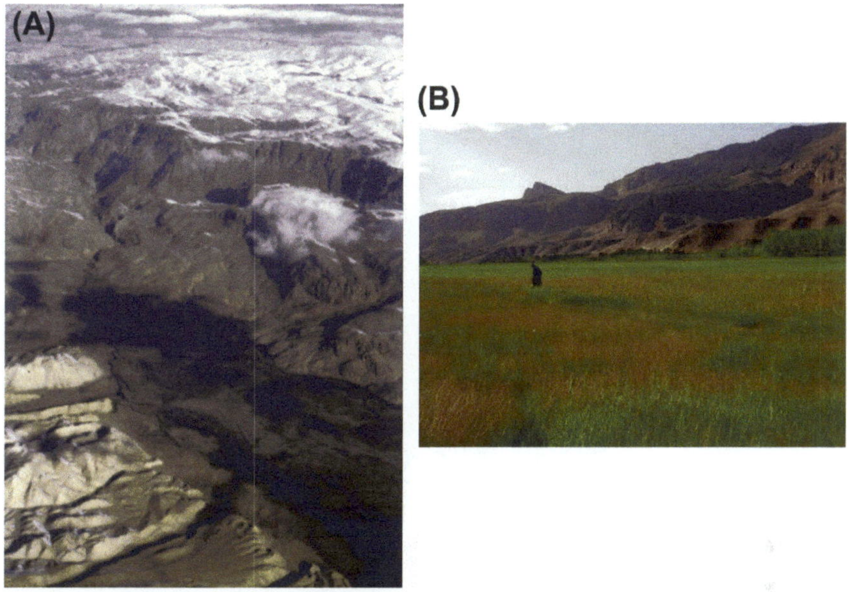

Figure 13.3 (A) Low-altitude oblique aerial photograph of Yakawlang graben area in central Afghanistan where the bedrock has been pulled apart by a bend in the right-lateral strike-slip fault. The river below is near the headwaters of the Balkh Ab. The photograph was taken from US Ambassador Theodore Elliot's aircraft in 1977. *Photograph by J. Shroder.* (B) Ground view of the same scene in approximately the same direction at the southeast end of the pull-apart rhombochasm. *From M. AliStani on Google Earth™.*

diplomatic pouch, which ensured their safe arrival. A few important rock specimens were shipped that way too, although a later request for a greater weight of such was politely rejected. I myself was placed under house arrest for several months in the fall of 1978, before being deported to Germany on a Lufthansa jet in mid-November of that year. At least the plane ticket was free.

The next big project involved with satellite imagery of Afghanistan was an effort by Ritchie Williams and Jane Ferrigno at the US Geological Survey (USGS). They had decided that the world-wide coverage of snow and ice with satellite imagery in the 1970s would be able to give a good measure of the health of the glaciers of the world in the face of the long-predicted climate change since the nineteenth century. Accordingly, they were casting about to find those willing to take on the challenge of assessing ice in the Hindu Kush and Himalaya, and discovered my ongoing work on such glaciers. That led to provision of plentiful and free imagery from the USGS in the late 1970s and early 1980s, which coupled with the plentiful and large-scale (1:100,000) topographic maps from the US Department of Defense (DOD) of high mountain areas in Afghanistan that were generally denied to most other people, led to the very beginnings of a first glacier inventory of the Hindu Kush (Shroder, 1980) (Figure 13.4).

Figure 13.4 Return-beam vidicon (RBV) scene of glaciers in the Little Pamir of the Wakhan Corridor of Afghanistan (top third of image), as well as glaciers in the Hindu Raj mountains of northwest Pakistan in 1977. Such temporally valuable imagery was provided by the USGS to work on glacier configuration and change, but all RBV images were subsequently erased from storage media at the EROS Date Center in Sioux Falls, SD, in an ill-advised cost-cutting measure during the Reagan presidency in the late 1980s.

Some of the imagery provided included that of the Return Beam Vidicon (RBV) imaging device that was put into orbit in the late 1970s. In a subsequent cost-saving measure by the EROS Data Center in Sioux Falls, SD, during the President Reagan years, the RBV image files were destroyed, which left the author with the only known copies of such imagery of Afghanistan and Pakistan. Later these became important when the author's postdoctoral researcher, Dr Umesh Haritashya, assessed the glacier changes in the Wakhan Pamir (Haritashya et al., 2009). A paper to describe the glaciers of Afghanistan was submitted in the early 1980s, the benchmark time period for the Satellite Image Atlas of Glaciers of the World, USGS Professional Paper 1386A-K, but because there were subsequent decades of delays in publishing, the work needed to be updated by adding references to more recent work while intentionally retaining the original benchmark information (Shroder and Bishop, 2010).

After the full Soviet invasion of Afghanistan in late 1979, the war between the Mujahideen and the Soviet troops escalated, and I returned to Peshawar, Pakistan, to work with Pakistani geologists, as well as to aid the Afghan resistance in education. At the same time, the US government was funding a variety of projects to aid the progress of the war against the Soviets (Crile, 2003; Coll, 2004; Tomsen, 2011). One of the issues was to get many forms of assistance (books, school supplies, guns, ammunition, and Stinger missiles) through the mountain passes from Pakistan into Afghanistan. To do this, the US Central Intelligence Agency (CIA) made a new set of mosaicked satellite images of Afghanistan that were printed at a scale of 1:500,000 (Shroder, 2008). The seven sheets (Sheet I, Herat; Sheet II, Mymanah; Sheet III, Kabul; Sheet IV, Islamabad; Sheet V, Zaranj; Sheet VI, Qandahar; Sheet VII, Ghazni) were false color images (red, green, and blue, with the red being used for the portrayal of vegetation spectral signatures). The large-sized sheets were overlain with useful designations and a legend portraying the international borders, internal administrative boundaries, Federally Administered Tribal Areas Boundary in Pakistan, district of Agency boundaries, Internal administrative capitals, railroads, paved roads, unpaved roads, tracks and trails, airfields, and spot elevations in meters (Figure 13.5(A and B)). Passes through the mountains along the borders of the country were numbered. On the reverse side of each sheet was a variety of bilingual gazetteer information, with names of the passes if known; otherwise, left blank for insertion of a name if found out by a covert operative (Figure 13.6). Many of the American civilians living in Peshawar during the war got to know some of these people who went over the passes during the war, even though otherwise everyone officially was not supposed to do this; a few ignored the warnings anyway and came back with interesting stories to relate over drinks in the American Club and elsewhere.

After the Soviets went home in some ignominy and the Second Afghan civil war erupted in the 1990s, work on satellite imagery of Afghanistan at UNO was limited to an occasional look at glaciers and landforms for various reasons. In the late 1990s, when the USGS was first looking for scientists worldwide to work on the GLIMS Project (Global land Ice Measurements from Space), in connection with a new NASA rocket and satellite program, the Earth Observing System's (EOS) flagship satellite "Terra," named for Earth. Terra collected data about the Earth's changing climate and carried five state-of-the-art sensors that studied interactions among the Earth's atmosphere, lands, oceans, and radiant energy, but the one that was used for Afghanistan the most in this context was Advanced Spaceborne Thermal Emission and Reflection Radiometer (ASTER). ASTER was a cooperative effort between NASA, the USGS, Japan's Ministry of Economy, Trade and Industry (METI), and Japan Space Systems (J-spacesystems). ASTER data were used to create detailed maps of land-surface temperatures, reflectance, and elevation. The coordinated system of EOS satellites, including Terra, has been a major component of NASA's Science Mission

(A)

(B)

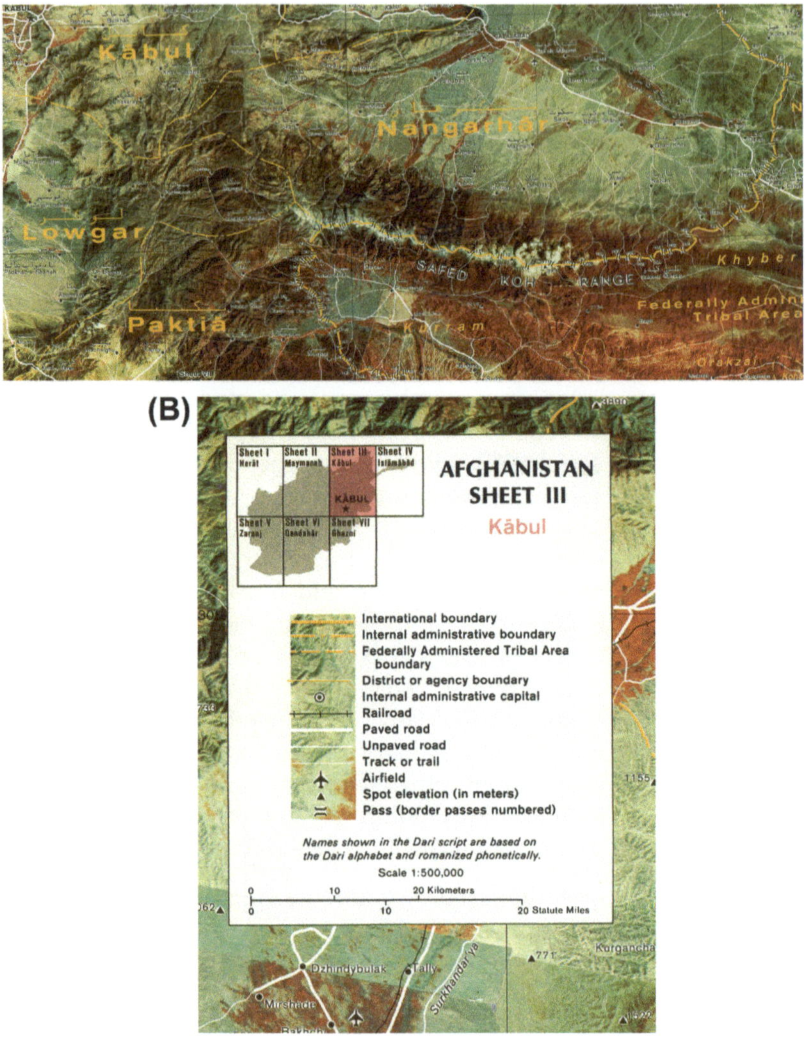

Figure 13.5 (A) Portion of the Kabul III satellite-image mosaic sheet produced by the US CIA in the 1980s as a part of the effort to aid the Afghan resistance against Soviet occupation. Part of Kabul city is located in the extreme upper left and the Safed Koh (Spin Ghar) mountain range occurs, running east–west through the lower center of the map, north of the North–West Frontier Province of Pakistan running along the bottom quarter of the map. Passes through the mountains that are used for smuggling all manner of people, guns, weapons, and untaxed consumer goods are enumerated. (B) Title part of Kabul III sheet showing the numbers and names of the other six sheets, as well as the standard legend for all sheets.

Directorate and the Earth Science Division. The goal of NASA Earth Science has been to develop a scientific understanding of the Earth as an integrated system, its response to change, and to better predict the variability and changing trends in climate, weather, and natural hazards. Afghanistan, more than most

111	Tora Tigga Kandao	توره تیکه کندو	111
112	Kot Mohmand Kandao	کوت محمد کندو	112
113	Oghaz Kandao (Owghaz Kandaw, Afghanistan)	اوغاز کندو اوغاز کندو	113
114	Papin Kandao	پاپین کندو	114
115	Agam Pass	آگام کوتل	115
116	______	______	116
117	Shah Hussaini Kandao	شاه حسینی کندو	117
118	______	______	118
119	Rishak Kandao	ریشک کندو	119
120	Spinkai Raghai	سپینکی رغائی	120
121	______	______	121
122	Zarai Khande Kandao	زراعی خنده کندو	122
123	Gawai Tangai	گوائی تنگی	123
124	Kirkai Tangai	کیرکی تنگی	124
125	______	______	125
126	Bargawi Kandao	برگاوی کندو	126
127	______	______	127
128	Peiwar Kotal (Peywâr Kandaw, Afghanistan)	پیوار کوتل (پیوار کندو)	128
129	______	______	129
130	Sursurang Kandao	سورسورنگ کندو	130
131	Shautalo Chhapri Kandao	شاوتلو چهاپری کندو	131
132	Zhâwar Kandao (Zhâwar Kandaw, Afghanistan)	ژاور کندو (ژاور کندو)	132
133	Margho Kandao	مرغو کندو	133
134	Darwezi Kandao	دروزی کندو	134
135	Mandatt Kandao	مندت کندو	135
136	Dre Drang Kandao	دره درنگ کندو	136
137	Swaro Gawai	سوارو گاوائی	137

Border passes have been numbered for ease of reference. Blank spaces have been left for adding the English and Dari names as they become known. Unless otherwise indicated, pass names are those used on the Pakistan side of the border.

Figure 13.6 Small portion of the back of the Kabul III sheet with the passes enumerated and named, if known, but a blank left for the name to be filled in later if discovered by covert operatives.

countries, exists right at the edge of major environmental problems and climatic livability, so that some of their environmental problems, which might be less extreme elsewhere, can take on more severe import that satellite monitoring is able to detect more rapidly or easier than most other places. It is a country on the margins, where the brink may not be far away between livability, and something more extreme. ASTER satellite data provided hosts of new information to enable new looks at the problems (Shroder and Bishop, 2010; Shroder, 2012).

In the course of working with the ASTER data on Afghanistan and Pakistan, for which UNO became the Southwest Asia (Afghanistan and Pakistan) Regional Center for GLIMS, terabytes of reasonably high (30 m)-resolution, satellite-image data of Afghanistan were collected, which enabled detailed looks at many features, not only ice. In 2000, the Shuttle Radar Topographic Mission (SRTM) was flown by NASA onboard the Space Shuttle Endeavor to obtain elevation data on a near-global scale. This gave a new ability to see the topography of the Earth recreated in three dimensions on the computer screen with 90 m resolution. ASTER also had stereographic overlap side-to-side with its image acquisitions, which with its higher resolution added to the topographic data or digital elevation models (DEMs) available for the whole of Afghanistan. By draping a single ASTER image over the new DEM, a simulated relief map could be generated that gave the appearance of actual topography as viewed from any angle above the surface. This gave an unparalleled ability to see and interpret landforms in three dimensions (3D) at a range of scales. Then when Google Earth™ came online in 2005, an even greater ability to map the physical character of Afghanistan was made available. For the first time, the whole physical character of the country was revealed from space and could be accessed at a moment's notice on almost any modern computer. This enabled unprecedented and rapid access to the surface of Afghanistan, at least in a virtual sense. Many hours of pouring over apparent low-altitude, high-resolution Google Earth™ imagery revealed many newly recognized features. From these data sets, a variety of new applications were then made concerning glaciers, water resources, and natural hazards (Hagen et al., 2010; Haritashya et al., 2009; Sarikaya et al., 2012; Shroder, 2009; Shroder and Weihs, 2010; Shroder et al., 2007, 2011a,b).

Continual research progress made in the remote-sensing community ultimately enabled comprehensive mapping of about 70 percent of Afghanistan using an advanced remote sensing technique known as *hyperspectral imaging* (Brozena et al., 2007). Such work is quite expensive because it requires specially equipped aircraft, massive equipment expenditures, and hundreds of hours of flight time and data-processing work, but was possible for a country like Afghanistan because the budget could be justified through US national security funding channels. Hyperspectral remote sensing collects and processes information across much of the electromagnetic spectrum, including visible light as well as the near infrared. Hyperspectral sensors deployed from aircraft were viewed as an ideal tool for mapping the mineral provinces of the country because of its rugged topographic relief, lack of substantial ground cover of soil and vegetation that can obscure the bedrock, and the substantial security issues that precluded normal geological exploration on the ground. Such hyperspectral sensors measure light spectra reflected from the Earth and can be interpreted to identify mineral compositions, man-made materials, snow and ice, and vegetation. Areas can be mapped quickly to show mineral resources, natural hazards, agricultural conditions, and development of infrastructure.

If the Afghanistan Geological Survey had attempted the standard geologic ground reconnaissance that has been characteristic of such work all over the world for several centuries, it would have taken 15–20 years to do the work, the time that the country cannot afford. Accordingly, the USAID, the Naval Research Laboratory, and the US National Geospatial Intelligence Agency provided about $11 million to Operation Rampant Lion I and II in the summers of 2006 and 2008 (McKinney, 2009), and the HyMap Data Collection Program by the USGS in summer 2007. All three of the summer flight operations were hampered to a certain extent by dust-storm weather that is characteristic of the *badi sado bist roz* or wind of 120 days in summer in the west of Afghanistan. Such windstorms occur when air masses rush south from Central Asia through the lowlands around the western edges of the Hindu Kush to meet the monsoonal low-pressure intertropical convergence zone (ITCZ) coming in from over Pakistan.

The US government picked the Navy's Scientific Development Squadron One (VXS-1; "Warlocks") to conduct the Rampant Lion Operations in Afghanistan, where they deployed one four-engine turboprop NP-3D Orion aircraft to the Kandahar airfield to execute the research flights (Figures 13.7 and 13.8). Forty research survey flights were flown 37,000 km over the course of 43 days and 28 flights, totaling 226 flight hours covering over two-thirds of the country (440,000 km^2). This excluded most border areas, where buffers of about 46 km were established around all borders except Iran, where the buffer zone was a larger ~65 km. The higher altitudes in the center and northeast of the country were excluded as well because of ceiling limitations of about 8.6 km for the aircraft. This meant that the aircraft could not fly over ground above much above 3.7 km altitude and had to maintain a relatively high flying altitude to avoid weapons fired from the ground (Figure1 3.9).

Approximately 50 military personnel and civilian scientists were involved in the Rampant Lion efforts, keeping the 38-year-old aircraft ready to fly in spite of extremely hot temperatures ranging from 40–54 °C (105°–130 °F). The

Figure 13.7 Specially equipped Lockheed Orion NP-3D aircraft taxiing at the Kandahar International Airfield in June 2006 in connection with Operation Rampant Lion to gather hyperspectral and other valuable geophysical data over Afghanistan. *Photo by USGS.*

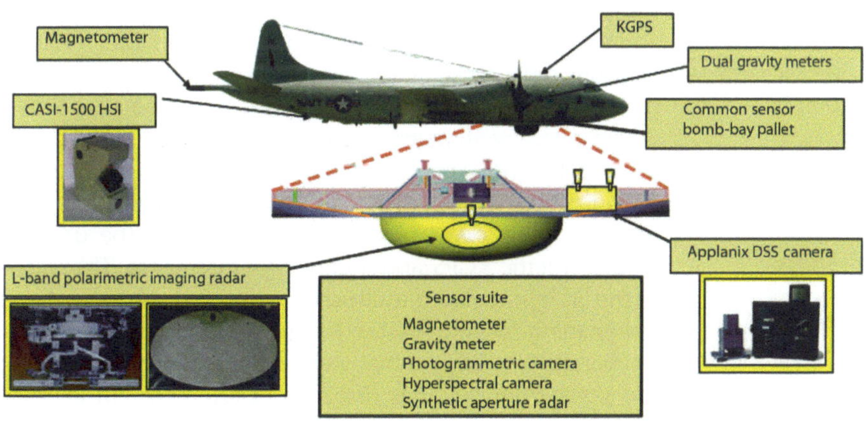

Figure 13.8 The sensor systems installed on the Lockheed Orion NP-3D research aircraft for the aerogeophysical survey of Afghanistan. CASI-1500 HIS is the hyperspectral imaging camera. *Photograph is after Brozena et al. (2007).*

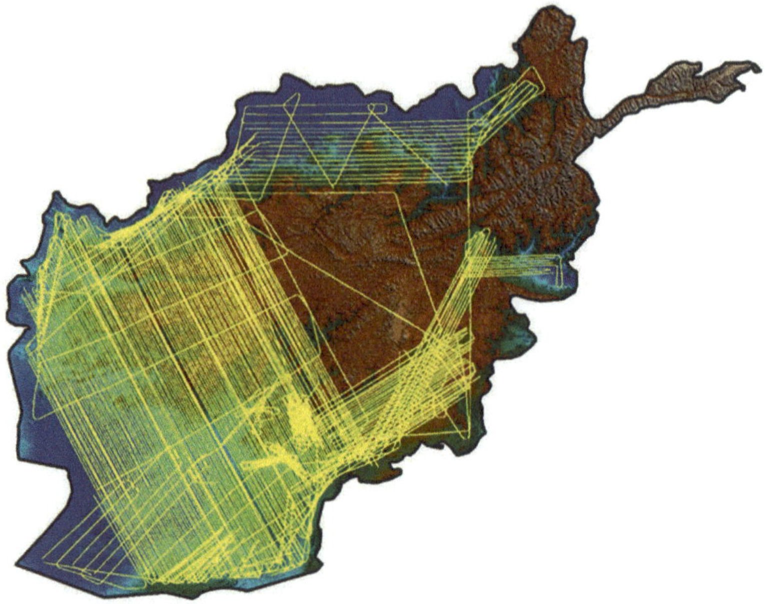

Figure 13.9 Map of the 40 Rampant Lion I mission flights over Afghanistan that produced more than 125,000 line km of airborne survey tracks and some 330,000 km^2 of imagery. *After Brozena et al. (2007).*

environment was not optimal for scientific research where maximum control of takeoffs, landings, navigation, and altitude are required because the region was an active combat theater, with avoidance of other aircraft, tanker traffic, and ground fire being necessary as well (Burgess, 2006, Brozena et al., 2007). The plane required outfitting with the addition of a flare/chaff dispenser, optical launch warning systems to counter any possible firing of surface-to-air missiles, infrared strobes, secure communications channels, a low-observable paint scheme (gray), as well as foamed fuel tanks to reduce the danger from ground-weapons fire.

The aircrafts in 2006 and 2008 were specially equipped with a unique suite of state-of-the-art remote-sensing technology, including dual gravimeters, scalar and vector magnetometers, a hyperspectral imager, L-band polarimetric synthetic aperture radar, and digital photogrammetric camera (Figure 13.8). A kilometric global positioning system (KGPS) that enabled pinpoint location of the aircraft within a few centimeters established a stable air platform from which to control the experiments. The combination of all these sensors in one aircraft was a unique configuration that enabled high comparative precision between data sets. Data from all sources was precisely coregistered to the ground by a combination of interferometric KGPS and inertial measurements. The mission advanced the state of the art in integrated multisensor airborne remote sensing as a result. Hyperspectral imaging for Afghanistan Rampant Lion I divided the spectrum into 288 spectral bands (0.4–1.0 μm) that were averaged down to 72 bands in order to improve the signal-to-noise ratios (Brozena et al., 2007). In Rampant Lion II, the spectral

Table 13.1 Comparison of Published Rampant Lion I and II Operations in 2006 and 2008, Respectively *(After Brozena et al. (2007); Anonymous (2008); McKinny (2009).)*, and USGS HyMap Flights in 2007 *(Kokaly et al., 2008; Anonymous, 2012)*

Operation	Flights	Flight Hours	Altitude	Data Amount	Data Collected
Rampant Lion I	37	226	6–9 km	50 TB	1.5×10^5 km^2 hyperspectral coverage
					3.3×10^5 km^2 stereo true color images
					1.13×10^5 line km magnetic data
					7.3×10^4 line km gravity data
					1.1×10^5 line km SAR (radar) data
USGS HyMap	28	–	~16 km	–	39,546 line km hyperspectral cover
					438,012 line km area coverage
Rampant Lion II	40	335	6–9 km	30	Not released

The sources do not agree on all parameters but a general consensus for some data is possible; much is not released as well, probably because of the necessity of military security and the uses in fighting the Taliban to which some of the information was put. TB = terabytes.

bands were extended from 0.4 μm up to 2.5 μm, a thermal imaging camera was added, as well as a high-altitude scanning topographic LiDAR system (Table 13.1).

Some 65,000 high-resolution photogrammetric images were collected over Afghanistan with a 4000 × 4000 pixel, three-color Applanix Digital Sensor System with a pixel resolution of 85–135 cm on the ground, and a locational positioning accuracy on the order of 1–2 m or less. A 60 percent along-path overlap allowed stereographic viewing to see the topography in 3D and stereo models for elevation and slope estimates could be made as well. Such scenes are capable of considerable enlargement "zoom" to larger scales without losing detail. These images were ortho-corrected to remove parallax distortion errors, and are unclassified so that they can be used freely by ground troops in the field, who are otherwise prohibited from carrying sensitive and classified maps and images. In fact, some images were provided to ground combat forces within 48 h of landing, which is an unprecedented rate of delivery of information from aerial information acquisition in a combat theater (Brozena et al., 2007) (Figure 13.10).

In contrast to the Rampant Lion missions, the USGS HyMap operation in 2007 utilized an NASA WB-57F high-altitude research jet that was deployed to

Figure 13.10 A photograph from part of a mosaic of high-resolution digital image acquisition over the Kajakai Dam and reservoir on the Helmand Province in southern Afghanistan. The dam was built by the USA in the 1950s but is now a hotbed of insurgent Taliban fighters, as well as opium farming. This and other images were provided to ISAF personnel fighting Taliban in the region in 2006. *Photograph after Brozena et al. (2007).*

Figure 13.11 Photograph of NASA WB-57F aircraft 928 on the ground at the Kandahar International Airport, working for the US Geological Survey without its usual NASA logo on the tail. Originally built as a Martin RB-57 Canberra tactical bomber in the 1960s, it was rebuilt by NASA for high-altitude weather research but has also been put to a number of other uses (communications, hyperspectral image capture, etc.). Apparently when on non-NASA missions, the logo is removed.

Afghanistan with the NASA logos removed, probably because it was being used by USGS to collect Airborne Visible Infra Red Imaging Spectrometer (AVIRIS) for assessments of mineral assemblages, rather than a strictly NASA-oriented project (Figure 13.11). The HyMap imaging spectrometer had 512 cross-track

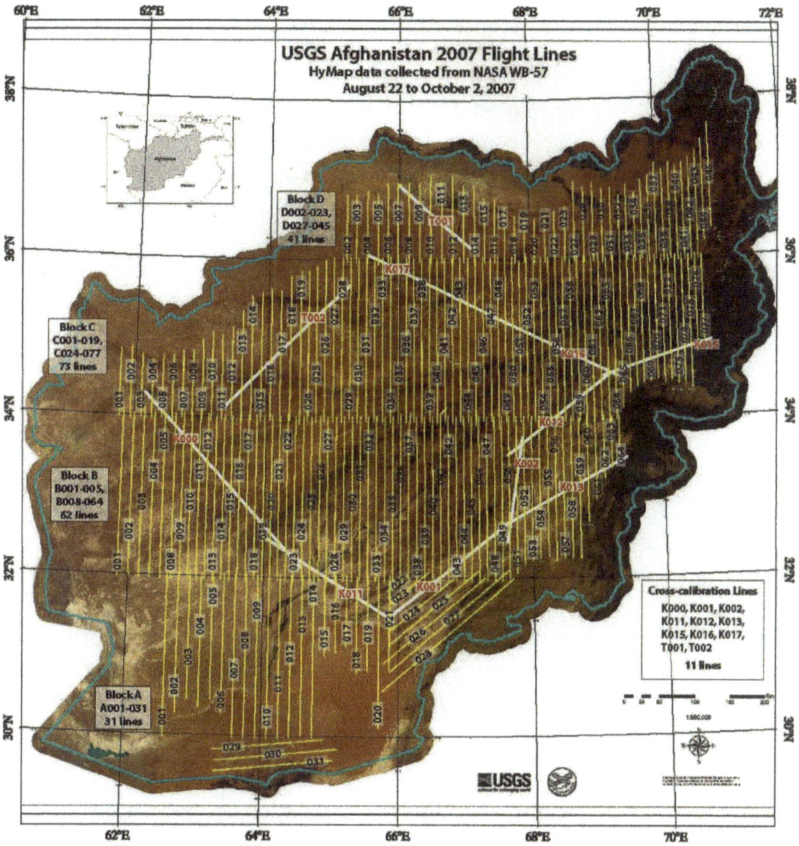

Figure 13.12 Plot of the actual flight lines used by the NASA aircraft WB-57F during the USGS HyMap mission of 2007 over Afghanistan to collect hyperspectral data. The cyan line is the border of Afghanistan with the neighboring countries. *Map after Kokaly et al. (2008).*

pixels and covered the wavelength range 0.43–2.48 microns in 128 channels with four detectors in visible and near-infrared, near-infrared, shortwave infrared 1, and shortwave infrared 2 in the sensor. This would allow production of thematic maps showing distributions of certain selected minerals and vegetation, with applications in detecting commercial mineral deposits, natural hazards, hydrology, and infrastructure assessments (Kokaly et al., 2008). Unlike the propeller-driven Orion aircraft of the Rampant Lion operations that had a lower altitude flight ceiling of about 6–9 km, the WB-57F could fly at about 16 km altitude over the higher altitudes so that much more of the high Hindu Kush could be imaged (Figure 13.12).

The data from the three missions have not all been released, but enough has been interpreted and given out so that one can observe the general geophysical foundations of much of Afghanistan in a remarkable fashion that is actually more complete than that of the USA, or indeed, most other countries of the world. For example, the gravity-anomaly material (Brozena et al., 2007; Jung et al., 2012) allows the interpretation of structural relations (faults, folds) between rocks of different types, as well as between uplifted mountains, sedimentary plateaus, plains, and sedimentary basins. Gravity anomalies are the differences between the observed acceleration of Earth's gravity and the values predicted from some model of how the gravity would be predicted to appear. Positive anomalies show more gravity than expected, whereas negative anomalies exhibit less. The free-air anomaly is the rate of change of gravity with distance up from the datum, which is generally about sea level. The free-air gravity anomaly map of Afghanistan (Figure 13.13) shows the mountains as relative gravity highs (positive), presented as warm reddish colors, and compared to the relative lows (negative) in the plains of the sedimentary basins (cool blues). Salt domes, which can be excellent oil traps, generally show gravity low because of the low density of the mobile salt (cool blues), compared to the surrounding sedimentary rocks they intrude, which themselves can be bowed up into oil-trapping anticlinal domes, which without salt can be the reddish positive highs. Both types show in the north of Afghanistan at this scale as interesting potential oil traps, but at the large scales also possible to produce from these data sets, the resolution will be even more detailed and interesting to guide hydrocarbon exploration. At larger scales, local positive anomalies might otherwise indicate metallic ores of higher density.

The complete Bouguer gravity anomaly is a correction for the height at which it was measured, plus the attraction of the terrain itself. It is actually the free-air anomaly, plus the mass of the material that exists above the same datum,

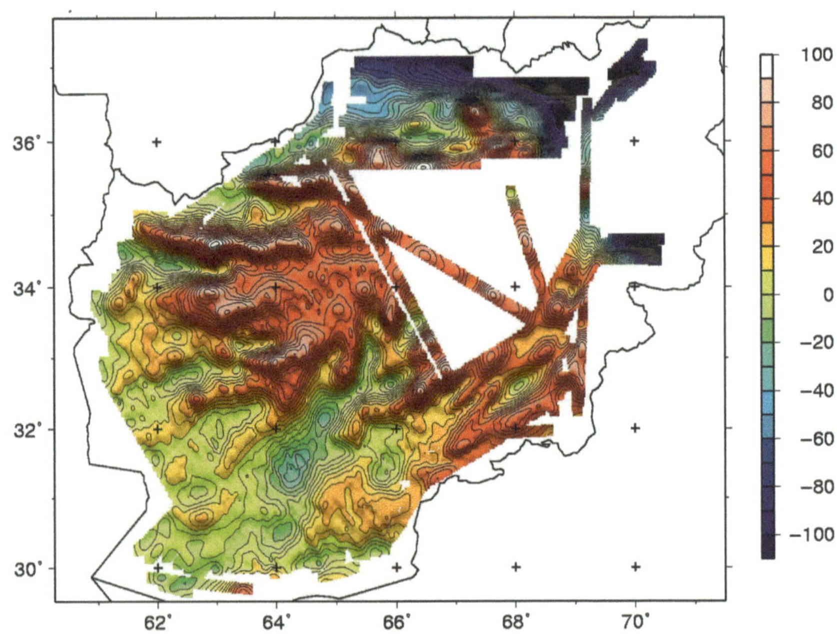

Figure 13.13 Plot of the free-air gravity anomaly map of Afghanistan showing the relative gravity highs produced by the mass of the mountains (warm reddish colors), in contrast to the relative lows of the plains (cool blue greens). *After Brozena et al. (2007).*

generally sea level. Bouguer gravity anomalies are generally more negative in mountains because of isostasy wherein mountains of lower density crustal rocks tend to have deep roots of lower density rocks, rather like low-density icebergs floating in water, where 9/10 of the berg mass is below the waterline. The crustal roots of mountains have a lower rock density "floating" and surrounded in a sea of higher density mantle material down at depth. They are ideal for geophysics because they show many effects of different rock densities concealed below the surface.

The complete Bouger gravity anomaly map of Afghanistan (Figure 13.14) is produced by removing the estimates of the mass effects of the topography and the altitude. Subsurface mass variations, plus deviations produced by some of the assumptions used in the initial modeling control the overall residual Bouguer gravity anomaly. At a small scale like this, the Bouguer gravity anomaly is indicative of crustal thickening in the mountain regions where millions of years of accreting, largely lower density sedimentary rock terrains have piled up into mountains to produce a negative anomaly (cool blue colors). This is compared to thick sections of even less-dense less-consolidated sediments that fill the Helmand sedimentary basin to the south and the Afghan–Tajik basin to the north (warm red colors) indicating a positive anomaly where a sedimentary basin has been strongly downwarped along some east–west trending crustal flexural zones, which show clearly on both types of gravity maps (Figures 13.13 and 13.14).

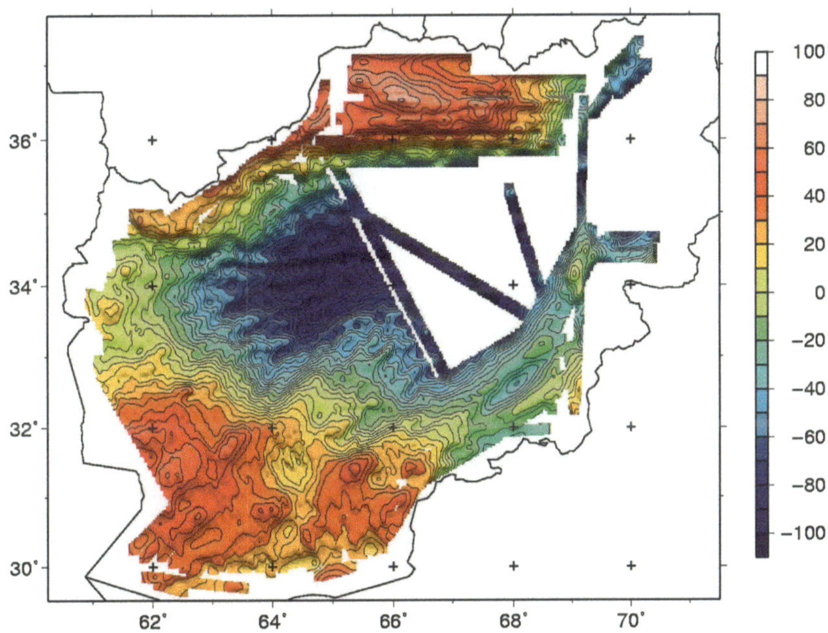

Figure 13.14 Plot of complete Bouguer gravity anomaly map produced by removing the estimated effects of the mass variations of topography and of altitude. The resulting residual anomaly is an indicator of subsurface mass variation as well as some assumptions made in the modeling used as a base. *After Brozena et al. (2007).*

Geophysical manipulation of the data even allow estimation of the sedimentary rock thickness in the basins, with perhaps a 5 km thickness of sediments north of the Hindu Kush, which is a good indication for more potential oil traps in that region. Most high mountain valleys probably have sediment thicknesses of <1 km. The Helmand basin on the other hand does not have the large gravity low that might be expected if it was a thick sedimentary basin with oil and gas potential. It does have a number of positive anomalies that might be either batholiths or small deeper basins (Finn, 2007). The presence of the Khaneshin carbonatite (low density) volcano that deflected the Helmand River at about 64° east longitude and 30°30′ north latitude shows a slightly negative free-air anomaly and a slightly positive Bouguer anomaly, which lends credence to the notion of possible batholithic variations at depth, also thereby precluding oil and gas because of the possible presence of hydrocarbon-absent igneous rocks.

Magnetic anomalies occur in rocks that have magnetically susceptible minerals such as iron ore with magnetite minerals, or those with many ferromagnesian minerals typical of mafic igneous or metamorphic rocks. Areas occur with either a higher (positive) or a lower (negative) magnetic anomaly than the average magnetic field of an area. Thus the igneous and metamorphic rocks are generally far more magnetic than sedimentary rocks where any iron-bearing ferromagnesian minerals tend to have never been there in the first place, or have weathered away. The result is that positive anomalies can be created by irregularities in the buried basement rocks beneath a cover of sediments. Similarly, negative anomalies may form where sedimentary basins or troughs occur, as well as where faults have dropped down the basement in grabens or other structures. The Earth's magnetic field is a composite of magnetic anomalies of different and varying frequencies. High-frequency magnetic events are created by geological variations in the shallow subsurface, whereas the lowest frequency events are caused by contrasting magnetic properties at or beneath the surface of the crystalline basement rocks. Faulting, which abounds in a country such as Afghanistan, juxtaposes rocks of contrasting magnetic variation, and important lateral variations can result from some combination of faulting, deposition, and secondary mineralization emplaced by groundwater along structural displacements such as fault planes (Figure 13.15).

Solar energy causes diurnal (daily) magnetic currents in the Earth's crust that must be removed to see the true background magnetism of the rocks. For the best results, a base station must be located on the ground near the center of the area to be flown for detecting magnetic anomalies. The range of the base station is about 100 km. Accordingly in Afghanistan, five base-station magnetometers were established in Kandahar, Kabul, Herat, Faizabad, and Shiberghan, and the midnight readings averaged over 22 nights were used to calculate the diurnal variation to be removed. Other variations of data spikes and noise in the base-station records were removed with statistical filters, and additional processing removed line-leveling noise and elevation variations. Everything was projected to a nominal 5000 m above the terrain and the final magnetic anomaly grids were

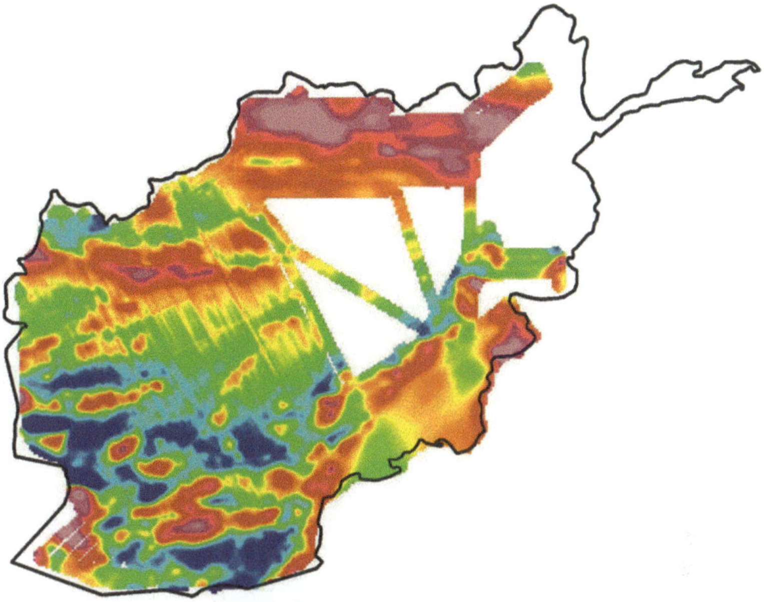

Figure 13.15 Plot of the preliminary data set of magnetic anomalies of Afghanistan that has only been corrected for diurnal variation at the Kandahar International Airfield. Other diurnal variation can be detected northwest of the Kandahar location where diurnal artifacts occur as streaks following the flight lines. *After Brozena et al. (2007).*

Figure 13.16 Three engineer-geophysicists from the AGS operating a proton-precession magnetometer along the Ghorband River in central Afghanistan in September, 2006. *After Shenwary (2011).*

merged with the ground magnetic survey data that were run in September 2006 by the Afghanistan Geological Survey along the Ghorband and Hari Rud river valleys through the central, east–west running fault-trained valleys (Shenwary et al., 2011). The magnetic data were reduced to the pole to center the anomalies over their geologic sources (Finn, 2007) (Figure 13.16).

The results of these surveys constitute a first detailed look at the magnetic anomalies of Afghanistan, in which a number of interesting and informative anomalies show prospective mineral and hydrocarbon locations possibly

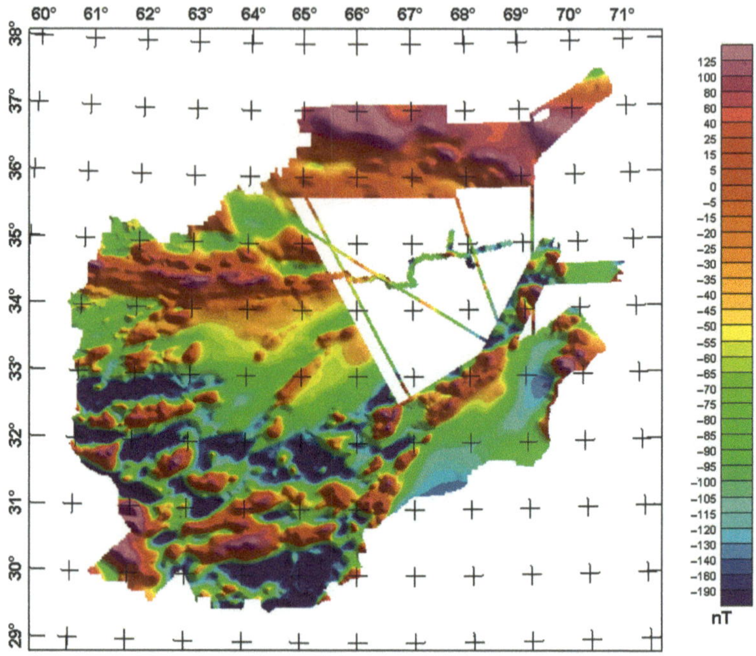

Figure 13.17 Plot of the corrected and merged multiple Afghanistan magnetic anomaly grids showing that most of the diurnal variation of the magnetic fields has been removed. The serpentine magnetic values through the central Hindu Kush where only three linear flight line readings can be seen are the ground survey data collected by personnel of the AGS. The simulated elevation of the displayed magnetic grid is 5000 m above terrain, and the illumination is from the northeast. *After Shenwary et al. (2011).*

worthy of exploitation (Figure 13.17). These magnetic data can detect accreted island arc terranes and other Precambrian crystalline basement blocks of interest because of their mineral resource potential. In addition, the magnetic highs associated with the Hari Rud, Helmand, and Chaman strike-slip fault systems separate highly magnetic from nonmagnetic rocks, which may allow modeling of fault dips that may in turn aid in assessing the seismic hazard potential.

The geophysical data sets in combination commonly reinforce each other as well. For example, gravity and magnetics in combination (Figure 13.18) show that the Katawaz area close to the border with Pakistan has low magnetic anomalies that could have oil and gas prospects, which supposition is reinforced by low rock densities on a gravity manipulation that also indicates hydrocarbon prospects (Aikins, 2010). An additional representation of a far more sophisticated combination of data sets is that done by Hubbard et al. (2011), in which a number of priority areas based on overlapping anomalies indicate the need for further work, particularly in the field on the ground. The anomalies are (1) geophysics (gravity and magnetics—GEO); (2) ASTER satellite

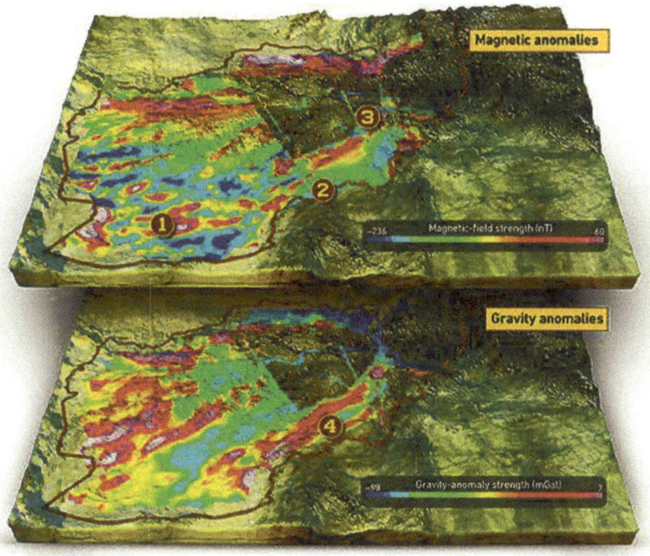

Figure 13.18 Comparison of magnetic to gravitational anomalies over Afghanistan. 1, Magnetic anomaly high at Khanneshin on the lower Helmand indicating the presence of rare-earth metals; 2, Low magnetic anomaly that could indicate hydrocarbon potential (oil, gas, and coal); 3, Magnetic anomaly high indicating presence of copper in the Aynak region; 4, Gravity low that further supports the hydrocarbon possibility of item 2. *After Aikins (2010).*

spectral signatures (AST); and (3) HyMap hyperspectral data (HSD). Mineral anomaly locations number 334, with 225 overlapping anomalies, of which:

- 55 are AST and HSD;
- 107 are HSD and GEO;
- 34 are AST and GEO;
- Several are AST, GEO, and HSD.

Other influences on prioritizations are the following:

- Number of data sets providing overlapped agreement.
- Other known and nearby mineral occurrences.
- Proximity to roads, infrastructure, and major population to provide work support.
- General knowledge base about particular mineral types and associations.

Five higher priority areas of interest (AOIs) were designated from this work (Figure 13.19):

1. Tourmaline AOI (tin vein, tungsten, precious and base-metal sulfides)—overlap HSD, AST, and GEO.
2. Zarkashan AOI (lode gold)—overlap HSD, AST, and GEO.

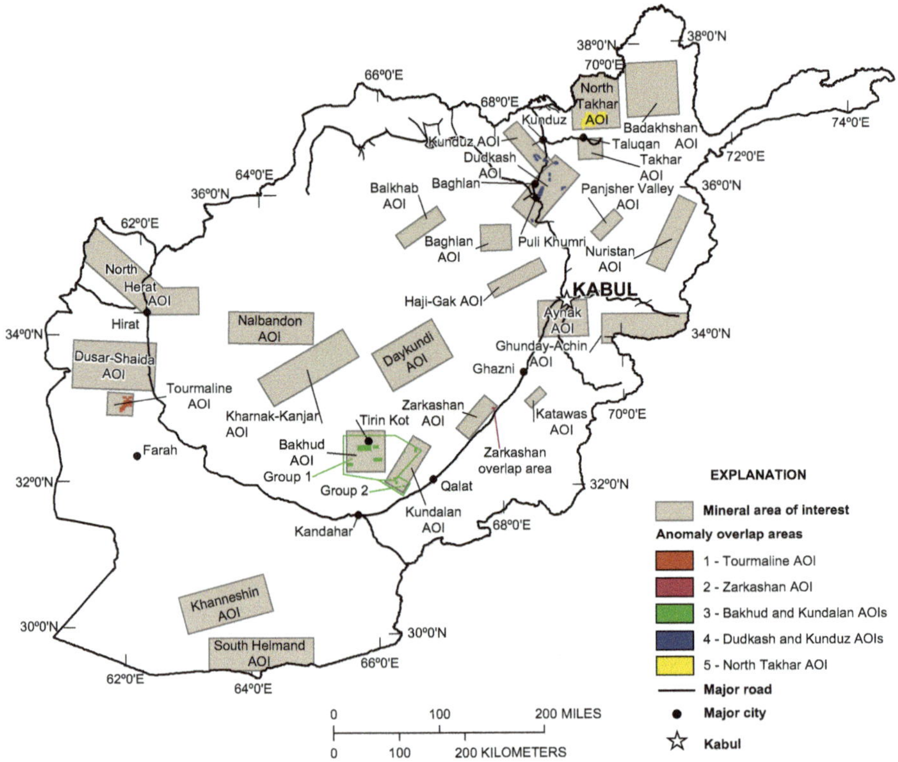

Figure 13.19 Location map showing selected overlapping anomalies in mineral AOI defined by the source data (geophysical, GEO; ASTER spectral, AST; and HyMap hyperspectral, HSD. 1, Toutmaline AOI; 2, Zarkashan AOI; 3, Bakhud and Kundalan AOIs; 4, Dudkash and Kunduz AOIs; 5, North Takhar AOI). *After Hubbard et al. (2011).*

3. Bakhud and Kundalan AOI (fluorite and porphyry)—overlap HSD and GEO with gravity lows and magnetic highs, coupled with strong surface alteration.
4. Dudkash and Kunduz AOI (coal and/or bauxite)—overlap AST and HSD.
5. North Takhar AOI (placer gold)—overlap AST and GEO with magnetic high probably in basement and not relevant to more surficial placer deposits above it.

The HyMap hyperspectral results obtained from the 2007 efforts are huge data sets that do take considerable time to process; nonetheless, many results have been released and reasonable observations can be made about what has been discovered (Brozena et al., 2007; King et al., 2011a,b,c, 2012; Kokaly et al., 2011). The HyMap results included more than 8×10^8 pixels of imaging spectrometer data, which after conversion to reflectance and georegistration, were compared against a library of 97 standard reference spectra of various surficial materials. The resulting maps have a fairly massive and bewildering breakdown of recognized mineral types but are useful to the professional, especially in combination with the other imaging data sets. These imaging spectrometer data are presented in two maps: the first is the distribution of iron-bearing minerals with absorption features in the visible and near-infrared wavelengths

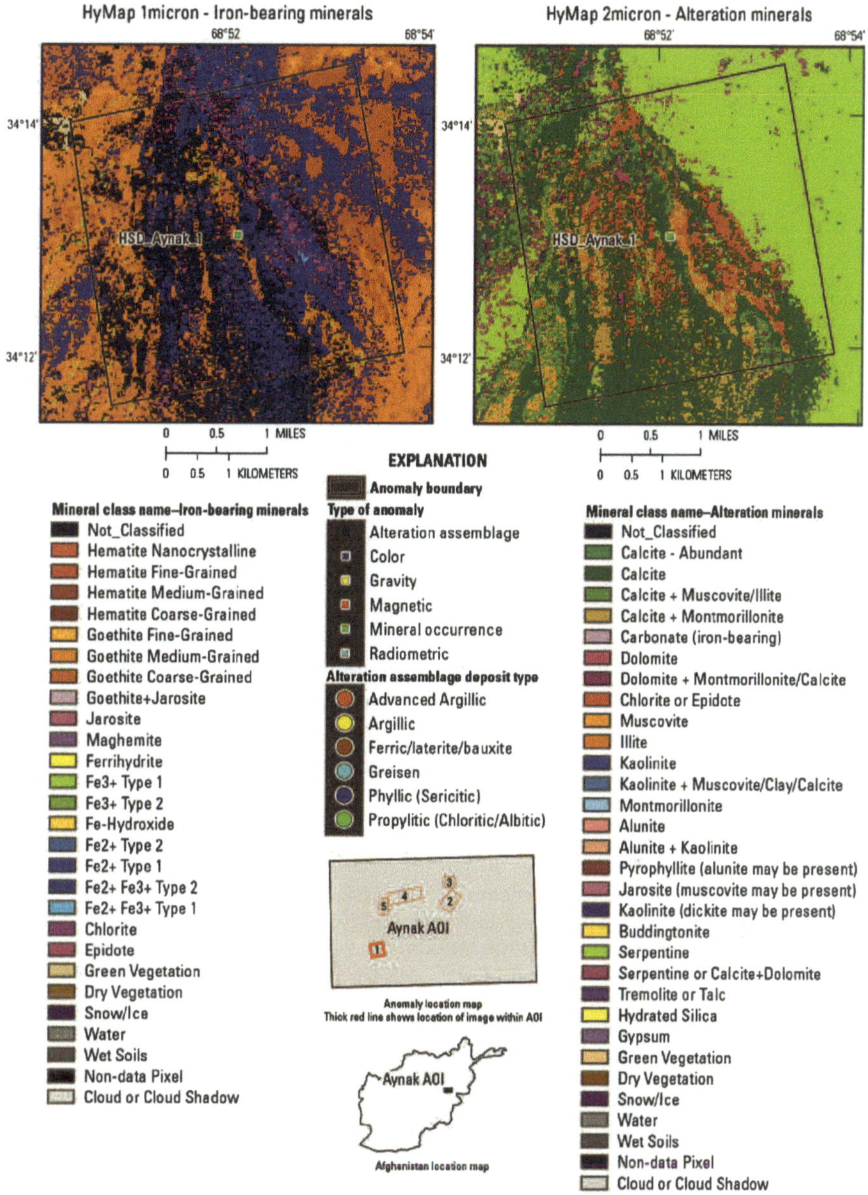

Figure 13.20 Hyperspectral anomaly located in the Aynak AOI, which is based upon the combined presence of chlorite or epidote, serpentine, illite clay mineral, dolomite minerals, ferric iron, and goethite minerals. *After King et al. (2011b).*

(1 μm), and the second is the carbonates, phyllosilicates, sulfates, altered minerals, and other minerals that have their primary absorption features in the short-wave infrared wavelength region (2 μm). These designations were used for exposition of selected AOIs such as the Aynak copper site (Figure 13.20), as

well as for the country as a whole (Kokaly et al., 2011; King et al., 2011c). The significance of these particular mineral designation types is that many of these minerals are indicative of zones of hydrothermal alteration, which is where good ore mineralization can be located; thus they are another prospecting guide, but ultimately the boots-on-the-ground geologist with hammer in hand is an essential element as well. This activity, however, is difficult and expensive when coupled with the necessary military security required for elemental survival in the war-torn region.

For this reason, the Task Force for Business and Stability Operations (TFBSO) was set up by the US Department of Defense (DOD) to provide economic stabilization in order to reduce violence, enhance stability, and try to restore economic normalcy in those areas where unrest and insurgency have produced downward spiraling cycles of economic hardship and violence (Anonymous, 2013). TFBSO tries to create stabilization by developing economic opportunities through a range of efforts including their own capabilities (water-well drilling, road building, etc.), as well as (1) encouraging investment in the mineral resources by US and international companies; (2) developing the resources in economically sound and environmentally responsible fashions; (3) assisting in industrial development; and (4) agricultural revitalization. This is nation building at its finest, if it can possibly work in benighted Afghanistan. TFBSO's only real problems are bureaucratic snarls typical of so many efforts directed from Washington and the Pentagon, coupled with some probable cultural or legal corner cutting to accomplish objectives in trying to operate from within the rather stultifying or Byzantine legal frameworks of Afghanistan.

References

Aikins, M., September 2010. The Treasure of the Safit Chir. Popular Science, 6 p. http://www.popsci.com/science/article/2010-08/treasure-safit-chir.

Anonymous, 2008. Airborne Gravity Survey and Ground Gravity in Afghanistan: A Website for Distribution of Data. US Geological Survey OF 2008-1089, 3 p. http://pubs.usgs.gov/of/2008/1089/Afghan_grv.html.

Anonymous, July 17, 2012. Afghanistan the First Country Mapped Using Broad Scale Hyperspectral Data. US Geological Survey, New release; 2 p. http://www.usgs.gov/newsroom/article.asp?ID=3280.

Anonymous, 2013. TFBSO Task Force for Business & Stability Operations. TFBSO Report; 2 p. http://tfbso.defense.gov/www/index.aspx.

Brozena, J., Childers, V., Gardner, J., Liang, R., Jarvis, J., Bowles, J., 2007. A Multi-Sensor Aerogeophysical Study of Afghanistan. Naval Research Laboratory (NRL), Review: 107–117 http://mapserver.cmf.nrl.navy.mil/FA7rev.pdf.

Burgess, R.R., 2006. Beneath the Surface: A Navy Flight Squadron Conducts a Geological Survey of Afghanistan. Navy League of the United States, News release; 5 p. http://www.navyleague.org/sea_power/nov06-20.php.

Coll, S., 2004. Ghost Wars. Penguin Press, New York, 695 p.

Crile, G., 2003. Charlie Wilson's War: the Extraordinary Story of the Largest Covert Operation in History. Atlantic Monthly Press, New York, 550 p.

Dimarzio, K., Shroder Jr., J.F., Rundquist, D., 1977. Satellite-Image Mosaic of Afghanistan. Remote Sensing Applications Laboratory. University of Nebraska at Omaha. first published as a frontispiece in Debon, F., Afzali, H., Le Fort, P., Sonet, J. and Zimmerman, J.L., 1987. Plutonic Rocks and Associations in Afghanistan: Typology, Age and Geodynamic Setting. Mémoire No. 49, Mémoires Sciences de la Terre; Nancy, France; 132 p.

Finn, C.A., 2007. Regional Gravity and Magnetic Data Help Map Subsurface Geology of Afghanistan. Geological Society of America, Abstract #156–9; GSA Denver Annual Meeting https://gsa.confex.com/gsa/2007AM/finalprogram/abstract_129281.htm.

Glicken, M., 1960. Making a map of Afghanistan. Photogrammetric Engineering 26, 743–745.

Hagen, E., Shroder Jr., J.F., Lu, X.X., Teufert, J.F., 2010. Reverse engineered flood hazard mapping in Afghanistan: a parsimonious flood map model for developing countries. Quaternary International 226 (1–2), 82–91.

Haritashya, U.K., Bishop, M.P., Shroder, J.F., Bush, A.B.G., Bulley, H.N.N., 2009. Space-based assessment of glacier fluctuations in the Wakhan Pamir, Afghanistan. Climatic Change 94, 5–18.

Hubbard, B.E., King, T.V.V., Johnson, M.R., Drenth, B.J., 2011. Chapter 6: discussion of anomaly overlap areas and potential for economic development. In: King, T.V.V., Johnson, M.R., Hubbard, B.E., Drenth, B.J. (Eds.), Identification of Mineral Resources in Afghanistan – Detecting and Mapping Resource Anomalies in Prioritized Areas Using Geophysical and Remote Sensing (ASTER and HyMap) Data. US Geological Survey Open-file report 2011 – 1229.

Jung, W., Brozena, J., Peters, M., 2012. Predicting gravity and sediment thickness in Afghanistan. Geophysical Journal International Geophys 192, 586–601. http://dx.doi.org/10.1093/gji/ggs038. http://gji.oxfordjournals.org/content/192/2/586.full.pdf+html.

King, T.V.V., Johnson, M.R., Hubbard, B.E., Drenth, B.J. (Eds.), 2011a. Identification of Mineral Resources in Afghanistan – Detecting and Mapping Resource Anomalies in Prioritized Areas Using Geophysical and Remote Sensing (ASTER and HyMap) Data. US Geological Survey Open-file report 2011 – 1229. 327 p. http://pubs.usgs.gov/of/2011/1229/.

King, T.V.V., Kokaly, R.F., Hoefen, T.M., Dudek, K.B., Livo, K.E., 2011b. Surface Materials Map of Afghanistan: Iron-Bearing Minerals and Other Materials. U.S. Geological Survey Scientific Investigations Map 3152–B, one sheet, scale 1:1,100,000, Also available at http://pubs.usgs.gov/sim/3152/B/.

King, T.V.V., Hoefen, T.M., Johnson, M.R., Kokaly, R.F., Livo, K.E., 2011c. Chapter 5: mapping anomalous mineral zones using HyMap imaging spectrometer—data for selected areas of interest in Afghanistan. In: King, T.V.V., Johnson, M.R., Hubbard, B.E., Drenth, B.J. (Eds.), Identification of Mineral Resources in Afghanistan – Detecting and Mapping Resource Anomalies in Prioritized Areas Using Geophysical and Remote Sensing (ASTER and HyMap) Data. US Geological Survey Open-file report 2011 – 1229, pp. 158–320. http://pubs.usgs.gov/of/2011/1229/.

King, T.V.V., Kokaly, R.F., Hoefen, T.M., Johnson, M.R., 2012. Hyperspectral remote sensing data maps minerals in Afghanistan. Eos, Transactions American Geophysical Union 93 (34), 325. http://onlinelibrary.wiley.com/doi/10.1029/2012EO340002/abstract.

Kokaly, R.F., King, T.V.V., Livo, K.E., 2008. Airborne hyperspectral survey of Afghanistan 2007: Flight line planning and HyMap Data Collection. USGS Afghanistan Project Product No. 186. USGS Open-file report 2008 – 1235, 14 p. http://pubs.usgs.gov/of/2008/1235/downloads/OF08-1235_508.pdf.

Kokaly, R.F., King, T.V.V., Hoefen, T.M., Dudek, K.B., Livo, K.E., 2011. Surface Materials Map of Afghanistan: Carbonates, Phyllosilicates, Sulfates, Altered Minerals, and Other Materials. U.S. Geological Survey Scientific Investigations Map 3152–A, one sheet, scale 1:1,100,000, Also available at http://pubs.usgs.gov/sim/3152/A/. http://pubs.usgs.gov/sim/3152/A/.

McKinney, D., January 26, 2009. NRL's Rampant Lion II Survey Builds on Success of Rampant Lion I. Naval research Laboratory, News Release 2 p. http://www.nrl.navy.mil/media/news-releases/2009/nrls-rampant-lion-II-survey-builds-on-success-of-rampant-lion-I.

Sarikaya, M.A., Bishop, M.P., Shroder, J.F., Olsenholler, J.A., 2012. Space-based observations of eastern Hindu Kush glaciers between 1976 and 2007, Afghanistan and Pakistan. Remote Sensing Letters 3 (1), 77–84. http://dx.doi.org/10.1080/01431161.2010.536181.

Shenwary, G.S., Kohistany, A.H., Hussain, Sardar, Ashan, Said, Mutty, A.S., Daud, M.A., Wussow, M.W., Sweeney, R.E., Phillips, J.D., Lindsay, C.R., Kucks, R.P., Finn, C.A., Drenth, B.A., Anderson, E.D., Abraham, J.D., Liang, R.T., Jarvis, J.L., Gardner, J.M., Childers, V.A., Ball, D.C., Brozena, J.M., 2011. Aeromagnetic surveys in Afghanistan—an updated website for distribution of data. U.S. Geological Survey Open-file Report 2011 – 1055, 8 p. http://pubs.usgs.gov/of/2011/1055/.

Shroder Jr., J.F., 1978. Remote sensing of Afghanistan. Afghanistan Journal 5, 123–128.

Shroder Jr., J.F., 1980. Special Problems of Glacial Inventory in Afghanistan. World Glacier Inventory Proceedings Reideralp Workshop, September 1978. (IAHS-AISH) Publication No. 126. Hydrological Sciences Bulletin: 142–147.

Shroder, Jr., J.F., 1983. The U.S.S.R. and Afghanistan Mineral Resources. In: International Mineral Resources a National Perspective, ed. A. Agnew, American Association for Advancement of Science, p. 115–153.

Shroder, J., 2008. Remote sensing and GIS as counterterrorism tools for homeland security: the case of Afghanistan. In: Sui, D. (Ed.), Geospatial Technologies and Homeland Security, 94. Springer Science & Business Media B.V.; Geojournal Library, pp. 11–33.

Shroder Jr., J.F., 2009. Saving Afghanistan: redevelopment one resource at a time. Earth 54 (7), 38–47.

Shroder Jr., J.F., 2012. Afghanistan: rich resource base and existing environmental despoliation. Environmental Earth Sciences. http://dx.doi.org/10.1007/s12665-012-1638-7.

Shroder Jr., J.F., Weihs, B.J., 2010. Geomorphology of the Lake Shewa landslide dam, Badakshan, Afghanistan, using remote sensing data. Geografiska Annaler 92A (4), 471–486.

Shroder Jr., J.F., Bishop, M.P., 2010. Glaciers of Afghanistan. In: Williams Jr., R.S., Ferrigno, J.G. (Eds.), Satellite Image Atlas of Glaciers of the World – Asia. U.S. Geological Survey Professional Paper 1386–F3, pp. 167–199.

Shroder Jr., J.F., Weihs, B., Schettler, M., 2011a. Mass movement in northeast Afghanistan. Journal of Physics and Chemistry of the Earth 36, 1267–1286. http://dx.doi.org/10.1016/j.pce.2011.03.003.

Shroder, J.F., Schettler, M.J., Weihs, B.J., 2011b. Loess failure in northeast Afghanistan. Journal of Physics and Chemistry of the Earth 36, 1287–1293. http://dx.doi.org/10.1016/j.pce.2011.03.001.

Shroder Jr., J.F., Bishop, M.P., Bulley, H.N.N., Haritashya, U.K., Olsenholler, J., 2007. Global land ice monitoring from space (GLIMS) project regional center for Southwest Asia (Afghanistan and Pakistan). In: Baudo, R., Tartari, G., Vuillermoz, E., Shroder Jr., J.F. (Eds.), Mountains, Witnesses of Global Changes: Research in the Himalaya and Karakoram. Under the Auspices of SHARE-Asia (Stations at High Altitude for Research on the Environment in Asia). Developments in Earth Surface Processes 10. Elsevier.

Tomsen, P., 2011. The Wars of Afghanistan: Messianic Terrorism, Tribal Conflicts, and the Failures of the Great Powers. Public Affairs, Perseus Books, New York, 849 p.

In this Chapter

14

Resource Rushes in Afghanistan

Abstract

The idea of a resource rush in Afghanistan begins with military nation building wherein an entity such as the Task Force for Business and Stability Operations (TFBSO) from the US Department of Defense sought business opportunities with new resource extractive efforts to jump start the Afghan economy. Investment opportunities were established by obtaining all the old Soviet maps and reports, utilizing modern remote-sensing techniques, and new ground sampling programs to build up mineral-bidding packages that could be used to attract external mining companies. An extractive industries transparency initiative was also established to reduce corruption. Neighboring powerful countries such as China and India were allowed to bid successfully on, and assume control of a number of important resources.

Keywords: Extractive industries transparency initiative (EITI); Mineral-bidding packages; Mineral-investment opportunities; Task Force for Business and Stability Operations (TFBSO).

The gold rushes of the nineteenth century all over world (California, Alaska, South Africa, Australia, New Zealand, etc.) have nothing on what is transpiring in Afghanistan at the present time, except perhaps the lack of mining camps springing up with all of the apocryphal excesses of drinking, gambling, and prostitution, although Afghanistan has its own variants on those storied excesses. But the presence of armed hostiles all about in the nineteenth century who objected to the mining process by outsiders is, of course, well represented in Afghanistan. The next steps will be to see who gets rich in the process, and the most operant question of all is, will the Afghan people make it into the twenty-first century with a prosperous share of the riches, or will their near neighbors run off with the goodies, so to speak? To investigate possible answers to this problem, it is necessary to see what has been done in the recent past, and where and how things might go in the future.

Natural Resources in Afghanistan. http://dx.doi.org/10.1016/B978-0-12-800135-6.00014-3

Military Nation Building in Afghanistan

Armed nation building has been greatly criticized as it is considered not possible to build society at the barrel of a gun, but in fact, the USA has had a long history of such activities in its own domains, and has necessarily been fully engaged in such activities in Afghanistan. From the founding days of the American republic, the early military was involved in nation building; they cut down trees and farmed, they built schools and hospitals, and by 1830, they had built 1900 miles of roads. They dredged harbors, erected lighthouses, dug canals, and built bridges, as well as surveying and mapping the new lands to the west. President Jefferson's founding of the West Point Military Academy in 1802 led quickly to the best engineering education in the USA, which thereafter participated in nation building by the military (Tierney, 2010). Something like this is being tried in Afghanistan as well, but it did not start easily and its ending may not be simple (Shamoo, 2011).

During President Bush's election campaign in 2000, he declared that he was opposed to nation building and wanted to pursue a policy of isolationism. The 9/11 attack on US soil quickly jolted the country out of its isolationism and when the bombing campaign by the US began in Afghanistan, Bush famously reiterated that "We're not into nation building, we're into justice." But this limited thinking was soon overturned when it was realized that a major gear shift was required to get Afghanistan back on its feet. In short order in late 2001, preparations were already being made to rebuild the country. Bush himself even began to tout a kind of Marshall Plan to rebuild Afghanistan, but in spite of the overblown rhetoric, nation building barely materialized and few resources were committed to it as resources were instead diverted to Iraq. Bush's so-called new Marshall Plan for Afghanistan was actually the most poorly resourced American venture into nation building in 60 years. Only under the later Obama administration were even marginally adequate resources brought to bear on development problems in Afghanistan (Greenblatt, 2011; Cabe, 2002; Gentile, 2010; Bergen and Gelb, 2009).

The very concept of nation building has come under considerable academic and semantic scrutiny and criticism, in that the terms, *nation building* and *state building* are used differently by political scientists than in the popular vernacular. Nation building to the political scientists refers to the national identity, whereas state building refers to the institutions of the state. The two terms tend to be used interchangeably in North America. Furthermore, the media tends to portray state building as an interventionist action by foreign countries, whereas increasingly international institutions tend to see true state building as a mostly indigenous process. The terminology confusion has come to mean especially that the use of armed force in the aftermath of a conflict underpins a transition to democracy. Unfortunately the deliberate efforts by a foreign power to construct or install the institutions of a national government generally follow a model that may be more familiar to the foreign power, but can be considered too foreign for the culture of the new government, and may even be destabilizing. In this instance in Afghanistan especially, the state building has been characterized by massive investment, military occupation, transitional government, and the use of propaganda to communicate governmental policy. The result in the face of an ongoing resistance by the Taliban may not be a pretty sight or result in a very durable entity.

Criticisms of nation building in Afghanistan have long been pronounced by outsiders who were unfamiliar with the country, that Afghanistan was just a collection of warring tribes and bitter ethnic factions that could not tolerate working with each other to have a state. But the first Afghan state, the Durrani Empire of 1747, occurred long before the US was a state in its own right, and prior to the past three decades of war, the Kingdom of Zahir Shah was fairly successful, albeit quite poverty stricken and with only a weak central government. Two contrasting points of view have arisen in recent years, that the state must be built at the point of a gun, or that the state needs to be abandoned by the western Coalition in an unwinnable war, and left to the citizens to build their own state if they are able. The problem with this thinking is that the so-called "nation-building light" of the Bush administration has already been tried and resulted in a resurgent Taliban. As has been pointed out, if a pared-down counterterrorism strategy works no better the second time around, will the US and Coalition have to invade a second time in the event of a new Taliban takeover after the 2014 drawdown of troops? In fact a third choice is probably the most realistic projection of what will happen in the present conflict.

To help Afghanistan recover from its violent politics of the last three decades, the US must paradoxically engage with the country over the long term, yet also engage only from a careful distance so that the Afghan people can build their own new nation. Having overthrown the Afghan government, such as it was in 2001, the US has an obligation to leave to the Afghans a country that is at least somewhat stable. A possible way to do this is to help the Afghans take advantage of the trillions of dollars of mineral resources that are in the ground that have never been exploited (Risen, 2010). World-class deposits in copper, iron, rare earths, and lithium for batteries, along with plentiful oil, gas, and coal are certainly enough to jump start any small economy such as Afghanistan, but do take time to "come on stream", as the mining, extraction, transportation, refining, and smelting get underway. To this end, the US Geological Survey has spent considerable effort and money getting all the old Soviet resource information translated and revised into the highest quality possible and explained in modern economic terms. To help Afghanistan become much more than the failed state, it has been for the past several decades, the USA must help them keep the peace if it can, while the economy grows by its bootstraps with its own resources so that the Afghans can arise from the traumas of their long wars and create a new nation.

Task Force for Business and Stability Operations

The Task Force for Business and Stability Operations (TFBSO) was set up by the US Department of Defense (DOD) in 2006 to leverage American and international economic power as a strategic tool to promote economic stabilization (TFBSO, 2011). This process is seen to reduce violence, enhance stability, and restore economic normalcy in areas where unrest and insurgency have created a synchronous downward spiral of economic hardship and violence. The TFBSO model creates stabilization by developing economic opportunities through a considerable range of efforts, including encouraging investment by US and international businesses to develop a country's natural resources in a way that is economically sound and environmentally responsible, as well as assisting industrial development and agricultural revitalization.

The Task Force initiated its effort on Afghanistan with the assumption that the absence of a viable economic base was one of the primary factors which historically had led to state failure. In 2009, with a nominal gross domestic product of just over $10 billion, with much of that attributed to foreign assistance, Afghanistan was seen to need industries that could produce sustainable wealth, create jobs, and reduce the country's overreliance on international donors to fund its treasury and armed forces. The answer was thought to be the untold mineral wealth lying beneath the ground that could provide the foundation needed for modern, thriving communities above it. The Task Force used the USGS (Peters et al., 2007) preliminary nonfuel mineral report to estimate the gross in-place value of minerals in Afghanistan at more than $900 billion. By leveraging the power of the market, as opposed to simple foreign assistance, the Task Force considers that it works differently than many other international government and nongovernmental organizations who tend to try and remain "pure" and above the corruptive fray of the marketplace. One of the Task Forces primary methodologies first identifies investment opportunities, then hosts private sector investors where the identified opportunities are located, and then finally, works to ensure that transparent business transactions occur between the investors and the local communities to help reduce corruption and exploitation. To this end in Afghanistan, the Task Force recruited private-sector exploration geologists to join with the other geologists of the USGS and Afghanistan Geological Survey (AGS) to assess the Soviet-era geological data and maps to identify and assess the highest priority areas for further exploration and assessment. The Task Force then: (1) coupled the troves of data collected by the Soviet geologists, together with; (2) cutting-edge hyperspectral remote sensing data (Chapter 13); and (3) on-the-ground sampling work, to thereby create the multidimensional, resource-assessment construct required by international investors before entering an emerging market such as Afghanistan. These three primary sources of data were then integrated to complete the investment packages.

1. *Soviet-era geologic reports and maps.* The Soviet Union spent several decades and untold millions supporting some of its best geologists to survey Afghanistan, who produced hundreds of reports and maps, but in untranslated Russian language and only single-paper copies at the AGS offices. The Task Force digitized and translated all these reports, which then allowed a leap-frogging approach to several stages of mineral exploration that generally otherwise, have to start from scratch. These legacy data sources could then be used to reassess the resources, as for example, the economics of gold figured at the $35/troy ounce of the 1960s, versus the >$1000/troy ounce of 2010s dramatically changes the value of an assessed property. Similarly, the value of rare earths, which was negligible half a century ago, has now escalated into rarified high-technology territories not imagined previously.
2. *Modern remote sensing techniques used by scientific and private-sector exploration communities.* The hyperspectral, ASTER, geophysical data, and other techniques outlined in Chapter 13 now have given Afghanistan more information on its surficial geology than probably any other area of that size in the world, including the USA. This high-tech data trove has enabled advanced prospecting in areas not sampled by the Soviet-era geologists.
3. *New ground sampling.* Because of the great danger from the Taliban to the ordinary field geologist, the DOD Task Force relied upon military protection to allow collection of essential ground samples by USGS, AGS, and private geologists in essential areas. As noted in Chapter 12, some of these comprehensive sampling programs by full helicopter and military support teams resulted perhaps in some of the most expensive rock samples, barring those from the Moon, ever collected. Nonetheless, these samples allow quantitative evaluation of resources in Afghanistan, and result in a clearer, more accurate, and compelling assessment of the economic potential.

This overall strategy of the Task Force was designed to minimize risk for investors, at least in terms of the resource itself, because they know that the data are of the highest quality, from a reliable source, and complies with the high standards that financial backers require to risk venturing into the Afghanistan milieu, with all of its other political and insurgency hazards. The results of the effort have shown recently that many of the deposits analyzed have turned out to be more compelling, more valuable, and more attractive to international investors than realized heretofore (TFBSO, 2011).

The selected findings by the Task Force in 2010 singled out six key findings selected from among 27 total sites surveyed or analyzed across Afghanistan, each of which demonstrates economic potential for development (Table 14.1; Figure 14.1).

- Zarkashan gold and copper porphyry deposit in Ghazni Province, estimated to be worth $30 billion and capable of production within 1–2 years.

Table 14.1 Mineral Resource Project Areas of the US Department of Defense Task Force for Business and Stability Operations (2010–2011)

1. Khanneshin carbonatite
2. Balkhab copper
3. Hajigak iron
4. Takhar placer gold
5. Aynak copper
6. Badakshan lode gold
7. Takhar evaporate
8. Kundalyan gold and copper
9. Dusar-Shaida copper and tin
10. Tourmaline tin
11. Karnak-Kanjar mercury
12. Nalbandon lead and zinc
13. Panshir Valley emerald
14. Katawaz gold
15. Chagai Hills travertine, copper, and gold
16. Baghlan clay and gypsum
17. Zarkashan gold and copper
18. Dudkash industrial minerals
19. Kunduz celestite
20. Herat barite and limestone
21. Bakhus fluorite
22. Uruzgan tin and tungsten
23. Ghunday Achin magnesite and talc
24. Nuristan pegmatites
25. Namaksar lithium salts
26. Godzareh (Gaudi Zireh) lithium salts
27. Dashti Nawar lithium salts

TFBSO (2011).

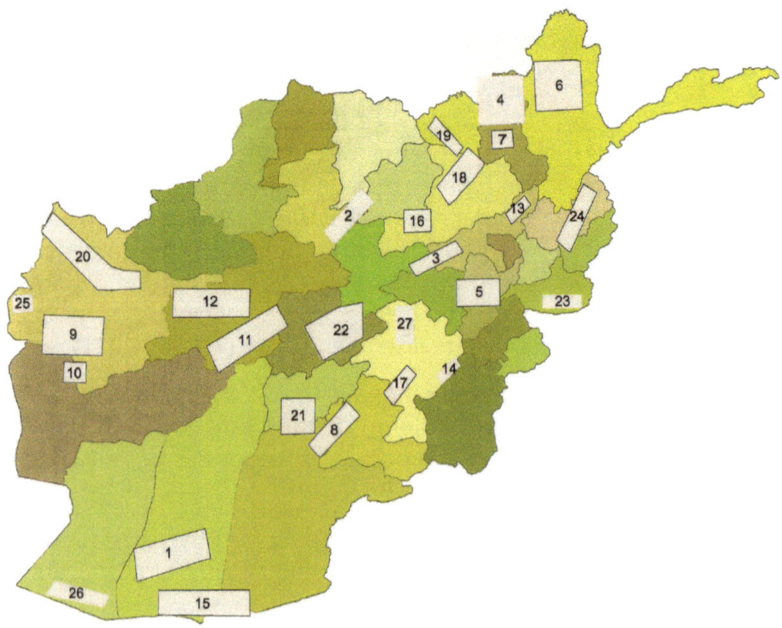

Figure 14.1 Mineral resource project areas of the US Department of Defense Task Force for Business and Stability Operations (2010–2011). See Table 14.1 for an enumerated list of the 27 names of each project area that is keyed to numbers in the figure. *After TFBSO (2011).*

- Lithium production from evaporate deposits in Herat, Ghazni, Nimruz, and Farah provinces worth $60 billion and capable of initial production within 1 year, with large-scale production 2–4 years later.
- Khanneshin rare-earth production in Helmand Province worth $89 billion that is crucial to development of steel alloys, lasers, and super magnets. With 99 percent of world production owned and controlled by China, alternate sources such as this one are a highly strategic resource for Afghanistan that will lead to jobs from low-skilled labor to chemists, physicists, and engineers.
- North Aynak copper and cobalt in Logar province only 30 km south of Kabul inferred to be equal to, or larger than the Aynak copper award to China, such that underground cobalt production could begin within 1–2 years, followed within 5 years with an open-pit copper mine.
- Dusar-Shaida copper deposit in Herat province thought to contain $29 billion in copper alone, far greater than previously recognized.
- Balkhab high-grade copper deposit located in politically and militarily stable area of Balkh Province that could become a significant mining operation in less than 5 years.

With only one of these deposits brought into production, the impact on the economy of Afghanistan will drastically change life for the Afghan people in a completely transformational way. These crucial assets that are so essential to the future of Afghanistan hold the promise of thousands of good jobs for the people of the country, many of whom have known only abject poverty and subsistence living for decades. Furthermore, the royalties and fees paid to the government by the developers of these resources will not only pay for the police and military to ensure security but also lead directly to the construction of highways, railroads, schools, hospitals, and public facilities at all levels that are the very bricks and mortar of any successful society.

Mineral Bidding Packages

The USGS and the DOD TFBSO worked together to summarize their results and to come up with the means to present summaries and data packages of important areas for mineral investment and production opportunities in Afghanistan (Peters, 2011; Peters et al., 2011). Twenty-four areas of interest (AOIs) were identified, leaving three areas of lithium salts from the TFBSO report, but otherwise substantially the same (Table 14.2; Figure 14.2). Each AOI has digital data in the form of geographic information system databases, as well as archival, and non-USGS reports. The data packages contain from 20 to 50 digital layers of data, such as geology, geophysics, and hyperspectral and other remotely sensed imagery. Existing archival reports and maps are mainly from the AGS and are translated Soviet-era reports from the 1960s and 1970s, which in addition are available on several websites as well. Large amounts of technical data are available in these reports from the Soviet Union and its Eastern European

Table 14.2 List of the 24 AOIs for Mineral Resources in Afghanistan and Their Contained Commodities, Together with Minor or Possible Other Commodities, as well as Comments on the Deposits. ?: Probable

AOIs	Subareas	Deposit Models	Main Commodities	Minor or Possible Commodities	Comments
Aynak copper–cobalt (Logar chromite)	Yes	Sediment-hosted copper, podiform chromite	Copper, cobalt, silver	Marble, chromite, asbestos, talc	Reserves at Aynak, Jawhar, and Darband deposits
Badakhshan gold	Yes	Gold-quartz veins, gold and iron skarn	Gold, iron	Silver, copper, uranium	Weka Dur deposit contains 958 kg gold
Baghlan clay–gypsum	No	Bauxite, clay, gypsum	Bauxite, clay (kaolin), gypsum	–	Resources contained in Tala Barfak bauxite deposit, extensive clay deposits
Bakhud fluorite	No	Sediment-hosted fluorite, fluorite vein, polymetallic skarn	Fluorite	Zinc, lead, silver, antimony, barium	Main fluorite district in Afghanistan
Balkhab copper	Yes	Volcanogenic massive sulfide	Copper	Lead, zinc, coal	Balkhab copper prospect; over 3-km strike length
Chagai Hills travertine	No	Travertine, porphyry copper–gold	Travertine (onyx)	Copper, gold, molybdenum	Travertine production, porphyry copper–gold deposits in adjacent Pakistan
Daykundi tin–tungsten–lithium	No	Greisen tin–tungsten, tin–tungsten skarn, lithium pegmatite	Tin, tungsten, lithium	Copper, lead–zinc	Taghawlor lithium pegmatite field, also contains tin and base metals
Dudkash industrial minerals	No	Limestone cement, dolomite, bedded celestite, gypsum clay	Limestone cement, dolomite, celestite, gypsum clay	Coal	Located in Pul-e-Khumri area and in Tangi-Murch celestite deposit

Table 14.2 List of the 24 AOIs for Mineral Resources in Afghanistan and Their Contained Commodities, Together with Minor or Possible Other Commodities, as well as Comments on the Deposits—cont'd

AOIs	Subareas	Deposit Models	Main Commodities	Minor or Possible Commodities	Comments
Dusar-Shaida copper–tin	Yes	Porphyry copper, volcanogenic massive sulfide, tin–copper skarn	Copper, tin	Lead, zinc, tungsten	Shaida porphyry copper prospect
Ghunday-achin magnesite–talc	Yes	Metasomatic magnesite–talc, ultramafic-hosted asbestos	Magnesite, talc, asbestos	Graphite, coal, marble	Acbin and Ghunday magnesite–talc deposits near Tora Bora
Haji-Gak iron	Yes	Volcanogenic iron (?)	Iron	Barium, marble, sandstone, uranium, copper	1.7 billion tonnes + iron ore
Kharnak-Khanjar mercury	Yes	Epithermal mercury, base-metal skarn	Mercury	Antimony, arsenic, gold, silver (?)	Mercury belt
Katawas gold	Yes	Epithermal gold–silver	Gold, silver (?)	Mercury, tungsten (?)	ASTER anomaly
Khanneshin carbonatite	Yes	Carbonatite	REE, uranium, phosphorus	Thorium, barium, strontium, limestone	Significant REE prospect and uranium resources
Kundalan copper–gold	Yes	Porphyry copper–gold and skarn	Copper, gold, molybdenum	Silver, lead	Copper and gold resource; multiple occurrences
Kunduz celestite	No	Bedded celestite, oil and gas, bedded phosphate deposits	Celestite ($SrSO_4$)	Oil and gas	About 1 million tonnes of celestite in speculative resource and Katar oil occurrence
Nalbandon lead–zinc	Yes	Sediment-hosted lead–zinc	Lead, zinc	Silver (?)	Extensive mineral field
North Herat barium–limestone	No	Bedded barite, vein barite, chemical limestone, marble, clay, iron skarn	Barium, limestone, marble, clay, iron	–	Major barite field, marble factory, industrial center

Continued

Table 14.2 List of the 24 AOIs for Mineral Resources in Afghanistan and Their Contained Commodities, Together with Minor or Possible Other Commodities, as well as Comments on the Deposits—cont'd

AOIs	Subareas	Deposit Models	Main Commodities	Minor or Possible Commodities	Comments
North Takhar gold placer	No	Gold placer	Gold	–	Past production, gold resources
Nuristan pegmatite	No	Pegmatites	REE, lithium, tin, and mica	Tantalum, niobium	Paron (Jamanak-Pasgushta) and Pachigram pegmatite fields; lithium resources
Panjsher Valley emerald–iron–silver	No	Emerald, sedimentary iron, and silver	Emerald, iron, silver	–	Major emerald mining area
Takhar evaporite	No	Salt dome, clay, sandstone	Salt, clay silica	Coal, oil and gas	Rock salt deposit at Namakab; porcelain and pottery clay
Tourmaline tin	No	Tin–tungsten vein, placer tin	Tin (tungsten)	Copper (?)	Previous mining for placer tin
Zarkashan copper–gold	Yes	Porphyry copper–gold and skarn	Copper, gold	Silver, lead	Copper and gold resource; multiple occurrences, gold placers

AOIs, areas of interest; ASTER, Advanced Spaceborne Thermal Emission and Reflection; REE, rare-earth element.
After Peters (2011), Peters et al. (2011).

satellite nations of the time, and these have been enhanced by new USGS interpretations and quantitative assessments of the old data. Most of the important AOIs were field checked by USGS and TFBSO geologists from 2009 to 2011, who produced newly generated data that are included in the aggregated data packages for each AOI (Figure 14.3). These comprehensive technical information packages for the individual AOIs were then given to the Afghanistan Ministry of Mines for the preparation of an individual Mineral Bidding Package for the AOIs to be distributed to potential bidders from internal sources, multinational companies, or foreign countries.

The highest priority AOIs were identified by the USGS (Peters, 2011; Peters et al., 2011) to be Badakhshan Gold, Balkhab Copper, Haji-Gak Iron, Aynak

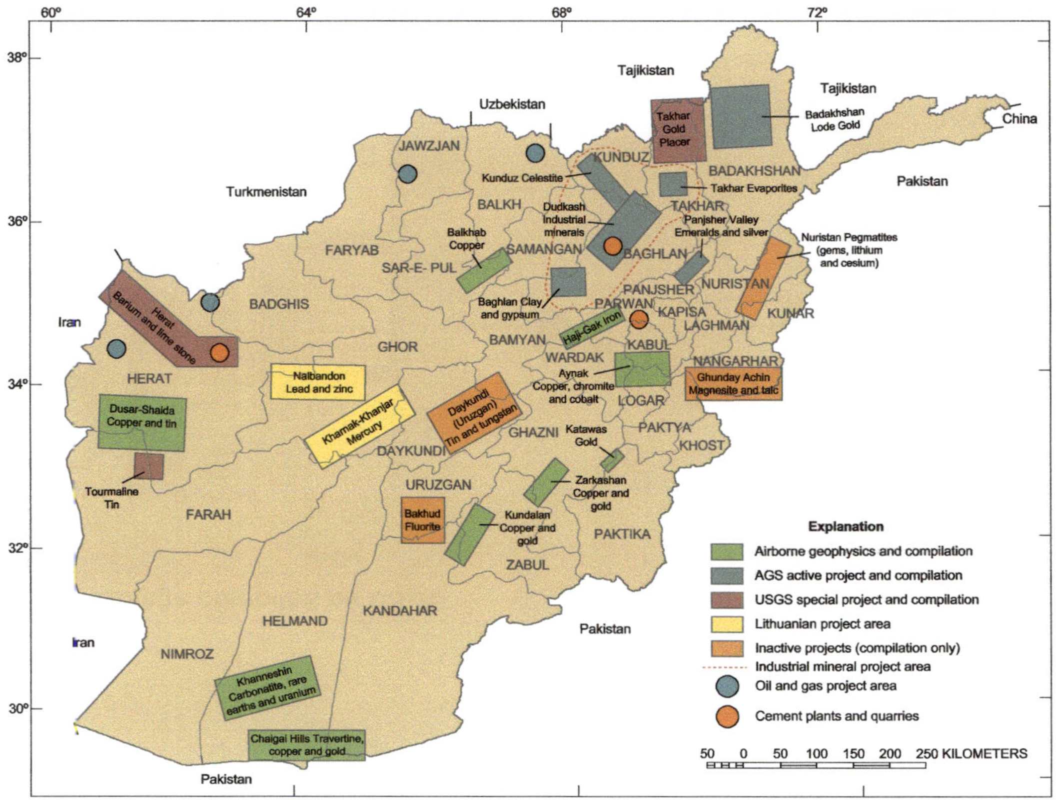

Figure 14.2 Mineral resource project areas that is similar to those in Figure 14.1 but includes additional information concerning the dominant mineral or element type, ancillary data sources, as well as oil and gas project areas, cement plants and quarries, and some notice as to who is responsible for certain projects. *After Peters et al. (2011).*

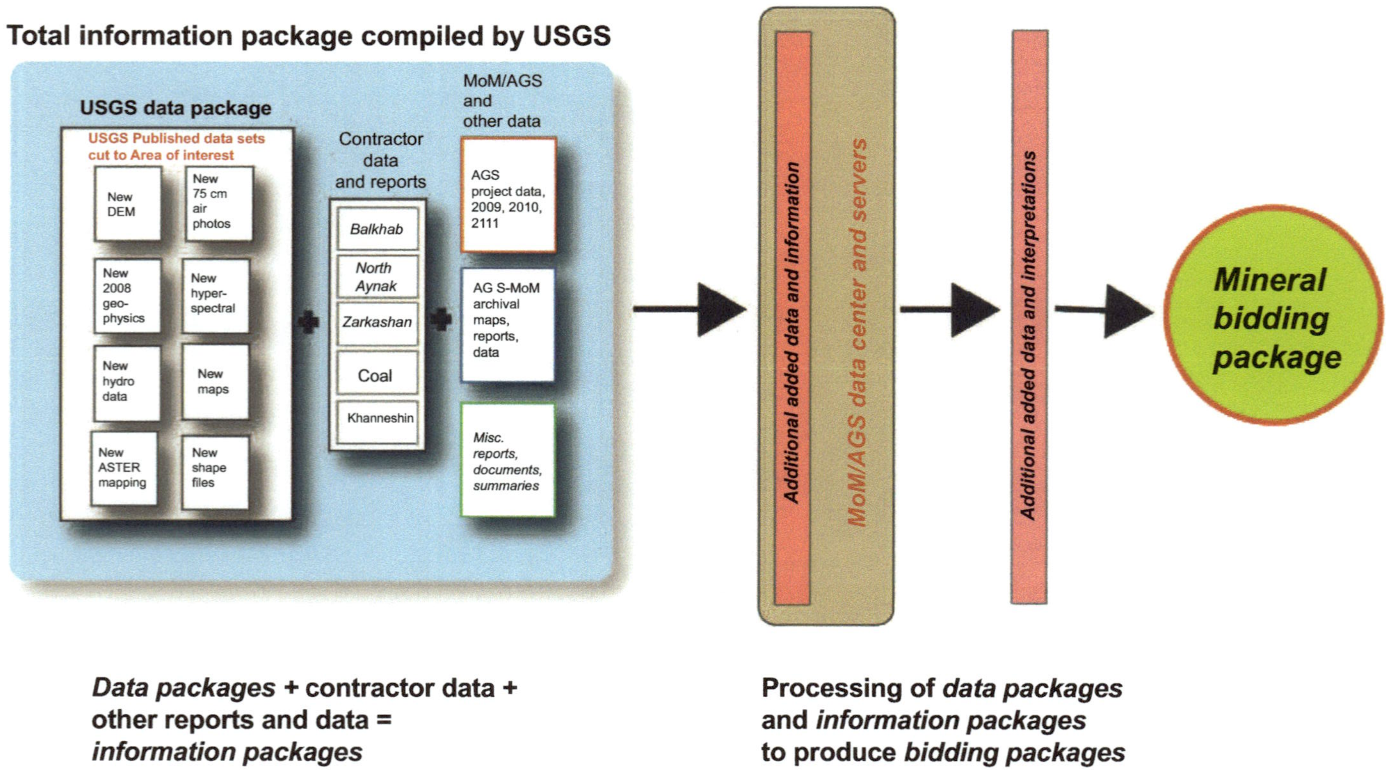

Figure 14.3 Flow sheet of total information data packages compiled by the USGS for bidding on mineral resource project areas. *After Peters et al. (2011).*

Copper–Cobalt, Zarkashan Copper–Gold, Kundalan Copper–Gold, Khanneshin Carbonatite, and Dusar-Shaida Copper–Tin. These AOIs were judged as the most important because they display outcropping mineralization with favorable geometries and reasonably simple metallurgical ore types, which are thought to translate into lower capital costs, shorter lead times to get the deposit producing ore pay, and therefore shorter periods of payback of investment. Furthermore, many of the selected AOIs may contain ore deposits that could be medium to world class in size and tenor. The result of this intensive effort then has been to provide the best available information that has already enabled the Afghanistan Ministry of Mines to attract some major bids from foreign companies and countries such as China and India, for example, on the Aynak copper and the Hajigak iron, with others in the works.

Extractive Industries Transparency Initiative

As the mineral bidding packages were prepared and given out to prospective bidders, other young private citizens in Kabul were watching the process with great interest to ensure that the reported endemic corruption in the ministries would not destroy the prospects for their country to finally prosper. I helped a member of Integrity Watch Afghanistan (IWA) (http://www.iwaweb.org/news.html) to edit both their Aynak as well as their Hajigak reports that were designed as citizen watches to ensure that the government managed the mining and its environmental and economic issues in a beneficial and sustainable way for the future of the country (Nichols, 2011). Because public sector corruption in Afghanistan is seen as worse than any other country except for Myanmar and Somalia, according to corruption watchdog Transparency International, the pressure is on to make sure that foreigners do not corrupt the government any more than they already have, as President Karzai is quick to point out. But a serious concern also exists that the warlord military commanders who built up private armies and fortunes during the years of civil war, and who still command the loyalties of their supporters and gunmen, will seek to profit from the country's mineral resources, which would be a sure trigger for additional violent conflict. In addition, the warlords are quite likely to want to invest in these mineral sectors in order to set up the means to money launder and thereby legalize profits from drugs and other illegal activities. IWA feels that the mining sector in Afghanistan has become too exposed to war profiteering so that the capacity and experience of proposed investors needed to be closely scrutinized to weed out those who were not really capable of performing what they proposed.

Implementing the Extractive Industries Transparency Initiative (EITI) principles and criteria that were developed by the World Bank in 2008 (Anon, 2008) has been seen by IWA and others as critical to the future of Afghanistan to ensure the purity of the bidding and all elements of the resource extractive process. The 12 principles and 6 criteria to ensure such transparency (Tables 14.3 and 14.4),

Table 14.3 EITI Principles Developed by the World Bank (Anon, 2008)

The EITI Principles (edited and summarized)
1. Prudent use of natural-resource wealth should be an important engine for sustainable economic growth that contributes to sustainable development and poverty reduction, but if not managed properly, can create negative economic and social impacts.
2. Management of natural-resource wealth for the benefit of a country's citizens is the domain of sovereign government to be exercised in the interests of national development.
3. The benefits of resource extraction occur as revenue streams over many years and can be highly price dependent.
4. Public understanding of government revenues and expenditure over time could help public debate and inform choice of appropriate and realistic options for sustainable development.
5. The importance of transparency by governments and companies in the extractive industries cannot be overemphasized, as well as the need to enhance public financial management and accountability.
6. The achievement of greater transparency must be set in the context of respect for contracts and laws.
7. Financial transparency almost certainly brings an enhanced environment for domestic and foreign direct investment.
8. The principle and practice of accountability by government to all citizens for the stewardship of revenue streams and public expenditure is a high value.
9. The encouragement of high standards of transparency and accountability in public life, government operations, and in business is a commitment to be undertaken.
10. A broadly consistent and workable approach to the disclosure of payments and revenues is required, which is simple to undertake and use.
11. The payments' disclosure in a given country should involve all extractive industry companies operating in that country.
12. In seeking solutions, all stakeholders have important and relevant contributions to make—including governments and agencies, extractive industry companies, service companies, multilateral organizations, financial organizations, investors, and nongovernmental organizations.

Extractive Industries Transparency Initiative (EITI) principles are developed by the World Bank for application in countries where the extractive industries are clearly in need of a great deal of transparency.

Table 14.4 EITI Criteria Developed by the World Bank (Anon, 2008)

The EITI Criteria (minimally edited)
1. Regular publication of all material oil, gas, and mining payments by companies to governments, and all material revenues received by governments from those companies made to a wide audience in publically accessible, comprehensive, and a comprehensible manner.
2. Where such audits do not already exist, payments and revenues are the subject of a credible, independent audit, applying international auditing standards.
3. Payments and revenues are reconciled by a credible, independent administrator, applying international auditing standards and with publication of the administrator's opinion regarding that reconciliation, including discrepancies, should any be identified.
4. This approach is extended to all companies, including state-owned enterprises.
5. Civil society is actively engaged as a participant in the design, monitoring, and evaluation of this process and contributes toward public debate.
6. A public, financially sustainable work plan for all of the above is developed by the host government, with assistance from the international financial institutions where required, including measurable targets, a timetable for implementation, and an assessment of potential capacity constraints.

Extractive Industries Transparency Initiative (EITI) criteria are developed by the World Bank for application in countries where the extractive industries are clearly in need of a great deal of transparency.

if applied with some degree of rigor in Afghanistan, will go a long way to look after the details of such attempted beneficent actions in a country not greatly known for such heretofore. Such rigor can be ensured because the World Bank supports these efforts through making EITI advisors and consultants available to governments to assist them in implementation, as well as sharing international best practices, and providing grants to governments to help support EITI implementation. The EITI implementation is a means by which the World Bank works with governments on these issues as part of broader Bank-supported programs on reform of extractive industries, natural resource management, and good governance/anticorruption.

On the other hand, people with extensive experience working with the Ministry of Mines in Kabul are more sanguine about the prospects for success of the EITI programs in Afghanistan (Stephen Metcalf email communication, 07/14/2011). The EITI program was not felt to be a very good mechanism for Afghanistan where good governance was thought to be decades away. Metcalf felt that not much hope could be pinned upon it, at least as things stood in his eyes several years ago.

Neighboring Countries Desiring Access to Resources

Two neighboring countries, China and India, that desire access to the resources of Afghanistan have already bid successfully on copper and iron deposits in the country. This indeed augers well for the future of mining in Afghanistan because both countries seem committed to production of the necessary road and rail infrastructure to get the ore and commodities to market, while also maintaining the peace. Furthermore, because the economies of the two huge countries have been experiencing robust growth and the two are in highly competitive rivalries to develop their own nations, as well as to achieve overall greater hegemony in Asia, their beneficial influence on the future political–military situation in Afghanistan will better enable them to help control things. With each looking carefully at competitive and corruptive machinations of the other, the EITI effects will be enhanced, which will help Afghanistan curtail the endemic corruption, as well as the violence. Furthermore, with China so strongly in the resource-extraction effort in Afghanistan, their considerable friendship with Pakistan will help bring pressure on that country toward greater control of the Taliban insurgency against Afghanistan, which will thereby also help India.

Iran does not yet figure much in this resource-extraction effort in Afghanistan, largely because of its position in the Middle East as a spoiler, with economic sanctions against it that make it less competitive in world resources. It does have considerable influence in the Shiite parts of Afghanistan, however, and could perhaps be expected to enter the resource-extraction effort in Afghanistan in future, once the Coalition troops depart. The Central Asian countries north of Afghanistan, Turkmenistan, Uzbekistan, and Tajikistan, similarly do not yet have much at stake in Afghanistan resources, but that too could change, especially insofar as transportation of resources by pipeline and rail across their territories is concerned.

Multinational companies with positions far outside Afghanistan already are involved in Afghanistan resources, and more can be expected the longer time passes and others are seen to be prospering in the extractive industries in the country. The more the neighboring countries and those far from Afghanistan are seen to be making progress in extracting resources and in developing Afghanistan, the greater the beneficial influence will be, as long as the EITI efforts can maintain some semblance of fairness. Still, corruption is everywhere in this commercial world so the bribes will be paid, but the big neighbors looking on who are competitive with each other will be likely to force greater transparency by necessity. Everyone has much to gain and a great deal to lose if it were to be otherwise.

References

Anon, 2008. Implementing the Extractive Industries Transparency Initiative: Applying Early Lessons from the Field. The World Bank, 107 p. http://siteresources.worldbank.org/INTOGMC/Resources/implementing_eiti_final.pdf.

Bergen, P., Gelb, L.H., October 5, 2009. Two Arguments for What to Do in Afghanistan. Time Magazine. http://www.time.com/time/magazine/article/0, 9171, 1927290,00.html.

Cabe, D.K., 2002. Nation Building. Kennedy School Bulletin, Spring. http://www.hks.harvard.edu/ksgpress/bulletin/spring2002/features/nation_building.html.

Gentile, G.P., July 12, 2010. Petraeus's Impossible Mission: Armed Nation Building. The Christian Science Monitor. http://www.csmonitor.com/Commentary/Opinion/2010/0706/Petraeus-s-impossible-mission-in-Afghanistan-armed-nation-building.

Greenblatt, A., November 24, 2011. US Easing Out of Nation-Building Business. NPR. http://www.npr.org/2011/11/23/142699506/u-s-easing-out-of-nation-building-business.

Nichols, M., August 1, 2011. Warlords, Lack of Government Capacity Threaten Afghan Iron Mining: Watchdog. The China Post. http://www.chinapost.com.tw/business/middle-east/2011/08/01/311764/Warlords-lack.htm.

Peters, S.G., 2011. Summaries and Data Packages of Important Areas for Mineral Investment and Production Opportunities in Afghanistan. Fact sheet 2011-3108 US Geological Survey. http://pubs.usgs.gov/fs/2011/3108/fs2011-3108.pdf.

The US Geological Survey-Afghanistan Ministry of Mines Joint Mineral Resource Assessment Team, 2007. In: Peters, S.G., Ludington, S.D., Orris, G.J., Sutphin, D.M., Bliss, J.D., Rytuba, J.J. (Eds.), Preliminary Non-fuel Mineral Resource Assessment of Afghanistan 2007. Prepared in cooperation with the Afghanistan Geological Survey under the auspices of the U.S. Agency for International Development; USGS Open-file report 2007-1214. http://pubs.usgs.gov/of/2007/1214/.

The U.S. Geological Survey Afghanistan Mineral Assessment Team, 2011. In: Peters, S.G., King, T.V.V., Mack, T.J., Chornack, M.P. (Eds.), Summaries of Important Areas for Mineral Investment and Production Opportunities of Nonfuel Minerals in Afghanistan: U.S. Geological Survey Open-File Report 2011–1204. 1810 p. plus appendixes on DVD. Available at: http://pubs.usgs.gov/of/2011/1204/.

Risen, J., June 13, 2010. U.S. Identifies Vast Mineral Riches in Afghanistan. The New York Times. http://www.nytimes.com/2010/06/14/world/asia/14minerals.html?pagewanted=all.

Shamoo, A.E., November 30, 2011. Nation-Building in Afghanistan. Foreign Policy in Focus, Washington, DC. http://www.fpif.org/articles/nation-building_in_afghanistan.

TFBSO, 2011. Mineral Resource Team 2010 Activities Summary. 48 p. http://tfbso.defense.gov/www/attachments/TFBSO_AFGHAN_MINERALS.pdf.

Tierney, D., November 11, 2010. Jefferson's Army of Nation Builders. The New York Times. http://www.nytimes.com/2010/11/11/opinion/11tierney.html.

In this Chapter

15

Resource Corridors

Abstract

The ideas of developing "resource corridors" in Afghanistan were patterned after similar situations elsewhere in the world. These have been set up as effective combinations of mineral resource locations, means of transport, and associated infrastructure to bring a country up to modern economic mining standards. The advanced scheme developed for Afghanistan, the National and Regional Resource Corridors Program, will depend upon development of infrastructure (road, power, water sources, and railways), livelihoods, governance, and associated social and environmental issues. Four main resource corridor segments are envisioned, with three main extensions later if the concept works well for Afghanistan. The entire concept plan has also been developed with four levels of possible development from full flowering if all goes very well, all the way down to full economic collapse of the country in a renewed war. In this fashion, planning for any future eventuality can be accounted for and worked with more effectively.

Keywords: Cross Hindu Kush resource corridor; Governance; Livelihoods; National and Regional Resource Corridors Program (NRRCP); Northern hydrocarbons resource corridor; Resource corridors; Social and environmental issues; Southeast copper resource corridor; Steel and energy resource corridor; Transport infrastructure.

Finally after some 40 years of writing and lecturing on the resource riches of Afghanistan by a variety of people (Abdullah et al., 1980; Ali and Shroder, 2011; Shareq et al., 1977; Shroder, 1983, 1989, 2007, 2012; Shroder and Assifi, 1987; Shroder and Watrel, 1992; Vejlupek, 1980), critical mass seems to have been reached at last, and Afghanistan may actually leap forward with its new resource corridor (RC) approach to development. The somewhat clunky nomenclature notwithstanding because of possible confusion with transportation corridors (NRRCP, National and Regional Resource Corridors Program) (Anon, 2011b; Shahrani, 2013), the RC ideas are now on the table with some sense of urgency in the middle of considerable enthusiasm. The urgency, of course, is engendered by the looming Taliban as the Western Coalition troops prepare to withdraw their forces; the enthusiasm has occurred infectiously as the World Bank has stepped in with vigor to aid the Government of the Islamic Republic of Afghanistan (GIRoA) because most can see the great attractiveness

Natural Resources in Afghanistan. http://dx.doi.org/10.1016/B978-0-12-800135-6.00015-5

and eminent sensibility of the NRRCP. After decades of shifting politics, false starts, dashed hopes, and millions of war deaths, hope is actually on the horizon and could really have a chance.

The concepts of RCs have been around for a while, with earlier examples established in Chile, Mozambique, Zambia, and elsewhere (Anon, 2012a; McMahon and Lokanc, 2011), but applied to the uniquely diverse and potentially quite rich country of Afghanistan, some startling innovations may yet be possible. One of the problems that has developed in any number of prior efforts at resource extraction elsewhere, for example, is that mining enclaves develop to extract the resource, which is then shipped out of the country, leaving behind not much more than a hole in the ground, commonly to the detriment of the local people in many ways. What is planned in Afghanistan, on the other hand, is something quite different. An Afghanistan RC, most simply defined, is a sequence of actions and investments to leverage a large extractive industry (EI) investment in infrastructure, goods, and services, into a viable economic development and diversification process along a defined geographic strip or region. In Afghanistan, the infrastructure part of the equation, particularly the transportation (Chapter 7), is the most critical, so an integrated and staged building and overhaul of roads, railroads, and aviation facilities is planned and underway (Shahrani, 2013; Kuchins, 2011). The GIRoA has recognized the absolutely vital importance of such an effort for any sort of viable or decent future for the country as a whole and so they have made the NRRCP a national priority program, which elevates its status and helps to fast-track the procedures. The process aims to facilitate and leverage private sector investment through sequenced, prioritized investments across five dimensions (infrastructure, livelihoods, governance, environmental impact, and social affects) (Figure 15.1).

The RC approach in Afghanistan is set up according to the EI value chain (EIVC) that is promoted by the World Bank (Anon, 2011a). The EIVC is a framework for coordinated and integrated interactions that work across broad public and private partnerships. These associations can leverage the broader economic growth that has eluded so many other countries and regions elsewhere for various reasons. The basic EIVC plan begins with: (1) the awarding of contracts and licenses; which are followed by (2) regulation and monitoring of operations; (3) collection of taxes and royalties; (4) revenue management and allocation; and finally, (5) implementation of sustainable development policies and projects. In Afghanistan, each step of these processes will be subject to the vicissitudes of endemic corruption, ethnic competitions, and other problems so that monitoring by the World Bank and other GIRoA and private watchdog agencies will have to be especially vigilant to ensure success. Nonetheless, this is the mechanism that has been tried and true for any number of the developed nations, to which Afghanistan aspires to join; indeed, with sufficient attention to such detail, and probably a fair dose of good luck, it just may be possible. "Inshahallah", as the Afghans say.

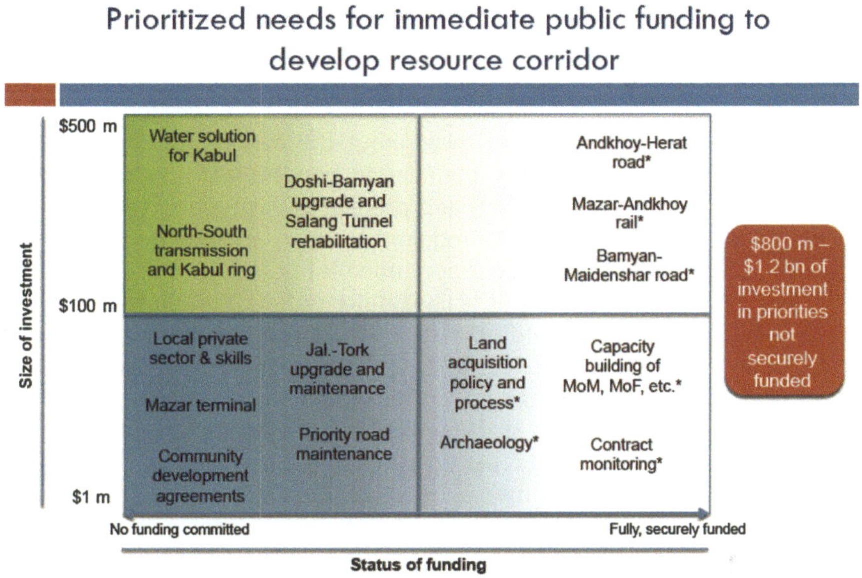

Figure 15.1 Prioritization of different looming needs in the public sector to upgrade facilities, especially in urban water and land use, transportation (road, rail, and air), as well as protection of antiquities to prevent the rapacious looting that has been underway for some time. m, million; bn, billion. *After World Bank Tokyo presentation, 2012.*

If the mining sector in Afghanistan is going to become a true engine of growth, the infrastructure and the jobs created servicing this nascent mining industry will eventually need to have successful development of a large infrastructure, as well as an increase in the skills of people (including learning by doing) for creation and expansion of the other sectors. Large mining projects and supporting physical and social infrastructure can be set up to maximize the possibilities that these various collateral synergies will actually occur.

All modern mines need supportive infrastructure but the amount and type vary greatly depending on the type of mineral, the size and location of the mine to obtain that mineral, and the degree of processing of the ore that needs to take place on site. In the case of large mines that are not near a seaport, such as all those in Afghanistan, the type of mineral is the driving factor with respect to transport infrastructure. For example, in a typical gold mine in Afghanistan, there may be only 1–8 g of actual gold product for every metric ton of ore processed. At the other extreme, the rich iron ore deposit of Hajigak can contain over 60 percent of solid metal in any ore produced. Accordingly, while the gold can even be flown out, the iron will generally need to be shipped by train, or perhaps trucked out to market. Such base metal (iron and copper) operations in Afghanistan will require railroad shipping to China, India, Iran, and Pakistan, or elsewhere if the price is right. The amount of power required by mining operations also tends to rise exponentially with processing amounts, so that will have to be accounted for in the development process.

Construction of railways in Afghanistan has always been a conundrum since the nineteenth century for reasons of difficult topography and justifiable Afghan fears of invasion by foreigners by railroad. Nonetheless, the country is ringed today by railroads in the neighboring countries and plans are afoot to project connecting rail throughout the country to open it up for ease of export and import; in fact, the railroad from Hairatan on the Amu Darya to Mazari Sharif is already in action and presently construction on south is underway (Chapter 7). Railways will be required to open up such deposits as the Hajigak iron deposit but the Aynak copper mine is likely to be underway long before the arrival of a railroad near it. Railways are optimal for high-volume point-to-point transport, but can be disruptive in that a production of 15 metric tons per year (mt/year) of ore equates to 41,000 mt/day, which implies 40–50 trains/day, or roughly one each half hour. This would require excellent maintenance of rail, roadbed, and rolling stock, which is not something one would expect could ever be a characteristic performance in a place such as Afghanistan. In addition, railways next to highways, however, generally capture only 15–20 percent of the traffic because for most private enterprises, the flexibility of highway truck traffic overwhelms the cost advantage of rail. It seems most likely that both railways and highways will have to be utilized for the RCs in Afghanistan.

Paradoxically, the more infrastructure required by a mining operation, the better it can be for the country's economic development because of the collateral externalities generated and economies of scale in provision of infrastructure, the latter allowing for opportunities to form private–public partnerships (PPPs). In the Afghanistan context, the known copper and iron deposits by themselves provide an opportunity to provide transport and power infrastructure to regions of the country desperately in need of such energy sources for other purposes as well. If properly designed, this refurbished or new infrastructure could lead to the development of an RC that would provide large incentives for the discoveries of other minerals and the development of industries both related and unrelated to mining. It must be noted, however, that the concession process for PPPs is not a simple operation and will generally require world-class advice, skilled administration, and considerable financial support to be done properly. These are not particularly things that Afghanistan was ever known for in the past so will have to be worked on assiduously for any prospect of real success. Furthermore, infrastructure concessions have long tenures and it is important to do them correctly from the start, including provisions for strong renegotiation triggers for unforeseen future developments, such as environmental issues, noncompliance with enabling legislation, issues with corruption, and other such destabilizing occurrences.

It should be emphasized that a RC is generally developed to take advantage of some opportunity; it does not create the opportunity. The mines act as an anchor for the infrastructure that underpins the viability of the other sustainable activities and the infrastructure is intended to attract or bring in other private investment. If the mines and the infrastructure were largely in sparsely inhabited areas with little potential for other regional industrial groupings, or agriculture, tourism, or other forms of development, it would be quite unlikely that the mining operations alone could act as an anchor for an RC. The progression of a large mining operation from an undesirable enclave operation unrelated to the local population, to a fully functioning RC is complex and requires a multiphased approach over some considerable period of time. The following enumerations illustrate the various stages that must be dealt with in the case of two large mining operations such as Aynak and Hajigak, each of which gives a strong potential to the regions in between them.

1. Make a baseline assessment of existing socio-economic, environmental, cultural, and political conditions, and governance capacities.
2. Set up a coordinated development by governments, companies, and civil society on an action plan to leverage the infrastructure requirements of the mining companies and the new opportunities that will be afforded. Train the appropriate local people and administration to implement this development plan.
3. Develop other nondedicated infrastructure for the mine with expansions or routing to provide access for other users, either as part of the mining agreements or through the PPPs.
4. Train individuals about the mining operations, and in the skills needed for mining and other identified opportunities in the region or along the corridor.
5. See to development of enterprises in the region of the mines to be able to supply goods and services needed for mine, including financing and skill training or education.
6. Develop the local governance capacity to be able to regulate and guide the development of the RC, including ensuring that small- and medium-sized enterprises (SMEs) obtain the required assistance.
7. Develop a mining-related growth pole around one of the operations or at a suitable location between them (such as a medium-sized city).
8. Make more geological exploration along the corridor and in nearby regions. Make further extension of the infrastructure when discoveries occur.
9. Feeder roads—generally using fiscal revenues—are built to link up otherwise stranded opportunities with the corridor as business possibilities increase in both mining-related and other activities (sometimes referred to as densification of the corridor).
10. Skills that are developed in the mining and related industries are transferred to other industries; in-migration of entrepreneurs and other skilled individuals takes advantage of the newly expanding business opportunities.
11. The timespan between the first and last events in this list will clearly be both lengthy and vary significantly from one corridor to another. It can take several decades.

Minerals and hydrocarbon development can be a pillar of future economic growth in Afghanistan by creating both direct and indirect employment and income, by developing transport and other infrastructure that will help open up areas for broader economic development, and by generating considerable domestic revenue as well as trade and balance of payments benefits. Mines also contribute indirectly. In other countries, it has been shown that every direct job created by a mining operation can result in as many as 5–10 indirect jobs by providing contracts for numerous small businesses and services that supply the mine. Each of these jobs in turn produces tax revenues and other expenditures that pass through the economy.

The Aynak copper and Hajigak iron ore mines, with Kabul as a central node, provide the opportunity to develop an initial resource growth corridor in Afghanistan. These copper and iron deposits can provide transport and power infrastructure to regions of the country that need this and that in turn can stimulate economic development parallel to the mining industry.

Railways, roads, and power infrastructure built to service Aynak and Hajigak and export their production could be integrated with other infrastructure plans and routed through areas of high potential for other geological resources, agriculture and, eventually, back to the tourism that once in prewar times, was beginning to prosper. If this infrastructure is properly designed, an RC could emerge providing incentives for the discoveries of additional minerals and the development of industries both related and unrelated to mining.

The NRRCP is the flagship program of the Afghan Government to focus upon taking advantage of the investments being made in the extractives sector, including supporting infrastructure, to develop the Afghanistan economy more broadly. A **resource corridor** in this context is a sequential action to turn large investments in extractive resources into broader economic development and diversification in a specific geographic area (Figure 15.2). A **development corridor** (DC) is a well-known economic initiative that has been successfully employed in many other countries, helping to develop local economies by leveraging opportunities offered by mining operations, which includes opening landlocked countries such as Afghanistan to new world markets by giving them access to ports and other trade highways.

The Afghanistan NRRCP will be implemented through a dedicated Secretariat in the GIRoA that will keep track of all the disparate elements and ensure that some efficiencies of good government are brought into the DC and RC processes. Classically in the government bureaucracy of any developing country like Afghanistan, a certain amount of in-coordination and official lethargy can prevail, unless methods are implemented to cut through the red tape; it is envisioned that the Secretariat of the NRRCP will facilitate such endeavors.

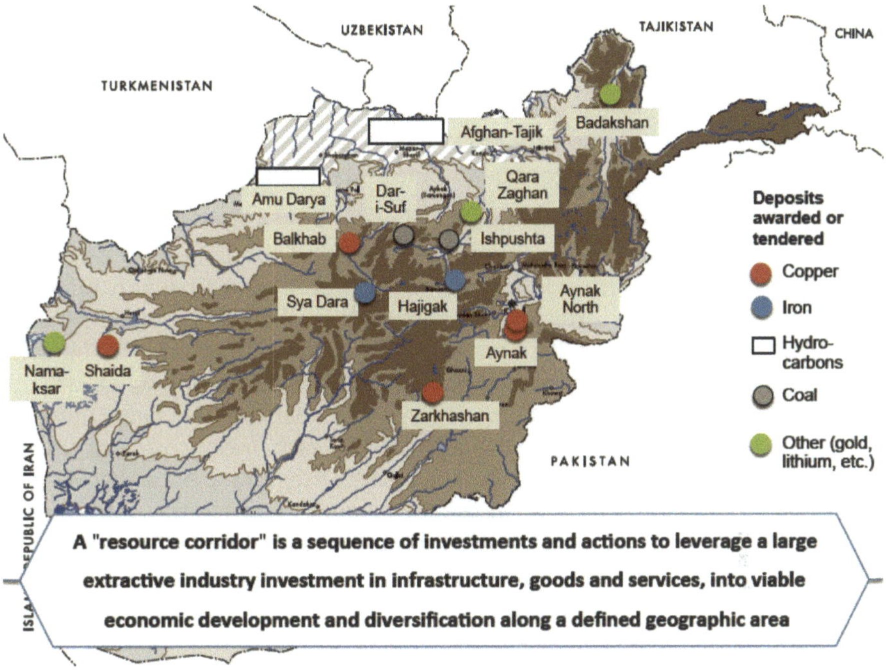

Figure 15.2 The resource corridor approach to Afghanistan. *After World Bank Tokyo presentation, 2012.*

The established NRRCP Secretariat will help coordinate the various efforts among a diverse group of relevant organizations in support of the RC concept. Its key roles are to the following:

1. To support the Afghanistan ministries in conducting much needed analytical, diagnostic, design, and feasibility work for realization of the RC concept over the mid- to long term.
2. To assist with improving intragovernment (central and subnational), regional, and donor coordination in support of the RC concept.
3. To advocate and facilitate clear policy decisions in sectors such as land acquisition, benefit sharing, PPP, trade, transport, roads, mining, energy, SME development, and information communications technology.
4. To encourage public–private dialogue by establishing consultative groups with the private sector and civil society, at national and provincial levels (in corridor provinces).
5. To serve as an information focal point for government, the communities, and donors.

In Afghanistan, the key implementing Ministries for the NRRCP that will be looked after by the Secretariat are the: (1) Ministry of Public Works (MoPW), (2) Ministry of Transport and Civil Aviation (MoTCA), (3) Ministry of Commerce and Industries (MoCI), and (4) Ministry of Energy and Power (MoEP). The (5)

Ministry of Mines (MoM) will support coordination of NRRCP implementation through the NRRCP Secretariat, whereas the (6) Ministry of Finance (MoF) will facilitate its monitoring and evaluation. Other key stakeholders include minerals-sector investors as well as the local private sector, civil society, local communities, and special development partners who will be vetted through the World Bank or other quasi-independent and therefore less corruptible agency.

The need for development of RCs in Afghanistan is exceptionally high because the country is entering a period of difficult transition away from foreign aid and money pouring in to support the Coalition forces that has been driving the economy. This very high level of aid is not sustainable at these former levels after 2014 when the Coalition troops depart. The coming slowdown in aid will exert a drag on the economy at a time when growth is vitally needed to offset depredations by the Taliban. At the same time, some 300,000 people are entering the job market each year, corrosive underemployment already at 48 percent is expected to rise, and a huge impact of the transition will be the projected fiscal (or revenue) gap for the GIRoA, estimated at 25 percent of GDP, which will then put the government under enormous pressure to provide basic services for citizens and pay the salaries of its government workers. The country will need to find new sources of growth to ensure the success of the economic transition. At this time when investment in the minerals sector is accelerating, the natural resources sector offers a window of opportunity around which an RC can emerge. The mining sector by itself alone, however, will not generate the changes and the financial resources necessary to ensure Afghanistan's economic transition. The good news, however, is that the investments in mining can trigger additional business opportunities and social benefits with the potential to transform the Afghan economy.

The NRRCP's vision is to advance Afghanistan's economic and financial independence through strategic and prioritized infrastructure development that will enable investment by the private sector. The hard infrastructure referred to in the documents is the building of roads, railways, and power plants. Soft infrastructure, on the other hand, includes community development, enterprise development, and institutional capacity training and education, among other things. The NRRCP's purpose is to connect Afghanistan to itself, to the region and the rest of the world through promotion of interstate RCs as a means to realize Afghanistan's natural resource, growth, revenue, and employment potential.

The NRRCP plan has four main dimensions that must include the following:

- *Infrastructure (road, power, water sources, and railways)*: The idea is to leverage road, rail, and power infrastructure built by the mining investors. This infrastructure will then support other sectors and expand their reach through incremental catalytic investments, for example, in feeder roads that will also support the development of agriculture. It also means integrating water and power plans more broadly.

- *Livelihoods (jobs)*: Develop the capabilities of the local private sector to supply goods and services to the mines, or to develop refineries and steel-reprocessing facilities, thereby capturing benefits for local businesses along the mining-sector value chain; building the skills of local businesses to ensure that they become trusted partners for mining operations; and developing the capability of the agriculture and agribusiness industries to take advantage of the mine-related, power, and transport infrastructure.
- *Governance*: This is about strengthening public financial management and monitoring revenues from natural resources; building the capacity of key Ministries to develop and implement policy and regulation; and establishing the means for effective delivery of services by both national and local Government. Endemic corruption and bureaucratic lethargy must be rooted out through constant attention to the problems, and equally perseverant efforts for its suppression through education and an enhanced media and whistle-blower campaign.
- *Social and environmental issues*: This includes the provision of benefits (economic, education, and health) to communities around the corridor, including them through community development agreements; it also means improved land administration including accurate and noncorrupt cadastre delineation for all land within the corridor, revised legal frameworks for land acquisition, resettlement, and compensation; the protection of physical and cultural heritage through archaeological preservation and museums; and the mitigation of environmental impacts occurring from poor development practices to ensure environmental sustainability, particularly with regards to water quality and quantity.

To ensure the implementation of RCs in the multiethnic and highly fractious Afghanistan, special precautions have to be implemented to ensure that moderate equality is maintained between north and south, east and west, mountainous and flatland, and desert and well vegetated. In general, a superb interplay between the diverse and commonly mutually hostile elements has been achieved in the NRRCP plans. In fact this has been greatly facilitated by the fortuitously excellent spatial diversity exhibited by the various resource distributions; almost all of the diverse ethnicities have an excellent prospect to be included in the plans for expansion and growth because they generally all have a significant resource contribution to provide from their different geographic regions. This has the potential to greatly inculcate a sense of belonging to a more integrated national space, in a country where any sense of national patriotism is most generally lacking.

All development of RCs takes a certain amount of time for the mine to "come on stream", the railroad to be built, or the personnel to be trained, so scenarios of building the various segments can be short term (few years), medium term (decade), or long term (several decades). Accordingly in Afghanistan, four main RCs have been developed comprehensively: (1) Northern hydrocarbons

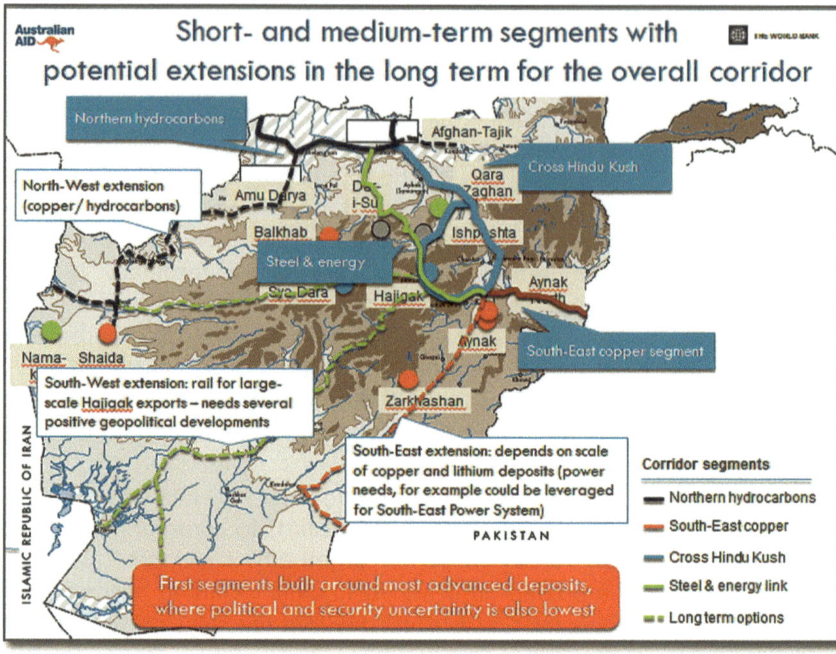

Figure 15.3 The Afghanistan Resource Corridor plan as conceived in 2012. *After World Bank Tokyo presentation, 2012.*

corridor segment; (2) Southeast copper segment; (3) Cross Hindu Kush segment; and (4) Steel and energy link segment (Figure 15.2). Long-term options are also attached to all of these segments to further open out the prospects for more people; these options also include a northwest extension that includes copper and hydrocarbons, a southwest extension that includes important rare earths and a transportation nexus, a southeast extension that has lithium and more copper, as well as transportation and power, and northern extensions into more hydrocarbon zones in the Afghan-Tajik and Amu Darya basins. The first segments are being built around where the most advanced development of resource deposits exist already, which is also where the political and security uncertainties are considered the lowest or easiest to deal with (Figure 15.3).

Northern Hydrocarbons Resource Corridor Segment

An example of short- and medium-term investments to be made in the Northern Hydrocarbons Resource Corridor Segment (NHRCS) are maintenance and upgrading of the Shiberghan-Mazari Sharif road; construction and operation of the new multimodal rail terminal at Mazari Sharif; Shiberghan gas and power; construction of a midscale (~15,000 barrels per day) refinery at Mazari Sharif; and a new railway from Mazari Sharif west to Andkhoy (with spur north to Aqina to connect to a new railroad coming to the border across Turkmenistan). All of these projects are scheduled to take place over the next 2–4 years.

Southeast Copper Resource Corridor Segment

The all-important and ethnically delicate, southeastern copper segment needs to start with robust and public financial management, coupled with transparency for the projected Aynak revenues. National and local capacity needs building to plan and implement this RC. Next it needs to have the capacity strengthened of the new Afghanistan Land Authority (ARAZI) of the Ministry of Agriculture, Irrigation and Livestock (MoAIL). The road network must be enlarged, rebuilt, and maintained with effective means. Needed also are the improvement of the north–south electrical transmission lines, a new power ring and "grid code" around Kabul, and the setting up of a mining supply chain of investments such as drilling services, foundries, and related small industries. A new water system for Kabul, including Shatoot dam, needs to be established, perhaps by bringing new water directly south from the Panshir River. Livelihoods of people will be augmented by development of new enterprises and skills matching that will multiply creation of jobs. Environmental and social impacts in this segment will be critical, so community development agreements must be implemented around Aynak, as land acquisition is strengthened with local buy-ins, water impacts are controlled and constantly tested, and archaeological excavation of the impressive Buddhist sites are accelerated and afforded greater protection. Later the new power and transport system should be extended to the Zarkhashan gold and copper mining district near Ghazni, and thence to Kandahar.

Cross Hindu Kush Resource Corridor Segment

The CHKRCS is set up to connect Hajigak iron with coking coal nearby and then create options for export north into NHRCS transportation system, or southeast into the SECRCS transport system. This involves bringing the railway south from Mazari Sharif through the Shibar Pass to Bamiyan and the coal mines, as well as to Hajigak and on to Kabul through the Ghorband River Valley and the Shomali Plain to Kabul to connect to the SECRCS.

Steel and Energy Resource Corridor Link Segment

This SERCLS example is established especially to add to the linkages in railways, highways, and energy delivery systems between the northern, southeastern, and cross Hindu Kush segments. It is anchored in the iron and coking coal mining, with a projected steel mill and several options for power and transport connections outward.

Southeast Extension

This extends the transportation and power grid south from Kabul to Ghazni, to the Zarkashan gold and copper mine, and on to Kandahar and the border with Pakistan at Spin Boldak. It depends upon the scale of the lithium deposits available and the grade and accessibility of any additional copper deposits, with power needs being leveraged for a southeast power system as well.

Southwest Extension

Extensions of a railway to the southwest from the CHKRCS and SERCLS to the borders of both Iran and Baluchistan, Pakistan, taking advantage of the rare-earth mine to be opened at Khaneshin on the lower Helmand, and perhaps also the decorative travertine and marble stone quarries in the Chagai Hills on the south border as well.

Northwest Extension

Extensions from the NHRCS east to the Afghan-Tajik hydrocarbon fields and especially southwest to Herat, the Iranian border, and south from Herat to the Shaida copper location.

The timeline and contingencies for development of the RCs has been worked out to a certain extent, although uncertainties remain. The overall investment scenarios for all the segments have a range or continuum of possibilities based on unforeseeable futures, but which do allow for some contingency planning. These future possibilities can be divided into a basic four scenarios to include: (1) full flowering, (2) a reduced set of segments, (3) delayed or constricted growth, and (4) collapse (Figure 15.4). The full-flowering scenario would result from additional positive geological exploration results, positive commodity markets, conducive regional politics, and relative freedom from, or control of corruption. This would result in growth of the private sector and job creation that would be taking off all across the country. The second scenario of a reduced set of segments would result from disappointing exploration results for some deposits, mixed commodity markets in future, or turbulent regional politics, but which would still result in some private sector growth and job creation in key parts of the country. A tertiary delayed or restricted growth scenario would occur with too many delays by private sector investors, falling commodity prices, instability caused by the Taliban or other bad actors, and rampant corruption. This, however, would still mean that some long-term fiscal targets would be met with some job creation. Finally, full collapse could result from bad governance, extreme corruption, overwhelming instability from a strongly resurgent Taliban, and/or a commodity price collapse, in which case investments would be needed anyway to be ready to take off again at an eventual

Overall investment scenarios (for all segments)

		Investment (USD bn, 2012–2025)			
Scenario	Triggers for scenario	Public[1]	Private[2]	Total	Outcome
Full flowering	• Positive exploration results • Positive commodity markets • Conducive regional politics	4–4.5	14–16	18–20	Private sector growth and job creation taking off across country
Reduced set of segments	• Disappointing exploration results for some deposits • Mixed commodity markets • Turbulent regional politics	3.2–3.5	8–8.5	11–12	Private sector growth and job creation in key parts of the country
Delayed or constricted growth	• Pvt sector investment delays • Falling commodity prices • Instability	2.8–3[3]	2–2.5	4.5–5.5	Long-term fiscal targets met with some job creation
Collapse	• Governance • Instability • Commodity price collapse	0.05–0.5	0.6	0.7–1.2	Investments needed anyway ready to be made at recovery

1 Includes already allocated funds and provisions for O&M spending over the period
2 Includes private spending on public good infrastructure (e.g., power plant)
3 Includes spending on infrastructure whose economic value is independent of the resource investments

Figure 15.4 The four scenarios from full flowering of the Afghanistan Resource Corridors plan to its initial collapse under adverse conditions, together with estimates of investment money needed to effect some sort of reasonable progress for Afghanistan. bn, billion.

recovery. Furthermore, even in the case of a full takeover by the Taliban, most people think that the resources will come out of the ground anyway.

The timelines for these scenarios range from robust to risk and include sets of priorities and decision factors with interacting contingencies. For example, the priorities of 2012–2014 include: (1) making a variety of soft infrastructure investments such as improved land management and land acquisition practices, development of better communications, skill training, and private sector capabilities, greater empowerment of farmers, increases in sound governance and transparency in deal making to underpin parts of the first RCs, along with planning and coordination of the newer infrastructure; (2) doing priority road construction and maintenance works; (3) making institutional and regulatory reforms in hard infrastructure; and (4) making design and land surveys to prepare for hard infrastructure options such as power, roads, railways, and waterworks. This then will lead in 2015–2016 to new decisions based on: (1) the most essential political stability and security; (2) results of the geological resource exploration and proving out ore bodies; (3) the timing of mine development with appropriate investment inputs; (4) health of the regional and world commodity markets; and (5) regional economic integration factors. Depending upon how things turn out on the positive or negative sides, a number of upside and downside actions could then come into play several years later in 2017–2018. Upside actions would trigger hard infrastructure investments, expand the soft infrastructure programs to benefit the people directly, and set the stage for the next waves of hard infrastructure development. On the other hand, an offset to the negative side would put a pause to the hard infrastructure investments that are dependent upon the resources actually being there and becoming available; all of this in turn would allow only the selective pursuit of actions dependent upon the geography of the situation (ethnicity and culture, climate-change situations, etc.), and by the sector that would be affected (farming, shop keeping, manufacturing, etc.). The positive and negative decision factors arising in 2017–2018 would then depend upon political stability, commodity market prices, timing of the development of the new mines, any new discoveries or not, and the completion of some of the investments.

All in all, this design for Resource Corridors in Afghanistan is a very robust and brilliant new look by forward thinking (read, "young"), modern economists and regional planners to put together plans that have a high probability of actually working, providing that the hard working and much beleaguered Afghan people are actually allowed to attempt this really reasonable hope for their future. The alternative, of course, is that the illiterate and suicidal gunmen from the backward hordes of Taliban and al Qaeda enthusiasts supported by those Pakistanis, who want to control the resources and the people for their own purposes, actually continue to play the insurgency card they have been using so much of late. The fact that the Indian government has bid successfully on the Hajigak iron deposit and wishes to help the Afghans dam the Hari Rud and Kabul River makes the Pakistani government apoplectic about possible Indian machinations and successes on their western border, so that the Pakistanis see that they have their enemy surrounding them on both sides, with no strategic depth into which they can fall back. Even the Iranians want a piece of the action; more of the water, perhaps a little uranium. China will sit quietly to the north, willing to do business with almost anyone, and do its best not to be drawn too far into anything while there are resources to gain and money to be made. Their recent negotiations with India concerning the use of Afghanistan resources in the development of their own economies are an additional hopeful sign for Afghanistan's future.

But the Afghans also have a few other aces in the hole, providing they can be played with skill. The proposed gas pipeline from Turkmenistan, across Afghanistan into Pakistan, and perhaps on to India, would be an excellent start. With their potentially huge resource base in Afghanistan that far exceeds Pakistan's own, and Afghan plans for developing a completely new transport network across their own territory; this could all be to the considerable advantage of Pakistan, and might even be an impetus for them to refurbish their own dilapidated railways and roads (Anon, n.d.). The resources in Afghanistan might also result in a totally new railway access across Baluchistan to the new Gwador port on the Indian Ocean that China built for Pakistan, with perhaps a spur to Iran; these are factors to Pakistan's great advantage, if they can overcome their pathological fear and hatred of India just enough to let it happen. Alternatively the Pakistanis are just as likely to bring themselves down in the process, as they are caught between the hammer of their own jihadists on the west and the Indian anvil on the east.

In Afghanistan, the hopes for the establishment of robust RCs, coupled with their own augmented agriculture will be the key drivers of growth and poverty reduction. Private investments and regional integration are vital to the process, and the neighboring countries have much to gain as well where the RC nexus points of Afghanistan come to the border's edge to tie in with the neighboring country's own regional efforts. Only China's border with Afghanistan is too high, cold, dry, and short to offer much in the way of development possibilities, although even that could change with a well-engineered tunnel and rail system through the Pamir border range and into the Wakhan Corridor, with connections on south through the Panjsher Valley. In any case, the opportunities along the Afghanistan RCs are real and need both public and private sector investment and action to come to fruition. Obviously significant challenges and risks are attached to the difficulties at the governing that will be required, as well as the timing of the resource investments. Some or all of the pressure from the Taliban against the Kabul government likely can be attributed at attempts to derail these RC development processes. In some cases, public and private actions are attempting to move ahead against this negative tide anyway. The first actions identified in the RC plans, however, are realistic about these risks, and the Afghans seem staunchly prepared and able to deal with them, so they are likely to succeed to a certain extent. The GIRoA office of NRRCP has noticed benefits that have spilled over already and is setting up options for later such effects, although they have not yet specified what those might be. The GIRoA feels at this point in time that it is important to maintain a sense of balance, but that the RC concept can be a powerful means to avoid enclave mining investments that do little overall for the country, and that the program will most definitely generate more diversified growth, raise livelihoods, and create new jobs. The time is short, however, perhaps even too short given the geographic restrictions inherent in that difficult milieu. Nonetheless, the people of Afghanistan and Pakistan are strong at business, and good at seeing potential profit margins; one can only hope that they can also see how much good solid, mineral-resource potential, and personal wealth-building capacity that occur in this model.

Lest people may think that the plans that GIRoA have put into motion for the country-wide installation of the NRRCP are too much unrealistic pie-in-the-sky, given the ethnic antagonisms, they perhaps have a point. The reality, however, is that with several world-class ore bodies as anchor points, coupled with a reasonable host of other money making in precious metal mines, hydrocarbons, and other such; this all makes an exceptionally good prospect in the world of developing countries. The RC issues have serious stakes. Afghanistan's neighboring countries would very much like access to these resources for their own burgeoning development programs of growth and modernization. Any return to armed confrontation would be antithetical to their own programs, yet some of the countries may see shortsightedly that such actions might help ensure their own success in gaining an upper hand for access to those resources. Pakistan, in particular, needs to be strongly disabused of these risky notions, but whether or not this is at all possible in this volatile swamp is quite another matter.

All over the world, people of good will toward Afghanistan very much want this NRRCP to work, but they also want to be sure that any money invested is a safe bet; given the recent huge bank fraud in Kabul (Sommerville, 2011; Anon, 2012b; Rosenberg, 2012), however, plus the many marginal people who like to play in the wild west of new mining ventures, this may be a faint hope at best. The veracity of the NRRCP idea as a truly viable possibility in a country like Afghanistan is thus at long odds, with a possibly limited payback possibility given the stakes and players involved; we can only wish them the best in attracting venture capital and hopes for a better future (Anon, 2012c). The alternative roads to violence and chaos are perhaps not useful to contemplate here further.

References

Abdullah, S.H., Chmyriov, V.M., Dronov, V.I., 1980. Geology and Mineral Resources of Afghanistan. Geological Survey of Afghanistan, 2 vols. Kabul.

Ali, S., Shroder Jr., J.F., 2011. Afghanistan's Mineral Fortune: Multinational Influence and Development in a Post-war Economy. Institute for Environmental Diplomacy and Security, James Jeffords Center for Policy Research, University of Vermont, Burlington, VT, 24 pp.

Anon, 2011a. A Programmatic Approach to Leverage Extractive Industries for Economic Development (E4D): Background Note; Resource Corridors. Overview – EI Sourcebook; 12 pp. http://www.eisourcebook.org/cms/files/resource_corridors_background_note.pdf.

Anon, 2011b. National and Regional Resource Corridors Program. Infrastructure Development Cluster, Ministry of Mines, Islamic Republic of Afghanistan, Kabul, 52 pp. http://www.iwaweb.org/Docs/resource_centre/EIM/Extractive%20Industerial%20Fiscal%20Model.pdf.

Anon, 2012a. Afghanistan: resource corridor (presentation slides). In: Technical Summary at Tokyo Conference, 8 July; 14 Slides; World Bank and Australian AID. http://siteresources.worldbank.org/SOUTHASIAEXT/Resources/223546-1328913542665/8436738-1341156360475/Afghanistan-Resource-Corridors-Technical-Summary.pdf.

Anon, November 28, 2012b. Kabul Bank Fraud Profited Elite, Leaked Audit Says. BBC News Asia. http://www.bbc.co.uk/news/world-south-asia-20512713.

Anon, 2012c. Towards Self-reliance: Strategic Vision for the Transformation Decade. Islamic Republic of Afghanistan, Ministry of Finance, 25 pp. http://mof.gov.af/Content/Media/Documents/Towards-Self-Reliance-27-6-201216720121028258355325325.pdf.

Anon, n.d. Emerging Resource Growth Corridor Afghanistan (Power Point Presentation). http://www.eisourcebook.org/cms/files/attachments/worldbank/Emerging%20Resource%20Growth%20Corridor%20Afghanistan.pdf.

Kuchins, A.C., 2011. A truly regional economic strategy for Afghanistan. The Washington Quarterly. 34 (2), 77–91. http://csis.org/files/publication/twq11springkuchins.pdf.

McMahon, G., Lokanc, M., 2011. Mining Section for Afghanistan Growth Paper (Draft). http://www.nrrcp.gov.af/images/docs/2msa.pdf.

Rosenberg, M., November 28, 2012. Political Meddling Hampers Inquiry into Kabul Bank Debacle. The New York Times Asia Pacific. http://www.nytimes.com/2012/11/28/world/asia/inquiry-into-kabul-bank-fraud-hits-snags.html.

Shahrani, W., 2013. National and Regional Resource Corridors Program (NRRCP). Ministry of Mines, Islamic Republic of Afghanistan, Kabul, 3 pp. http://mom.gov.af/en/page/6355.

Shareq, A., Chmyriov, W.M., Stazhilo-Aleekseev, K.F., Dronov, V.I., Gannon, P.J., Lubemov, B.K., Kafarskiy, A.Kh., Malyrov, E.P., Rossovskiy, L.N., 1977. Mineral Resources of Afghanistan, second ed. United Nations Development Program. Project AFG/74/012, 419 pp.

Shroder, J., 1983. The U.S.S.R. and Afghanistan mineral resources. In: Agnew, A. (Ed.), International Mineral Resources a National Perspective. American Association for Advancement of Science, pp. 115–153.

Shroder, J., 1989. Afghanistan resources and Soviet policy in central and south Asia. In: Hauner, M., Canfield, R. (Eds.), Afghanistan and the Soviet Union: Collision and Transformation. Westview Press, pp. 101–119.

Shroder, J., 2007. Afghanistan's development and functionality: renewing a collapsed state. Geojournal 70, 91–107.

Shroder, J., 2012. Afghanistan: rich resource base and existing environmental despoliation. Environmental Earth Sciences. http://dx.doi.org/10.1007/s12665-012-1638-7.

Shroder Jr., J.F., Watrel, R.H., 1992. Mineral resources in Afghanistan (summary of report for USAID). AACAR Bulletin (Association for the Advancement of Central Asian Research) 5 (2), 9.

Shroder Jr., J.F., Assifi, A.T., 1987. Afghan mineral resources and Soviet exploitation. In: Klass, R. (Ed.), Afghanistan and the Soviet Union: The Great Game Revisited. Freedom House Pub, New York, pp. 97–134.

Sommerville, Q., February 2, 2011. Calls for Prosecutions Over Kabul Bank's Near Collapse. BBC News Asia. http://www.bbc.co.uk/news/business-12348045.

Vejlupek, M., 1980. Geologie a Nerostne Suroviny Afghanistanu (Geology and unexploited resources of Afghanistan). Geologicky Pruzkum, Prague, Czechoslovakia 22, 16–18.

Part III

Overview of Environment and Development of Afghanistan

The extraction industries are known to be problematic for the environment, commonly because they so regularly destroy land surfaces and contaminate soil and water sources unless special precautions are taken. In the process of development of natural resources, great care must be expended to look after many issues of land, water, and the extractive mineral resources. Accordingly, a slightly expanded look at environment and development issues has also been, if not in the actual forefront of much of the resource extraction efforts in the world, at least not too far back. Many issues have come up in dealing with the problems of land ownership. Major environmental issues deal with water availability and its propensity to contribute to so many disasters in Afghanistan, from too much of it in the all-too-common flooding, or too little of it in the ever recurrent droughts. The extractive mineral resources themselves have been viewed by many in the GIRoA, as well as by their external civilian and military advisers, as the main solution to all of Afghanistan's problems in a kind of "gold-rush" as hundreds of resource claims are granted or still being readied for bid tenders. Fortunately also, however, is the robust recognition of all the environmental hazards tied to resource extraction, which has led to creation of strong legal frameworks to try to combat the expected environmental despoliation that will come with it.

Environmental degradation in Afghanistan has already been rampant in the country for a very long time but modern recognition of the problem by the scientific community and independent watchdog groups has given us recognition of most of the chief problems. Water contamination and diminution from human-caused, as well as natural, processes is a most severe problem, with almost everything else paling beside them. In fact, in the last chapter, Section III is all about just water and climate change, with a focus on the old water treaties, as well as the coming climate shifts. In addition, the main environmental issues that are so common in Afghanistan anyway include all of the following: soil salinization from irrigation, soil erosion from flowing water and wind; as well as landslides, deforestation, dearbification shrub removal, biodiversity diminution, habitat loss and protected areas, desertification, pollution and environmental health, and the residuals of long-term warfare. As the natural resources are produced and come on commercial stream in Afghanistan, it will be essential to study and map these hazard-producing, environmental processes in detail so as to reduce their most harmful effects. Afghanistan's environmental despoliation from all these sources, on the other hand, had also received enough legal attention already that a new National Environmental Protection Agency (NEPA) has been formed and begun to train personnel and develop testing mechanisms to solve these serious problems. Five environmental pillars have been established as a basis around which to manage protection and remedial activities. These pillars include the formation of coordinated environmental institutions, as well as new laws and policies, impact assessments, environmental information and other assessments, and community-based natural resource management. Collectively, these would seem a good start to control and fix new problems, but there is a very long way to go and perhaps not enough time or funding to take care of all that will be required. In any case, whether or not Afghanistan can ever rise out of its status as a failed state seems to depend to a large measure on whether the newly recognized mineral resources are exploited in environmentally benign ways.

A most serious resource-utilization problem in all this is the development of Afghanistan's water resources for improved flood control and agriculture in the face of severe climate change wherein downstream deliveries across borders cannot be guaranteed in the future at all times. These facts of looming climate change must be recognized by all the riparian nations and they must come together over the negotiating table to work out the means to deal with this new reality. As the saying goes, if they do not hang together on this, they will all hang separately anyway and the results will not be pleasant. Thus, all the nations of the Indus and Amu Darya hydrologic basins of South and Central Asia, with Afghanistan as the central hydrological hub, must come together repeatedly in what may seem to become a hydrological hubbub, but which will be essential to smooth increasingly troubled waters. Without the means to discuss the hydrologic uncertainties and the certain coming droughts and water shortfalls, along with the increasing likelihoods of equally devastating floods, nations like Afghanistan will be caught without recourse and will fail to solve essential planning problems that will arise. Great attention to the details of the environment of Afghanistan are thus paramount issues that must be attended to if the country is to have any realistic future beyond increasing environmental despoliation and ever-increasing violence from a sorely stressed people.

In this Chapter

16

Afghanistan Environment and Development Issues

Abstract

Environment and development issues are a highly serious situation in Afghanistan because of centuries of despoliation. Beginning with the disruptions to the land-ownership records caused by the total misrule of the communists and Taliban in the 1980s and 1990s, a path back to normalcy has been the difficult reestablishment of proper cadastral or property-ownership information. Sharing of the critically essential water in Afghanistan is another major issue that is being addressed in a variety of ways. Issues with the extractive mineral resources and associated environmental problems are also of great concern and new legislation has been developed to control any potential problems.

Keywords: Environmental devastation; Extractive mineral resources; *Kuchi* nomads; Land conflict; Land titling & economic restructuring activity (LTERA); *Mirab* councils; Resource-based environmental management; Resource management; Rural land administration project (RLAP); Scarce natural resources; Variable land-use rights; Water inequities; Water management; Water rights.

In order to prosper best, any country in the world needs a reasonable natural environment from which to springboard its own economic development. In this second decade of the new millennium, a country like Afghanistan that is trying desperately to recover not only from nearly three decades of nearly constant war but also from the pressures on it from life in a marginal environment; things can be rather extreme, especially for a generally poverty-stricken people. Afghanistan is too hot and windy, too dry, too high, too cold, and has been lived in by so many poor humans and their rapacious grazing animals for so many millennia that its environment clearly shows the effects of long-term environmental degradation, depredation, and despoliation (Shroder, 2012). As the poorest country in Asia, Afghanistan needs a great deal of help to try to keep it from succumbing even further to violence and environmental destruction, something that the United Nations Environmental Programme (UNEP) seeks to do, along with many other nations who are concerned that the problems endemic in this country are a possible harbinger of similar

Natural Resources in Afghanistan. http://dx.doi.org/10.1016/B978-0-12-800135-6.00016-7

problems that other nations must also begin to face (Brown and Blankenship, 2013). Part of the answer in Afghanistan lies in wise management of its natural resources, both those on the surface and the ones deep underground, so that the chances for a far deeper peace can be more likely in this strongly bellicose place.

Decades of conflict and instability in Afghanistan have devastated the people, infrastructure, economy, and environment in the country. In fact at any given time, numerous conflicts are underway at several different scales and intensities, many of them having little at all to do with the standard Taliban versus the military forces of the government and international ISAF coalition, upon which the media has long been so focused. Instead, in this volume, attention has focused on the numerous smaller issues with natural resources as sources of internal friction in the country. These include such things as the division of water at the local and transboundary levels, disputes over land ownership, the regional trade in drugs, and the illegal harvesting and transboundary smuggling of high-value timber that are all the source of numerous fracture lines and bones of contention, not only in Afghanistan but also throughout other parts of South Asia as well. Basically, the management of the natural resources, or its lack, results in a strong interaction with the violent conflict underway in Afghanistan, and so the institutions, policies, and practices that govern all the land, water, forests, rangeland, minerals, and hydrocarbons are a form of conflict resolution and peace building that must be stressed if Afghanistan is to have any sort of reasonable future at all. Furthermore, any large mineral resource-development schemes, including resource corridors, large mining developments, and so forth, whether from the government or the international community, are neither far from immune to the environmental impacts of natural resource-related conflict nor for making the situation inadvertently worse. For these reasons, the natural resource management, the resulting environment, and the future development of Afghanistan are tied inextricably together with the very interrelationships that can serve as a bellwether of similar situations developing elsewhere, which may need considerable attention to stave off unpleasant consequences as climates change and populations increase.

In a limited survey of some provinces in Afghanistan, as reported by Oxfam in Brown and Blankenship (2013) in 2008, disputes over both water and land were the two most commonly reported reasons for violent conflict and accounted for some 55 percent of the cases assessed. Natural resources thus play a variety of roles already in conflicts at different scales, locations, and intensities in the country. Five major points about resources are critical to the country's prospects for a peaceful and prosperous future, where an estimated 70–80 percent of Afghans rely upon agriculture, animal husbandry, and artisanal mining for their daily survival:

- Communities, just as countries, do fight over scarce resources such as irrigation water and productive land.
- Powerful stakeholders do indeed use natural resources as coercion instruments to exert control over others, such as farmers in upstream watersheds who can control downstream water usage.
- Thievery of natural resources provides a source of illicit revenues that sustains corruption and does build powerful war economies.
- Natural resources act as incentives for peace spoilers at local, regional, and international, transboundary levels where neighboring countries have vested interests in maintaining security vacuums.
- Natural resources are a source of grievance, especially where endemic corruption delegitimizes the government.

Thus, Afghanistan needs to harness its natural-resource assets to create jobs and generate revenues to fund basic government services for the betterment of its people so that the country can be raised from its present position near the bottom of any index of human development (Brown and Blankenship, 2013). Effective resource management at its heart is a strong form of conflict prevention by creating structures and rules for managing and sharing in the fruits of the natural resources where order, predictability, and trust can be brought into situations that are otherwise rife with competition and conflicting interests. Under improved rules of resource management, the benefits from the wise exploitation of natural resources can be divided equitably, communities can be directly involved and informed about the decisions that affect them, contracts and payments can be conducted transparently, and communities can raise and settle their grievances. These are not ideas or procedures that Afghanistan has ever been known much for in the past, and the cynics will say, "Good luck with that", but still without attention to such prior insufficiencies, little progress into the future can be made effectively. Perhaps well-managed resource extraction ultimately will not ever work very well in the difficult and corruption-ridden environment of Afghanistan, but certainly without attention to these issues, future development most definitely will fail. It thus appears that only by building an awareness of the essential aspects of the environment into a society such as that of Afghanistan, will an even marginal success with development be capable of being achieved because the increasing synergistically reinforcing environmental degradation will increasingly overcome forward development steps. Clearly, one step forward and two back are not where Afghanistan can go with any degree of developmental sustainability.

In terms of natural resource management and development, the three most essential elements to consider are the land itself upon which an overwhelming number of Afghans depend for their agricultural livelihoods; the water, without which no agriculture is possible; and the mineral and hydrocarbon bonanzas that offer the possibility of such a robust future in revenues and consequent developmental potentials if they are handled properly. All three are fraught with so many problems for a country like Afghanistan with its problematic history and recurrent themes of seemingly unwarranted violence, which any cynic could perhaps be forgiven indeed for doubting whether these problems can be overcome on the road forward to more enlightened development by a sorely beset country. Instead, most will expect Afghanistan to fail at this, as it has done on so many other issues in the past. Nonetheless, a look at the issues of land, water, and resource extraction from the point of view of overall development can be instructive.

Land

The land of Afghanistan includes some 652,000 km^2, upon which about 80 percent of the population depends directly for sustenance and income. The ownership and rights to land in any country can vary considerably between individual owners, shared ownership, common land, government land, and so forth. Titles to land and different usages of land by legal owners, illegal squatters, itinerant nomad grazing with or without rights for that, and other different interpretations of rights to land use, have produced no end of squabbles and various levels of violent disputes in Afghanistan. An example of the long-term problems with land in Afghanistan goes back into the nineteenth century when in the course of their border delineations of Afghanistan, the British granted the common-land pastures of Hazarajat customarily used by the Shia Hazara people, to the Sunni *Kuchis* (*Powindahs* in Pakistan), who are nomadic, border-crossing Pashtun pastoralists. This issue for over a century has been at the heart of every contested interethnic relation that has flared into violence on many past occasions. In 2009, for example, reports were made of Hazara communities constructing trenches at strategic entry points to the central highlands against *Kuchi* arrivals, while Taliban arms were provided to the *Kuchi*, and Shia Iranian support was given to the Hazara (Brown and Blankenship, 2013).

The chaotic land conflict in Afghanistan has occurred also because so many houses were destroyed in the second civil war or were taken over and occupied by others, official records were inaccurate or destroyed, and ownership and transfer records were commonly forged. People had been forced to sell property under duress, or long-term residents might have no official title because their land tenure was in the form of a social ownership, or recognized only through customary or common law. A solution is a cadastre, which is normally a parcel-based system of up-to-date land information that contains a record of interests (e.g., rights, restrictions, and responsibilities). The cadastre normally contains a geometric description of land parcels linked to other records that describe the nature of the interests, the ownership or control of those interests, and commonly the value of the parcel and its improvements. Property rights are critical in nation building in Afghanistan (Table 16.1) and include the following:

Table 16.1 Registering the Human Terrain or a Valuation of Cadastre in Afghanistan

- Required for an effective land-administration system.
- Provides a platform upon which nation building can take place.
- Establishes a platform for securing land tenure, which is fundamental to building peace and stability within the society.
- Provides a platform for solving local land disputes before they escalate into violent conflict.
- Is sustainable because it is embedded in the societal fabric of an area.
- Guarantees ownership and security of tenure.
- Provides the basis for land and property taxes.
- Provides security of credit.
- Guarantees results of judicial procedures relating to land rights, including repossession.
- Reduces land disputes.
- Develops and monitors land and mortgage markets.
- Protects government lands.
- Facilitates land reform.
- Promotes improvement of land and buildings.
- Facilitates reliable land-use records.
- Improves urban planning and infrastructure development.
- Supports environmental management.
- Produces statistical data as a base for social and economic development.

Registering the human terrain or a valuation of cadastre in Afghanistan is a measure of land use there by the people. The value of a good and manageable cadastral system is manifold.
Batson (2008); www.ndic.edu/10278.htm.

- Use rights, in which the land is used for grazing, subsistence crops, and minor forest products.
- Control rights wherein there is custodianship of the land use, control over which crops get planted, who gets the money from the sale of crops from the land, and so forth.
- Transfer rights, which are the right to sell or mortgage the land, pass it on through inheritance, or reallocate its use and control rights.
- Security of tenure, in which the various land-use rights are recognized by others and are protected by the government.

Land is commonly a significant factor in widespread violence and is also a critical element in peace building and economic reconstruction in postconflict situations. Cadastral information (land and property records) can predict threats to regional stability, world peace, and national sovereignty. Beyond its application to the refugee situation in Afghanistan's reconstruction, cadastral data can also aid in recovery from natural disaster or the long-term war from which the coun try suffers.

Under the largely illiterate Taliban of the 1990s, no interest or knowledge by them of cadastral land ownership was exhibited in Afghanistan, with the result that property records that had gotten into a shambles from the prior foray into communist ideology under the Soviets, were subsequently destroyed, lost, forged, or otherwise completely mismanaged and left in an entirely wretched state. These records required much work to rehabilitate in so far as possible after the invasion by the ISAF Coalition in late 2001 (Batson, 2008). In practice even up to contemporary times, the lands of Afghanistan have been governed largely by a quite unpredictable mix of statutory and traditional institutions and practices, which have forced far too many land disputes into the courts. The slow response times of the judiciary, their unfortunate reputations for susceptibility to corruption and coercion by powerful interests, or tendency to favor males, the wealthier elites, and the dominant ethnic groups, detract from the judiciary's credibility and lead to the avoidance of the courts by most Afghans insofar as possible.

In recognition of the considerable importance of proper land ownership in the fractured country, the US Agency for International Development (USAID) established a project in Afghanistan in 2004 entitled, "Land Titling and Economic Restructuring Activity (LTERA)". LTERA (Figure 16.1) was set up to improve the management of property records, formalized land tenures, restructured and privatized previously state-owned entities, as well as implementing modern land mapping and surveying methodologies where little had existed previously. LTERA also worked at the policy level to improve the legal and regulatory framework for establishing property rights, property registration, and economic restructuring. Land-titling activities included improving the legal framework, developing a modern land policy based upon community recognition of property rights within informal settlements, implementing an efficient property

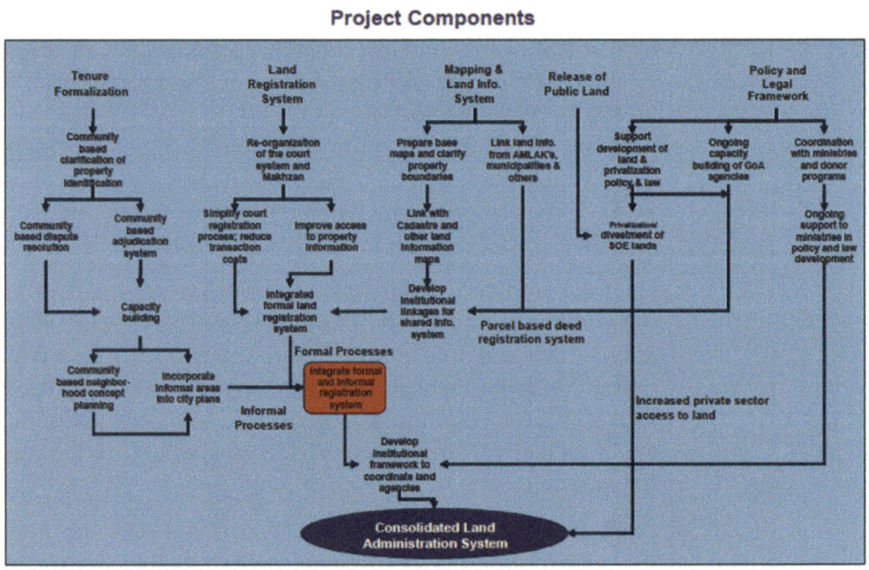

Figure 16.1 Project components of the LTERA project sponsored by the US Agency for International Development that are aimed at improving security in land tenure for millions of Afghans, as well as assisting the government of Afghanistan to undertake a comprehensive land privatization program to support the Afghan National Development Strategy and international donor programs in the country.

registration system, and reactivating more normal cadastral activities. LTERA was able ultimately to introduce a comprehensive land administration system throughout the country, as well as producing maps of land and buildings held by the various state-owned entities so that they could be restructured or privatized. This economic restructuring assisted the Ministry of Finance and 11 other ministries with the restructuring and privatization of some state-owned entities through international tenders, long-term leases, corporatization, or the sale of certain assets to the private sector.

The results of the LTERA effort in management of cadastral records were that some 6.5 million property records were reorganized in 21 out of what are now 37 provinces, with more than 700,000 deeds digitized. At the same time, the previously labyrinthine and opaque bureaucracy on immovable property registration was cut down from an original 25–34 steps, to only 3 steps for urban land and 4 for rural. The formalization of land tenure was approved by the Cabinet of Ministers in September 2007. At the same time, LTERA was able to help about 2 million informal settlers in Kabul, Mazari Sharif, Kunduz, and Taluqan to gain formal ownership of the land upon which they were living. In the process, personnel from USAID and the ISAF military provincial reconstruction teams (PRTs) were assisted in navigating the land-acquisition process so that they would understand how to help their Afghan clients do the same. In the economic restructuring and privatization process, 25 liquidation, restructuring, and corporatization proposals for state-owned entities were approved by the Afghanistan government, which started the process of transferring their assets to the private sector. This identified 1320 state-owned entity land parcels and buildings for privatization and ultimately allowed the transfer of assets worth almost $12 million to the private sector through 38 public auctions. In the process of this activity, LTERA also was able to develop a curriculum for land surveying and mapping at Kabul Polytechnic University, as well as creating a cost-effective technique to survey and map property boundaries, which supports the protection of property rights.

The Rural Lands Administration Project (RLAP) was also an initiative that used the cadastre to help rebuild the shattered nation. Where the project was begun, community cooperation was requested by the government and was then given by the communities. The borders of rangeland parcels were then delineated and agreements were prepared concerning who were to be the legitimate users of the rangeland parcels. The community stakeholders met, discussed, and approved the agreements and delineations on satellite images, while plans were also made for improvement of the rangeland parcels.

The government of Afghanistan has recognized all the problems with land in the country and has attempted to modernize the land management system by legislation. Starting in 2003, the Policy and Strategy for Forest and Rangeland Management act was promoted and enacted by 2006, as was the National

Land Policy in 2007, and the Rangeland Law in 2008. Since 2011, the Afghanistan Land Authority (*Arazi*) has been put in place to deal with managing land rights but it is so new that it is not yet well recognized nationally or embedded at the local level. Because the government of the country owns some 6 million ha of unused land, the Ministry of Agriculture created the Arazi to develop and implement a strategy to lease some of this land for agriculture, agroprocessing, and industrial purposes. The Arazi has been given the responsibility of inventorying all of the government-owned land, establishing fair and transparent procedures for its leasing, and setting up a one-stop procedure window that makes such leasing attractive to both domestic and foreign investors. Via the means of this new proactive and responsive authority, the processes of land lease and land use will be significantly eased for: (1) the population of Afghanistan directly through the increased access to government-owned land for lease and indirectly through activities that promote private-sector development; (2) international investors through access to governmental land for various legal business purposes; and (3) the government of Afghanistan through both increased revenues and a more positive reputation through the means of successful land reforms. With these understandings, the new Arazi organization hopes to contribute to poverty reduction and the creation of a more secure and sustainable future for Afghanistan.

The future success of the new structures and institutions managing the lands of Afghanistan hinges upon how well these hybrid and evolving mechanisms succeed in their wider efforts at peace building through being able to handle three major and most daunting challenges: (1) managing the growing demands for land; (2) stopping illegal land grabs; and (3) finding effective ways to resolve the disputes over land. These challenges are being exacerbated and forced by rapid population growth, large numbers of returning refugees who need some place to go, and environmental degradation that reduces the amount of productive land available, and increases the competition for land in rural areas for agriculture, and in urban areas for building locations. A telling statistic is that arable land per capita in 1980 was 0.55 ha, but that by 2007 this had fallen over 55 percent to only 0.25 ha per person (Brown and Blankenship, 2013).

Population growth in Afghanistan is quite out-of-hand, as the total fertility rate has skyrocketed to a horrific and unsustainable 7.48, whereas a normal replacement rate is 2.1 (Batson, 2008). Enormous rural-to-urban migration has forced the growth of Kabul from about 300,000 people in 2001 to around 4 million today, which is a rise of about 4.9 percent every year, one of the highest rates of urban growth in the world. This cannot be sustained, and is compounded by the return of millions of refugees, 4.6 million since 2002, all hoping to reestablish themselves and their families where they once lived, on land now occupied by others. Pakistan has even threatened to expel several million additional refugees living within its borders, which at one fell swoop would increase the total population in Afghanistan by about 5 percent, a most unsettling prospect.

Illegal land seizures have occurred repeatedly by powerful elite groups whom the local people refer to as the "land mafias" (Irvine, 2007). The urban development minister estimated in 2007 that as much as 2 km^2/day was being appropriated and sold off. In other cases, where long-running land disputes had been ignored for decades, new road-building projects suddenly increased the land value so much that it was seized from the local people (Sexton, 2012). Resolving land disputes is essential so that development is not inhibited or community relations are not poisoned further. And although considerable progress seems to have been made in implementing a land management and registry system in some places, country-wide implementation is still a long way off. Questions of central government management or its lack, versus local authority and traditional arbitration and mediation through *jirgas, shuras*, and tribal elder decisions based on sharia law, are paramount. Such systems will be strongly tested by expansion of the mining industry, which is likely to result in the need for both voluntary and forced population resettlement in connection with mining concessions. This will certainly be disruptive and likely to lead to considerable problematic behavior unless handled exceptionally carefully, something Afghanistan is not well-known for with its all too often inept or corrupt bureaucracy.

The questions of national versus local authority in matters dealing with land registration is quite important in terms of government credibility because prior to about 1979 when communism intervened, the central government was never involved in settling any sort of land disputes. This means that land administration and management would most likely have to be fully devolved to the local level before most land interests could be brought under some form of record-based ownership. Estimates are that it takes more than a generation (~25 years) to build any effective land administration in a postconflict situation, and Afghanistan has not yet even entered the postconflict stage (Brown and Blankenship, 2013). Some solutions to recurrent problems are to bring new lands that could be farmed into production by fixing irrigation and soil conservation procedures, building much better management structures to adjudicate land disputes by encouraging transparency and accountability, empowering marginalized groups, and using procedures with new safeguards designed to overcome all the myriad of mistakes or omissions that can occur in land and other conflict assessments (Table 16.2).

As discussed in more detail in the following chapter, soil erosion in Afghanistan is rampant because of millennia of deforestation, dearbification, and the lack of attention to range degradation in the presence of near universal sheet erosion. The results of such erosion have been a huge mobilization of soil particles downhill into dam sites, with an attendant loss of water-storage capacity in the filled-up reservoirs as well. In recent years, some limited success at prevention of soil erosion has been achieved by installation of some new berms along the contour on the outward convex noses of hills, as well as plentiful rock check-dams and berms in the hillside convexities to suppress such erosional

Table 16.2 Guidelines for Rapid Project-Level Peace and Conflict Assessment

1. Assessing the context
   a. What are the key natural resources in the project area?
   b. Who controls these natural resources?
   c. What are the main conflict risks in the area?
   d. How does natural resource management interact with conflict in the project area?
   e. What is the level of state capacity to extend its presence and authority into rural areas in order to enforce laws and resolve disputes?
   f. What previous peace-building initiatives have been deployed in the project area?
2. Assessing the intervention
   a. What is the nature of the project?
   b. Have there been other similar projects in the area?
   c. What institutions/organizations does the project seek to engage?
   d. What potential negative impacts does the project have on natural resources in the project area?
   e. What potential positive impacts does the project have on natural resources in the project area? How might this have an impact on conflict dynamics?
   f. What mechanisms does the project currently have in place to address conflict and deal with grievances?
3. Maximizing opportunities for peace building
   a. What sort of safeguards can the project put in place to avoid the identified risks?
   b. How can the project maximize the identified positive opportunities?
   c. How can the project assess its progress and learn from the experience?

Government and international-sponsored projects can never be fully insulated from potential conflicts over natural resources unless numerous guiding questions can be addressed by project members. Each question has several descriptive possibilities not presented here, but which need to be assessed to help ensure success.
After Annex 3; Brown and Blankenship (2013).

movement of soil. As an attempt by the US Agency for International Development to reduce soil erosion, such contour remediation is a welcome addition to range management to show people how to better handle their essential soil resource, but its present-day miniscule application would have to be taken up in a forthright manner nationwide to have any realistic and lasting effect. This would be an ideal employment opportunity for thousands of the illiterate unemployed but without a strong government plan and funding to be effective, the problem will not be solved any time soon.

Water

As go problems with the land in Afghanistan, water is also unequally shared and constitutes the second-most cited cause of conflict in the country so a discussion of its role in the future development of the country is essential as well. Sexton (2012) has detailed a number of case studies of water issues in Samangan and Balk Provinces, as well as Herat Province that can serve as paradigmic types of contentious problems that have occurred repeatedly in different forms and locations in the past, and that are highly likely to show up again and again in different places. The first such problem in Afghanistan is the limitations to downstream irrigation and drinking water wherein upstream users take what they need, leaving an insufficiency downstream, and the second is a situation in which a downstream country foments armed incursion upstream where plans for hydroelectric and irrigation damming are underway. Violent confrontation in both cases can serve as a warning of the understandable increased bellicosity associated with future water insufficiencies.

The past thirty-odd years of near-continuous conflict in Afghanistan have significantly degraded the country's water storage and irrigation network because of decades of underinvestment and inadequate management. In fact, in spite of being at the top of the five major watersheds of the region (Amu/Panj, Northern River Basin, Harirud-Murghab, Helmand, Kabul), Afghanistan has the lowest water storage capacity of any country in the region, and strikingly for such an arid country, one of the lowest storage rates in the world (<100 m^3/capita) (Brown and Blankenship, 2013). Furthermore, irrigation reaches only an estimated 1.8 million ha, which is down 40 percent from what was able to be irrigated in the mid-1970s, whereas the population increase over the same time period has nearly doubled. Similarly, the many years of conflict have significantly degraded the traditional water management systems and community relations so that the ever scarcer water resource is increasingly a source of instability.

Traditional water management in Afghanistan in any given area is done by a *mirab*, the water master, who may be assisted by one or more *chachbashi*, who help take care of the local water arrangements. If a river is large, *mirab* councils are formed with each *mirab* and his assistants supervising approximately 20 km segments of the river, making sure that the many communities, side canals, irrigation lands, and land owners receive the predetermined water amounts appropriately. The central government always sat on the outside of these local arrangements as the final arbiter to make sure that the arrangements were respected, and only used state force to intervene in water disputes when absolutely necessary. The past three decades of instability, however, have yielded in some cases to coercive and corrupt politicking that has undermined the *mirab* structures. In addition, problems with the social structures of water management have been compounded by inadequate water infrastructure, such that weak or poorly developed water distribution systems have led to periods of scarcity followed by devastating floods (Sexton, 2012). Land degradation in the headwaters of all the major water basins in Afghanistan, particularly by overgrazing, deforestation, and dearbification wherein shrubby vegetation is pulled up by its roots for fuel, has been a major factor in the rapid siltation of Afghanistan's irrigation and water-storage networks. In many cases, the soil erosion process is accelerated even more when the degraded upper watersheds fail in various kinds of landslides (Shroder et al., 2011a,b), which break up bedrock and overlying soil enough to pour it into the drainages, which is then deposited as overwhelming amounts of sediment onto agricultural fields and into the irrigation works lower down (see Figure 8.28).

The water available gives the land its value in Afghanistan; some studies have shown that reliably irrigated land can produce crops with a value 4–14 times greater than rain-fed land. Because some 36 percent of Afghans live in absolute poverty, the fluctuations of seasonal water availability can be devastating in the dry years when crop yields decline dramatically. Because Afghanistan is at the top of most of its watersheds, the average water availability for Afghans is a more reasonable 2775 m^3/person/year, whereas in neighboring Iran and Pakistan, the annual water availability per person is half that of the Afghanistan. If Afghanistan managed its water better than it does now, and in future transactions sent regular documented amounts across its borders to the other countries, the nation would be seen as a source of cooperation and mutual benefit, rather than the present more difficult situation wherein its neighbors machinate problems within its borders to get what they want. Part of the problem also is that water treaties with the neighboring countries are either old or quite out-of-date in not reflecting present-day climate realities and their changes, or in a few places, nonexistent as with the Hari Rud that flows into Iran and Turkestan. In this latter case, no formal treaty with Iran was ever made because Afghanistan never used very much of the water. With the new Bandi Salma dam, however, being built across the Hari Rud upstream from Herat with help from the Indian government, considerable unease has occurred in Iran, along with their plentiful support to the Taliban who have repeatedly attacked personnel building the dam (Sexton, 2012). Similarly, on the Panj–Amu river system in the north, Afghanistan only draws about 2 km^3 of water, in spite of a 1946 agreement with the Soviet Union to be able to use 9 km^3, but this prior treaty since the end of the Cold War has not been recognized or incorporated into any of the other agreements between Tajikistan, Uzbekistan, or Turkmenistan over the waters going into the Aral Sea basin (Brown and Blankenship, 2013). Other similar problems exist with Iran and the Helmand, or Pakistan and the Kabul River. Unless Afghanistan can act on these treaty issues with some knowledge and authority concerning coming likely discharges and timings of flood and drought variations with a well-managed series of impoundments, water releases, and helpful predictions to the neighboring states, problems will continue and violence will erupt (see further discussion in Chapter 18).

Internally in Afghanistan, some dissonance has occurred between traditional *abandaz* water allocation through the *mirab* and *chachbashi* systems, and the new 2009 Water Law enacted by the government. The new framework for the water sector is based upon three main tracks: (1) integrated water resource management; (2) river basin management and participation in decentralized decision-making via multistakeholder platforms or river basin agencies; and (3) river basin councils. On paper, this design for better water management makes sense but implementation can be quite another matter, as has been demonstrated amply in connection with the Panj–Amu River Basin Program that has been piloting the introduction of these imported concepts in northeastern Afghanistan (Thomas et al., 2012).

Issues of access to water in South Asia in terms of ethnicity, religious sect, and gender constitute largely hidden problems that should be brought out in open because the problems need to be known to a wider audience (A. Eschen, email communication, September 2013). For example, if people might be from the wrong tribe or different religious sect, many examples are known where water might not be made available. In the Ghazni area, the Hazara people live in the more arid parts of the province and the Pashtuns are located near the fertile land with the water. Here the water *mirabs* are supposed to be impartial, but this is all too often not the case. Throughout Afghanistan's history, the Pashtuns have been able to better position themselves where the land is more fertile and the water more plentiful.

Women in both Afghanistan and Pakistan can also have many problems in getting water for use in their houses (A. Eschen, email communication, September 2013). If water is available, men will have the women walk miles to fill buckets and water jugs. In a usual scenario at a well or other water source, the people of the community will know each other. Thus the women will cover themselves with head scarves of *chadris* if men are around and will quickly retrieve their water and go on their way. Women, however, have been assaulted when water is scarce or under control by another community. Only through education and dialog can these cultural norms be addressed.

In Afghanistan as a result of three decades of war, in many cases the men have been killed, which produces many problems for the women to do agriculture. As a result of the civil war and Taliban rule, Afghanistan lost some 70 percent of its agricultural sector. In many cases, the displaced women with the lost male family members had to move to the cities for other income. Sometimes this meant they were forced to become beggars. The women who stayed behind with the land would most likely become peasant farmers. They would not be able to have an agricultural business, but might grow enough to feed their family. Currently, however, no particular means exist to support women's agriculture although the global Partnership for Afghanistan is an NGO helping women farmers on a large scale (http://gpfa.org/programs/ruralafghanwomen/). Due to cultural restraints, women are not able to be in water management position or act as any kind of *mirab*. This may be changing in places like Kabul, but not in rural Afghanistan. Pakistan has the same situation because of their similar Islamic and social mores. The *mirabs* are generally actually hired by either a warlord, or other power broker within the community. Usually they are hired within the same ethnicity, which, of course, can lead to inequities.

Over the next several decades, Afghanistan will be facing three major and specific challenges regarding management of particular conflicts over water: (1) increased water demand; (2) coping with likely increased frequency and magnitude of water-related disasters such as floods and droughts; and (3) building or reconstruction of the water-related infrastructure for agriculture, water impoundment, and hydroelectric production. This will entail careful attention to well-considered procedures designed to keep local and cross-border tensions to a minimum by artful public discussions, and transparency in presentations. Diplomatic discussions will be required to ease the fears of local people, as well as the international audiences. Particularly telling will be data presented about the facts that three quarters of the Afghan population lacks any sort of safe drinking water, and that nearly a quarter of the entire huge spectrum of child mortality in Afghanistan is directly related to water contamination and poor sanitation. Development of the water resources of Afghanistan clearly requires great attention to these three pressing issues.

Increasing water demand is a fact of life worldwide as populations increase and climate changes, with Afghanistan perhaps becoming one of the world examples of this international type of problem for which the solutions are rather few. The water situation in Afghanistan is such that water availability per person in 2025 will drop by one-third from 2004 levels and be down by 50 percent by 2050, which is obviously unsustainable. By then also, the capital city, Kabul, will require six times as much water as it does now. Some parts of the country will drop below the absolute thresholds for water scarcity to the point where life will not be possible; under such conditions, keeping the peace will be equally impossible. At the same time, the mineral resource bonanza, which is expected to be ramping up greatly in production over coming decades, will require considerable new water usage as well, or will produce collateral issues of water contamination, situations that have not had to be dealt with significantly before.

Coping with water-related disasters is another issue of serious import for Afghanistan, particularly because the rate of flood deaths in the country relative to population exposure to floods is one of the highest in the world (Brown and Blankenship, 2013). Average numbers of flood fatalities in Afghanistan are at least 20 times that of USA, which has a similar exposure of population to the hazard. As an example, floods in 2009 in northeast Afghanistan took the lives of nearly 1200 people, killed 35,000 livestock, and affected almost 29,000 homes. The converse hazard of drought can be equally destructive to populations and livelihoods with displacements of people equivalent to that of the fighting associated with the Taliban in the late 1990s.

The government of Afghanistan has plans to build or reconstruct some 31 major infrastructural water projects in the country, including 15 large water-storage dams. Twelve dams are purportedly to be supported by India on the Kabul River that flows into Pakistan, and are a source of great unease in Islamabad, just as are tensions for similar projects on rivers to Iran. But with no new water treaties with either country, and no frameworks for cooperation in the offing because of regional intransigence, Afghanistan has been at a loss as how to proceed further. Two possibilities result; the water-starved neighbors are seeking ways to just take the water they want by fomenting violence, or they will just sit not so idly by and let the status quo of no treaty or impoundment activity accomplish the same purpose. The United Nations Environmental

Programme (UNEP) has proposed a number of solutions to maximize opportunities for peace building by reducing competition over scarce water resources through crop substitution, reforestation, and other methods of demand reduction, as well as improving water governance, preparing better for climate uncertainty, improving transboundary management, and introducing conflict-sensitive approaches by presumably well-meaning, but rather water-heedless donors (Brown and Blankenship, 2013).

Extractive Mineral Resources

Exploitation of the great trove of mineral resources detailed in Chapter 11 has been viewed by many as the answer to all the Afghanistan's problems. At the same time, however, some have feared that given the present security situation and the potential environmental and social risks of mismanagement of exploration and extraction, perhaps the country cannot afford to push ahead with wholesale mining at all. That domino has already toppled, as it were, however, as it is unlikely that once begun with the payment of millions of dollars into the Kabul treasury, this process will be unlikely to stop unless massive violence is unleashed, and few want that. The attendant problems are many, of course, and include the need for a dramatically improved transport grid already discussed in Chapter 7, mining profits potentially funding warlords, and the seemingly endless opportunities for corruption in spite of the many attempts to develop methods of ensuring transparency in dealings already discussed in Chapter 14 and elsewhere. The development of the mining sector is likely to remain a strategic priority of the government for the foreseeable future because of its strongly lucrative nature, with the result that as of late 2012, some 212 resource extraction contracts had already been signed, with many more in the works. This "gold-rush" mentality detailed in Chapter 14 is unlikely to abate any time soon and will certainly lead to considerable development, but at what risk to the environment remains to be seen, although if the past is any sort of reliable guide to the future, perhaps the environmental results will not be that pleasant.

To the credit of the government of Afghanistan, however, reform of the mining sector has been set out in several important documents (Oil, Gas, and Mining Sector Vision, National Extractive Industries Excellence Programme; Anon, 2010) that if successful will help considerably. The overall goal is to work toward financial independence of the country with reforms that support growth of the extractive industries through a strong regulatory environment, modernization, and strengthening of the Ministry of Mines, enabling of the business environment to attract investment of the private sector, and development of reliable and updated geological information to support sector growth. All of this effort is expected to provide 42–45 percent of the GDP of Afghanistan by 2024, as well as creating jobs, reducing reliance on foreign aid, and helping to diversify and enlarge the economy of the country. At the same time, the passing of the National Environmental Protection Act in 2005 and the formation of the new National Environmental Protection Agency (NEPA) (Table 16.3) to monitor environmental performance are essential actions to protect what is left of the environment, but they will have to have legions of

Table 16.3 Functions and Powers of the Afghanistan National Environmental Protection Agency (NEPA)

1. Maintenance of environmental integrity and promotion of sustainable use of natural resources.
2. Promotion of conservation and rehabilitation of environment.
3. Coordination of environmental affairs at local, national, and international levels.
4. Development and implementation of national environmental policies and strategies in order to integrate environmental issues and sustain developmental approaches into legal and regulatory frameworks.
5. Provision of environmental management services in areas of environmental impact assessment, air- and water-quality management, waste management, pollution control, and permitting of related activities.
6. Establishment of communication and outreach for environmental information to ensure improved awareness of environmental issues.
7. Implementation of bilateral or multilateral environmental agreements to which Afghanistan is a Party.
8. Implementation of Convention on the International Trade in Endangered Species of Fauna and Flora (CITES).
9. Signing on behalf of the government agreements regarding the protection and rehabilitation of environment.
10. Promotion and management of Islamic Republic of Afghanistan's accession to and ratification of bilateral and multilateral environmental agreements.
11. Coordination of preparation and implementation of a national program for environmental monitoring and effective utilization of the data provided by that program.
12. Preparation every 2 years in relation to urban areas and every 5 years in relation to rural areas, a State of the Environment report for Islamic Republic of Afghanistan for submission to President's Office.
13. Preparation of an interim State of the Environment report on emerging issues relevant to environment in Afghanistan not less than every 2 years.
14. Within a period of 3 years of Act promulgation, development of a national environmental action plan, which assesses urgency and importance of actions that should be taken in short-, medium-, and long term to prevent, eliminate, and reduce adverse effects as described in most recent State of the Environment report, and, in consultation with relevant ministries and institutions, determination of coordinated strategy and schedule for implementation of those actions.
15. Compilation periodically and publication of reports on significant environmental indicators.
16. Compilation and publication on annual basis of report detailing authorizations granted and activities undertaken by NEPA.
17. Assessment of effectiveness of implementation of NEPA Act and any regulations made under it to improve sustainability of use and management of natural resources and conservation and rehabilitation of environment.
18. Development and implementation of plans for environmental training, environmental education, and environmental awareness raising in cooperation with relevant ministries and public bodies.
19. Active coordination and cooperation with ministries, Provincial Councils and District and Village Councils, public bodies, and private sector on all issues related to sustainable use of natural resources and conservation and rehabilitation of environment.
20. Monitoring implementation of objectives and provisions of this law.
21. Fulfilling any other functions that may be assigned by Council of Ministers.

After Shroder (2012).

professionals trained quickly to ensure compliance and clean-up, which will be a tall order in an endemically corrupt country such as Afghanistan. The new transparency initiatives discussed in Chapters 14 and 15 will help, but much will be required to translate the rhetoric into reality on the ground. Required also will be core challenges so as not to create or exacerbate conflict by minimizing negative social affects, human rights, and environmental impacts of the mining, while also ensuring that the mining benefits the people by managing the revenues effectively, diversifying the economy, strengthening the accountability and transparency to reduce corruption, and maintaining responsive security around the mining sites. The government of Afghanistan has taken a number of steps to ensure that these criteria for success are met with their mining endeavors, and UNEP has responded with carefully reasoned assessments to ensure success. Results must await later analyses in future, but for now much has been done to facilitate the processes of maintaining a reasonable environment in the course of greatly ramped up development in Afghanistan.

References

Anon, 2010. National Extractive Industry Excellence Programme. Infrastructure Development Cluster, Ministry of Mines, Government of Afghanistan, 24 p. https://www.acbar.org/admin/publication/02_neiep_project_document_revised_29march2011.pdf.

Batson, D.E., circa 2008. Registering the Human Terrain: A Valuation of Cadastre. National Defense Intelligence College Press, Washington, DC, 62 p. http://www.ni-u.edu/ni_press/pdf/Registering_the_Human_Terrains.pdf.

Brown, O., Blankenship, E., 2013. Natural Resource Management and Peace Building in Afghanistan. United Nations Country Team in Afghanistan, United Nations Environment Programme (UNEP), 62 p. http://postconflict.unep.ch/download/NRM&peacebuilding_Afg_Draft_02.pdf. http://postconflict.unep.ch/download/poster_preview_draft_04_A4_low-res.pdf.

Irvine, S., September 6, 2007. Powerful 'Grab Afghanistan Land'. BBC News. http://news.bbc.co.uk/2/hi/south_asia/6981035.stm.

Shroder Jr., J.F., Weihs, B., Schettler, M., 2011a. Mass movement in northeast Afghanistan. Journal of Physics and Chemistry of the Earth 36, 1267–1286. http://dx.doi.org/10.1016/j.pce.2011.03.003.

Shroder, J.F., Schettler, M.J., Weihs, B.J., 2011b. Loess failure in northeast Afghanistan. Journal of Physics and Chemistry of the Earth 36, 1287–1293. http://dx.doi.org/10.1016/j.pce.2011.03.001.

Shroder Jr., J.F., 2012. Afghanistan: rich resource base and existing environmental despoliation. Environmental Earth Sciences. 67 (7):1971–1986. http://dx.doi.org/10.1007/s12665-012-1638-7.

Sexton, R., 2012. Natural Resources and Conflict in Afghanistan: Seven Case Studies, Major Trends and Implications for the Transition. Afghanistan Watch, Kabul, 85 p. http://inec.usip.org/resource/natural-resources-and-conflict-afghanistan-seven-case-studies-major-trends-and-implications.

Shroder Jr., J.F., 2012. Afghanistan: rich resource base and existing environmental despoliation. Environmental Earth Sciences. http://dx.doi.org/10.1007/s12665-012-1638-7.

Thomas, V., Mumtaz, W., Azizi, M.A., 2012. Mind the Gap? Local Practices and Institutional Reforms for Water Allocation in Afghanistan's Panj – Amu River Basin. Afghanistan Research and Evaluation Unit Case Study Series, 150 p. http://www.areu.org.af/Uploads/EditionPdfs/1209E-Mind%20the%20Gap%20CS%202012.pdf.

In this Chapter

17

Afghanistan Environmental Degradation

Abstract

The environmental degradation of Afghanistan is extreme and includes ground- and surface-water degradation, diminution and contamination, soil salinization from irrigation, soil erosion from watershed and range mismanagement, deforestation, and dearbification wherein shrubs are uprooted, biodiversity reduction and habitat loss, widespread desertification, urban and industrial pollution, and the residuals of warfare. Each of these problems has now been recognized and described, with new efforts implemented to solve many of these problems.

Keywords: Biodiversity loss; Dearbification; Deforestation; Desertification; Environmental pollution; Groundwater drawdown; Habitat loss; Range mismanagement; Soil erosion; Soil salinization; Surface-water diminution.

At myopic first glance, Afghanistan might seem to be in good environmental condition because of the exceptionally low levels of development and the great natural beauty of the soaring mountains, great canyons, turquoise lakes, and vast expanses of rolling hills of grassy and flower-covered terrain of wind-blown loess dust. First appearances are deceiving, however, because beneath the veneer of initial landscape beauty are the realities of many kinds of environmental degradation and despoliation that have only recently begun to be addressed and remediated in a limited way or in any coherent fashion (Figure 17.1). The environmental problems in Afghanistan are profuse but the amount of detailed technical, quantitative, and mapped information about the environment in the country is limited; nevertheless, an overview of the chief environmental problems is possible. A list of these environmental topics for discussion includes the following:

1. Water diminution and contamination
2. Soil salinization from irrigation
3. Soil erosion (sheet erosion and soil landslides)
4. Overgrazing rangeland degradation
5. Deforestation
6. Dearbification or "deshrubification"

Natural Resources in Afghanistan. http://dx.doi.org/10.1016/B978-0-12-800135-6.00017-9

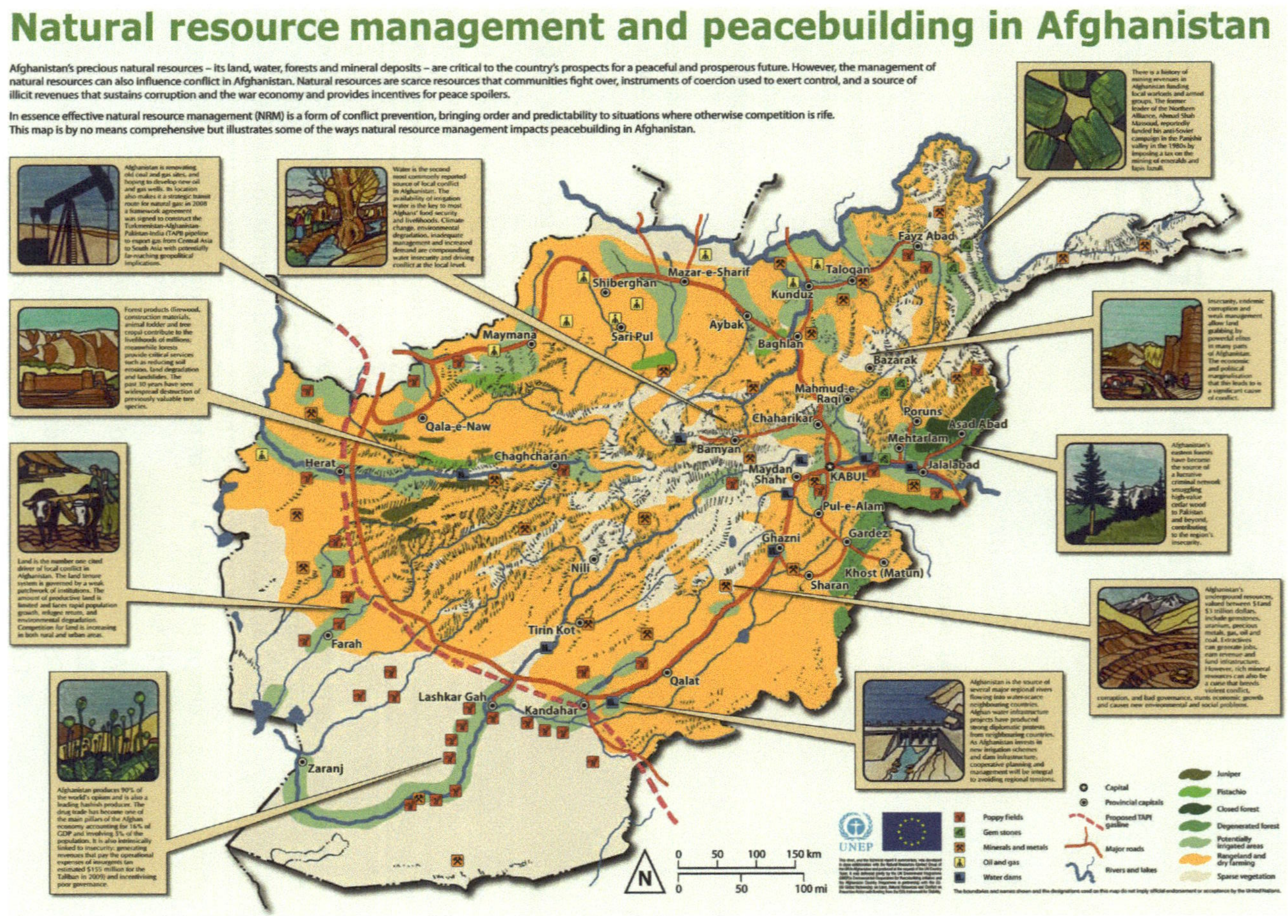

Figure 17.1 Map of natural resource management associated with peace-building efforts in Afghanistan by the United Nations Environment Programme (Brown and Blankenship, in press).

7. Biodiversity, habitat loss, and protected areas
8. Desertification
9. Pollution and environmental health (urban and industrial pollution)
10. Residuals of warfare

Following the ouster of the Taliban in late 2001 during the invasion of Afghanistan by the USA and the Western Coalition troops, personnel from the United Nations Environment Programme (UNEP), together with Afghan authorities, local communities, and international and national experts, set out to assess the environment in 35 rural locations and 38 urban sites throughout the country. Hundreds of interviews were conducted and 60 field samples were obtained of air, water, and soil to test for contamination in a spot-sample assessment of general environmental health. The results were presented in UNEP's (Anon, 2003) report on the postconflict assessment, which specified a long list of 163 environmental recommendations whose enumeration gave a good overview of the dimensions of the environmental problems to be addressed (Tables 17.1–17.3).

The UNEP (Anon, 2003) report found that the long-term results of three decades of war and long-term overexploitation of the country's once rich, natural-resource base have created numerous grave environmental threats. Not only the 10 environmental threats listed above are serious, but more detailed environmental degradation includes many places of interlinked, or complex, synergistic linkages between environmental stressors, such as severe surface- and groundwater depletion and water-table drawdown without any recharge attempts, coupled with water contamination, desertification, and urbanization of important wetlands, and increased soil erosion from combinations of deforestation and overgrazing. Thousands of people have become environmental refugees, which when combined with the return of many of the war-time refugees from Pakistan, where 3 million have resided, and Iran, where an additional 1.5 million took refuge; these factors have resulted in increased environmental problems. When more than 3 million refugees returned home following political stabilization after the Coalition invasion of late 2001, leaving 2 million behind in Pakistan and 900,000 in Iran, the prospect of the eventual return of these people as well leaves UNEP and Afghan government personnel highly concerned (Anon, 2008) (Figure 17.1).

As explained by the author previously (Shroder, 2012), the existing environmental despoliation in Afghanistan is obvious to even untrained eyes, but so little real data exist (Anon, 2008) that it is most difficult to obtain a clear picture of the current state of the environment in the country. The lack of systematic observation and recording of the rivers, forests, lands, wildlife, climate, or atmospheric conditions over the years prevent accurate comparisons 5, 10, 30, or 50 years ago, as would normally be done in most contemporary studies elsewhere in the developed world. Much uncertainty exists about the sources of environmental pollution, its extents and transfers, or its connections to public health and disease transmission. Little is known about the scope of all the many forms of soil erosion and deposition elsewhere over other useful land, or into sediment sinks where it is not important, or

Table 17.1 Recommendations by the United Nations Environmental Programme, Postconflict Environmental Assessment (UNEP, 2003), First Part. Cross-Cutting Environmental Management Recommendations

Environmental Legislation and Enforcement	
1.	Recognize environmental rights in national constitution
2.	Consult stakeholders on framework of environmental law
3.	Introduce routine use of environmental resource assessment
4.	Strengthen enforcement mechanisms
5.	Establish participatory processes
6.	Introduce environmental permits
7.	Employ economic incentives
8.	Share information
9.	Coordinate environmental monitoring
10.	Develop "soft" enforcement measures
11.	Develop environmental budgeting
12.	Debate multilateral environmental agreements
Capacity Building for Environmental Management	
13.	Provide basic infrastructure
14.	Provide professional training
15.	Strengthen mandate of Ministry of Irrigation, Water Resources, and Environment
16.	Introduce an interministerial coordination mechanism
17.	Establish interagency technical committees
18.	Adopt decentralized, community-based management approaches
19.	Strengthen regional cooperation
20.	Join other international processes
Environment and Job Creation	
21.	Support labor-based rehabilitation projects
22.	Prioritize hiring of Afghan professionals and staff
23.	Afghanistan conservation corps
24.	Transform and add value to products from natural resources
25.	Maximize economic benefits from nature tourism
Environmental Planning	
26.	Clarify land tenure and begin land-use planning
27.	Develop a National Environmental Action Plan (NEAP)
28.	Prepare emergency response and contingency plans
Environmental Impact Assessment (EIA) Procedures	
29.	Apply EIAs across the board
30.	Establish an interministerial task force
31.	Establish a dedicated EIA Unit
32.	Apportion EIA costs
33.	Build professional expertise to undertake EIAs

Table 17.1 Recommendations by the United Nations Environmental Programme, Postconflict Environmental Assessment (UNEP, 2003), First Part. Cross-Cutting Environmental Management Recommendations—cont'd

Industry and Trade
34. Develop best environmental practice in the private sector **35.** Prevent illegal trade in hazardous waste, raw materials, and products **36.** Promote cleaner production **37.** Assess privatization impacts **38.** Develop and market environmentally friendly products and services **39.** Minimize impacts in food security from new international trade **40.** Diversify industrial and agricultural production base **41.** Transform and add value to primary products for export **42.** Mitigate potential environmental and societal impacts of trade
Public Participation, Training, and Environmental Education
43. Establish rights to environmental information and access to justice **44.** Improve access to government processes **45.** Establish a resource center for environmental information and best practice **46.** Increase media awareness **47.** Develop environmental education materials **48.** Develop an environmental education curriculum **49.** Train teachers to deliver environmental education **50.** Provide environmental training **51.** Inventory traditional ecological knowledge **52.** Unite environment and religion

about the economic consequences of any of this. Because of the largely illiterate population, little understanding of the myriad environmental problems exists at the local or any other level, with the result that although people may understand that situations can be unpleasant for them, they will generally not know why, or will ascribe it vaguely to "Allah's will". With only insufficient data, it is most difficult to predict such natural hazards as different kinds of floods, massive, soil-eroding windstorms, droughts, or other natural hazards. Assessment of the quality and quantity of surface- and groundwater are varied and generally rather out-of-date. Because of these problems, the need for data collection and data exchange is consequently the major priority for effective environmental management. Commonly, the most that can be done is to present photographs of various environmental problems and let these serve as slim measures of the various environmental situations. The exceptions are those cases where repeat satellite imagery can be used to capture changes in deforestation, lake-water fluctuations, or in flooding, but other more subtle environmental changes on repeat satellite imagery that might be measured, such as soil salinization, range degradation through soil erosion from water or wind, or dearbification uprooting of shrubs generally require more ground control and quantitative sampling techniques to obtain reliable data.

Table 17.2 Recommendations by the United Nations Environmental Programme, Postconflict Environmental Assessment (UNEP, 2003), Second Part. Recommendations for Sectoral Environmental Management

Water Supply
53. Test and monitor water quality 54. Identify and eliminate cross-contamination 55. Stop unmanaged well drilling 56. Protect water supply sources 57. Properly equip the water supply sector 58. Promote water conservation 59. Develop a National Water Resources Management Strategy 60. Develop a National Water Authority 61. Develop groundwater monitoring programs 62. Promote transboundary water-resources management 63. Study hydrological contributions of glaciers, permanent snowfields, and snowfall 64. Consider impacts of climate change
Waste
65. Segregate and properly dispose of medical waste 66. Properly equip waste management section 67. Identify pollution hotspots 68. Develop a national strategy for waste management 69. Expand wastewater treatment in all cities and industrial sites 70. Use septic tanks as temporary solutions 71. Compost wastewater sludge
Hazardous Wastes and Chemicals
72. Investigate chemical storage sites 73. Minimize pesticide use 74. Monitor pesticides in the environment 75. Adopt a strict chemicals policy 76. Plan for management of hazardous and industrial wastes 77. Investigate war-related chemical damage
Open Woodlands
78. Undertake immediate soil-stabilization measures 79. Begin community-based reforestation 80. Reinstate a community-based forest-warden system 81. Allocate woodlands to communities 82. Employ grazing management and rotation systems 83. Establish community-based woodlots 84. Establish woodland management legislation 85. Establish seed banks 86. Establish woodland protective areas 87. Develop alternate incomes

Table 17.2 Recommendations by the United Nations Environmental Programme, Postconflict Environmental Assessment (UNEP, 2003), Second Part. Recommendations for Sectoral Environmental Management—cont'd

Eastern Conifer Forests
88. Rehabilitate community control and government influence
89. Improve transboundary cooperation
90. Introduce a timber cooperative
91. Seek "win–win" outcomes to use of "Green forces"
92. Control road access to forests
93. Estimate future timber demands
94. Establish forest management legislation
95. Undertake gradual implementation of export controls
96. Institute grazing management and rotation systems
97. Develop demonstration sites showing the viability and benefits of forest conservation
98. Promote strict protection of forest sites as components of a national protected areas system
Energy
99. Increase use of all renewable energy resources
100. Use best available technologies when rebuilding the energy sector
101. Encourage small-scale hydroelectric development
102. Save and conserve energy
103. Assess possibilities to construct combined heat and power plants
104. Manage community woodlots
Air Quality
105. Protect workers
106. Improve public transportation systems
107. Use cleaner fuels
108. Increase use of natural gas
109. Regularly inspect the transport sector
110. Institute a central system of heating
111. Stop uncontrolled burning of wastes
112. Manage industrial air pollution
Protected Areas Network and Wildlife Conservation
113. Identify biodiversity hotspots and set protection targets
114. Introduce protected areas legislation
115. Establish protected areas as a managed resource
116. Explore feasibility of restoring degraded wetlands
117. Monitor protected areas
118. Manage wildlife populations
119. Develop transboundary protected areas
120. Protect green spaces, green belts, and trees in urban areas
121. Promote tourism development
Desertification
122. Stabilize sand dunes and soils
123. Reseed highly degraded rangeland
124. Reduce grazing and dry-land cultivation in vulnerable areas
125. Map areas vulnerable to desertification
126. Create community-based rangeland assessment and management plans
127. Establish representative rangeland areas where grazing is excluded or experimentally controlled

Continued

Table 17.2 Recommendations by the United Nations Environmental Programme, Postconflict Environmental Assessment (UNEP, 2003), Second Part. Recommendations for Sectoral Environmental Management—cont'd

Plant Resources for Food and Agriculture
128. Reestablish gene banks 129. Promote community-based orchards 130. Promote local varieties of cereal 131. Promote sustainable agricultural practices
International Environmental Conventions (Multilateral Environmental Agreements—MEAs)
132. Strengthen the focal point for MEAs 133. Prioritize international environmental convention 134. Activate and implement existing agreements a. International Convention to Combat Desertification (CCD) b. Convention on biologic diversity c. Basel Convention on Control of Transboundary Movements of Hazardous Wastes and their Disposal d. International Convention for Protection of World Cultural and Natural Heritage e. Convention on international Trade in Endangered Species of Wild Flora and Fauna (CITES) f. United Nations Framework Convention on Climate Change (UNFCCC) g. Develop clean development mechanism (CDM) projects h. Convention on Wetlands of International Importance (Ramsar) i. Convention on Protection and Use of Transboundary Watercourse and International Lakes j. Convention on Conservation of Migratory Species (CMS)

Water Diminution and Contamination

The decrease in water supplies as a result of changes in climate has already been discussed to some extent in Chapters 5 and 8 but bears some repeat because of its tremendous seriousness; indeed without sufficient water, life in any region is simply not possible. Rampant contamination of the resource accomplishes much the same as diminution because the attendant lack of potability tends to lead to sickness and death as well (Figures 17.2 and 17.3).

Many laypeople do not realize that surface waters and underground waters are basically the same thing because they are directly connected through the pores and fracture spaces of the earth and rock over and through which they run. The result is that when droughts diminish surface water supply, they also diminish groundwater, which needs to be replenished, not only in later years of reasonably good precipitation but also through artificial recharge wherein surface water that might go on outside a country's borders could be diverted into recharge basins on a controlled basis to refill the underground aquifers. It will be very hard, however, for Afghans to accomplish this because of the outsider Taliban in the pay of the surrounding downstream competitors who want the water for themselves. This is why Afghanistan's participation in regional and global water partnership discussions leading to new treaties is a prerequisite for the future use of international waters. If Afghanistan does not have prior

Table 17.3 Recommendations by the United Nations Environmental Programme, Postconflict Environmental Assessment (UNEP, 2003), Third Part. Recommendations for Site-Specific Areas of Afghanistan

Waste Management
135. Herat 136. Kabul 137. Kandahar 138. Mazari Sharif
Wastewater
139. Herat
140. Kabul
Water Supply
141. Herat 142. Kabul 143. Kandahar 144. Mazari Sharif
Hospitals
145. Herat Regional Hospital 146. Mazari Sharif Civil Hospital 147. Mirwais Civilian Hospital
Industrial Sites
148. Shiberghan oil refineries 149. Mawlawy oil storage facility 150. Sari Pol crude oil terminal 151. Zhora plastic recycling/shoe factory 152. Brick factory 153. Zuri battery company 154. Asphalt factory
Protected Areas
155. Bandi Amir 156. Wakhan Corridor and Pamiri Buzurg 157. Seistan Basin 158. Abi Istda 159. Dashti Nawar 160. Koli Hashmat Khan 161. Imam Sahib and Darqad 162. Ajar Valley 163. Nuristan and Kunar

records kept of its annual downstream flow, or they have been destroyed by the past three decades of war, however, the neighbors like Pakistan can claim whatever they like in any international negotiation. For example, Pakistan could assert that it needs more water which Afghanistan must send their way just as

Figure 17.2 **Deeply polluted Kabul River: note people on left gathering to extract water in buckets.** *From Fareidullah Masi; Anon (2008).*

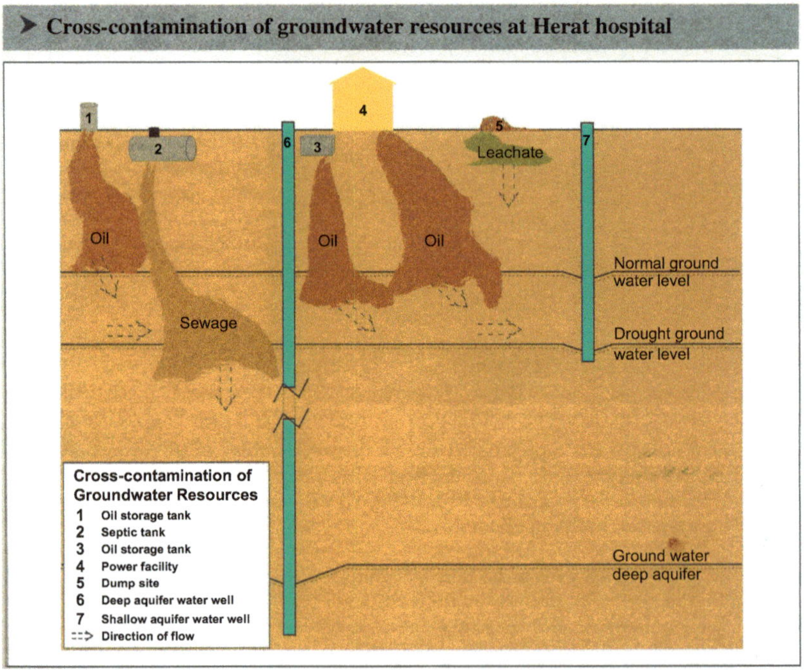

Figure 17.3 Diagram of cross-contamination of groundwater resources at Herat Hospital is actually an exceptionally serious problem that occurs repeatedly throughout other areas of Afghanistan. The conditions of ground and groundwater contamination, as well as the "normal" and drought-caused water table levels are shown clearly. If the lowered water tables in the current drought phase are ever restored, even greater groundwater contamination will occur. *After Anon (2003).*

it has always done in the past, even if the country did not know about it or keep track of it. Afghanistan must thus develop an appropriate and comprehensive data base concerning the quantity and quality of all its water resources to use in their water-treaty negotiations. The government of Afghanistan must also soon carry out a comprehensive assessment of current and future water needs so that it can better predict the future and begin to make allowances for more coming water shortages, rather than waiting for "Inshahlah" or whatever "Allah is willing". As long as the neighbors like Pakistan or Iran, who do not do business this way, keep better records and measure more water quantities, they always will have an upper hand at the negotiating table.

Soil Salinization from Irrigation

Wherever ancient cities occur as archaeological remnants in arid regions, extensive nearby saline ground generally occurs as well. For example, a simple glance at Google Earth™ in the region of ancient Balkh near Mazari Sharif shows plentiful saline ground where fewer plants grow in the western parts of the Balkh irrigation area (Figure 3.49). These salts accumulate as a result of long-term irrigation where drain tiles were not also installed to drain off the salty wastewater and thereby flush away the salinity. In some places, the ground may also become swampy because the irrigation may not be able to drain away due to "hardpan" calcrete (caliche) at depth or variable sediment sizes or mineralogies that slow down or stop the water drainage. The Nadi Ali and Marja areas of the Helmand project, for example, had considerable salt on and in the ground 40 years ago, as well as water logging from an impenetrable calcrete layer below the thin surficial soils (Figure 17.4). In recent years, however, Marja has been a prominent opium poppy growing area and appears to have considerably less salt at this time, apparently largely as a result of better drainage and improved cropping techniques as the formerly

Figure 17.4 Saline ground from excessive flood irrigation with only limited drainage in the Helmand Delta region. *Photograph taken by J. Shroder in 1978.*

kuchi nomadic herders who were newly settled there became better farmers and learned how to use the land properly (R. Scott, email communication, 23/04/2013). Perhaps the new drug profits enabled some remediation of the saline soils of past decades as well.

Soil Erosion (Sheet Erosion, Soil Landslides, Deflation)

The erosion of the weathered fine particulate soils of Afghanistan is a huge factor in the ecology of the country, and its presence is mainly the result of the particular hilly and mountainous geography and semiarid to arid climate of the region wherein plant cover thereby has been reduced anyway (Figure 17.5). Except in high-altitude alpine topography where other processes are prevalent, all soil erosion in most mid-latitude places in the world tends to be a situation where close to the divides between drainage basins, any precipitation that is rain or melt-water runs off in sheets everywhere and strips away millimeter- to centimeter-thin layers of soil in what is referred to as *sheet erosion*. Further down slope, the running water begins to collect together more and form small runnels a few centimeters deep that are referred to as *rills* or *rill erosion*. Through the process of rill deepening and rill coalescence eventually, multiple meter-deep gullies (*sareh, farkand, karaa, wadi, nullah*, etc.) form and transport much sediment out of the watershed. As gullies grow and intersect, they may consume the land surface and develop an extensively eroded badland

Figure 17.5 Formerly forested, now denuded hillsides near Farkhar District in Takhar Province of northeastern Afghanistan. Some 30 years ago, these slopes were forested with valuable pistachio trees but now they are cut down with resulting landslides and other forms of soil erosion, which is redeposited as alluvium at lower elevation.

Figure 17.6 Wind-deflated ground in the lower Helmand valley near Zaranj in the spring of 1978.

type of topography if there are plentiful and thick soft sediments into which the profuse gullies can dissect the terrain, or if bedrock is not too far below the soil cover, eventually the softer sediments are all stripped away, leaving only exposed and barren bedrock.

An additional serious process of soil erosion in Afghanistan is deflation wherein soil sediments are stripped away by strong winds. In the process, the landscape becomes pitted and grooved with linear deflation basins or blowouts, with the intervening streamlined ridges sculpted into yardang landforms (Figure 17.6). The sand grains from the soil that are mobilized by the wind form sand dunes downwind, whereas the finer grained dust from the soil eventually settles out of the wind and forms the extensive loess deposits that cover so much of Afghanistan.

Many other processes, as described further below, can initiate or accelerate soil erosion and any place in the world where mankind has managed flocks of grazing animals for many thousands of years tends to be severely eroded as a result of these processes; parts of Africa, the Middle East and South Asia in particular. Only where special restrictions to certain land-use practices are maintained by government control, can this soil erosion be slowed or stopped. In general, once the processes of rampant soil erosion get underway, changing or stopping the process is an exceedingly difficult procedure. Afghanistan is perhaps a world exemplar of strong soil erosion (Figure 17.7).

Overgrazing Rangeland Degradation

In general in Afghanistan, any area within reach by shepherds and hooved animals is commonly overgrazed, mostly by sheep and goats, as well as exploited for fodder plants (Breckle et al., 2010) and uprooted shrubs for fuel that are transported offsite (Figure 17.8). This also applies to remote deserts, as well

Figure 17.7
Decorative minaret built by Mahmud of Ghazni about 1000 years ago with a rough foundation that is now exposed down to a depth of about 3 m by soil erosion around it. *Photograph was taken by the Niedermayer-Hentig Expedition of 1914.*

Figure 17.8
Clear-cut conifer forest near Kamdesh in Nuristan Province in fall 1977.

as the high mountain valleys and alpine meadows that are used seasonally as rangeland by the nomadic *kuchis* and seminomadic shepherd tribes. Overgrazing of any landscape is a direct result of what has come to be known as the "tragedy of the commons" (Hardin, 1968), wherein individuals neglect the wellbeing of the shared or common grazing land in pursuit of their own personal gain. Everyone naturally has to share in the overgrazing depletion of the common-land resource whether or not they want to, but only the individual gains from adding just one more animal to his flocks until the common resource is depleted. As a consequence of this pervasive grazing pressure, the structure and floristic composition of most rangelands in Afghanistan have been changed, with an increase of unpalatable, poisonous, or spiny species, even in the better-watered monsoonal border areas with Pakistan.

Two additional factors of overgrazing pressure are the twin problems of the manner in which sheep and goats graze, and in their small, sharp hooves that penetrate into the ground surface. For example, unlike the larger horses and cows that tend to chew off the tops of the grasses they graze, sheep and goats more commonly clamp their teeth down on the plant and jerk it out of the ground, which of course removes the roots so that regeneration is less likely. Their trampling effects also puncture the soil surface, which enables it to be mobilized more effectively by erosional agents. In some parts of the Western USA, sheep and goats are commonly referred to by knowledgeable range managers as "four-footed locusts" as a result of their depredations of the land cover. Most developed countries with grazing animals control the amount of animals on the land, or the time that they are allowed to graze in certain areas, but nothing like that ever occurs in Afghanistan.

As overgrazing thins out the protective ground cover of plants, the more fertile upper layers of the soil grow more exposed to wind and slope-wash, water erosion. During drought years, the resulting erosion can be mainly by wind deflation, but strong episodic rains or heavy snowmelt will cause dramatic loss of soil by erosion as well, with a complete and partially irreversible change of habitat conditions. After many centuries of this activity, as has happened over the many millennia in Afghanistan, soils above bedrock that have taken many thousands of years to form are progressively striped away to leave only the underlying rock exposed on the surface. Little will continue to grow there with no soil into which new vegetation can become rooted. The extensive rocky wastelands of much of Afghanistan are the result.

Deforestation

The removal of trees in Afghanistan is a quite serious situation. We do know from archaeological studies and writings in antiquity (Emperor Babur's memoirs, Babur Nama, among others) that parts of Afghanistan were covered with deep, cedar-rich forests well into the Mughal period, which dates back about 500 years ago (Ponting, 1990; Anon, 2003). Only a few centuries in the past, deciduous and coniferous forests covered some 5 percent of the current land area of the country, with about 1 million ha of oak and 2 million ha of pine and cedar, growing mostly in the eastern portions of the country (Anon, 2008).

Figure 17.9 **Piles of uprooted shrubs that are used for fuel or fodder.** Such dearbification breaks the soil and mobilizes it for later transport by water and wind.

Figure 17.10 **Trucks in Afghanistan loaded with squared-off log beams of deodar cedar and other woods for export to Pakistan.** *Photograph from Anon (2003); by permission.*

These trees grow mainly between altitudes of 2500–3300 m (Figure 17.9), being replaced mainly by juniper above that. The highly fragrant wood of the deodar cedar is the main tree that has been extracted from these forest areas, mainly for illegal export to Pakistan (~66 percent). During the times of the Mujahideen and Taliban in the 1980s and 1990s, local government officers reported that up to 200 timber trucks per day were operating on the main road in Kunar Province (Figure 17.10). Because each truck could carry a volume of timber representing 5–10 trees, or ~44 m^3 of wood, this represents an off-take of up to 200 ha/day (assuming a density of 20 mature trees per hectare). By 2003 even with government decrees against uncontrolled logging, a range of 25–50 trucks/day were still removing logs. Forest officers and local people had estimated that the loss of forest was in the range of 50–70 percent, so that UNEP used satellite imagery for 1977–2002 to obtain better estimates. The results were that the figures were 71 percent for Nangarhar Province, 53 percent for Nuristan, and 29 percent for Kunar. Similar losses were estimated for the provinces of Paktya, Khost, and Paktika (Anon, 2003).

In the northern provinces, open savanna woodlands with pistachio, almond, and juniper trees originally covered some 32 million ha for about 48 percent of the country as well. Tree numbers were something like 20–100 trees/ha. At the present time, most of the forests have retreated or gone entirely in places where they once flourished. For example, some 40 years ago, the northeastern Badghis, Takhar, and Badakhshan provinces held the most continuous and productive pistachio woodlands in the country. Above 1500 m altitude, the pistachio woodlands increasingly gave way to juniper. In these areas, an estimated 50–70 percent of the trees have been lost through combinations of fuel-wood stockpiling during the Soviet occupation, and military forces cutting trees indiscriminately for fuel wood, as well as to reduce hiding and ambush opportunities for opposing forces. UNEP (Anon, 2003) has estimated that woody shrubs and dung meet about 50 percent of a family's fuel needs for a year in that region, with pistachio trees supplying the remaining percentages. Calculations show that this amounts to 20–40 trees per family annually, which because the nuts are harvested also, as well as the land being heavily grazed over; collectively, this eliminates regeneration possibilities so that no new seedlings have been observed in recent years. With help from USAID and the Afghanistan government, some 18 forest regeneration sites were established by 2003.

Dearbification or "Deshrubification"

The natural ecosystems of Afghanistan have been heavily degraded except for a few inaccessible or uninhabitable areas. As noted, the forests have been relentlessly exploited for timber and fuel. Woodlands, shrublands, and even the scarce vegetation of the arid semideserts were also thinned out over the years, and then increasingly devastated and destroyed as the search for firewood and brushwood has gone on unceasingly for cooking and warmth. Because of the increasing shortages of suitable coarser woody fuel, even slightly woody dwarf shrubs such as *Artemesia* (Asian sagebrush) and a number of other lignified herbs are uprooted for fuel in a process that can be called *dearbification* after the Latin for "shrub" (Figure 17.8). The result of rampant dearbification, of course, has been the ever-increasing disturbance and mobilization of the soil so that it is more easily eroded subsequently by the wind and slope wash.

Biodiversity, Habitat Loss, and Protected Areas

As noted by the many observations of wildlife reported by Emperor Babur in his memoirs of 500 years ago, the country was long known for its rich wildlife.

Since that time, a number of these species have been driven to extinction in the country, including the Caspian tiger (*Panthera tigris virgata*), the cheetah (*Acinonyx jubatus*), and the wild ass or onager (*Equus hemionus*). Many other large mammals are threatened, including snow leopards (*Uncia uncial*), wild goat (*Capra aegagrus*), markor (*Capra falconerie*), Marco Polo sheep (*Ovis ammon polii*), urial (*Ovis orientalis*), and the Asiatic black bear (*Ursus thibetanus*). Many of the bird species in Afghanistan are also threatened globally, with the Siberian crane (*Grus leucogeranus*), white-headed duck (*Oxyura leucocephala*), marbled teal (*Marmaronetta angustirostris*)), Pallas's sea eagle (*Haliaeetus leucoryphus*), greater spotted eagle (*Aquilla clanga*), imperial eagle (*Aquilla heliacal*), lesser kestrel (*Falco naumanni*), corncrake (*Crex crex*), sociable lapwing (*Vanellus gragaria*), and the pale-backed pigeon (*Columba hodgsonii*). Of these birds, the Siberian cranes are the most endangered because only a few thousand are left and the group that used to use the lake of Abi Istada (see Chapter 5) as a stop-over point between summering in Siberia and wintering in India has been decimated by the lake drying out in the recurrent droughts, as well as by hunting because the hungry Afghan people hunt and eat these and many other birds. Flamingos that used to breed there as well have also not been seen in Afghanistan for decades.

Many other species of mammals and birds exist in Afghanistan and are not particularly threatened as a species because they may occur in many other places as well, but wherever the environment is as strongly threatened as is that of Afghanistan, one can be fairly certain that a number of these other species may be threatened also with extirpation of their normal habitats or hunted mercilessly for food to eat or fur to sell to foreigners.

The genetic resources that exist in Afghanistan should be mentioned as well because a number of important plants such as common bread wheat (*Triticum aestivum*) appear to have been first cultivated in Afghanistan over 5000 years ago. More other native varieties of wheat occur wild in Afghanistan than any other place in the world and they could have residual genetic variation with great resistance to disease or pests that could be exploited, providing habitat destruction does not drive them into extinction as well (Anon, 2008).

Protected areas were first introduced in Western countries primarily to protect landscapes, wildlife, and habitats of particular value, generally by exclusion of people and regulation of access or use. With suitable design and management, twin goals of protection and engagement can be realized by the local populations for their own benefit. In past years in Afghanistan, royal hunting preserves were established, and nominal attempts by prior governments have also been to set aside parts of the Wakhan Corridor, as well as one-supposed national park at Bandi Amir, three waterfowl sanctuaries (Abi Istada, Dashti Nawar, and Koli Hashmat Khan), and two wildlife reserves (Ajar Valley and Pamiri Buzurg) (Figure 17.11). Unfortunately, all the wildfowl lakes have begun to dry out

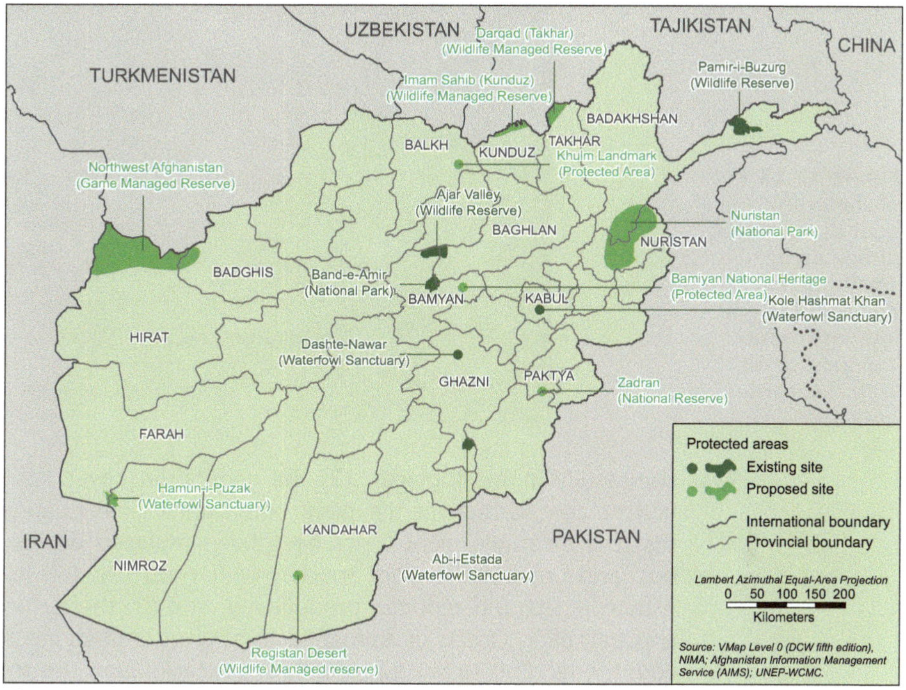

Figure 17.11 Map of protected areas in Afghanistan. *Map from Brown and Blankenship (in press).*

repeatedly in the recurrent droughts, which makes life difficult or impossible for the many birds who seek refuge in them. Between 1977 and 1992, an additional 10 sites were proposed for protected status, including in the southwest of the country, where supposedly exists the Registan Desert Wildlife Management Reserve, and Hamuni Puzak Waterfowl Sanctuary, the Northwest Afghanistan Game Management Preserve, and in the north, the Imam Sahib (Kunduz) Wildlife Preserve and Darqad (Takhar) Wildlife Managed Reserve. A national park has also been proposed for Nuristan as well (Figure 17.1). Most of these prior designations were never formally gazette in the government bureaucracy, however, and have never had any real enabling legislation or reasonable financing, in past governments or even at the present time, so their future is far from assured.

Desertification

Altogether, these multiple sources of environmental degradation as explained above have multicative reinforcing effects where combined with each other, as they most commonly are. The multiple impacts have had disastrous effects in Afghanistan: (1) the regeneration of trees and shrubs has stopped almost

Figure 17.12 Pile of sheep and goat dung pellets picked up by small children and saved for later use as fuel. *Photograph by author in Khash Rod valley in spring 1978.*

completely in most places; (2) the vegetation cover has decreased, and the floristic composition of the plant communities has changed fundamentally in that the palatable plant types have been replaced widely with spiny, poisonous, and unpalatable plant species, which has resulted in a heavy loss of floristic diversity and reduced productivity; and (3) the reduced vegetation cover and trampling effects of animal flocks have intensified erosion in all parts of the country, leading to widespread bare bedrock. Even the small pellet droppings of the sheep and goats are commonly collected by children in some areas and used to cook with later by the family, which removes vital nutrients from the soil (Figure 17.12). The addition of salt loads brought in by long-term irrigation without attention to flushing of salts in wastewater drains compounds the destruction of the land (Figure 17.4). The overall result of all these environmental degradation processes is the desertification of a region that may have once had a quite different ecological character. UNEP (Anon, 2008) has noted that the global assessment of soil degradation (GLASOD) shows some 16 percent of Afghanistan has been severely affected by anthropogenic activities, whereas the country's vulnerability to desertification is one of the highest in the world, with about three-fourth of the nation existing under that threat. In addition, <1 percent of the land lies within the nominal protected areas, and none of these cover any of the coniferous forest areas.

Pollution and Environmental Health (Urban and Industrial Pollution)

Waste management is completely out of hand in almost the whole Afghanistan, especially in the urban areas where all the plentiful trash dumps are routinely grazed over by animals and picked over by people looking for scraps to recycle in some fashion (Figure 17.13). At the present time, solid waste management in the country is not being handled at all properly, inasmuch as a great

Figure 17.13 Children scavenging in Qamar Qalla dumpsite in Herat, Afghanistan. Such completely uncontrolled garbage sites are common in Afghanistan and are generally grazed over by sheep, goats, and cattle, at the same time that humans seek to extract whatever they can reuse.

deal (up to 50 percent) of the waste is inert dirt (soil, gravel, stone, and dust), that is needlessly mixed with hazardous, medical, household, and industrial waste materials. Because everything has to be carried to dump sites all mixed together, this adds greatly to transport costs and vehicle emissions. The result of this is that outside many cities, only dump piles of dirt, broken glass, medical waste, and bits of unusable plastic bags are heaped. Furthermore, the whole country of Afghanistan has almost no sanitary landfills that hold the waste on-site safely, but instead open dumps are quite common. The capital city, Kabul, is considered to be an open garbage area, and is thought to be one of the most polluted cities in the world (Ahsan, 2012). Some of the dumps near Kandahar and Herat are completely inappropriate, in that the dumps are in dry river channels that at certain seasons of the year sweep the waste downstream and right back into the cities it came from. In some places, the dump sites have plumes of toxin leachates moving underground to contaminate the groundwater. In Kandahar, the burning of plastic wastes releases poisonous dioxins and furans into the air (Anon, 2003), which adds to the already problematic air pollution from other sources such as open cooking fires and completely uncontrolled vehicle emissions. The overall result of all these inappropriate waste situations is that diseases, bacteria, and viruses are also added to the mixes of poisonous chemicals in the water and air.

Open sewer ditches (*juies*) that are highly contaminated with numerous pathogens are disturbingly common throughout the urbanized areas of Afghanistan because by and large no piped sewer systems exist anywhere in the country except for a small area in Kabul that superficially treats the wastewater from about 100,000 people. Out of an urban population of several million people, this is a disturbingly low number. Household latrines and toilets commonly discharge through the wall and out onto the street because septic tanks are not generally available except to a certain extent in Kandahar and a few other places. Vendors commonly wash their produce in effluent wastewater and children play in the effluent ditches as well. Outside the cities, human waste (night soil) is spread directly on fields, in many areas by donkey-drawn, two-wheeled *sud* carts, or by tanker trucks where people or municipalities are wealthy enough to afford them. Cross-contamination by sewage-born *Escherichia coli* bacteria introduced into well-water supplies from shallow groundwater wells occurs in many, if not most urbanized areas in Afghanistan. Persistent pesticide chemicals such as DDT and benzene hexachloride (Lindane) also were used extensively for several decades in Afghanistan in the past to control locusts, among other pests, and have been stored unprotected in many sites, so they will have accumulated in ecosystems and perhaps the groundwater as well.

Reliable access to clean and safe drinking water sources, which is a fundamental requirement of any normal society, is quite nonexistent in the urban environments of Afghanistan. The results are severe threats to public health so that a full spectrum of intestinal (amoebic dysentery, giardiasis, and bacillary dysentery) disease is rife, as well as parasites and other water-borne diseases. UNEP (Anon, 2003) estimates that only 12–23 percent of the population can access safe water supplies in the country. Foreign assistance agencies (UNICEF, ICRC, etc.) have been working steadily to provide greater piping and chlorination of water supplies in Kabul, Herat, and elsewhere.

Air quality, especially in cities in Afghanistan, has been progressively deteriorating over the years as vehicles, smoke stacks, brick–kiln emissions, and blowing dust add to the pollutant load. Late autumn and winter are the worst, particularly during days of temperature inversion when warm air aloft traps lower, and colder air beneath as in a stable lid in the mountain-bounded valleys that are so common in Afghanistan. At these times, the common wood cooking fires further decrease the air quality to the point where more and more people choose to wrap their nose and mouth in masks or layers of cloth to breathe through. Turban ends are commonly used as well. Polyaromatic hydrocarbons (PAHs) are the chief worrisome culprit in air pollution as they are implicated in lung cancers. The PAHs bind with dust particles, which prevent them from escaping upward into the higher atmosphere.

Residuals of Warfare

Anywhere wars are fought for any length of time, a tremendous amount of chemical and explosives residues of many varieties tends to be left behind, to the obvious detriment of the people in whose land the war was fought.

In Afghanistan, the armies have come and gone many times, but the refuse remains; probably the Americans and Coalition forces will take away a great deal more than did the Soviets, but the residues will still be there in various fashions—destroyed armor, crashed planes, bullets, bombs and mines, perhaps unneeded chemicals such as perchlorate, depleted uranium; the list goes on (Anon, 2013). Both the garbage at the many military sites in Afghanistan that is disposed of in smoky burn pits, and other military operations with huge dust plumes behind vehicles are reported to have exposed soldiers and civilians to dangerous levels of toxic dust. With the Bradley fighting vehicle getting only less than half a kilometer per liter (~1 mile/gallon), some two-thirds of the enormous fuel consumption is spent just getting the fuel to the battlefield. The military vehicles have emitted many hundreds of thousands of tons of air contaminants (CO, CO_2, SO_2, and nitrogen oxides) that add to the pollution loads already there. The war-despoiled environment is overprinted across an already basket case of an environment that is thoroughly degraded by the general competition for land and life. This problematic environment will not be much of a problem to the latest militaries to return home after experiencing the Afghanistan grist mill, but for the people of Afghanistan, it will be another story altogether. One can only hope that their continued experience past the departure of the American and Coalition troops is not too extreme if they can start to rebuild with their rich resources, but many things must go exactly right in order for them to overcome the long odds.

Conclusion

The upshot of the tremendous and multifaceted environmental despoliation of so much of Afghanistan is that the country is so stressed environmentally, as well as militarily, politically, and socially, that the place can serve perhaps as a bit of a bellwether to alert people of potential future problems elsewhere in the world (Shroder, 2012). This is not a position that the government of Afghanistan or its people are particularly pleased with, of course, but it certainly reflects existing realities such as the fact that UNEP (Anon, 2008) thinks that the health status in the country is perhaps the poorest in the world. For example, the infant mortality and the under-age-five death-rate categories are some of the world's highest, with overall life expectancies at birth of 43.5 for females and 43 for males, which is well below world averages of about 67 years. At the same time, Afghanistan's vulnerability to hazards and disasters is increasing; while also having food shortages, all of which add further pressures to the beleaguered environment.

The environment of Afghanistan is caught in a precarious balance between low levels of precipitation and its primary production (Anon, 2009). Traditional and reasonably balanced, rather low-level agricultural practices have been disrupted by decades of insecurity, migration, breakdown of social structures, drought, and extreme poverty that have produced overexploitation of the natural resource base. The results have been situations of advanced loss of biodiversity, degradation of the land by both natural and anthropogenic causes, and the diminution of biophysical agencies of protection resulting in an acceleration of wind and water erosion to produce a real lack of productivity in the arid zones. The impoverished soils have a reduced crop-carrying capacity, which leads to attempted overstocking, the cultivation of unsuitable land for cash and subsistence crops, and an increased exposure of soils to erosion. The consequences have also produced severe flooding, slope wash and wind erosion, deforestation, dearbification, reduced pasture quality, decimation of wildlife numbers, air pollution, decreases of quantity and quality of water for drinking or irrigation, and so on, all of which are further compounded by climate changes, especially precipitation. All of these competing land uses and attendant environmental problems have exacerbated and led directly to violent confrontations. The extreme pressures for simple survival at the local level, coupled with the need for economic growth at the national level, even with the recognition by the government of the potentially ultimately fatal consequences of biodiversity loss and overall desertification, little substantive action has been taken to address the many issues.

A number of recent authors (Brown, 2009; Fagen, 1999, 2003; Diamond, 2005) have detailed the hosts of environmental problems in the middle of ever-changing climates that have brought down past human civilizations over and over. The list of damaging factors for the world, with many reinforcing feedback loops for each, reads like a comprehensive poster presentation for environmental problems in Afghanistan. These include overplowing, overgrazing, loss of topsoil through soil erosion, drought, flood, and water shortages in a warming world, as well as deforestation, dearbification, habitat destruction, soil salinization, fertility reduction, and surface and subsurface water management problems, all of which are extreme in places in Afghanistan. When a country facing all these problems also cannot provide basic services or even security to its citizens within the crisis of the failed state, which Afghanistan became some years ago (Rubin, 1995), the resulting social chaos generally spreads beyond its own borders, with spreading violence, refugees into neighboring states, political, and religious extremism, smuggling, and sales outward of plentiful drugs and weapons, sanctuary to international terrorists, and increasing disease. Afghanistan, of course, as detailed in these pages, already has almost all these causes and consequences of the failed state syndrome, and is not likely to be able to tolerate many more environmental stressors that would be added by the significant resource extraction discussed elsewhere in this volume (see Chapter 15). Furthermore, because the country has already been in the top tier of failed states for some time, any further diminution of environmental quality caused by the proposed extensive development of resource corridors will only compound the mechanisms to increase more failure of the state in the future. Afghanistan seems balanced on razor's edges of environmental instability, with nasty tipping points lurking everywhere; only by great attention to the environmental details will the country avert an unrecoverable crisis.

References

Ahsan, M., 2012. Afghanistan's Environmental Challenges. Daily Outlook Afghanistan, Olivier Immig & Jan van Heugten; Dutch Research on Afghanistan and Pakistan; 3 p. http://www.immigvanheugten.nl/.

Anon, 2003. Afghanistan: Post-conflict Environmental Assessment. United Nations Environment Programme (UNEP), 177 p. http://postconflict.unep.ch/publications/afghanistanpcajanuary2003.pdf.

Anon, 2008. Afghanistan's Environment. United Nations Environment Programme (UNEP). National Environmental Protection Agency (NEPA) of the Islamic Republic of Afghanistan, 30 p. http://reliefweb.int/sites/reliefweb.int/files/resources/37FB695F399DC788C1257582004CA11E-Full_Report.pdf.

Anon, 2009. Afghanistan: National Capacity Needs Self-Assessment for Global Environmental Management (NCSA) and National Adaptation Programme of Action for Climate Change (NAPA). United Nations Environment Programme (UNEP); National Environmental Protection Agency, Afghanistan; Global Environmental Facility. 125 p.

Anon, 2013. Costs of War (Environmental Costs: Afghanistan, Pakistan). Watson Institute, Brown University, Providence, RI, 3 p. http://costsofwar.org/article/environmental-costs.

Breckle, S.-W., Dittmann, A., Rafiqpoor, M.D., 2010. Field Guide Afghanistan: Flora and Vegetation. Scientia Bonnensis, Bonn, Germany, 861 p.

Brown, L., May 2009. Could food shortages bring down civilization? Scientific American 300, 50–57.

Brown, O., Blankenship, E., 2013. Natural Resource Management and Peacebuilding in Afghanistan. United Nations Country Team in Afghanistan; United Nations Environment Programme; 58 p. http://www.unep.org/disastersandconflicts/portals/155/countries/Afghanistan/pdf/UNEP_Afghanistan_NRM.pdf.

Diamond, J., 2005. Collapse. Penguin Books, NY.

Fagen, B., 1999. Floods, Famines, and Emperors: El Niño and the Fate of Civilizations. Basic Books, NY.

Fagen, B., 2003. The Long Summer: How Climate Changed Civilization. Basic Books, NY.

Hardin, G., 1968. The tragedy of the commons. Science. 162 (3859), 1243–1248. http://www.sciencemag.org/content/162/3859/1243.full.

Ponting, C., 1990. Historical perspective on sustainable development. Environment 32 (9), 4–9.

Rubin, B.R., 1995. The Search for Peace in Afghanistan: From Buffer State to Failed State. Yale University Press, New Haven, CN.

Shroder Jr., J.F., 2012. Afghanistan: rich resource base and existing environmental despoliation. Environmental Earth Sciences 67 (7), 1971–1986. http://dx.doi.org/10.1007/s12665-012-1638-7.

UNEP, 2003. Afghanistan: Post-Conflict Enbronmental Assessment. 180 p. http://postconflict.unep.ch/publications/afghanistanpcajanuary2003.pdf.

In this Chapter

18

Afghanistan Environmental Protection

Abstract

New means of environmental protection in Afghanistan are now fairly well developed in the law of the country, with five main environmental pillars to provide fundamental underpinnings to their new National Environmental Protection Agency. These pillars that have been established are: (1) environmental institutions and coordination; (2) environmental law and policy; (3) environmental impact assessment; (4) environmental information and assessment; and (5) community-based natural resource management. A set of so-called equator principles have been established to set standards for determining, assessing, and managing social and environmental risks in the financing of resource-exploitation projects.

Keywords: Community-based natural resource management; Environmental impact assessment; Environmental information and assessment; Environmental institutions and coordination; Environmental law and policy; Environmental pillars; Equator principles; National Environmental Protection Agency (NEPA).

The bad news or pessimistic environmental picture about Afghanistan that was presented in Chapter 17 is offset by the examination here of the strong attention that the environment has gotten from the international donor community and the many good things since the Coalition invasion of Afghanistan in late 2001 that have happened to improve the environmental situation since then. Once the government of Afghanistan was alerted by UNEP (Anonymous, 2003a,b) to the serious environmental problems in the country, it rose directly to the occasion to request key follow-up activities (Jensen and Zaidi, 2006). This would quickly develop into a program to provide an integrated package of capacity-building activities that would help evolve from a department of Environment in the Ministry of Irrigation, Water Resources and Environment, into a stand-alone and self-sufficient National Environmental Protection Agency (NEPA) for Afghanistan by 2005. NEPA serves as the country's environmental policy-making and regulatory institution. Its role has been to regulate, coordinate, monitor, and enforce the new environmental laws that were promulgated in 2007. The NEPA agency is expected to play a major role in protection of the environment, as well as to be the central point in dealing with

Natural Resources in Afghanistan. http://dx.doi.org/10.1016/B978-0-12-800135-6.00018-0

the management of the environment of Afghanistan, so that it benefits all of the Afghanistan's people. To achieve this goal, UNEP trained and mentored counterpart staff from NEPA, provided technical assistance and advice on an as-needed basis, as well as supplying basic office and field equipment. NEPA was able to take the lead role for all major decisions to ensure that Afghan ownership is retained and reinforced at all times. Thus soon after, it was founded that the NEPA was able to obtain the required technical and legal capacity to implement environmental mandates of the government. Five environmental pillars were established around which to manage protection and remedial activities (Jensen and Zaidi, 2006):

- Pillar 1: Environmental institutions and coordination
- Pillar 2: Environmental law and policy
- Pillar 3: Environmental impact assessment (EIA)
- Pillar 4: Environmental information and assessment
- Pillar 5: Community-based natural resource management.

Pillar 1: Environmental Institutions and Coordination

In order to achieve sustainable management of environmental resources, national- and local-level institutions in Afghanistan established clear mandates, effective and coordinated structures, technically skilled staff, and sufficient financial resources. UNEP provided technical support to NEPA for an ongoing reform and reorganization of their structure so that the several hundred staff members could be set up into eight regional hubs, with separate provincial desks. A new advisory group on the environment (AGE) was established to share information between stakeholders and help in the coordination of the different environmental activities that were proposed. The new Public Investment Program (PIP) for Environmental Management was established to enable external input to better share information. UNEP established a counterpart staff training program with 15 new employees, and provided environmental technical support, a computer laboratory, and gender-sensitivity training to be able to help women in both UNEP and NEPA.

Pillar 2: Environmental Law and Policy

Along with the establishment of the NEPA organization to look after the nation's environmental business, a realistic environmental legal framework had to be established as well. The setting up of an integrated legal and environmental regulatory framework for the country required training and capacity building in NEPA, as well as several of the line ministries so that people would know better what to do and why. Several environmental phrases were inserted into the new constitution such that the people of Afghanistan were to be ensured of a prosperous life and sound environment for everyone, and that the state would adopt the measures necessary to safeguard forests and the environment.

The Environmental Act of 2005 that became law in 2007 (Afroze and Karyab, 2007; Anonymous, 2007) was based upon international standards and recognized the current poor state of the environment, while laying a framework for the progressive improvement of governance about it to ultimately lead to effective and sustainable environmental management. Its stated primary purposes were to improve the livelihoods and protect the health of humans, fauna, and flora; maintain ecological functions and evolutionary processes; secure the needs and interests of present and future generations; conserve natural and cultural heritage; and facilitate the reconstruction and sustainable development of the national economy. The people of Afghanistan were granted the rights to use natural resources in accordance with customary but sustainable practices; establish and participate in environmental nongovernmental organizations; have access to information held by government about the environment; participate in environmental meetings and demonstrations, and participate in environmental decision-making processes that affect local communities; lobby the government regarding environmental issues; voice concerns regarding environmental acts of infrastructure and development projects; and refuse to undertake environmentally hazardous work, or work that threatens human health without appropriate safeguards in place. At the same time, the people of Afghanistan were given the legal duty to comply with the provisions of the Environment Law, to use environmental resources in a sustainable manner, to avoid causing pollution or harming the environment, and to obtain a permit or license before undertaking activities regulated under the Law.

All these new laws and regulations were a very long way and a major step against previous practices, so their implementation would have to be handled carefully as little by little the government began to exert greater control over its quite out-of-control environmental situation. No one thought the process would be easy, but at least a start could be made at exerting some control over environmental despoilations where absolutely none seems to have existed previously. Afghanistan would still have a long way to go to achieve even some sort of limited environmental control but at least a start had been made.

Pillar 3: Environmental Impact Assessment

Afghanistan is entering a phase of large-scale, multi-million-dollar investments in many different programs of development and infrastructure overhaul, some of which will have significant impacts on the environment. Clearly an absolute functional system of environmental screening and impact assessment of all investment decisions was seen to be essential to ensure that all decisions were made with access to data and full understandings about impacts and options to avoid future problems. Thus, the objectives under this pillar are to contribute to the institutionalization of EIA through training and technical support in the development and effective implementation of the EIA guidelines, policies, procedures, and legislation.

Among the first acts under this pillar were to find out about all the different projects with potential environmental impacts that were underway in the country and get them registered and identified for appropriate actions if any were needed. Interministerial meetings were held to identify goals and further clarify the current state of EIA in Afghanistan. EIA screening guidelines were developed, and a gift-giving roundtable was held to develop appropriate levels of support from such donors as the Asian Development Bank, USAID, the European Commission, and other agencies. Training of EIA counterpart staff was undertaken, technical assistance and technical papers on EIA and pollution control was provided, and EIA pollution control policy papers were written.

Pillar 4: Environmental Information and Assessment

The availability of timely and high-quality environmental information is the only way to ensure effective environmental decision making. The loss of so many maps, reports, records, reference materials, and essential expertise during the three decades of near-constant war has taken its toll on being able to make reliable comparisons of present to past conditions. Under these circumstances, developing the capacity for better environmental management is difficult, so special efforts have been made to increase public awareness and participation through educational campaigns, environmental reporting, and public participation in relevant decision making. Environmental library materials have been developed, computer laboratories established, environmental awareness-raising materials translated and printed, posters made, environmental events held, management and media training programs established, and even a clean-up of the Kabul River was undertaken to raise urban environmental awareness.

Pillar 5: Community-based Natural Resource Management

Strong traditions of provincial and local-level management of natural resources occurred in some parts of Afghanistan in prewar times. Uncontrolled extraction of many resources was a direct result of the decades of war when anyone with the means could just take whatever they wanted in the war-torn and lawless environments that abounded. During what the locals call, *topakiyaan*, the time of men with guns in recent decades, the local governing structures collapsed so much that NEPA has identified a most urgent need to reestablish community-based, natural-resource management and rehabilitation practices. Training and technical support courses and workshops were designed to contribute to reestablishment of linkages between national, provincial, and district or local levels of government, as well as through implementation of pilot projects that encouraged these goals.

Equator Principles

The world minerals extractive industry is capital intensive by its very nature where exceptionally large financial sums are required to get the resource "on-stream" to market. The financing of any given project is a function of many things but in many cases, the lender of the funds expects to be paid back directly by the funds generated by the project, as well as security for the financial exposure. This process has to be overseen according to a series of strong guidelines to control against important and complex social and environmental issues that can be quite challenging, especially in such emerging economies such as that of Afghanistan's. Thus in the context of resource extraction in Afghanistan, the equator principles (EPs) that were developed in 2003 at the behest of the World Bank (Anonymous, 2006; Watchman et al., 2007) do an excellent job of setting standards for determining, assessing, and managing social and environmental risk in the financing of resource exploitation projects (Table 18.1). These principles were developed especially for countries with developing or newly emerging legal frameworks where little or none had existed before. Any project done under the scope of the EPs is evaluated on the basis of certain fairly rigorous performance standards, with the result that the EPs have become a world-wide industry standard for environmental and social risk management. As Tables 18.1–18.4 amply demonstrate, the comprehensive nature of the many strictures that applied to any given project has the capacity to ensure fairly rigorous environmental and social controls for all resource extraction projects in Afghanistan. Integrity Watch Afghanistan (Anonymous, 2010, 2011) is doing an excellent job on maintaining public scrutiny and attention on environmental issues and adherence by GoA to the EPs that are relevant to the country.

In spite of the good news about new environmental laws and concerted attempts to alert the Afghan citizenry about all the problems with their environment, ecological concerns historically have been of only minor importance to most people in the country (Sexton, 2012). To most Afghans, the short-term issues of simple survival and livelihood generally far transcend longer term issues of environmental integrity. During periods of open conflict, of course, little attention is paid by anyone to environmental issues, but in the past dozen years of the post-Taliban era, a number of notable decisions on natural

Table 18.1 The Equator Principles (EPs) Set Up to Apply to all Project Financing Globally with Total Capital Costs of >$10 M across All Industry Sectors

Number	Principle Characteristics
1	**Review and Characterization**—Assess magnitude of potential impacts and risks according to whether or not project will have impacts that are (A) significant; (B) limited; or (C) minimal or none.
2	**Social and Environmental Assessment**—All (A) significant and (B) limited risk projects need an environmental assessment report done at sufficient standard and able to satisfactorily address key environmental and social issues according to the issues listed in Table 18.2.
3	**Applicable Social and Environmental Standards**—Environmental assessment reports will use the applicable IFC performance standards (Table 18.3), and the industry-specific EHS guidelines (Table 18.3) to establish acceptable EPs. Assessment reports will address compliance with relevant laws, regulations, and permits that pertain to environmental and social matters.
4	**Action Plan and Management System**—With the environmental assessment (EA) as a guide, the borrower constructs an action plan that describes and prioritizes the actions necessary to manage the impacts and risks identified in the EA. The social and environmental management system that is developed incorporates the social and environmental assessment, a management program, organizational capacity, training, community engagement, monitoring, and reporting.
5	**Consultation and Disclosure**—Borrowers consult with stakeholders to provide information about risks on projects in structured and culturally appropriate manner so that free, prior, and informed consultation is provided to affected communities.
6	**Grievance Mechanism**—The management system of any project will have a built-in grievance mechanism to allow appropriate airing and resolution of concerns from affected individuals and communities.
7	**Independent Review**—Assessment of EPs compliance and all relevant documents will be made by an independent social or environmental expert.
8	**Covenants**—A number of covenants are linked to compliance such that if the borrower does not comply with the agreed terms, or take appropriate corrective action, then the loan may be rescinded and full restitution of the loan made.
9	**Independent Monitoring and Reporting**—An independent environmental and/or social expert will be hired to monitor and report on and verify the monitoring information over the life of the loan.
10	**EPFI Reporting**—Each Equator Principles Financial Institution must commit to reporting publically about its EPs implementation processes.

resources were made. Banning and at least partial enforcement against extraction of deodar cedar logs was an excellent first move, for example, whereas the rangeland management restrictions in Nangarhar and Hazarajat (Sexton, 2012) show equally compelling narratives from both sides of the issue concerning marginalization of the combative sides, human-rights problems, faulty but intractable perceptions of the "will of Allah," and the bad behavior of the

Table 18.2 Illustrative List of Potential Social and Environmental Issues to be Addressed in Documentation of Social and Environmental Assessments to Meet EPs Requirements

1. Assessment of baseline social and environmental conditions.
2. Consideration of feasible environmentally and socially preferable conditions.
3. Requirements under host country laws and regulations, applicable international treaties, and agreements.
4. Protection of human rights and community health, safety, and security (including risks, impacts, and management of project's use of security personnel).
5. Protection of cultural property and heritage.
6. Protection and conservation of biodiversity, including endangered species, and sensitive ecosystems in modified, natural, and critical habitats, and identification of legally protected areas.
7. Sustainable management and use of renewable natural resources (including sustainable resource management through appropriate independent certification systems).
8. Use and management of dangerous substances.
9. Major hazards assessments and management.
10. Labor issues (including the four core labor standards), and occupational health and safety.
11. Fire prevention and life safety.
12. Socio-economic impacts.
13. Land acquisition.
14. Impacts on affected communities, and disadvantaged or vulnerable groups.
15. Impacts on indigenous peoples, and their unique cultural systems and values.
16. Cumulative impacts of existing projects, the proposed project, and anticipated future projects.
17. Consultation and participation of affected parties in the design, review, and implementation of the project.
18. Efficient production, delivery, and use of energy.
19. Pollution prevention and waste minimization, pollution controls (liquid effluents and air emissions), and solid and chemical waste management.

This list is not exhaustive and may or may not identify all the issues or be relevant to every project.
After Anonymous (2006).

Table 18.3 International Finance Corporation (IFC) Performance Standards on Environmental and Social Sustainability in the Equator Principles

Performance Standard 1—Assessment and management of social and environmental risks and impacts.
Performance Standard 2—Labor and working conditions.
Performance Standard 3—Resource efficiency and pollution prevention.
Performance Standard 4—Community health, safety, and security.
Performance Standard 5—Land acquisition and involuntary resettlement.
Performance Standard 6—Biodiversity conservation and sustainable management of living natural resources.
Performance Standard 7—Indigenous peoples.
Performance Standard 8—Cultural heritage.

Each of the performance standards has a lengthy set of Guidance Notes to accompany it with instructions on how to measure and achieve the standard. Equator Principle Financial Institutions (EPFI) or borrowers can use the Guidance Notes as useful points of reference when seeking either more information or interpretations of the performance standards.
After Anonymous (2006).

Table 18.4 Industry-Specific Environmental, Health, and Safety (EHS) Guidelines That are Used in Meeting the EPs Performance Standards of Relevance to Afghanistan

World Bank Guidelines (PPAH)	IFC Guidelines
Base metal and iron ore mining	Electrical power transmission & distribution
Cement manufacturing	Forestry operations: logging
Coal mining and production	Gas terminal systems
Coke manufacturing	Geothermal projects
Copper smelting	Hazardous materials management
Foundries	Office buildings
General environmental guidelines	Rail transit systems
Iron and steel manufacturing	Roads and highways
Lead and zinc smelting	Telecommunications
Mining and milling (underground, open pit)	Waste management facilities
Nitrogenous fertilizer plants	Wastewater reuse
Oil and gas development	Wildland management
Petroleum refining	Wind energy conversion systems
Thermal power plants	Wood products industries

Two complementary sets of guidelines are contained in the World Bank's Pollution Prevention and Abatement Handbook (PPAH) and the International Finance Corporation (IFC) Guidelines that will ultimately be combined and rewritten.
After Anonymous (2006).

enemy. Both the political leaders and the local combatants view their opposing causes as just, whereas those of the opponent are considered nefarious and in bad faith. These nuances of cultural behaviors and understandings are devilishly complex in most cases, but must be faced in most situations in a place like Afghanistan if there is to be any resolution of the environmental problems.

Natural resources are the obvious sources of a great many fracture lines in Afghanistan and its nearest neighbors. The timber smuggling, disputes over land ownership, division of water rights at both local and cross-border levels, the regional trade in drugs, all generate enormous tension and conflict that will not go away anytime soon, even with the best and most concerted efforts at increasing public awareness, education about the environment at all school levels, and even more draconian, or strong court-based systems of attempted eradication. The new investments for strongly ramped up mineral and hydrocarbon extraction, if not managed most carefully, will no doubt generate many new and perhaps equally intractable problems to reducing conflict in Afghanistan.

At the same time, however, to not give the Afghan people the chance to try to overcome their undeveloped backwardness is to already admit defeat by these apparently intractable issues, which is not a pragmatic viewpoint to work with to try to fix things, environmentally or any other way. Effective management of natural resources is already a form of conflict prevention and peace building, but no big development program, whether from the government of Afghanistan, or from international donors or corporations can remain entirely free from the impacts of the conflicts arising from the resource extraction, nor from making the situation inadvertently worse (Bowden and Haysom, 2013).

The key in Afghanistan, of course, is poverty alleviation, sustainable development, and improving the life of the people through the restoration and sustainable use of those renewable natural resources on the surface, and large-scale industrial development of those other nonrenewable, natural resources below the surface. The problem, of course, if these things are to happen, the state has to be stable enough to allow mining to proceed, and without the protection of the Western Coalition troops, the existing Afghan Army and police forces have to be good enough to hold their own against the Taliban opposition. Keeping the state stable and the military in control enough to get the nonrenewable resources out of the ground to pay the government's bills and raise wages without too much environmental destruction is a somewhat doubtful proposition, given all the prior despoliations at smaller scales in the past. Thus, Jensen and Zaidi (2006) have noted that sustained financial assistance and technical support from the international community will be needed during the entire process of restoration and until sustainable use of resources in Afghanistan is established. This is a strong order but is possibly conceivable if sufficient funding is available from the resource extraction.

Taliban, of course, encourages the continued maintenance of a failed state so that it can take over and run things again in an oppressive way, and Pakistan encourages, or at least does not seek to greatly manage, the Taliban strongholds into which Taliban retreats in the tribal areas. The result, of course, is that the continued instability in Afghanistan supports the failed-state aspect, so that Pakistan can have easier access to the resources. But India is already present on the other side in Afghanistan, and Iran controls some violent events in Afghanistan near its border, so the results will be interesting as the accommodations are made and the external powers decide how to divide up the pie, so to speak. Perhaps educated people in South Asia understand this quite well, or perhaps not so much, but the results of how this will play out have a number of different outcomes, given the history of the region.

If there is any good news in any of these assessments, perhaps it lies in the fact that Afghans are remarkably agile in hedging their bets, or "flipping" their alliances and points of view in order to gain ground on the opposing side. Regarded as inconstancy or unreliability in most of the West perhaps, hedging one's bets in Afghanistan is a fundamental strategy for survival in uncertain circumstances (Sexton, 2012; Fotini and Semple, 2009). The upshot in the context of environmental protection at least is that possibly while the resources are coming out of the ground and the royalties and fees are being paid into the local and national economies, the ever so pragmatic and bet-hedging Afghans may detect that these circumstances could be to their considerable long-term benefit. That could result in considerable attention paid, in turn, to protecting the environment so that more people will live happily to enjoy the benefits of the resources.

References

Afroze, Z., Karyab, A.Q., 2007. A guide to Afghanistan's 2007 Environment Law. UNEP Capacity Building and Institutional Development Programme for Environmental Management in Afghanistan. National Environment Protection Agency, 13 p. http://postconflict.unep.ch/publications/afg_env_law.pdf.

Anonymous, 2003a. Afghanistan: Post-conflict Environmental Assessment. United Nations Environment Programme (UNEP), 177 p. http://postconflict.unep.ch/publications/afghanistanpcajanuary2003.pdf.

Anonymous, 2003b. Afghanistan's Environmental Recovery: A Post-conflict Plan for People and Their Natural Resources. United Nations Environment Programme, 6 p. http://postconflict.unep.ch/publications/UNEP_afghanistan_lr.pdf.

Anonymous, 2006. The Equator Principles. http://www.equator-principles.com/index.php/about-the-equator-principles/38-about/about/17.

Anonymous, 2007. Environment Law (Islamic Republic of Afghanistan). Official Gazette No. 912, dated January 25, 2007; As approved by the National Assembly (Unofficial English translation). 39 p. http://www.afghan-web.com/environment/afghan_environment_law.html.

Anonymous, 2010. Perspectives for Development – Mineral Riches in Afghanistan. http://www.ecc-platform.org/index.php?option=com_content&view=article&id=2377&catid=139&Itemid=158.

Anonymous, 2011. Integrity Watch Afghanistan's March Newsletter. http://www.iwaweb.org/WebsiteMaterial/Newsletter/Newsletter%2024-03-11.pdf.

Bowden, M., Haysom, N., 2013. Foreword. In: Brown, O., Blankenship, E. (Eds.), Natural Resource Management and Peace Building in Afghanistan. United Nations Country Team in Afghanistan; United Nations Environment Programme (UNEP). 62 p. http://www.unep.org/disastersandconflicts/portals/155/countries/Afghanistan/pdf/UNEP_Afghanistan_NRM.pdf.

Fotini, C., Semple, M., 2009. Flipping the Taliban: How to Win in Afghanistan. Foreign Affairs, 8 p. http://www.foreignaffairs.com/articles/65151/fotini-christia-and-michael-semple/flipping-the-taliban.

Jensen, D., Zaidi, A., 2006. Progress Report on the Capacity Building and Institutional Development Programme for Environmental Management in Afghanistan 2003 – 2005. United Nations Environment Programme, UNEP Post-Conflict Branch Kabul, 31 p. http://postconflict.unep.ch/publications/afg_PR_jan06.pdf.

Sexton, R., 2012. Natural Resources and Conflict in Afghanistan. Afghanistan Watch; Rockefeller Brothers Fund, 85 p. http://www.watchafghanistan.org/research-&-publications/research-indicates-natural-resources-are-a-major-source-of-conflict-in-afghanistan.html.

Watchman, P.Q., Delfino, A., Addison, J., 2007. EP2: The Revised Equator Principles: Why Hard-Nosed Bankers are Embracing Soft-Law Principles. http://www.equator-principles.com/resources/ClientBriefingforEquatorPrinciples_2007-02-07.pdf.

In this Chapter

19

Afghanistan Water and Climate Change

Abstract

The situation and state of water in Afghanistan in the face of obvious climate change is exceedingly serious. Afghanistan's old and out-of-date water treaties with surrounding governments that no longer exist are very important because they are so antiquated, misleading in terms of modern understandings, and in need of strong new negotiations to reflect changing international water law, as well as to establish scientific understandings within the neighboring governments in light of the climate change that is shifting water availabilities. Changes in transboundary water management and transparency in all dealings are essential to avoid predictable coming conflicts or new water wars.

Keywords: British India; Climate change; Downstream riparian nation; Flood control; Indus water treaty; International water law; Irrigation; Political instability; Soviet Union; Transboundary water management; Upstream riparian nation; Water storage; Water treaties.

Kabul be-zar bashad, be-barf ne-bashad!

May Kabul be without gold, but not without snow!

This ancient and well-known admonition is reflective of the great value that Afghans place upon their water, now even more than ever in the face of climate change. The details of water in Afghanistan have been discussed to a limited extent in Chapter 5 where the basic surface hydrology and groundwater of the country were described, and some mention was made of the problems of water as an international issue, as well as the problem of coming climate change. As a country at the top of most of its main watersheds, Afghanistan hosts the sources of its major rivers, all of which then pass through international borders to empty into neighboring countries of Pakistan, Iran, Turkmenistan, Uzbekistan, and Tajikistan where the local people there absolutely must rely upon that water which is delivered to them from across the border, and without which life is not possible. At the same

Natural Resources in Afghanistan. http://dx.doi.org/10.1016/B978-0-12-800135-6.00019-2

time, premonitions of climate change can already be felt by the people of Afghanistan in the form of the recurrent droughts that have been increasing in frequency and duration for at least the past half century (Williams, 2009). Similarly in neighboring Pakistan, the effects of climate change have been showing up in the huge and repetitive floods, perhaps as the atmospheric, air-mass transport (Hadley) cells change their sizes and locations, with profound consequences for the people who live there. Water issues have repeatedly emerged as a serious international problem that must be addressed (Figure 19.1). This chapter discusses some of the implications of these factors.

Water Treaties

The history of water treaties and Afghanistan goes back into the early and middle twentieth century when all the earlier transboundary water treaties were negotiated between Afghanistan and the neighboring countries of British India in 1921, Iran in 1950 and 1973, and Russia or the Soviet Union in 1873, 1946, 1958, and 1978. Of these treaties, only the 60-year-old bilateral treaty with Iran could even be considered marginally applicable from the time of the government of the Shah to the present rule by the ayatollahs. Obviously, the old treaties with British India and with the Soviet Union are now completely invalid because those governments are long gone. No new treaties are in view anytime soon either, but clearly must come before much more time passes or trouble will occur. In modern times, the issue of

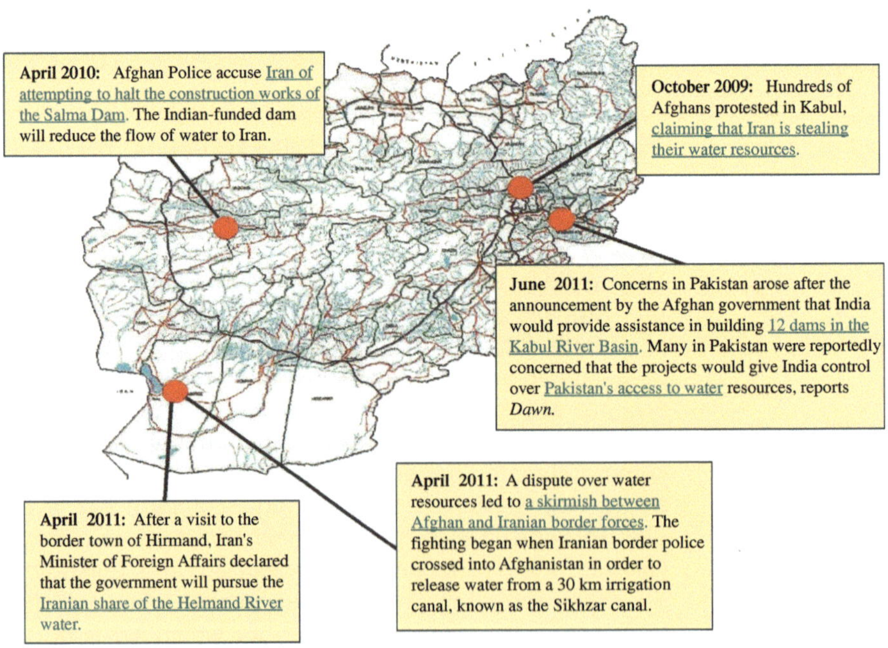

Figure 19.1 Relevant water-related incidents that have occurred in Afghanistan in the past few years *(after Palau, 2011)*. This is a small number in comparison to the decadal totals, which sooner or later if not attended to by serious diplomacy, will result in far more widespread violence and bloodshed.

developing new water treaties with the neighboring countries must overcome a series of challenges on the main rivers, as well as a number of the smaller basins (Table 19.1).

Table 19.1 Potential challenges and sources of friction between countries that are shared by Afghanistan with its neighbors

Kabul River Tributary to Indus River System in Pakistan
1. No water-sharing agreements exist between Afghanistan and Pakistan. 2. Pakistan benefits from river flow from Afghanistan but provides no technical or financial support for flow-control structures, measurement, or management of the rivers. 3. India is assisting Afghanistan to construct 12 new dams needed on the Kabul River for irrigation diversion, municipal water supply, hydroelectric power, and improved sanitation, which will affect flow into Pakistan. 4. Lack of waste collection or wastewater treatment in the heavily populated areas, especially Kabul but also Jalalabad and elsewhere, is a serious issue. Large quantities of diluted and floating pollutants fill the river so that water quality is well below international drinking standards. 5. Lack of both essential hydrological data as well as little technological capacity concerning rivers cripples monitoring of water use or the ability to begin to enforce old agreements or to reach new water-distribution arrangements.
Helmand River Basin
1. Wars and changes of government in both Kabul and Tehran mean that the original bilateral treaty between the two countries has rarely been applied. 2. Iran's government has noted that the Iranian Plateau has been becoming uninhabitable because their ground water has decreased and a negative water balance is widespread (Friedman, 2013). 3. Iran has overused the water resources and also has not compensated Afghanistan as the treaty specifies. It has also violated national sovereignty by crossing the border repeatedly and excavating channels to divert water for its benefit. Overpumping of groundwater in Iran has lowered water tables and water flows in Afghanistan. 4. The Afghanistan side of the river has been covered with new irrigation works that are impacting water flow downstream into Iran. Additional dams, reservoirs, and control structures could increase amounts of water available for irrigation and economic development. 5. Lack of historical monitoring of surficial and groundwater flow in both countries inhibits water management and water sharing. 6. Over a million Iranians depend upon Lake Hamun for drinking water and as a major driver of the local economy, yet recurrent drought upstream in Afghanistan has periodically lowered and dried up the lake repeatedly. 7. Declining water availability on the Afghanistan side has led to considerable production of drought-resistant opium poppies.
Harirud—Murghab River Basin
1. No treaties have ever been developed with either Iran or Turkmenistan with whom Afghanistan shares these rivers, and no records are kept of historic flows. 2. Neither Iran nor Turkmenistan have never compensated Afghanistan for the water that flows into their countries, nor have they paid any costs for water management or for dams, reservoirs, and flood control structures upstream in Afghanistan, nor have they informed Afghanistan about their own dam construction as is expected under international water law. 3. Development of necessary new dams and irrigation networks in Afghanistan will affect downstream water flows, and Iran repeatedly protests dam construction in Afghanistan. 4. The Harirud feeds water into the Dootsi Dam in Iran, which is the main water source for pumping to the city of Mashad, so any diminution in downstream flow will be a major problem.

Continued

Table 19.1 Potential challenges and sources of friction between countries that are shared by Afghanistan with its neighbors—cont'd

Amu Darya River Basin (Part of Aral Sea Basin)
1. Almost 92% of water demand on the downstream Amu Darya is irrigation so upstream damming in Afghanistan can have serious implications for the dramatically declining Aral Sea (Micklin, 2007). 2. Seasonal water use (energy production in winter which diminishes storage, irrigation in summer which uses storage) poses major challenges to balance. 3. Agriculture and irrigation development should greatly increase in northern Afghanistan, which will strongly affect river flow downstream. 4. Turkmenistan and Uzbekistan are highly dependent only upon Amu Darya and Syr Darya rivers so any changes in Afghanistan are serious to those countries. 5. Lack of waste facilities or wastewater treatment along the whole Amu Darya are problems. 6. Lateral avulsion by the river into the banks has shifted the channel many kilometers away from the international border, mostly to Afghanistan's disadvantage because of its lack of ability to armor the banks against erosion.

After Fipps, (2007a,b,c) and Palau (2011).

Afghanistan and all its neighboring countries obviously need water for domestic purposes, reliable irrigation supplies, flood control, and hydropower generation. The times would seem ripe for new negotiations on water treaties between the nations, yet centuries of cross-border incursions and other rapacious behaviors against each other have destroyed trust and made such talks most difficult. To date, no foreseeable or reliable means has been found to share water between the nations; indeed a certain fear exists against even discussing water issues across borders and has been a paramount concern in many cases. Water-sharing agreements take much time and discussion to bring to fruition, even in the best of times. In the case of Afghanistan, the technical capacity even to hold open discussions across the table have been gravely lacking because of three decades of turmoil and violence during which time education about important water issues was completely lacking, along with any sort of reasonable record keeping about the resource. If the Afghans cannot have any certainty about whether agreed-upon dams, power plants, and infrastructure will even be built, they have great difficulty in allocating or dedicating their limited resource (Vick, 2010, 2013). In fact, one of their missed opportunities in the past decade has been Afghanistan's failure to legislate a comprehensive water law for the country (Aman, 2013). The existing statutes do not define water rights as they should. Land owners are also the owners of the water on and beneath the land and landless residents have no official rights to water as one might expect they should. The institutions of governmental water management are highly ineffective and of little real use to people. This is compounded by the lack of databases on the natural resource and little government collection of additional data, which lack is a major obstacle to planning and development. With the absence of regional cooperation, Afghanistan will undoubtedly come to face largely unresolvable challenges after the Coalition forces depart in 2014, unless remedial measures are taken soon.

When the borders of Afghanistan were drawn in the late nineteenth century by the British in consultation with the Russians, and to a more limited extent with the rulers of the Persians and the Indian residents of the subcontinent, little attention was paid to the divisions of the water supplies. This was true except in the lower Helmand where an artificial jog of the border to the east was drawn by the British to accommodate the Persians with more access to lower Helmand water than the Afghans. The reason for this accommodation to the Persians over the Afghans was that the British had just fought two major wars, the First and Second Anglo-Afghan Wars (Chapter 1), and a certain amount of punishment for implicit transgressive behavior by the Afghans was no doubt assigned to them by the British. In any case at the end of the nineteenth century, the Afghans were able to overcome their previous foreign policy restrictions by the British Raj, although they did not get back any of the land they had once enjoyed. Their rights to the water resources that originated inside their borders, however, were not at the time seen by them or anyone else as any sort of impediment to getting on with their lives. Now however, as the populations of Southwest Asia have skyrocketed upward and irrigated lands in all the regions of the area have taken on a tremendous necessity to support people, water issues are paramount.

Water availability as a source of conflict has thus reached a necessity for negotiation, but it is so critical that many government officials do not wish to even put the issue on the table for discussion as it seems too intractable to solve. In a conference on the major rivers of the Hindu Kush—Himalaya in 2010 in Kathmandu, Nepal, that was hosted by the Chinese and Australian governments, for example, the author was asked not to discuss water treaties for these reasons of complexity and seeming intractability. Nevertheless, some experts do think that many of the problems are soluble and should receive fairly high-level attention to bring into fruition (King and Sturtewagen, 2010), which is now happening to a certain extent (Table 19.2).

A major problem is that upstream nations like Afghanistan quite logically claim the right to benefit from their own natural resources, whereas downstream states claim the right through international water-law doctrine of prior appropriation to all the water that has flowed through their territories for thousands of years (Ahmad, 2010). In addition, other applicable principles that apply are the long acceptable ideas of negotiations in good faith, no harm to anyone, and ensuring equitable utilization of waters. Nonetheless, few opportunities have come up in many decades for government-to-government discussions of transboundary water issues like this. The lack of bilateral treaties between Afghanistan and the countries that border it has not been possible in recent years, much less for the more desirable multilateral treaties with more than two countries such as those along the Panj-Amu Darya (Tajikistan, Uzbekistan, and Turkmenistan). In spite of considerable effort by the US Embassy and other agencies such as the East–West Institute to get Afghanistan and its neighbors to

Table 19.2 Steps toward effective and sustainable cross-border cooperation on water

1. Improve the hydrometeorological knowledge base in Afghanistan and the surrounding region.
2. Establish a formal confidence-building framework to share water policies between Afghanistan, its neighbors, and donor communities.
3. Mobilize support from the international community to more toward regional water strategies rather than existing national water schemes only.
4. Launch multilateral dialogue process to build confidence and establish agenda for cross-border water management mechanisms and intergovernmental, river-basin-based security watchdogs.
5. Establish regular meetings of senior water experts in the region.

After King and Sturtewagen (2010).

the negotiating table, these limited opportunities for discussion have been far too few to yet build any sort of trust and cooperation on water.

Afghanistan has had a strikingly low amount of hydrodevelopment for a dry country; in fact, the water-starved nation has one of the lowest water-storage capacities in the world (Kugelman et al., 2011). Clearly, Afghanistan needs to have constructed considerable water-storage dams in many of its deep and largely barren mountain valleys, yet both Pakistan and Iran regard the construction of any hydroelectric dams on the Afghanistan side of the basins they share as a direct security threat (Hanasz, 2012). This should not be tolerated much longer by Afghanistan, because all over the world treaties on transboundary water flow provide unique design requirements for run-of-the-river dams that ensure the steady flow of water, whereas at the same time providing power generation through hydroelectricity. An intelligent situation would be for Afghanistan to be providing electrical power generation for sale to Pakistan, which is chronically short of sufficient power resulting in their all-too-common, "load shedding". At the same time, however, the rampant soil erosion that occurs all over the country, in the absence of general recognition of the need to protect watersheds against it though the building of pervasive berms and check-dams on hillsides, will prematurely fill in reservoirs behind dams. Strong educational programs must be implemented simultaneously with any dam building to try to overcome such problems.

Pakistan and Afghanistan came close to drafting water treaties in both 2003 and 2006, but they ultimately failed. A conference on water between the two countries in about 2009–2010 was canceled by President Karzai over the nonsensical Durand-Line squabble, who apparently did not know about the understandings in international water law concerning legal prior appropriations. In fact, the World Bank and other lending agencies has threatened to cut off all future development funding in the event that a nation like Afghanistan fails to adhere in the future to keep up the same water deliveries downstream as it has done in the past. Nevertheless, from a strategic viewpoint, the timing is actually excellent for establishing a water treaty between the two countries because the pressure is building on many fronts for resource use and development in Afghanistan, as these pages have revealed. This should occur as a part of the integrated extraction and transportation schemes in the resource-corridors ideas of Afghanistan, being worked out between India, China, Pakistan, and perhaps Iran as well.

New treaties could be designed that would also factor in such contemporary concepts as climate change and integrated transboundary river-basin management. Historically, many past water treaties allocate measured water discharges on the basis of riparian river banks, but few if any account for the possibility of diminishing flow through time, or at critical times of the year. Furthermore, few if any treaties consider the possibility of intense floods that may increase as climates change (Kugelman et al., 2011). Thus, most transboundary water treaties in the past have focused primarily on the supply side of the water rather than the demand side. This attention to the supply side has been primarily focusing on developing water-control infrastructure, rather than on changing patterns in water use. New treaties could instead include ideas and rules on decreasing water wastage, and shifting from traditional flood irrigation to more technically advanced and environmentally conservative, furrow irrigation, or even very conservative drip-irrigation techniques.

The equitable sharing of upstream versus downstream water users in the distribution of waters that begin in the mountains of Afghanistan would solve many problems, including possibly turning adversaries into partners (Salman, 2010). It is fairly obvious to all concerned that upstream riparian (river-bank) users of a river can harm downstream users by interfering with the flow of the shared river, or by polluting the water. Far less obvious, however, is the fact that downstream riparian nations can also greatly harm the upstream country by foreclosing on their future usages of the water because of the prior appropriation of such water, and the claiming of such waters in perpetuity so that upstream development can never occur. In fact in the history of world development, most downstream water users were established first on the best lowlands, whereas development in the less agriculturally hospitable uplands has mostly come much later. Thus, if downstream always foreclosed on the uplands, the mountain water towers would remain perpetually backward, which is not acceptable in the modern world. Most powerful downstream states would prefer the status quo of such situations that seem to most profit them, except for the alternative, which is a completely unchecked river that is free to flood or dry up indiscriminately, with no possibility of dammed flood control or regulated releases of impounded water in times of drought.

Downstream riparian users, with their legally binding prior appropriations, thus may think that they are in control of the upstream state's waters through political and military intimidation, closing of trade routes into the uplands as Pakistan has done to Afghanistan on several prior occasions, and manipulation of political and developmental mechanisms against the upstream state. Conversely, if the upstream state feels that it can do as it pleases with its upstream watersheds against the wishes of the downstream state, a control on this high-handed behavior would include termination of development financing from the World Bank and other donor organizations unless proper procedures were utilized and the world water law doctrines of equitable utilization and no harm to downstream users were adhered to.

Downstream riparian states such as Pakistan have long championed the principle of "no significant harm" against upstream Afghanistan and used the established concept as a bulwark against claims by their upstream neighbor who aspires to develop and exploit their own riparian resources. The result of so little upstream development in Afghanistan, however, is that floods from the mountains travel unchecked downstream to Pakistan, and furthermore during droughts with no dams in place to provide controlled water releases, no drought relief is possible.

Both upstream and downstream states generally have the potential to derive considerable mutual benefit from the rational, equitable, and scientifically valid utilization of the shared water resource through the application of the rules of international water law (Paisley, 2002). But simply having international law on the side of the developing nation is unlikely to get them very far in negotiations with a more powerful downstream state without considerable knowledge of their own natural resources, and the ability to shrewdly negotiate from a position of resource acumen. Thus the upstream country such as Afghanistan must strongly and articulately advance the idea of their entitlement to the benefits of their water resources so that they can face down and overcome the resistance to any change coming from the downstream, more developed nation.

In order to manage the water resources properly, as well as to establish a strong negotiating position for the less developed upstream state, robust records would need to be kept; something that underdeveloped and war-torn Afghanistan does not have, and cannot yet do. One of the solutions to this dilemma is to begin by building trust and cooperation between the upstream and downstream nations by mutual calculation of benefits to both nations. For example, assumptions about unmeasured flows and the proposed heights of any planned dams or other water-control structures may be the key stumbling blocks (Paisley, 2002). Simple gathering of new data is unlikely to easily resolve such problems, but instead new understandings might be reached regarding the assumptions by both groups of scientists from the opposing states before real downstream benefits could be calculated with any degree of certainty.

Progress in this situation could be made by close collaboration between Afghanistan and Pakistan by joint field work and setting up of measurement stations in upstream and downstream locations. An ideal test collaboration area for the two countries would be the Kunar River, which rises in the high mountains of the Hindu Kush and Hindu Raj in northwest Pakistan as the Chitral River, before flowing into Afghanistan as the Kunar, where it is further downstream a tributary to the Kabul River that flows back out into Pakistan. With Pakistan as both the upstream and downstream riparian country and Afghanistan as the midstream nation, planning for jointly designed and managed dams in the upstream and midstream locations would be an ideal mechanism to ensure collective collaboration that would benefit both countries substantially.

Shared or joint planning, data collection, and measurement by both countries on the Kunar River would prepare them and enable better understandings of the effects of climate change on the flow of the river. The purpose of international water treaties is generally to enable easier construction of upstream and downstream water-retention structures, but a fixed, perpetual water supply and flow allocation regime is generally assumed. Classically, no provision is generally made for future altered circumstances, such as can occur with climate change. In the case of joint management of Chitral–Kunar, climate-change-induced, water-discharge variation could be an ideal training mechanism for the two countries to work out their differences on water deliveries downstream, especially if the potentially more problematic Kabul River were to act differently than its Kunar tributary, which is quite likely, given their vastly different source areas. Studying the details with collaboratively and honestly collected data could cause former adversaries to come to act more as partners (Paisley, 2002).

If Pakistan, for example, were to propose several run-of-the river dams on the Chitral to generate electricity and to help in flood control downstream, it might then also be able to credibly propose to Afghanistan, several similar dams on the Kunar downstream that could assist in the interests of useful cross-border planning and development. This would give Pakistan more of an air of verisimilitude and an entrée into the world of realistic local water politics in the region that would be looked upon favorably by the UN, World Bank, and USAID, who so much control the large finances required for such projects. This could bring useful pressure on the Kabul government, who has otherwise been penalized recently by international finance for not negotiating in good faith on trans-boundary water resources. In addition, if Pakistan were to propose a multinational Indus River Water Commission, to include all of the Indus Basin riparian nations (Afghanistan, China, India, and Afghanistan), with UN observers as well, Pakistan could be seen as forward looking and a new supporter of international water law. This would put Pakistan internationally in a good light, in that all nations who share water across borders are encouraged to have regular meetings of their government water specialists to discuss issues and work out differences. First, however, would come the meetings between the national and international scientists to define the water problems and share knowledge about all possible data sources on the waters.

Climate Change

In spite of considerable political distaste to admit that human industrial emissions into the atmosphere have affected the climate of the world, responsible top international scientists well recognize that fact; there is no dispute that global warming is occurring. The results of these climate changes in Afghanistan have been the diminution of winter snows and summer rains, the melting away of glacier ice where long-term, late-season melt-waters were stored (Shroder and Bishop, 2010; Williams, 2009; Haritashya et al., 2009), and recurrent drought cycles that lower water tables and cause large rivers to run dry for at least part of the year. For an already arid country such as Afghanistan, this recognition is a disheartening addition to the long list of environmental shortcomings the country must deal with as best it can.

A US national intelligence estimate about 5 years ago indicated that climate change will become a significant source of political instability; indeed the evidence is that climate change in Afghanistan has been underway for some time and is already a factor in the instability in the country. Long-term climate change and reduction in overall precipitation has, of course, dramatically reduced the lakes of the Seistan since the Pleistocene (Figure 5.26), but many people have assumed incorrectly that this trend was irrelevant to the future. Instead severe droughts that occurred in 1970–1973, 1998–2006, and 2008–2009 have made Afghans realize the strong implications this has for their future. In fact, Hsiang et al. (2013) have conclusively shown what many have intuited that increasing temperatures and extremes of rainfall (drought) for many millennia are causally associated with changes in interpersonal violence and civil war; this of course augers unpleasantly for the future of Afghanistan and elsewhere.

Computerized climate models indicate that over the next several decades, increases in the global carbon dioxide and other greenhouse gases will significantly alter precipitation patterns worldwide. The reason this is occurring in Afghanistan, Pakistan, and Iran is because the advancing global warming is increasing the size of the north-to-south atmospheric Hadley cells where moist and warm air rises from the equatorial Indian Ocean and travels north toward Afghanistan where it subsides and dries out to add to desertification. As the planet continues to warm overall, these zones are expected to expand and shift dry areas northward toward the poles. As the dry areas expand, the already arid regions of Afghanistan will grow even drier. At the same time, seasonal extremes will likewise intensify, as moisture accumulated in the atmosphere in the drier seasons, will be shed in downpours to produce flashfloods in cooler times, which will lead to seasonal flooding in regions that are otherwise prone to droughts.

With this as a simplified explanation of coming climate changes, a variety of quantitative methods can be used to predict future climates. In general, almost all methods rely upon past measured or recorded weather variations as a control for the equations to project onward into the future. A number of different equations to perform these calculations have been made in different countries. I, for example, was introduced to 15 different climate models produced by Chinese scientists, each first run separately with the historical data, and then combined to achieve the best results for looking into the future.

The Intergovernmental Panel on Climate Change (IPCC), for example, has developed a "Special Report on Emissions Scenarios" (SRES) that investigated six different pathways of future emissions of greenhouse gases, but only the A2 scenario of a heterogeneous or fragmented world almost exactly like of today was viewed as the most realistic. The underlying theme of this scenario was self-reliance for countries, and preservation of national identities. Increasing populations were assumed, with primarily regionally oriented economic development under a more fragmented and slower per capita economic growth and technological change. With these constraints for the quantitative climate-change modeling, Savage et al. (2008) developed a scoping study on the socioeconomic impacts of climate change in Afghanistan. With reservations appropriate to the profuse data unknowns and model uncertainties, certain trends became evident, such as a 0.6°C increase in mean annual temperature over that of the past 50 years. Most significantly, however, is a 20–30 percent reduction in precipitation in the south of the country by the year 2090, especially due to decreases in spring rainfall (March–April–May).

Only the world climate model from the NOAA Geophysical Fluid Dynamics Laboratory (GFDL) at Princeton University is shown here, however. The GFDL CM 2.1 model, for example, uses the measured period 1951–2000 as control and then projects forward in time for the whole of the twenty-first century to assess likely precipitation changes. These changes were computed in this model as the difference between the 2081–2100 20 year averages and the modeled 1951–2000 50 year average, for example. The results were quite varied but in terms of projected precipitation for subtropical dry regions such as Afghanistan, decreases in precipitation were projected for the region, especially on the poleward flanks of the Hadley cells. These projected climate changes were caused by the atmospheric response to increasing greenhouse gases and changing atmospheric aerosols, which NOAA GFDL based on a "middle of the road" estimate of future emissions.

The map presented here (Figure 19.2) of an NOAA GFDL run of about 10 years ago used precipitation change from the 1971–2000 average to the 2041–2070 average to show what climate change might result in about 30 years. The results were binned by percentages of projected change in precipitation ranging in color from positive wetter blues to negative drier browns. As can be seen in the figure, southern Afghanistan is projected to become considerably drier, whereas the north is expected to become slightly wetter.

In terms of a comparison of this map's results to actual recent changes, recent moisture increases in northern Pakistan have been observed by me and others

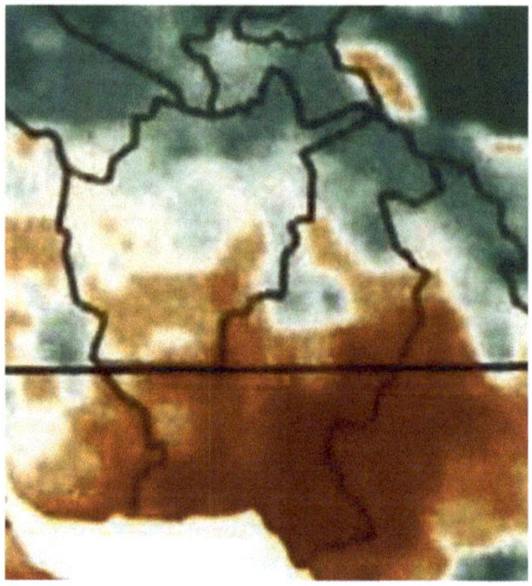

Figure 19.2 Climate change in Afghanistan and surrounding environs about 30 years hence with browner tones being 20–30 percent negative percentage change from measured 1971–2000 average precipitation control to 2041–2070 modeled changes. Bluer tones indicate estimated water surplus percentages *(after NOAA Geophysical Fluid Dynamics Laboratory, Princeton University)*. Blown-up larger-scale image from prior much smaller-scale world map. The black line running east–west across the image indicates the line of 30° N latitude.

(Hewitt, 2005; Copland et al., 2009, 2012) and have been observed as glacier increases in the higher elevations. The same has not been observed in the lower altitudes of the Afghanistan Pamir and Hindu Kush (Haritashya et al., 2009), however. My colleagues and I have presumed that the modeled increase in northern Afghanistan in Figure 19.3 has not occurred in the Afghan Pamir because of lower altitudes on the glaciers there than those in the higher altitude glaciers and mountains of the nearby Karakoram Himalaya, where any increase in snow or glacier advance is more noticeable because of higher altitudes and less melting.

The cumulative effects of increasing drought frequency in Afghanistan will obviously threaten the water supplies of entire communities in the country and thereby lead to a greater range of humanitarian crises, including population displacement, disease, and violent conflict. The lack of water availability will force Afghanistan and surrounding states to try to claim the greatest shares of the total regional water supply, which will further add to regional violence (Savage et al., 2008). Disputes over water have plagued central and South Asia for centuries and will most likely escalate in the face of these obvious climate projections.

The GoA has recognized these dire predictions and has attempted to respond as best it can in whatever limited ways would seem to make the most sense: (1) ratifying the Kyoto protocol that will require drawing up plans to develop low carbon energy footprints in transport and industry; (2) improved water management and use efficiency; (3) community-based watershed management; (4) improved terracing, agroforestry, and agro-silvo pastoral systems; (5) climate-related research and early-warning systems; (6) improved food security; and (7) better rangeland management. Education about the need for these factors, and the development of vocational skills for communities are also a focus of the attempts by GoA to respond in a logical fashion to these newly recognized climate threats.

Transboundary Water-Resource Management

The country of Afghanistan has actually been withdrawing less water from its transboundary rivers that had been legally allocated in the old international agreements and treaties with governments that no longer exist (Palau, 2011). The ecosystems in the downstream countries, however, are already undergoing significant pressures that have nothing to do with Afghanistan. Nevertheless, as economic development, resource extraction, and construction of new infrastructure in Afghanistan progresses with the new resource corridors, such pressures on the ecosystems will likely increase downstream. As a result, therefore, experts have highlighted the considerable importance of fostering development in manners that are sensitive to all the regional tensions that are occurring over the shared water resources. This is particularly relevant right now with Iran and Pakistan, and is certain to become more relevant later with Turkmenistan, Uzbekistan, and Tajikistan.

The Indus Water Treaty between India and Pakistan is an example of a surprisingly successful legal agreement between two otherwise rather belligerent nations (Leb, 2008). The 1947 partition of British India put most of the headwaters of the Indus River in the new India, along with much of the irrigation infrastructure as well. This resulted in almost all of Pakistan's main breadbasket region of the Punjab being left completely dependent upon water deliveries from across the border. Following partition of the two countries in 1947, over a decade of careful negotiations between the two nations resulted finally in a workable treaty in 1960. The treaty apportioned the tributary waters equitably between the two countries, as well as including a third-party guarantee of financing for additional water infrastructure to increase water-flow availability to both countries. The dispute resolution provisions of the treaty and the appointment of a neutral expert has allowed reasonably easy resolution of problems.

The channels of cooperation established by the Indus Water Treaty are also expected to prove useful in resolving new issues of hydrologic variability caused by climate change. The negotiation mechanisms established by this treaty show clearly that something similar could be established between Pakistan and Afghanistan, providing that both counties could be induced to come back to the negotiating table. This will be far from easy, however, especially when an intransigent Afghanistan will apparently not negotiate in any good faith or sever, the linkage between negotiations over transboundary waters and their ideas about the Durand Line as it should (M.J. Vick, oral comm., May 17, 2013). The problem with this is, of course, that the long-term failure of Afghanistan to recognize the Durand Line means that they would like to see the Pashtun part of Pakistan to be sutured back onto Afghanistan, which is quite unacceptable to Islamabad. This would spell the effective end of the country of Pakistan and the creation of a new greater Afghanistan, which would most likely become a most dangerous and bellicose power in an unstable part of the world. Certainly the implied partition of Pakistan in water negotiations with Afghanistan does not allow good-faith discussions, which is obviously to the detriment of both countries.

Ahmad (2010), expressing a Pakistani point of view, has advocated a position wherein Pakistan as the more developed nation must develop a new mindset to assist its neighbor in building their water infrastructure, rather than opposing it. A Kabul River water treaty would be a win–win for both countries, although the necessity was obvious for delinking the water issues from the Durand Line and other historic grievances. Acknowledging the past and then moving beyond it to new understandings would actually not be a sign of weakness but of statesmanship, something that seems in rather short supply in both countries.

Up to the present time, the external international community and individual nations in South Asia have primarily, although not exclusively, focused on countries as individual entities in the water resource management sector. This is obviously not the kind of region-wide strategy that should instead be implemented in a comprehensive, hydrological, basin-level approach to the process (Figure 19.3). A shift in focus to a regional approach is thus now being advocated by all the experts, and should be based upon (1) improved understanding of the hydrometeorological dynamics in the region; (2) setting up of transparent, confidence-building frameworks for the differently involved stakeholders; and (3) the new development of supreme or overarching, regional water-management mechanisms. This, of course, will require better resources and capacity or training for the personnel available to the national and regional authorities to become good enough to do this, especially in data collection and analysis. It is complex but soluble, given attention to detail.

The people of Afghanistan, but also those in the surrounding countries, are going to have to get smarter about their water management, including using and obtaining good measures of incoming precipitation, the outgoing surface and subsurface flow measures, and this information needs to be sent out across borders, something that all too commonly has tended to be left unmeasured or unknown, or has been hidden or is unavailable as a state secret. This is no way to be running hydrological systems upon which quite literally, millions of people absolutely must depend.

International Water Law

Afghanistan's position at the top of the main watersheds in that part of southwest Asia with the rivers that emerge from its borders (Amu Darya, Hari Rud—Murghab, Helmand, and Kabul), puts the country in an enviable position

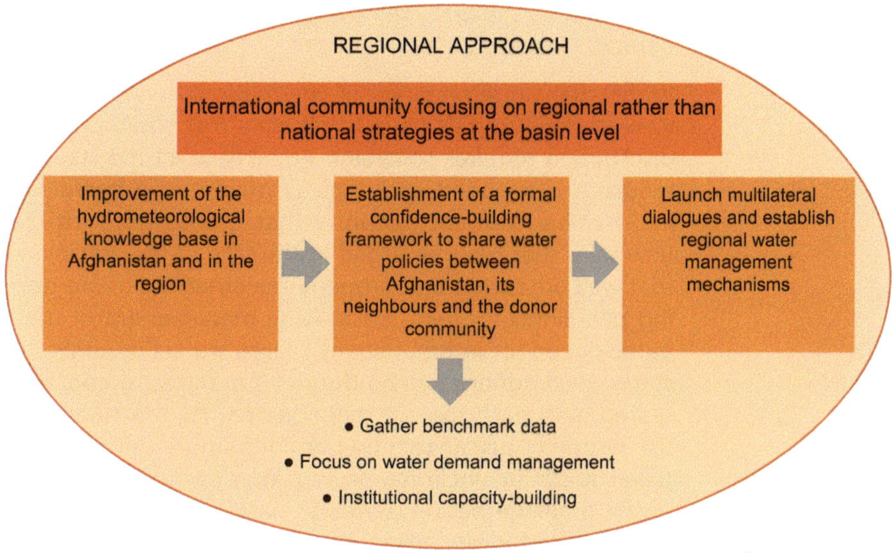

Figure 19.3 Regional approach framework toward effective transboundary management of water resources that is recommended for Afghanistan *(after Palau, 2011)*. The first large box on the left is the essential data-collection phase, which particularly needs newly trained, scientifically oriented water managers with environmental training to look after a declining resource in an already stressed environment.

in some ways, although as the country with the least developed waterways in the world, part of the reason its rivers are so underdeveloped is because the neighboring states want no change. But change is coming anyway, in the form of global climate variation, so given these facts, new ways to look at transboundary rivers are required.

Accordingly, international water management and allocations face severe readjustments in the face of climate changes that so affect the quantity and timing distributions of water discharges downstream. Some international rivers have been allocated by old treaties that created firm property ownership rights but were not designed to adjust to changed conditions and were created when people were generally unaware of the direct connections between ground water and surface water. The purposes of any international water treaty are in general to allow the construction of upstream and downstream dams, regardless of the locations of international borders. In addition, in treaty construction, the assumption has long been to assume a fixed perpetual water supply and flow allocation regime, with no provision for future changed circumstances such as upstream irrigation development, or most importantly, climate change. The great problem with this, of course, is that many people want whatever status quo there was to be maintained, no matter how inefficient, inequitable, or environmentally destructive that might be. Ecosystem protection is unfortunately generally quite subordinate to multipurpose regional water development (Tarlock, 2000).

Within the underpinnings of international water law, however, a simplified view of the complex issue is basically that the foundation of equitable water distribution requires sovereign equality of all the states involved, as well as a regular and required exchange of quantitative hydrologic data and collateral information from each country that is absolutely required. The basics of the new rules that have emerged in the face of the advancing climate change are that well organized and properly governed states are developing new governing paradigms for joint, basin-wide management of the rivers. This equitable participation for internationally shared waters requires a number of basics that states must adhere to: the duty of the states to cooperate; conjunctive, integrated, and participatory management by the states; equitable utilization of the resource; sustainable use; and minimization of environmental harm. Equitable utilization includes paying attention to the particular natural features of the drainage basin, as well as past, present, and foreseeable future needs, seeking alternative means for satisfying needs and sustainability of use, the avoidance of unnecessary harm, and reasonable compensation for injuries. Participatory management means that the people in any area have a right of access to water, the right to a voice in the decisions affecting their lives, access to education about their shared resource, protection of particularly vulnerable communities, the right to compensation, and access to legal remedies in case of dispute.

These ideas on evolving international water law applied to Afghanistan show clearly that the country needs a good deal of external support to get the neighboring states to come to the negotiating table. Recognition of the realities that climate change is affecting everyone is paramount, and Afghanistan must be allowed to develop its water impoundment systems to at the least provide for downstream flood control. With a finite and changing resource that is not guaranteed to provide dependable amounts of water downstream every year, the means to apportion equitable deliveries of certain percentages downstream during droughts is required. This will take quite robust measures to effect, and indeed, ultimately may not be possible, because of the strong resistance to transparency of water measures where water usage can be considered a state secret. But if this resistance cannot be overcome, then the results will be most unpleasant for Afghanistan as downstream users who will not negotiate, instead retreat into intransigence and support armed insurgency against dam building by the Afghans upstream. This attitude by the downstream riparian nations will backfire, of course, because the dams will be built anyway and then the upstream incentive will be to cut off downstream supplies altogether. Clearly strong diplomacy is required and more equitable distribution is a necessity.

References

Ahmad, S., 2010. Towards Kabul Water Treaty: Managing Shared Water Resources – Policy Issues and Options. International Union for Conservation of Nature and Natural Resources (IUCN), Karachi, Pakistan; Kingdom of the Netherlands, 11 p. http://cmsdata.iucn.org/downloads/pk_ulr_d3_1.pdf.

Aman, F., 2013. Afghan Water Infrastructure Threatens Iran, Regional Stability. AlMonitor; The Pulse of the Middle East, 1 p. http://www.al-monitor.com/pulse/originals/2013/01/afghanwatershortageiranpakistan.html.

Copland, L., Pope, S., Bishop, M., Shroder, J., Clendon, P., Bush, A., Kamp, U., 2009. Glacier velocities across the central Karakoram. Annals of Glaciology 50 (52), 41–49.

Copland, L., Sylvestre, T., Bishop, M.P., Shroder, J.F., Seong, Y.B., Owen, L.A., Bush, A., Kamp, U., 2012. Expanded and recently increased glacier surging in the Karakoram. Arctic, Antarctic, and Alpine Research 43 (4), 503–516. 2011. 10.1657/1938-4246-43.4.503.

Fipps, G., 2007a. Water Sector Briefing and Recommendations. US Embassy – Kabul briefing documents, 12 pages. http://gfipps.tamu.edu/Afghanistan/Water%20Sector%20Briefing%20and%20Recommendations.pdf.

Fipps, G., 2007b. Transboundary water policy of Afghanistan. US Embassy – Kabul briefing documents, 5 pages. http://gfipps.tamu.edu/Afghanistan/Draft%20Transboundary%20Policy26-04-17.pdf.

Fipps, G., 2007c. Appendix: Transboundary Water Issues. US Embassy – Kabul briefing documents, 10 pages. http://gfipps.tamu.edu/Afghanistan/Transboundary%20Water%20Issues26-04-17.pdf.

Friedman, T.L., September 22, 2013. Mother Nature and the Middle Class. The New York Times.

Hanasz, P., 2012. The Politics of Water Security between Afghanistan and Iran. Independent Strategic Analysis of Australia's Global InterestsFuture Directions International; Strategic Analysis Paper, 6 p. http://www.futuredirections.org.au/publications/food-and-water-crises/416-the-politics-of-water-security-between-afghanistan-and-iran.html.

Haritashya, U.K., Bishop, M.P., Shroder, J.F., Bush, A.B.G., Bulley, H.N.N., 2009. Space-based assessment of glacier fluctuations in the Wakhan Pamir, Afghanistan. Climatic Change 94, 5–18.

Hewitt, K., 2005. The Karakoram anomaly? Glacier Expansion and the 'Elevation Effect,' Karakoram Himalaya. Mountain Research and Development 25, 332–340.

Hsiang, U.K., Burke, M., Miguel, E., 2013. Quantifying the influence of climate on human conflict. Science 341, 1212 + Supplementary Materials.

King, M., Sturtewagen, B., 2010. Making the Most of Afghanistan's River Basins: Opportunities for Regional Cooperation. EastWest Institute and Gerda Henkel Stiftung, 13 p. http://www.ewi.info/making-most-afghanistans-river-basins.

Kugelman, M., Alam, A.R., Bakshi, G., 2011. Peace through Water. The AfPak Channel. Foreign Policy. December 2. http://southasia.foreignpolicy.com/posts/2011/12/02/peace_through_water#sthash.352dZtEj.dpbs

Leb, C., 2008. International Water Law and Climate Change: Avoiding Disputes over Water. Bridges, Embassy of Austria, Washington, DC, 18: 4 p. http://www.ostina.org/index.php?option=com_content&view=article&id=3231&Itemid=1006.

Micklin, P., 2007. The Aral Sea disaster. Annual Revue Earth Planetary Science 35, 47–72.

Paisley, R., 2002. Adversaries into partners: International water law and the equitable sharing of downstream benefits. Melbourne Journal of International Law. 3 (2), 280–300. http://www.law.unimelb.edu.au/files/dmfile/download22881.pdf.

Palau, R.G., 2011. Afghanistan's Transboundary Water Resources: Regional Dimensions. Civil Military Fusion Center, 8 p. https://www.cimicweb.org/Documents/CFC%20AFG%20Infrastructure%20Archive/CFC_Afg_Monthly_Transboundary_Water_Resources_July2011.doc.pdf.

Salman, S.M.A., 2010. Downstream riparians can also harm upstream riparians: the concept of foreclosure of future uses. Water International. 35 (4), 350–364. http://www.tandfonline.com/doi/abs/10.1080/02508060.2010.508160#preview.

Savage, M., Dougherty, B., Hamza, M., Butterfield, R., Bharwani, S., 2008. Socio-economic Impacts of Climate Change in Afghanistan. Department for International Development (DFID CNTR 08 8507); Stockholm Environmental Institute (SEI), 6 p. http://www.necsi.edu/afghanistan/pdf_data/2007447_AfghanCC_ExS_09MAR09.pdf.

Shroder Jr., J.F., Bishop, M.P., 2010. Glaciers of Afghanistan. In: Williams Jr., R.S., Ferrigno, J.G. (Eds.), Satellite Image Atlas of Glaciers of the World – Asia. U.S. Geological Survey Professional Paper 1386-F3, pp. 167–199.

Tarlock, A.D., 2000. How well can international water allocation regimes adapt to global climate change? Journal of Land Use and Environmental Law. 15, 423–449. http://www.law.fsu.edu/journals/landuse/Vol153/tarlock.pdf.

Vick, M.J., 2010. Developing the Transboundary Waters of Afghanistan. Infrastructure Working Group, United States Embassy Powerpoint, 19 p. https://ronna-afghan.harmonieweb.org/Lists/Submitted%20Content/Attachments/98/IWS%20US%20Embassy%202%20Mar%202010%20Noon.ppt.

Vick, M.J., 2013. Sharing Central Asia's Waters: The Case of Afghanistan. International Water Law Project Blog, 3 p. http://www.internationalwaterlaw.org/blog/2013/01/19/sharing-central-asias-waters-the-case-of-afghanistan/.

Williams, D.B., 2009. Finding Water in the Heart of Darkness: Afghanistan's Ongoing Water Challenges. Earth, July; 4 p. http://www.earthmagazine.org/article/finding-water-heart-darkness-afghanistans-ongoing-water-challenges.

Part IV
Afghanistan in Future

Overview

Predicting the future of a place like Afghanistan is probably not possible because so many good and bad possibilities pile up to reinforce each other in both positive and negative directions that a middle-road prediction of things not getting too wonderful or too awful either, may be more likely and what they end up with in the end. But increasingly, disastrous scenarios loom the largest, but are countered because ever so slightly, possible optimistic scenarios also glimmer on the horizon. As a scientist having no real pretensions or ability at either historical analysis or sociological or cultural assessment, the author's lot is to attempt a summing up of the possibilities for Afghanistan. Will the place continue on its long descent into the cultural and political swamps, helped on by climate change in all the wrong directions for the country, or will instead, the usual resource curse actually bring about a resource miracle? It seems too soon to tell but the possibilities for the either/or, upward or downward spirals are certainly there to see.

This book has made an attempt to show a number of recent facts and past occurrences that can be called upon to make some better sense of whatever is coming. Pessimistic scenarios would have it that Afghanistan is doomed, if only by its ethnic and religious hatreds and past histories of big-power machinations, border fixings, brutal wars, ethnic cleansings, and intransigent or corrupt politics, and then by the collapse of ecosystems, catastrophic water losses, and desolate, nonecologic, stripping away of so much macroscopic life. Alternative thoughts are that Afghanistan may actually have a chance simply because many considerable powers arrayed around the periphery of the country have too much to lose and a fair amount of gain if they work together to use the resources of Afghanistan to help them all develop their economies, thereby controlling the unpredictable violence, which if allowed to escalate as it has in the past, can only lead to mutual ruin across several borders.

At the same time, in terms of what is coming to us all, not just to the Afghans, according to the novelist Rich (2013), the US Army Corps of Engineers sees it coming, and scientists and government officials know what will happen when the larger storms come. As Rich says, "the future, once an amorphous, distant plain, is rapidly articulating itself." Dystopian stories about environmental apocalypses tend to contain warnings—this is the hell we all shall share if we do not change our ways—and quickly. But as Rich points out, we already know that it is too late to change the coming planetary transformations, and we also know what is coming next, namely more of the same—only worse, such as super-droughts, mega-windstorms, monster-pandemics, rampant desertification, and all the rest. Does Afghanistan have a chance at all with the arriving climate change, Hell-on-Earth for that eternally benighted country, or does it go down under the renewed onslaught of Taliban and neighboring intransigence operating against an incompetent government? Things do not look good, but at the same time, hope exists that a few wise heads will prevail. Perhaps the resource riches will rule the day, and life will become better in Afghanistan; the people certainly do deserve rather better than they have now. After all, not all Afghans are fundamentalists or ignorant and violence-prone. By far, the majority of Afghans just want to just lead their lives and take care of their families, free from oppression by their government, the Taliban, or anyone else.

Reference

Rich, N., April 21, 2013. Writing the End (Book Review). New York Times. p. 31.

In this Chapter

20

Pessimistic Scenarios: Incessant War in Afghanistan

Abstract

Pessimistic scenarios for Afghanistan's future rest particularly on the prospects for ongoing incessant war in the country because of a long list of such things as Pashtun cultural characteristics; neo-environmental determinism with a harsh physical environment leading to environmental and developmental problems; charismatic or "mad" mullahs who encourage bad behaviors; strong ethnic patchworks leading to ethnic cleansing, and the like; repeated foreign invasions; Afghan xenophobia; endemic corruption; land-ownership disputes; Islamicist desire for endless jihad; decades of war, destroyed educational systems, reduced capacity, widespread illiteracy, drug culture, and the purdah culture wherein valuable women are excluded from normal and useful problem solving. In addition are the paranoia, insecurity, and inability of nearby Pakistan to gain the upper hand over the ISI-driven and protected Taliban insurgency enjoying largely unmolested safe haven there, coupled with similar but not as extreme difficulties with fanatically religious Iran as well: the existing environmental basket case of Afghanistan only contributes further fuel to the pessimism. None of these problems is sufficient by itself to keep Afghanistan down, but in aggregate, they seem to have been enough to spoil the country up to now.

Keywords: Charismatic mullahs; Destroyed educational systems; Drug cultures; Endemic corruption; Environmental basket case; Ethnic cleansing; Islamicist jihad; Land-ownership problems; Neo-environmental determinism; Pashtun cultural bellicosity; Pakistan ISI; Purdah cultural removal of female contribution; Taliban insurgency.

Afghanistan seems cursed by near perpetual war; the only operant question that would seem to make any sense is why this fact of repeated and incessant bellicosity is so common. What human condition or natural occurrences would force war upon so many people for so long? It is quite understandable to seek answers to questions like these, even though the answers may not be so clear or the results too satisfying. Furthermore, it is not really the role of a natural scientist like me to try to understand deep human psychologies, historical vagaries, or complex sociologies, but still, the nagging question remains as to why

Natural Resources in Afghanistan. http://dx.doi.org/10.1016/B978-0-12-800135-6.00020-9

so much war has continued to plague the people of Afghanistan for so long. A few suggestions might help advance perspectives on why Afghanistan has had so many wars for so many years. Not all these issues may be directly germane to the problem of incessant violence, but many are, and a greater exposition may be useful to increase discussion and enable the seeking of possible solutions. Thus the reasons for incessant war would seem to include a mixture of internal and external factors that interact at unknown levels of significance, such as (1) Pashtun cultural characteristics; (2) neo-environmental determinism; (3) charismatic mullah movements; (4) ethnic patchworks; (5) foreign invaders; (6) corruption; (7) Afghan xenophobia; (8) land-ownership disputes; (9) endless *jihad*; (10) illiteracy; (11) drug cultures; and (12) the *purdah* culture. Probably other causes, or more scientifically or culturally, sophisticated interpretations of all these factors can ultimately be interpreted or discovered, but for now these factors might seem to add some understandings.

Pashtun Cultural Characteristics

As has been noted by a number of people (Fletcher, 1965; Caroe, 1958), Pashtuns have long had a reputation with many people for their generally rather less than worthy behavior in many very essential aspects, even while a certain, largely British contingent who had fought them many times, began to tout "a noble people", which no doubt elicited much local pride. In any case, outside visitors noticed combinations of extreme friendly hospitality coupled by the opposite with generational blood feuds. People characterized the Pashtuns as turbulent, warlike, predatory, and revengeful (Fletcher, 1965); Ibn Batuta in 1333 CE thought them mostly highway men, and Timuri Leng (Tamerlane) concurred, which is high praise indeed, from a fellow traveler. Khushal Khani Khattak, the great Pashtun warrior-poet of the seventeenth century, was reputed to have described his people as malevolent, ruthless, and contentious (Raverty, 1867; Fletcher, 1965). Even some early European explorers (Olearius, 1669) must have had some encounters with the difficult people, which led to comments of the Pashtuns being a self-conceited, insolent, cruel, and barbarous people. They were noted to slight others even if for no other reason than that they were not quite so rash themselves in hazarding their own lives without any conceivable necessity. An exaggerated warrior bravery was conceived by some who met them as another kind of "noble savage", above the general run of local savages (Ahmad, 1978).

When the British set up the Durand Line to divide and rule the Pashtuns, the new border complemented a cultural strategy of pacification known as the Pathan (Pashtun) Renaissance, in which the British colonial agents imagined that they were aligning their own interests with those of their ostensible tribal allies (Cullather, 2002). The Pashtuns to the British overlords were "our chaps", or natural rulers, equals to the British and therefore worthy of elevation to ruling status under British tutelage. Racist overtones also prevailed on the part of both ruler and ruled that the Afghans, that is, the Pashtuns were actually whites amidst a general population of "black-skinned" Mohammedans' and Hindus. Pashtuns were therefore entitled to financial subsidies, to rank in the armies of the British Raj, and to a direct relationship with the Crown. The British glorified the Pashtuns as a martial people, much like their own Highland Scots, peoples who had special codes for living life. The code of Pushtunwali speaks to issues of revenge and retaliation (*badal*), sanctuary (*nanawati*), honor, tribal values, blood money for wrongs (*saz*), and great hospitality (*malmasti*). Pashtun values of protecting women (*zan*), gold (*zar*) or money, and the land (*zameen*) were especially strong, and were stressed in this thinking.

When Nadir Shah crossed the Durand Line in 1929 to seize Kabul back from the Tajiks, he was establishing a monarchy based on Pashtun pride and nationalism, with overtones of scientific racism (Cullather, 2002) to the effect that Pashtuns were really the natural rulers of Afghanistan; and they would brook no interference with that natural order of things on into the future. The recognition of this historical distance between Pashtuns and the other ethnicities of Afghanistan, particularly the Turkic and Persian speakers, is partly why President Karzai wears the karakul hat and long-sleeved chapan coat in public, as these are the dress of the Central Asian peoples, not traditional Pashtun dress such as one might expect from him. Karzai is thus seeking a healing of traditional ethnic rifts with sartorial symbolism.

For all their cultural difficulties or flaws in dealing with themselves or outsiders, still some of the cultural tendencies and values of Pashtuns were seen by foreigners more realistically than the customary pejorative viewpoints of enemies. For example, the early British explorer Mountstuart Elphinstone was a scholar and linguist who saw through some superficialities. He noted in particular that Pashtuns were fond of liberty, faithful to their friends, kind to their dependents, hospitable, brave, hardy, frugal, laborious, and prudent. He thought that they were less disposed than the other neighboring nations in South Asia to cultural aspects of lying, intrigue, and deceit (Elphinstone, 1815; Fletcher, 1965).

On the other hand, some of these good cultural traits that would seem to help curtail violence can be offset by other far less beneficent behaviors such as a form of childhood prostitution known as *bacha bazi*, or "boy play", for example, which is the kind of cultural pediphilia or pederasty undertaken in many strongly purdah societies, where women are so commonly excluded from external society outside of the home. The early British explorers called this cultural pederasty the "Uzbeg disease" (Londoño, 2012) because presumably that was the first ethnic group they noticed using the practice, but in fact *bacha bazi* occurs all over Afghanistan and surrounding countries, not just among the Pashtuns. It is especially prevalent in the Pashtun south, however, and particularly among the truck stops for the common Pashtun drivers. Where such practice is so usual, one cannot but think that a certain amount of associated distaste for the practice and associated violence might ensue as well, especially among people who find the practice repellent. Homosexuality is prohibited in Islam but many Afghans do not regard *bacha bazi* as homosexual behavior at all. This mental dissonance must increase the cultural frictions that can only add to the incessant violence.

Neo-environmental Determinism

Although environmental determinism, wherein human behavior was supposedly determined by aspects of the physical environment, has long been a repudiated philosophy of geography, some environmental stressors discussed here may be at least partially relevant. Afghanistan, for example, has a large share of harsh environments where it is difficult to eke out even a marginal existence. The land may be too dry, too rocky, too high, or too cold, or subject to high-velocity parching winds, recurrent droughts and blinding sand- and dust-storms, rampant floods, rapid wet debris flows and massive landslides, debilitating and deadly blizzards and snow avalanches, common and repeated deadly earthquakes; in such a land of extremes, the people must scrabble for a bare minimum livelihood. To compound these factors of environmental stress, centuries of environmental despoliation have occurred, which in many cases adds to the environmental problems through overgrazing, deforestation, dearbification, soil erosion, soil salinization, ground- and surface-water contamination, and overuse beyond natural recharge, threats to biodiversity, and climate change, both natural and human induced (Shroder, 2007, 2012). Afghanistan seems to be a "poster-child" exhibit of such environmental stressors, all of which contribute to major parts of the failed-state syndrome discussed by Brown (2009) as major factors that bring about threats to other parts of the world as well. When a nation's government, such as has occurred for so long in Afghanistan, can no longer provide basic services or security for its people, the resulting social chaos can have serious adverse effects beyond its own borders. Disease spreads, sanctuary is offered to terrorists, pirates, and thugs, the sale of drugs and weapons increases, political extremism is fostered, and violence is generated, which produces refugees who flee outward to neighboring states. A harsh land can breed a belligerent people, or at least people who are so stressed by their environmental problems that they may seek solutions at the end of a gun when there is little else to lose.

Charismatic Mullah Movements

The Pashtun tribes on both sides of the Durand Line in Afghanistan and Pakistan are prone to a periodic cultural phenomenon wherein a cyclical pattern of insurrection erupts, generally in the form of a charismatic uprising where some new religious figure promises some sort of religious or mystical solution to a problem and legions of followers unite around the figure. These uprisings were so common in fact that the British jokingly dubbed them "mad mullah movements". The phenomenon of Mullah Omar operating from hiding on both sides of the Durand Line is just the latest in a long list of such characters to emerge out of the rural Pashtun experience (Anon, 2012).

For example, a personage in the 1930s and 1940s remarkably similar to the present-day Mullah Omar, leader of the Taliban, was the so-called Fakir of Ipi who raised the cry that Islam was in danger and that every Moslem must rally to the Holy War. His real name was Mirza Ali Khan, who was a Tori Khel Waziri, and he led the British and then later the Pakistani government on a frustrating chase around the Northwest Frontier Province for 30 years because they never caught him. He was protected by successions of his supporters and finally died of old age in 1960. The Mullah of Hadda was another, who provoked the so-called "Great Pashtun Revolt of 1897" that convulsed the Northwest Frontier Province of British India when he used mysticism, parlor tricks, and promises to turn British bullets to water as he exhorted the faithful. The British Raj was troubled for decades by these "Hindustani Fanatics" uprisings. This susceptibility of the Pashtun people of the regions on both sides of the Durand Line to such religious insurgencies, and their resistance to external governmental control, seems to be mainly the result of illiterate religious naiveté combined with the strong martial tribal culture that evolved into a cathartic relief for many of these tribal people.

Most of the mad mullah movements of the past have been of relatively short duration and limited in territorial scope so that they had little long-term impact on tribal mechanisms or structures. The actions in 1994 of the politicized insurgent Mullah Omar in Kandahar, however, were a masterful act that seems to have had longer term repercussions. Standing atop a mosque in the city, he wrapped himself in an ancient cloak believed to have been worn by the Prophet Mohammad. This audacious act of political theater catapulted him into a high level of mystical power with the illiterate and impressionable tribesmen. The result was a leader for the Taliban of Afghanistan and Pakistan who has become a modern archetype of the charismatic mullah to the great detriment of those trying to bring a measure of peace to the fractious region (Johnson and Mason, 2007).

Ethnic Patchworks

Afghanistan is a land of multiple ethnicities, many with mutually unintelligible languages, all with histories of religious intolerance, the Sunni–Shia divide, past and present exploitation of weaker tribes, ethnic cleansing, and repeated offenses against dignity and honor. The country is a complete patchwork quilt of ethnic groups with a major fault line in Afghan society between the Pashtuns of the south and the Tajiks, Hazaras, Aimak, Turkmen, and Uzbeks (among others) in the north, but small patches of different ethnicities occur all over the country (Saddique, 2012). The *kuchi* nomads who migrate all over the region are ethnic

Pashtuns. Dari (Afghan Persian) and Pashto are the main languages, but more than two dozen other languages are spoken fairly commonly and 49 languages occur in total (Emadi, 2005). Tension and conflict that lead to violence are not solely the product of ethnic strife, however, because there may be more internal conflict within an ethnic group in some places over issues of power, than between ethnic groups. Barfield (2010) has pointed out that the primary loyalty of people in Afghanistan is, respectively, to their own kin, then to their village, tribe, or ethnic group, all of which are generally glossed together as *qawm*. The oversimplification of ethnic strife in Afghanistan is a muddy concept because ethnicity commonly overlaps in areas with mixed populations, and is hidden in the cross-cutting patterns of intermarriages, bilingualism, and unity through common geographies. People in a shared locality, the *manteqa*, may display more solidarity with their neighbors of different ethnicities than they do with coethnics from other regions in the country. The so-called ethnic cleansing, the term used in the former Yugoslavia, in Afghanistan in some cases was Pashtun Sunni against Hazara Shia, because the latter were regarded not so much as a different ethnic group or political affiliation, but as members of an apostate religious sect as well.

Loyalty to the tribe, the *qawm*, or the *manteqa* may be a ruling factor in the violence in some cases, rather than to any sense of national pride or of a mutual nationhood, but even that model may be oversimplified. For example, a strong sense of national unity was forged during the Soviet–Afghanistan War of the 1980s when so many refugees had to flee the country (Barfield, 2010). The sense of national unity was, however, not rooted in a sense of nationalism, but rather by the will of its people to persist together, united by this traumatic and eye-opening experience that transcended regional or ethnic differences. Even any kind of partition along ethnic lines seemed to be resisted by the Afghan people overall. Four good explanations prevailed (Barfield, 2010): (1) The persistence of an old central Asian view of political order that did not ever link nationalism with ethnicity. (2) Every ethnic group in Afghanistan felt secure enough in its own regions to cooperate with others as partners at the national level. (3) The negative consequences of disunion outweigh internal frictions. (4) Afghans have few illusions about the nature of national politics and the compromises necessary to engage in them. Thus in some sense, the ethnic patchwork can be seen as a trigger to unceasing violence, yet in others this concept is less clear.

Taliban is a dominantly Pashtun organization, with a distinctly Pakistan background and support structure when it started, which remains to the present. With the power vacuum that followed the destruction of the Taliban government in 2001 by the invasion of the Western Coalition, most of the major ethnic groups (Pashtuns, Tajiks, Uzbeks, and Hazaras) returned swiftly to localized predatory power plays, which resulted in a number of competing fiefdoms that pay only lip-service to the central government. As a result of the non-Pashtun, Northern Alliance's treaty with the Coalition invasion in 2001, the Tajiks are overrepresented in the fledging central government in Kabul, which is a source of friction with the southern Pashtuns, even though President Karzai is an ethnic Pashtun himself. This may play itself out in the next elections of 2014, or it may be a tinderbox source of additional violence—it is too soon to tell.

As of this writing in late 2013, Karzai is trying to hedge his bets with whatever political events and changes in government come along postelection, by trying not to sign an agreement to have the American military presence stay on in some form after the elections. This is a dangerous and naive game that is attempting to curry favor with the Pashtun-dominated Taliban and antigovernment forces so that the former president does not have to leave Afghanistan to preserve his life, once he steps down as president.

Foreign Invaders

As Afghanistan's history has shown us repeatedly, chaotic and weak rule in Kabul has invited continual attempts by external invaders to attempt to control the country. This is compounded by the naiveté of foreigners who think that because Afghanistan is easy to invade, that it will be tractable thereafter and become a kind of "mission accomplished". Any number of prior British, Soviet, American, and NATO commands appears to have imagined that their task would be straightforward. Alexander the Great probably thought this too. The British tried three times, with perhaps a forth in their presence in NATO and ISAF at the present time. This only speaks to the hubris of external people who imagine that they can actually control a country like Afghanistan for long. The freedom-loving spirit of the Afghans resists domination by anyone, and the result of the naiveté and hubris of the foreigners only adds more fuel to the flames of Afghan intransigence and ultimately to the recurrent violence.

Corruption

The "*baksheesh* culture" of graft, corruption, internal civil war, and strife continually breeds resentments and keeps all Afghans weaker and less effective as a national people. Decades ago, the common payment of minor small bribes in the government bureaucracy was an accepted means for the poorly paid bureaucrats to augment their miniscule incomes. The infusions of vast billions of foreign aid in the recent decade since the Coalition invasion, however, have skewed the baksheesh out of all proportion to the financial realities of the common people. When such things as a common driver's license or property deed require payment of $100–200 bribes, the corruption is quite out of control, and this is just for the small things. The country thus ranks 175th in the world in corruption, at the top of the list but effectively down on the bottom in the swamp with North Korea and Somalia (Transparency International, 2013). Corruption weakens the polity and makes the country subject to such arbitrariness at so many levels that the frustration breeds violence. This can even lead to attempts by outsiders to take over certain aspects of the country because they think any kind of a decision can be "bought". This is why the resource issue in Afghanistan is so sensitive, yet also so full of promise if the corruption can be navigated without too much bad behavior. The general lack of transparency in the Afghan bureaucracy would seem to breed some violence as well.

Afghan Xenophobia

A completely natural or understandable Afghan xenophobia would occur against foreigners who are commonly thought to have nefarious "designs" on the country. This perhaps would not allow easy assimilation of beneficial external cultural attributes. Any hopes for a reduction in violence, or the seeking of greater peace and tranquility by foreigners, might be seen in an unfavorable xenophobic light by some in Afghanistan who have so little else in their poverty. An illiterate gunman who does not know how to farm after more than three decades of war might fear being tricked by the external foreigner. This could lead to greater xenophobia, with attendant violence.

Land-Ownership Disputes

As a result of the disastrous experiment with communism and the social and political chaos thereafter, land conflict erupted in Afghanistan because of border disputes and cadastral confusion over ownership (Batson, 2008). Redistribution of land to peasants taken from wealthy land owners began the disruption as people had been compelled to give up or sell property under duress. Houses were destroyed or occupied by others during the three decades of war. Official records were inaccurate or destroyed and ownership and ownership transfer documents were forged. Long-term residents might have had no official title because the rights to land use were in the form of social ownership or recognized through customary or common law, as when *kuchi* nomads would bring their flocks onto government or common land. Since the invasion of Afghanistan by the Western Coalition forces in late 2001, various land grabs have been perpetrated by different groups who apparently felt that they could just get away with it in the absence of strong government authority (Mansfield, 2013). All this cadastral confusion has contributed to a great deal of unrest in different parts of the country. The result has been one more cause of the seemingly endless violence.

Endless *Jihad*

Most nonIslamic Westerners do not understand the concept of *jihad* very well. In fact, multiple types of *jihad* have been defined. The Koran itself has more than 150 verses about *jihad*. How many might exist in reality is another story, but the philosophy of much *jihadi* violence in the name of Islam is instructive, especially the more destructive kinds. Some forms of *jihad* are quite benign and actually quite useful in general, nonreligious terms; as for example, the *jihad* of knowledge and scholarship, or that of the inner being to be a better person according to the fundamental precepts of Islam (Nasiri, 2006). Other forms include the *jihad* of the tongue, for example, where protests, proselytizing, politics, propaganda, and sermons are all various meritorious forms. The *jihad* of actions would include the haj pilgrimage to Mecca or money to support certain Moslem causes. *Faridat al-jihad* is an obligatory or defensive *jihad* battle to reclaim the lands of the caliphate (Ottoman Empire) from the infidels, whereas a *kifayat al-jihad* is an offensive or preemptive *jihad*. Only a caliph can proclaim a *kifayat al-jihad* or the ultimate *jihad* (*kutila fi sabilillah*)—The Holy War, and caliphs ended with the Ottoman Empire (1922). Of course an illiterate gunman who cannot read or interpret the Koran for himself, can only accept what others with a salafist jihadism agenda for violence tell him about what he should do according to religious strictures. Suicide bombers and the slaughter and indiscriminant violence against innocents are easy under such circumstances, in spite of Koranic injunctions against just such actions.

Illiteracy

After 30 years of war and the destruction of so much of the originally rather limited educational system, of course, it is little wonder that a lack of educational sophistication and backward thinking has left a great many Afghan people open to attempted exploitation by the cynical Taliban. Illiteracy leads directly to the ability to exploit a largely uneducated, and thereby a politically naive people. Islamic militias, cultures of repression, and aggrandizement of the culture of *jihad* began in earnest during the war of national liberation from the Soviet repression and developed to an acute degree under the Taliban regime. The long-continued armed conflict among the various fundamentalist Islamic parties for domination of the country's politics led to the destruction of most of the social, political, and economic institutions throughout the nation, with the result that people despised their leaders as the culture was so shattered by the never-ending violence (Ehmadi, 2005). These factors thereby contribute greatly to the culture of violence.

Drug Cultures

The cultivation and use of derivatives of *Cannabis* (marijuana) and opium poppies (*Papaver somniferum*) in Afghanistan are well known and do not need further elaboration here beyond the recognition that that use of these products in the local society is clearly antithetical to reasonable civil behavior. The combination of drug-induced, addled thinking and paranoid behaviors, the destruction of the family attendant with the needs of the addict for more money to obtain drugs, the drain on local economies to deal with those whose only concern is the desire for more drugs, the criminal gangs who deal the drugs, and the violence required to control the drug–money supply; all the other societal problems associated with such behaviors must contribute some measure of considerable ill to Afghan society that add to the pessimism that the country can ever be remade to rise from its ashes of war.

Purdah Culture

Finally, the pre-Islamic codes of *purdah* and exclusion of women from so much of public life in Afghanistan have added their measure of cultural disruption and violence to the mix as well, especially because its emphasis was so much lower in the mid-twentieth century than later during the Taliban repressions. From a purely religious point of view, *purdah* is presented as engendering an aura of respect for women, but the actual ignorant practice under Taliban rule led to untold hardship and ill feeling. The overall result of resurgent *purdah* in Afghanistan has in recent decades seemingly only added to cultures of restriction and repressive violence, in this case against half of the productive population.

Promotion of Peace

Afghanistan has long recognized that it has a problem with incessant war; accordingly great efforts have been made repeatedly over the years to find ways to manage some sort of peace. Centuries of tribal conflicts, messianic terrorism, and the repeated failures of the great powers of the day in dealing with the fragmented country (Tomsen, 2011) have taught everyone, Afghan and foreigner alike, that efforts must be attempted to curtail, or at least to control the interminable violence. Many have given it up as hopeless; the late archaeologist and Afghan specialist, Louis Dupree, once told me that the repeated evidence in the archaeological ruins of war piled upon war gave little evidence of any possible lasting peace. Nevertheless, the indefatigable optimists of the country try anyway. In modern times, for example, the Afghanistan Peace and Reintegration Program (APRP) is a peace program designed, implemented, and executed by the government of Afghanistan (GoA) to provide a way for insurgents to stop their fighting and rejoin their communities with both dignity and honor. No less than 24 other important organizations (Table 20.1) have been created to add their efforts to the peace building as well. Given the historical evidence that peace is never for long in that benighted country, 25 peace-building organizations may not be enough, but commendable efforts are underway anyway.

APRP, for example, has been set up so that a fighter can make a commitment to renounce violence, sever all of his ties to the insurgency, and abide by the constitution of Afghanistan, which includes the acceptance of the new laws on

Table 20.1 Peace-Building Organizations in Afghanistan from Governmental, Nongovernmental, Domestic, and Foreign Sources

Afghan Civil Society Forum (ACSF)—Has the widest network of civil society members in Afghanistan, and uses this reach to promote mass engagement in the peace-building process.
Afghan Defense of Women's Rights of Balkh (ADWRB)—Formed to raise Afghan women's awareness of their constitutional rights and to provide advice, education, and occupational development to vulnerable women.
Afghan Development Association (ADA)—One leading the development organizations in Afghanistan with projects across the country, including in peace building and conflict resolution.
Afghan Organization of Human Rights and Environmental Protection (AHOREP)
Afghan Peace and Democracy Act (APDA)—A communication and co-ordination network of NGOs aimed at enhancing role of civil society organizations and assisting in coordination of Afghan NGOs and civil society organizations through direct support and collaboration.
Afghan Women Education Centre (AWEC)—A commitment to promote human rights and work for gender equality and abolishment of any kind of discrimination and violence against women and children.
Afghan Women's Skills Development Center (AWSDC)—Exists to reinforce the sustainable development process of women and children in Afghanistan.
Afghanistan Human Rights and Democracy Organization (AHDRO)—Established by Afghan human rights and civil society activists with a strong commitment to work with the most marginalized sectors of society.
Afghanistan Justice Organization (AJO)—A public policy research and training organization dedicated to the principles of rule of law, free markets, and individual liberty in Afghanistan.
Afghanistan Research and Evaluation Unit (AREU)—An independent research institute with a mission to inform and influence policy and practice through high-quality research.
All Afghan Women Union (AAWU)—Formed to oppose the flood of oppression and violence that surround women's lives.
Bureau for Reconstruction and Development (BRD)—The vision is a sustainable Afghan society, focusing on human rights, civil society, government, and the local economy.
Center for Conflict and Peace Studies (CAPS)—Dedicated to reducing threats of violence, building local capacity for creating a sustainable environment that can support peace and stability.
Coordination of Afghan Relief (CoAR)—Serves marginalized communities in food security, agriculture/irrigation, animal husbandry, health, women and child support, education, rural engineering/support, emergency relief, disaster risk reduction, management, and income generation.
Coordination of Humanitarian Assistance (CHA)—One of the largest national NGOs working on humanitarian assistance and social development in Afghanistan.
Cooperation Center for Afghanistan Mazari Sharif (CCA)—Works for promotion of human rights, the rule of law and democracy as well as sustainable development in Afghanistan.
Co-operation for Peace and Unity (CPAU)—Works toward creating a sustainable culture of peace in Afghanistan through active peace-building and peace education programs.
Education Training Center for Poor Women and Girls of Afghanistan (ECW)—Nongovernmental, nonpolitical, and nonprofit welfare organization organized in Taliban time when a degrading, depressing, and stifling atmosphere was imposed on Afghans.

Table 20.1 Peace-Building Organizations in Afghanistan from Governmental, Nongovernmental, Domestic, and Foreign Sources—cont'd

Health and Rehabilitation Program for Afghanistan (HRPA)—A vision for a free democratic, healthy, and green Afghanistan, based on social justice and absence of violence against women and children.
Humanitarian Assistance for Women and Children of Afghanistan (HAWCA)—Established to encourage the active participation of women in reconstruction, development, and peace in Afghanistan.
Sanayee Development Organization (SDO)—Works toward serving vulnerable and disadvantaged people in sectors of education, peace building, community development, and community health in rural Afghanistan. http://www.insightonconflict.org/conflicts/afghanistan/peacebuilding-organisations/sd/
Tribal Liaison Office (TLO)—An Afghan nongovernmental organization that emerged on good governance after an approach by community elders requesting assistance for participation in the peace and reconstruction process in their southeastern communities.
Voice of Women (VOW)—Established in 1998, and works toward improving conditions for women and girls in Afghanistan.
Welfare Association for the Development of Afghanistan (WADAN)—An indigenous, nonprofit, nongovernmental organization that envisions a peaceful, drug-free, democratic, developed, and prosperous Afghanistan.

Some organizations are quite large and focused; others are more diffuse in their approaches or much smaller. Individual web sites and Insight on Conflict http://www.insightonconflict.org/conflicts/afghanistan/peacebuilding-organisations/.

women's rights. Instead of being paid to stop fighting, the insurgent is shown that incentives such as grants to his community will be awarded instead. In that fashion, APRP enables a local community level of support for peace and reconciliation such that communities of groups of people can lend their support to those who choose the path of peace instead of violence because there is a direct tangible award to them to get behind the program. The effort is divided into three phases.

Phase 1: Social Outreach, Confidence Building, and Negotiation

Strategic communications efforts are made by community, district, and provincial leaders to reach out to possible or known insurgents. Once contact is made, the insurgents are encouraged to stop fighting and rejoin their community. These negotiations involve discovery and resolution of grievances though confidence building and securing of peace agreements between victims, former insurgents, communities, and the government. After more than three decades of war, great mistrust and many grievances have accumulated so the most important part of the process is identifying the actual grievances and convincing the insurgent that the government will actually deliver on its promises. This can be a most difficult process as Derksen (2011) has pointed out; one Taliban commander from Helmand Province noted that, "This is not a reintegration process, this is an American process. With whom should we join? With this corrupt government? I will never join this process and will not let any of my friends." At most only a few thousand fighters had joined the process, mostly from the north and west, and only $7.7 million had been expended by September 22, 2011 from a fund of $133.4 million. Few inroads in the process had been made in the most critical areas in the south.

Phase 2: Demobilization

This part of the process first involves vetting of the potential reintegree to ensure that the applicant is a bona fide insurgent rather than, for example, a member of a criminal gang or actually a member of a progovernment militia looking for a hand-out. If this process is successful, personal data are collected such as retinal scans and other information to make sure of actual identities later, and personal weapons are registered and returned, although heavy weapons and improvised explosive materials are confiscated if turned in. The former fighters are helped to write down their intent to reintegrate, their family histories, why they are fighting, and why they would like to stop. These materials are then reviewed and approved by five levels of officials (security, army, police, peace committee, and governor) at the provincial level, before being sent on to the central government at Kabul. Only then does the reintegree receive a stipend of $120/month for 3 months while going through disengagement training for a new job.

Phase 3: Consolidation of Peace

This phase focuses most upon the community that is willing to accept the former fighter back. Development projects to consolidate the peace are a preferred method and include agriculture and infrastructure processes, as well as demining programs. Education and training in literacy, religion, and vocational aspects can be provided.

A big problem in Phases 2 and 3, however, has been in maintaining realistic security for the respondents as they can easily become targets by the unreconstructed Taliban (Derksen, 2011). Many have joined the Afghan Local Police as a result, and inadequate vetting has led to abuses of authority by them, as well as the perpetuation and intensification of local powerbrokers who want to get their own people into the APRP program. For example in Baghlan Province, many Tajiks saw former Pashtun insurgents join the government, receive weapons, and become powerful locally so they wanted to increase their own strength in the same fashion.

Some measure of Afghan bellicosity is provided by the more than 20 organizations that are devoted to different aspects of seeking some measure of peace in the land, which is a rather high number for a small place; people seem very tired of incessant war, yet it never seems to stop for very long. Peace in Afghanistan is thus an elusive quantity but an issue where the sheer number of organizations trying to do something about it (Table 20.1) is a measure of the sincere interest on the part of many people of good will to try to effect change. The odds against success in their many individual endeavors are long but the payoffs should be worth it; we can only hope.

Future Guides from Past Performance

Given the many reasons suspected of giving rise to incessant war in Afghanistan that have been described above, time would seem rather limited for the country to succeed in its quest to rise from its past stews of invasions, civil wars, foreign interventions, and other detractions that keep the place from prospering. But there are many problems and pitfalls lurking once the Coalition forces withdraw to see where the possibilities for solutions may lie. The neighbors surrounding Afghanistan are known to want access to the country's resources, so the future is murky in that regard and seems capable of producing situations where Afghanistan will fall back into its unfortunate failed-state characteristics of dysfunctionality, backwardness, and violence. Byrd (2012) has pointed out too that it is important not to overlearn from history as, for example, the country's problematic experiences over the past half-century with political parties, which are actually essential to democratic political systems, and without such Afghanistan will be unlikely to prosper. So the point is that the history of Afghanistan can teach us useful things but will not predict the future or give us reliable templates for managing the country in the transition back to more complete home rule and internal military integrity. A look, therefore, at the salient points of Afghanistan's history should provide instructive background.

As the decades from 1933–1973 in Afghanistan demonstrate, the country can be stable and effectively governed but stability during that period was grounded in traditional local governance and a centralized but weak state. Both the local and central governance means and structures were severely damaged after 1978 during the Soviet–Taliban years. The country also has had chronic succession problems in its long history, with associated conflicts so that the next presidential election of 2014, if successful, will be the first peaceful transfer of leadership and power since 1933, and only the fourth in the past 267 years since 1747 when Amad Shah Abdali (Durrani) established the country. Continued and even more effective new leadership is critical during the transition from President Karzai to a newly elected leader. The difficulties of reaching peaceful solutions to an insurgency during military withdrawals are manifold, and the adverse consequences of withdrawing financial aid and other support were amply demonstrated during the period of time from 1986 to 1992 when the Soviets withdrew and the Najibullah regime finally fell as the Soviet Union also collapsed. Certainly any options of arming and paying militias though the transition are dangerous because this only opens doors to the instability and predatory behavior that Afghanistan has been known for several times in the past. The Afghan economy, however, is in much better shape than it was during and after the Soviet presence, so any deep fiscal contraction after the coming Coalition withdrawal must be avoided. Expectations about the pace of progress with the extraction of resources to answer all needs must be modest because of the dangers of overambitious reforms that can lead directly to violent reactions, which therefore must be recognized clearly and prepared for. Obviously then, Afghanistan's future will depend heavily upon outside financial support for many years, so this support must not be cut back abruptly or stopped in hopes that the much vaunted resource extraction is going to leap in to pay all bills immediately. Smart resource extraction takes time and establishment of the resource corridors is not a universal panacea. The shortness of time notwithstanding until the withdrawal is complete, the means must be found to support the Afghans through these coming, even more difficult years. This will be a difficult effort to press upon a tapped-out and Afghanistan-war-weary world.

Both Pakistan and Iran are also keys to much of what will happen in Afghanistan in the coming decade, and their own futures do look a bit grim as well. Furthermore, the two most poisonous relationships that these neighboring countries have are the toxicity of the Pakistan–India and the USA–Iran associations, neither relationship of which bode well for Afghanistan's future, but both of which have some glimmers of hope for a decline in tensions. Ensuring later peace in Afghanistan is very much doubtful if one considers the intractability of these problematic relationships insofar as they affect the internal Afghanistan situation. With India's new road through southeastern Iran from the Indian Ocean to western Afghanistan, and with the Indian presence at the Hajigak iron deposit not far from Kabul, as well as Iran's extensive development program in western Afghanistan where it is providing roads, electricity, and gas (Rashid, 2012), both Pakistan and the USA have become rather nervous about these factors, even if they do not discuss it much. Byrd (2012) has noted that the current unstable political situation in Pakistan shows a need to work around, or to make contingency plans, with respect to Pakistan and Iran playing spoiler roles to the plans for Afghanistan in Kabul and Washington. But spoiler roles have a way of blowing up in a perpetrator's face, and care must be taken to lest the plots backfire, something the educated people in both nations understand, even if their militaries and religious figures do not seem to.

Pakistan in fact, may be on the brink of seriously unpleasant developments, as Rashid (2012) has emphatically pointed out in his cogent assessment of the problems. But the USA, who seems to very much wish that not too much unpredictability will continue to play out in Pakistan, does not outline exactly or even vaguely what policies it will pursue in the region after 2014. Perhaps the US Department of State does not yet know what to do, but precise US policies actually may be quite essential to Pakistan's stability. A positive outcome for both Afghanistan and Pakistan would seem to depend upon a deliberate and carefully considered Western-withdrawal strategy from Afghanistan, along with a political settlement with the Taliban, as well as ensuring Pakistan's willingness to take on Islamic extremism and prevent the breakdown of the state in the spreading violence. As both the former Pakistani ambassador to the USA and author Rashid (2012) have pointed out, however, mending Afghanistan is just not possible if Pakistan continues to fall apart.

It would seem, indeed, that Pakistan should act more as a normal state would do, rather than as the paranoid, insecure, ISI-driven entity that uses extremists and diplomatic blackmail as operational norms. The ISI has far too much power as its own state-within-a-state that clearly needs to be brought back under civilian control under the new administration of Nawaz Sharif. Furthermore, Pakistan as a nation seems hesitant to carry out any self-analysis according to logical rationality instead of the customary emotion and prejudice. Few in the country take responsibility for their actions or mistakes and instead try to paint their country as the victim, maligned and treated underhandedly (Rashid, 2012). The USA and India are too often painted by Pakistan as trying to undermine, subvert, and destroy poor Pakistan but no logical explanation is ever attached as to why they might want to do this. On the other hand, to the USA, Pakistan is too nuclear to ignore, and has become the ultimate "frenemy", using its control over the Taliban to help Pakistan face down India in its contest over Afghanistan. But at the same time, any peaceful solution to the war in Afghanistan must include India at the negotiating table, something that Pakistan in all its paranoia will not yet countenance.

Some think as a result of all this that Pakistan actually faces a much more dangerous future than does Afghanistan. The Pakistani military and political parties may know that change and reform actually have to happen soon in that country before political control is lost for good. Growing violence between ethnic groups, Sunni against Shia with major killings, huge economic catastrophes for millions of people, enormous earthquakes and floods taking countless lives with little government help, and the ever-growing threat of militant Islam creeping throughout and exploding over the country could in time overwhelm the whole Pakistani security apparatus. Potential coup plotters inside the Pakistani military, with their linkages to outside extremists, could threaten the army and its nuclear weapons. At the present time, dealing with these existential eventualities seems to have posed such a serious threat that the USA appears either rather flummoxed since their emissary Richard Holbrook died, or they do not wish to make their intentions known, which if true would be a strange position to take. But Holbrooke had approached Pakistan more from a position of engaging civilian needs in Pakistan for a first time, rather than by assisting the military fixation on India that the Pakistanis have always had. Instead he had begun more discussions on energy, electricity, water, and women, with some reasonable results actually in evidence as the Pakistanis began to move more against their own Taliban and allow much more intelligence gathering and drone attacks by the US. The hardliner approach to Pakistan with threats and coercion favored inside the White House was not preferred by Holbrooke (Nasr, 2012, 2013), and after his untimely passing, these less subtle and more militaristic approaches resumed for a time.

American foreign policy in the past few years in regards to Afghanistan and Pakistan seems to have become completely subservient to tactical domestic political considerations. The reelection of President Obama seems to have occurred at least in part because the US public wanted extrication from its big wars, as well as a smaller global footprint. Furthermore, the foreign policy for Southwest Asia, such as it has been, seems to have been most driven by the Pentagon, intelligence agencies, and the short-term political calculus instead of the difficult and subtle art of diplomacy as preferred by the Department of State. The lack of support by the White House for Secretary of State Clinton in many of her South Asian endeavors appears in abeyance with new appointee John Kerry because one of his first tasks has been to meet with Afghan and Pakistani leaders (Gordon, 2013), but it is too soon to tell how this new mode will play out. As of this writing in late fall 2013, US policy in Afghanistan and Pakistan, in the absence of any coherent and well-reasoned alternatives for the greater region, has seemed little more than a desire to be well rid of it all, hoping against hope that a decent interval of calm will occur between withdrawal and the expected likely chaos to follow.

Although Pakistan greatly fears a renewed civil war in Afghanistan, it will not countenance a partition of that country because of the grave implications for its own integrity. The position of the Pakistani military apparently is that if they cannot get what they want in an Afghan settlement, no one will get any peace; it will not allow any real solution to occur except on its own terms (Rashid, 2012). If, however, Pakistan continues to follow its present chosen path of continued conflict with the USA and supremacy over Afghanistan, while also being inflexible and ignoring the interests of the other neighbors, it is quite likely to lose everything. Pakistani General Kayani, who ran an intransigent ISI for many years, had long maintained that the Americans not only were not going to win their war in Afghanistan, but they would never be able to transform the place in any substantive fashion. His point all along was that Afghanistan had devoured many empires before and would defy the Americans as well. He felt that the only possibilities were stopping the grandiose American plans, getting practical, and discussing the departure of the foreign militaries and deciding what sort of a state would be left behind that both the US and Pakistan could live with (Nasr, 2013). But the long-sought retirement of General Kayani has finely come to pass and the future with the ISI and the Taliban may change considerably; it is too soon to tell.

Before the 2014 withdrawal deadlines roll around too far though, the USA really does need to be able to tell its friends and nominal allies what relationships it would like to have in the region and how those relationships should be structured. But President Karzai's intransigence about signing a post-withdrawal agreement with the USA to keep on some troops on the ground in-country has obviously given Washington considerable uncertainty. Pakistan obviously has to open up more to the Indian economic powerhouse if it wants to participate in the coming economic growth that is keyed to grow outward from Afghanistan as the resources are developed. India needs to be invited to the negotiating table in spite of Pakistani objections. The USA needs to talk more to the people that it does not really understand very well, such as the Iranians, who should be at the negotiating table as well.

A point can be made (Yusuf, 2012) that Iran's fundamental interests overlap greatly with those of the USA, but the numerous other difficult issues at hand elsewhere in the Middle East mean that Iran will not help the USA or its Afghan client state as long as the quid pro quo can be played out. For the USA to engage Iran successfully again after such a long hiatus, and at the same time take the Pakistan–Indian equation head on in the process as well, obviously does not conform to the priorities and moods in Washington at the present time, but ultimately even marginal success on Afghanistan is likely to require it. In spite of Secretary Kerry's intrepid new agreements with Iran, nothing like anything really substantive seems likely to happen any time soon, however, unless a new rapprochement develops stronger ties as some have noted as a distinct possibility (Patrikarakos, 2013). Given the apparent lethargy of the White House to continue engagement in South Asia/Middle East issues, however, especially now that they seem to be midway into their pivot off toward a bigger Asia with China and India, realistically, not too much is likely to develop very quickly with those ideas. Meanwhile, of course, inside the fears of those most worried about losing American primacy in the world, China and Russia are powering in to fill any gap left in the Middle East, especially that part held inside the vice grip China seems to be applying in trade and development around Iran and Pakistan.

The success of Afghanistan following the departure of most of the forces of the Western Coalition in 2014 is far from certain given the many external negative factors arrayed against the county. In addition, other factors of internal negativity have been mentioned in passing throughout prior pages that are summarized here. For example, Afghanistan's well-known proclivity for endemic corruption has repeatedly worked against the best interests of the state. With the hand of nearly every bureaucrat, minister, governor, and other functionary out for an upfront payoff or a later percentage of the profits, it is little wonder that so many projects have foundered upon the shoals of dishonest dealing. The lack of transparency in the bidding process on government resource tender offerings, the general distrust of local provincial officials and their people against the desires of the central government, withholding of critical government information for personal gain, control of projects by single actors rather than more widespread shared viewpoints; all these factors add to the sense of negative thinking about any possible success that Afghans are trying to pull off before another collapse.

Added to this are the environmental issues to consider as well. In general, potential environmental problems would not really set up too many situations that would actually preclude success with resource extraction in a place like Afghanistan where the environment is already so marginal; rather the environmental situation there forms more a series of factors that should be considered carefully to ensure long-term sustainability of local environments, which if well-managed in future could contribute enough to the people's well-being and new prosperity so that they would want to support more mining efforts. With a protected environment while mining proceeds, fewer criticisms can be leveled against the government or the mining company that holds the mining permit. But given the local sensibilities that do not generally set much store in environmental protection, as well as the apparent lack of much concern by so many big multinational companies in the past, or nations such as China in the present, certainly does indicate a potential for increase of this poor behavior in Afghanistan. The result obviously will be a great detraction from the country's success unless great care is taken.

References

Ahmad, A.S., 1978. An aspect of the colonial encounter in the North-west Frontier Province. Asian Affairs. 9, 319–327. http://www.khyber.org/publications/016-020/colonialencounter.shtml.

Anon, 2012. Mad Mullahs, Opportunists, and Family Connections: The Violent Pashtun Cycle. Tribal Analysis Center, Ltd, Willamsberg, VA, 20 p. http://usacac.army.mil/cac2/call/docs/12-18/ch_10.asp.

Barfield, T., 2010. Afghanistan: A Cultural and Political History. Princeton University Press, Princeton, NJ.

Batson, D.E., 2008. Registering the Human Terrain: A Valuation of Cadastre. National Defense Intelligence College (NDIC) Press, Alexandria, VA, 162 p. http://ni-u.edu/ni_press/pdf/Registering_the_Human_Terrains.pdf.

Brown, L., May 2009. Could food shortages bring down civilization? Scientific American. 50–57. http://www.earth-policy.org/images/uploads/press_room/SciAm-final.pdf.

Byrd, W., 2012. Lessons from Afghanistan's history for the current transition and beyond. United States Institute of Peace; Special Report 314, 15 p. http://www.usip.org/publications/lessons-afghanistan-s-history-the-current-transition-and-beyond.

Caroe, O., 1958. The Pathans: 550 B.C. – A.D. 1957 (1973 Pakistan Reprint). Macmillan, London, 521 p.

Cullather, N., 2002. Damming Afghanistan: modernization in a buffer state. Journal of American History, 89 (2), 512–537. http://www.journalofamericanhistory.org/teaching/2002_09/article.pdf.

Derksen, D., 2011. Impact or Illusion? Reintegration under the Afghanistan Peace and Reintegration Program. Peace Brief 106; United States Institute of Peace, 4 p. http://www.usip.org/files/resources/PB%20106.pdf.

Elphinstone, H.M., 1815. An Account of the Kingdom of Caubul, 2 vols. A. Strahan for Longman, Hurst, Rees, Orme & Brown and J. Murray, London.

Emadi, H., 2005. Culture and Customs of Afghanistan. Greenwood Press, Westport, CN, 252 p.

Fletcher, 1965. Afghanistan: Highway of Conquest. Cornell University Press, Cornell, NY, 325 p.

Gordon, M.R., April 23, 2013. Kerry to Meet with Afghan and Pakistani Leaders. New York Times.

Johnson, T.H., Mason, M.C., 2007. Understanding the Taliban and Insurgency in Afghanistan. Foreign Policy Research Institute. http://www.nps.edu/programs/ccs/docs/pubs/understanding%20the%20taliban%20and%20insurgency%20in%20afghanistan.pdf.

Londoño, E., April 4, 2012. Afghanistan Sees Rise in 'Dancing Boys' Exploitation. Washington Post, 2 p. http://articles.washingtonpost.com/2012-04-04/world/35451705_1_bacha-bazi-afghans-pashtun.

Mansfield, D., 2013. All Bets are Off! Prospects for (B)reaching Agreements and Drug Control in Helmand and Nangrahar in the Run up to Transition. Case Study Series Afghanistan Research and Evaluation Unit, Kabul. 104 p. http://reliefweb.int/sites/reliefweb.int/files/resources/The%20Impact%20of%20Microfinance%20Programmes%20on%20Women%27s%20Lives%3A%20A%20Case%20Study%20in%20Kabul%20Province.pdf.

Nasr, V., 2012. The Dispensable Nation: American Foreign Policy in Retreat. Knopf Doubleday Publishing Group; Random House. 320 p.

Nasr, V., 2013. The Inside Story of How the White House Let Diplomacy Fail in Afghanistan. Foreign Policy, 12 p. http://www.foreignpolicy.com/articles/2013/03/04/the_inside_story_of_how_the_white_house_let_diplomacy_fail_in_afghanistan.

Nasiri, O., 2006. Inside the Jihad: My Life with Al Qaeda – A Spy's Story. Basic Books; Perseus Group, New York, 337 p.

Olearius, A., 1669. The Voyages and Travels of the Ambassadors. John Starkey and Thomas Basset, London, 548 p.

Patrikarakos, D., December 9, 2013. Iran, from Enemy to Ally. The New York Times.

Rashid, A., 2012. Pakistan on the Brink: The Future of America, Pakistan, and Afghanistan. Penguin Books, NY.

Raverty, M.H., 1867. Selections from the Poetry of the Afghans. Williams and Norgate, London.

Saddique, A., 2012. Afghanistan's Ethnic Divides. Sources of Tension in Afghanistan and Pakistan: A Regional Perspective. CIDOB Policy Research Project Barcelona Centre for International Affairs. 11 p. http://www.cidob.org/en/temas_regiones/asia/sources_of_tension_in_afghanistan_pakistan_regional_perspectives.

Shroder, J.F., 2007. Afghanistan's development and functionality: renewing a collapsed state. Geojournal 70, 91–107.

Shroder, J.F., 2012. Afghanistan: rich resource base and existing environmental despoliation. Environmental Earth Science, 16. http://dx.doi.org/10.1007/s12665-012-1638-7.

Tomsen, P., 2011. The Wars of Afghanistan. Public Affairs. Perseus Group, Philadelphia, PA, 849 p.

Transparency International, 2013. http://www.transparency.org/cpi2013/results.

Yusuf, M., 2012. Should We Hope for a Regional Solution to Afghanistan? United States Institute of Peace; Olive Branch Post, 2 p. http://www.usip.org/publications/should-we-hope-regional-solution-afghanistan.

In this Chapter

21

Optimistic Scenarios: Successful Resource Corridors

Abstract

Prospects for Afghanistan appear slim against all the negatives. Afghans are famously pragmatic, however, and can realign political positions on short notice to back a winning plan. Resource corridors thus offer serious consideration for guarded optimism. The better educated urban youth are especially interested in these prospects. This in combination with some recent evidence that the newly trained Afghanistan Army can hold its own against the Taliban insurgency does offer some reasonable hope for success. Pakistan's new nonmilitary government seems to be restoring civilian control of the government, which might enable encouragement of stabilization in Afghanistan. Thus the resource corridors in Afghanistan may work to get the country away from its resource curse and more into a modern economy, which could contribute to regional well-being and increase standards of living. Careful environmental control, with governing transparency and oversight can help curtail corruption. The industrial competition of the large local powers in Afghanistan, China, and India is essential in getting the resources out of the ground, and in maintaining regional oversight and control. Pakistan could share in resource benefits if they assisted in transboundary flood control and irrigation, while also encouraging transport across their lands. Proposed internal government reform in Afghanistan, with external oversight might actually be successful.

Keywords: Competitive neighboring states curtail violence; Mining-sector growth engine; Resource corridors.

The sorely beleaguered country of Afghanistan does indeed need a break from its dreadful past, and their new ideas about setting up robust resource corridors just might serve that possibility. Furthermore, after the expenditure of seemingly bottomless blood and treasure trying to extract that country from the grip of al Qaeda and the Taliban, the US and its allies might be forgiven for a bit of optimism if the huge resource base worth some $1–3 trillion actually succeeds in pulling the country out of its seemingly bottomless swamp of poverty and despair. On the other hand, the pessimists cannot be dismissed either,

Natural Resources in Afghanistan. http://dx.doi.org/10.1016/B978-0-12-800135-6.00021-0

given the centuries of turmoil and political machinations so characteristic of the region. The murky future for that bellicose country notwithstanding, assessment of the possibilities for recovery from decades of violence seem real enough for Afghanistan, at least from a geological-resource point-of-view. The possibilities for thousands of new jobs and great wealth enhancement in a chronically wage-starved country are very real now, given the finally recognized trillions of dollars of natural resources that lie in the ground there awaiting exploitation.

But the greedy neighbors on the borders also want to control the resources for themselves; the question is whether the neighboring states of China, India, Iran, and Pakistan will restrain each other effectively enough to keep the lid on the pot and enable mutually profitable resource extraction for everyone's benefit (Pant, 2013), or whether the spoilers instead will win. Can the violence be restrained so that resource extraction lifts the country by its bootstraps, or does the country continue to wallow in its pit of illiterate gunmen and underdeveloped backwardness? Still, the people of Afghanistan are very pragmatic; they might just be tired enough of interminable war to seek an alternative, with the resource-corridor (RC) approach providing that opportunity.

More than four decades of scientific disclosures about the rich resources of Afghanistan first discovered by Soviet geologists, and then further assessed by a number of US Geological Survey geologists as well as by the author and others, the facts about the astonishingly rich resources have attracted major attention from the US military, the government of Afghanistan (GoA), the neighboring countries, and a number of multinational companies who are not adverse to accepting some risk in their international ventures. The Soviets had paid some attention to the hydrocarbons (gas, oil, and coal), and made some investments to bring the Aynak copper deposits on stream, but little more had ever been done, other than minor mining of some industrial minerals such as cement, oil-drilling fluid additives, and a few others.

Following invasion of Afghanistan by the Western Coalition in late 2001, the USGS, the British Geological Survey, and some other NATO-member geologists obtained and translated all the Soviet geology reports, and went into the field while it was still safe enough with a minimum of security to protect the geologists. In the later part of the decade of the invasion, however, new information began to accumulate that the resource picture was actually considerably more robust than had been thought previously (Chapter 11). First the USGS deduced that three times more undiscovered gas and 18 times more oil existed than previously thought, and that extensive but unproven coal was there to exploit. The Aynak and other copper prospects were the first of the metal ores to receive major attention, albeit through a flawed and corrupt bidding process that when exposed, brought in new competent management. Thereafter, the new Minister of Mines and Industries, Wahidullah Shahrani, came in with a robust skill set for managing their resources and preparing their many deposits for bidding by credible companies. In partnership with the USGS, the US Agency for International Development and the US military's Task Force for Business and Stability Operations, new mineral-bidding packages (Chapters 14 and 15) were established with a fairly full complement of ancillary information, especially a variety of new geophysical investigations, including much of the country covered with high-resolution, hyperspectral remote sensing for mineral identification, and numerous geophysical data sets (magnetics, gravity, etc.) that companies need for producing competitive bids on selected deposits (Chapter 13). In the process of this new resource data acquisition, a number of new discoveries were made, such as lithium and rare earths, among others, that added further to the good news about natural resources in Afghanistan.

In recognition of the great problems Afghanistan has long had from corruption, new mechanisms to cope with the corrupt bureaucracy were put into place (Chapter 14). Escaping the so-called resource curse, wherein violent struggles to control the resource riches ensue, such as the "blood diamonds" in so much of Africa, became a most important issue to avoid a new resource war. Adherence to newly written resource laws, and to the extractive industries transparency initiatives was stressed in hopes of controlling at least some of the endemic corruption.

Because of the sky-high prices, more than five new gold prospects (placer and lode) were successfully let in the bidding process and opened up for production in the belief that success with a few new mines where the resource could even be flown out of contested areas would spur on additional, slower paced mines. The Aynak copper bonanza that had gone to a corrupt bid from China was reconfigured but kept in Chinese hands, along with several oil prospects as well. The planned railroad south from Tajikistan and Uzbekistan is already at Mazari Sharif, with plans to head on south to Kabul and Aynak. Eventually railroads are planned throughout the country, which will give incentives for new industries and plentiful jobs. Both discussions about transporting Hajigak iron out to India across Pakistan by various means, or failing that, other transport developments (road and rail) have taken place on the ground in western Afghanistan to perhaps move the iron out through Iran to India via the Chabahar seaport at the mouth of the Persian Gulf into the Indian Ocean. The important rare earths of the Khanneshin carbonatites on the lower Helmand were recently investigated from helicopters and on the ground by USGS geologists surrounded by much military security, in what has been joked about in the geocommunity as the most expensive recent field tripping short of the Moon landings, but which offers a potential "goldmine" source of rare earths to whoever bids on it successfully.

Critical mass on resource extraction in Afghanistan seems to have been reached at last, and the country may actually leap forward with its new RC approach to development. The RC ideas are now on the table with some sense of urgency in the middle of considerable enthusiasm. The urgency, of course, is engendered by the looming Taliban as the Western Coalition troops prepare to withdraw their forces, although early indications are that some nine bases will be maintained to help the Afghan National Army maintain some security in the face of ongoing attacks from the Taliban. Even so, enthusiasm for the RC idea has occurred infectiously as the World Bank and the United Nations have stepped in with vigor to aid the GoA because most can see the great attractiveness and eminent sensibility of the plan. After decades of shifting politics, false starts, dashed hopes, and three decades of war, real hope is actually on the horizon.

As noted in detail in Chapter 15, the concepts of RCs have existed for quite a while, with earlier examples established in most of the developed countries and elsewhere in the developing world. The RC idea applied to the uniquely diverse and potentially quite rich country of Afghanistan has the potential for some startling innovations because of the closeness of huge and growing markets and industries in China and India. A problem that has developed in many prior efforts elsewhere, for example, is that mining enclaves can be set up to extract the resource, with no other collateral development. The hole in the ground left behind in many places commonly becomes a toxic lake that poisons the groundwater. What is planned in Afghanistan, on the other hand, is quite different. The establishment of RCs in Afghanistan is planned as a sequence of actions and investments to leverage a large extractive industry investment in infrastructure, goods, and services, into a viable economic development along defined geographic strips. In Afghanistan, the infrastructure, particularly transport, is the most critical so an integrated and staged building and overhaul of roads, railroads, and aviation facilities is being planned and is underway as much as funds allow. The overall concept of RC development in Afghanistan has been made a national priority program to raise the status of the concept and help speed along many of its procedures, insofar as that is possible within the generally obstructive Kabul government bureaucracy. Private sector investment is viewed as essential, so its sequencing and prioritization have been facilitated and leveraged across a number of aspects of livelihoods, social effects, governance, environmental impact, and the infrastructure.

Most importantly in an optimistic scenario, Afghanistan has set up an approach to RC development according to the extractive industries value chain (EIVC) promoted by the World Bank. This EIVC is a framework for coordinated and integrated interactions that work across broad public and private partnerships. The broader economic growth can be leveraged by these associations, which has eluded many other nations and regions elsewhere for a variety of reasons. The essential plan of the EIVC starts in the awarding of contracts and licenses to extract the resource, followed by government regulation and monitoring of the extractive operations, collection of tax and royalty revenue, management and allocation of these monies, and lastly, the implementation of policies and projects of development that are sustainable into the future. Of course in Afghanistan, the cancer of endemic corruption will greatly plague the situation, and other problems with ethnic rivalries and incessant violence will short circuit the process, so that monitoring by the World Bank and other government and external watchdog agencies will have to be applied with a rigor to generate enough reasonable transparency to allow even marginal prosperity to result. The idea that this could be done even marginally successfully in a place like Afghanistan will give pause to some, but such a mechanism has been used by numerous other developed nations and Afghanistan has great aspirations to join them.

With the mining sector in Afghanistan planned to become a major engine of growth, the jobs and infrastructure established to service the growing mining industry will require much new training and building of additional or collateral infrastructure to enhance expansion of other parts of the economy. Development of RCs does take variable amounts time in construction, education, transport, sales, and general ancillary construction, with the result that a phasing in of the various RC aspects will occur in different geographic areas (Table 21.1). This will occur in a staged fashion depending upon levels of acceptance by local people, especially with regard to levels of violence and security capabilities. The first RC segments are under construction where the most advanced development of resource deposits exists already, which is also where the political and security uncertainties are considered the lowest or easiest to deal with, such as in the north.

The timeline and contingencies for RC development have been established as a range of possibilities that add to optimistic outlooks because of the flexibility of approaches between a full flowering to even a major collapse, wherein even this latter unpleasant possibility can be considered just a minor setback on the road to recovery for greater prosperity. Timing flexibility in future decades is required, given the always uncertain nature of Afghanistan circumstances, but even with the need for contingency plans, the general and widespread recognition of the possibilities provided by the rich resources presumably will not allow more decades of further stagnation as before.

The idea of reestablishing Afghanistan as a hub of transport and trade to link Europe and the Middle East with the Indian subcontinent and all the rest of East and South Asia is a robust notion that is certainly already bringing much positive change to the region. This Silk Road initiative is viewed by the GoA as the reemerging of the "Asian roundabout", a central point of interconnection of local resources, goods, ideas, services, and people in the rapidly expanding economies of East and South Asia (Kuchins, 2011). The essence of this Silk Road initiative is giving the regional neighbors a higher stake in stabilizing Afghanistan through mutual economic benefit derived from increased trade and more efficient transit of raw materials and finished goods, with Afghanistan as a source, a way station, and ultimately as a destination for finished goods as the economy grows, achieves constancy, and joins in the general developed world trade.

Table 21.1 Afghanistan RC Plan

Northern Hydrocarbon RC Segment—Short- and medium-term investment to maintain existing, and upgrade highway between Mazari Sharif and Shiberghan; construct and operate new railway terminal at Mazari Sharif; establish Shiberghan gas and power system; construct middle-scale (~15,000 barrels/day) refinery at Mazari Sharif; build new railway from Mazari Sharif west to Andkhoy, with spur north across border into Turkmenistan to connect with new railway there; all to be accomplished within next 2–4 years.
Southeast Copper RC Segment—Based on all-important world-class Aynak copper deposit; needs to start with robust and transparent public financial management; planning and implementation requires upgrading of national and local capacities to handle world exposure to production capabilities and associated problems, especially in the new Afghanistan Land Authority of the Ministry of Agriculture, Irrigation and Livestock that must deal with the local disruptions; access road networks must be enlarged, rebuilt, and maintained with effective means; north–south electrical transmission lines and new power ring and electrical "grid code" must be established to service the site; a mining supply chain of service industries (drilling services, core extraction and assessment, foundries, etc.) must be established; a new water system for both Kabul and Aynak (probably from Panshir) needs establishment; critical environmental and social impacts must be attended to in essential Pashtun tribal area directly adjacent to Kabul; opportune community development schemes and essential archaeological preservation schemes will buy off some local and world opposition; water impacts will need constant testing and control against adverse impacts in this first major test case; later developments will include extension of power and transport southwest to Zarkhashan gold and copper mining district near Ghaznir, and thence to Kandahar.
Cross Hindu Kush RC Segment—Establish connection of Hajigak iron with nearby coking coal; set up transport either into northern, southeastern, or western to southwestern transport systems depending upon engineering possibilities and politics.
Steel and Energy RC Segment—Adds multiple linkages to railways, highways, and energy-delivery systems across Hindu Kush and northern and southeastern transport systems; anchored in mining of iron and coking coal, with projected steel mill; multiple options for outwards transport and power connections.
Southeast Extension RC Segment—Depending upon the scale of lithium and additional copper, power and transport grids would be extended south from Kabul to Ghazni and the Zarkashan gold and copper mine, as well as on to Kandahar and south to Spin Baldak at the border with Pakistan. Power needs would be leveraged for a southeast power system.
Southwest Extension RC Segment—Railway extensions would take advantage of rare earths at Khanneshin in southern Helmand, perhaps also decorative travertine and marble stone quarries in Chagai Hills on the border with Balochistan, as well as to nearby Iran.
Northwest Extension RC Segment—Transport and power grid extensions east from the northern regions to Afghan–Tajik hydrocarbon fields, as well as southwest to Herat, the Iranian border, and south from Herat to the Shaida copper deposit.

All of the different resource-corridor (RC) segments are to be phased in progressively in coming years.

The limited governmental resources and technical capacities in the country have caused the GoA to begin to think more strategically about solid plans to promote economic growth in the private sector, while also seeking to engage friendly neighboring states as the best means to enhance political and economic stability. Because Pakistan is not perceived as much of a friend officially, but rather as a fairly constant spoiler through its apparent support of the Taliban, it could get left rather far behind in the exploitation of resources and the transport renaissance that is planned. This may be an unfortunate circumstance because Pakistan's own dilapidated infrastructure also needs great refurbishment to maintain even marginal standards. In fact, if the means can be found to help Pakistan out of its own economic doldrums, that country could also achieve far greater prosperity through the accelerant of Afghanistan's own resurgence in the coupled boon of the resource bonanza, together with the Silk Road initiative of greatly enhanced regional transportation.

For its part, Pakistan justifies its own double-standard policy and its support for armed opposition in Afghanistan in the name of ensuring and protecting its "legitimate" interests inside the country, although it officially claims that it has dropped the strategic depth issue visa vie India. Of course, it also does not want a strong Afghan government that is able to reassert its irredentist territorial claims to the Pakistani provinces of Khyber Pakhtunkhwa or Balochistan, or to support secessionist movements in either place that would ultimately fatally dismember Pakistan. Similarly, Iran's ambassador in Kabul, Abdul Fazal Zahrawand, has exerted pressure on the parliament of Afghanistan not to endorse the Strategic Partnership agreement with the USA. The other neighboring countries in the north, generally run by rather autocratic regimes in Turkmenistan, Uzbekistan, and Tajikistan seem unwilling or unable, or both, to yet engage in much political or economic cooperation with Afghanistan. Reasons for this lack of cooperation are not entirely clear, although part of the problem is the rather inward-looking attention of Turkmenistan, or the economic weakness of Tajikistan. Uzbekistan has, however, recently increased its trade with Afghanistan through the new railway facility from Hairatan to Mazari Sharif, and Tajikistan, Kyrgyzstan, and Turkmenistan are all making increased plans to run energy gridlines and pipelines south to Afghanistan and on to energy-hungry Pakistan and India as well.

Afghanistan feels that regional economic cooperation must be vigorously pursued for the sake of regional economic and political stability, while at the same time ensuring that the country's political and economic isolation must never be repeated for the sake of the stability of the whole region. Through regional cooperation, the nation of Afghanistan desires to improve trading opportunities, integrate itself firmly within the regional rail and road networks, be an important partner in regional energy markets, eliminate the trade in narcotics, and achieve its millennium development goals (MDGs). These laudable goals are rank ordered into the list below, although the last-ranked issue, which is that of security seemingly added on as an afterthought, might better be reconsidered the number one priority:

1. Eradicate extreme poverty and hunger.
2. Achieve universal primary education.
3. Promote gender equality and empower women.
4. Reduce child mortality.
5. Improve maternal health.
6. Combat HIV/AIDS, malaria, and other diseases.
7. Ensure environmental sustainability.
8. Develop a global partnership for development.
9. Enhance security.

Because Afghanistan started at such a low base in all of its development goals among all the countries in the world; in fact at the bottom in Asia and close to the bottom in the world, it had to opt for a time target of 2020 to even try realistically to meet these millennium (2000) goals. With only 6 years left, a great deal of more progress needs to be made with a national development strategy organized around three essential pillars: (1) security; (2) governance, human rights, and the rule of law; and (3) economic and social development. Each of the nine MDGs enumerated above is reflected inside one or more of these pillars and the associated underlying sector strategies that have been developed by the government. An accounting of progress toward these goals (Zakilwal, 2010) is frank but hopeful, especially where so much progress has been made already in education and the essential security. Clearly there is a long way to go but the establishment of the concepts of the time-staged RC development obviously offers the potential to greatly enhance the MDGs as well.

Afghanistan is a difficult place in which to live, and has long been a recurrent problem to its neighbors. Periodic eruptions of its warriors from within the Hindu Kush mountain fastness to prey upon the lowlanders beyond its nominal borders are legend in South Asia. Now the country is equally poised to provide new resource riches to those intrepid enough to attempt the risk. At the very least, the new Silk Road initiative will enhance transport through and beyond the region, with plentiful benefits to a great many. The future looks good and a spirit of guarded optimism does give hope to many who have been so doomed to oppressive negativity for so long. We can only wish them the best.

Bismullah!

References

Kuchins, A.C., 2011. A truly regional economic strategy for Afghanistan. The Washington Quarterly 34 (2), 77–91.

Pant, H.V., 2013. China Reaches Out to India and Afghanistan. Center for Strategic & International Studies, 3(2), 2 p. http://csis.org/publication/china-reaches-out-india-afghanistan.

Zakilwal, H.O., 2010. Afghanistan Millennium Development Goals Report. Ministry of Finance, Kabul, 60 p. http://www.undp.org/content/dam/undp/library/MDG/english/MDG%20Country%20Reports/Afghanistan/MDG-2010-Report-Final-Draft-25Nov2010.pdf.

Index

Note: Page Numbers followed by "*f*" indicate figures; "*t*", tables; "*b*", boxes.

FSC
www.fsc.org
MIX
Papier | Fördert
gute Waldnutzung
FSC® C083411

Kolibri 360 GmbH
Ferdinand-Jühlke-Straße 7
99095 Erfurt, Deutschland
produktsicherheit@kolibri360.de